THE SEA

Ideas and Observations on Progress in the Study of the Seas

THE SEA

Ideas and Observations on Progress in the Study of the Seas

EDITORIAL BOARD

TSUNAMIS

Edited by

EDDIE N. BERNARD   and   **ALLAN R. ROBINSON**
National Oceanic and   *Harvard University*
Atmospheric Administration

THE SEA

**Ideas and Observations on Progress in the Study
of the Seas**

Volume 15

Harvard University Press
Cambridge, MA
London, England
2009

Library of Congress Cataloging-in-Publication Data

Tsunamis / edited by Eddie N. Bernard, Allan R. Robinson. -- 1st ed.
    p. cm. — (The Sea, ideas and observations on progress in the study of the seas ; v. 15)
  Includes bibliographical references and index.
  ISBN 978-0-674-03173-9 (cloth : alk. paper) 1. Tsunamis. I.
  Bernard, E. N. (Eddie N.) II. Robinson, Allan R.
  GC221.2.T78 2009
  551.46'37--dc22   2008029240

CONTENTS

TSUNAMIS

Indian Ocean tsunami approaching Ray Leh Beach in Krabi, Thailand, around 11 a.m. on December 26, 2004. Although approximately 230,000 people died that morning, all the people in this photo miraculously survived. AFP/Getty Images.

PREFACE

The intent of The Sea is to provide a continuing, comprehensive, and timely synthesis of the state of knowledge of ocean science, and to overview frontiers as ocean science progresses. This volume treats a formidable force from the sea—tsunamis. The December 26, 2004 Indian Ocean tsunami was the most destructive tsunami in recorded history in terms of casualties (228,000) and economic impact ($10B). The world's response to this horrific natural disaster was an unprecedented $13.5B in international aid. Given the magnitude and impact of such a disaster, the demand for scientific and technical information on tsunamis well exceeded the supply of knowledgeable people and useable reference literature. The dissemination of inaccurate and incorrect information about tsunamis led to confusion and misinformation, revealing a huge gap in the tsunami field: *education.*

At the time of the 2004 tsunami, there was little formal tsunami education available anywhere in the world. The Sea Volume 15: *Tsunamis* is an effort to fill the education gap at two levels. The first level is for the aspiring researcher/specialist who wants to learn basic tsunami science, e.g. a first-year graduate student. The second level is for the veteran tsunami researcher who wants to have a handy, state of the science reference for tsunami research.

The scope and structure of the volume were developed with advice from an ad hoc Editorial Advisory Panel, to whom we are grateful. Chapters were reviewed both by authors of other chapters of this volume and by external reviewers, whose efforts have contributed significantly to this study. A list of Editorial Advisory Panel members and an alphabetical list of the external reviewers appear following the list of contributors. It is a pleasure to thank Ms. Gioia Sweetland of Harvard University and Mr. Wayne Leslie (HU and MIT) for their essential administrative and technical editorial assistance. We thank Mr. Michael Fisher, Editor-in-Chief of Harvard University Press, for his interest in this study and Ms. Anne Zarrella (HUP) for her valuable technical assistance.

We dedicate this volume to the victims of tsunamis.

EDDIE N. BERNARD
ALLAN R. ROBINSON

May 2008

CONTRIBUTORS

EDDIE N. BERNARD
National Oceanic and Atmospheric
  Administration
Pacific Marine Environmental
  Laboratory
7600 Sand Point Way N.E.
Seattle, WA 98115-6349

JOANNE BOURGEOIS
University of Washington
Earth and Space Sciences
Box 351310
Seattle, WA 98195-1310

URI S. TEN BRINK
U. S. Geological Survey
384 Woods Hole Rd.
Woods Hole, MA 02543

ERIC L. GEIST
U. S. Geological Survey
345 Middlefield Rd.
MS 999
Menlo Park, CA 94025

GALEN GISLER
Physics of Geological Processes
University of Oslo
P.O. Box 1048 Blindern
0316 Oslo
Norway

VIACHESLAV K. GUSIAKOV
Tsunami Laboratory
Institute of Computational
  Mathematics and Mathematical
  Geophysics, Siberian Division
Russian Academy of Sciences
Pr. Lavrentieva, 6
Novosibirsk 630090
Russia

FUMIHIKO IMAMURA
Disaster Control Research Center
Graduate School of Engineering
  Tohoku University
Aoba 6-6-11
Sendai 980-8579
Japan

UTKU KÂNOĞLU
Department of Engineering Sciences
Middle East Technical University
Ankara 06531
Turkey

HOMA J. LEE
U. S. Geological Survey
345 Middlefield Rd.
MS 999
Menlo Park, CA 94025

PHILIP L.F. LIU
School of Civil and Environmental
  Engineering
Cornell University
Hollister Hall
Ithaca, NY 14853

HAROLD O. MOFJELD
School of Oceanography
University of Washington
Box 357940
Seattle, WA 98195-7940

EMILE A. OKAL
Department of Geological Sciences
Northwestern University
1850 Campus Dr. #212
Evanston, IL 60208-2150

TOM PARSONS
U. S. Geological Survey
345 Middlefield Rd.
MS 999
Menlo Park, CA 94025

ALLAN R. ROBINSON
School of Engineering and Applied
    Sciences
Harvard University
29 Oxford Street
Cambridge, MA 02138

COSTAS SYNOLAKIS
Tsunami Research Center
Viterbi School of Engineering
University of Southern California
Los Angeles, CA 90089-2531

VASILY TITOV
National Oceanic and Atmospheric
    Administration
Pacific Marine Environmental
    Laboratory
University of Washington/JISAO
7600 Sand Point Way NE
Seattle, WA 98115-6349

PAUL M. WHITMORE
National Oceanic and Atmospheric
    Administration
West Coast/Alaska Tsunami
    Warning Center
910 Felton Street
Palmer, AK 99645-6552

HARRY YEH
School of Civil & Construction
    Engineering
Oregon State University
Corvallis, OR 97331-3212

Chapter 1. Introduction: Emergent Findings and New Directions in Tsunami Science

E. N. BERNARD

*National Oceanic and Atmospheric Administration,
Pacific Marine Environmental Laboratory*

A. R. ROBINSON

Harvard University

Contents

1. Tsunami Science after the 2004 Indian Ocean Tsunami

The December 26, 2004 Indian Ocean tsunami was the most destructive tsunami in recorded history in terms of casualties (228,000) and economic impact ($10B). The world's response to this horrific natural disaster was an unprecedented $13.5B in international aid. Given the magnitude and impact of such a disaster, the demand for scientific and technical information on tsunamis well exceeded the supply of knowledgeable people and useable referenced literature. The dissemination of inaccurate and incorrect information about tsunamis led to confusion and misinformation, revealing a huge gap in the tsunami field: *education.*

At the time of the 2004 tsunami, there was little formal tsunami education available anywhere in the world. The Sea Volume 15: Tsunami is an effort to fill the education gap at two levels. The first level is for the aspiring researcher/specialist who wants to learn basic tsunami science, e.g. a first-year graduate student. The second level is for the veteran tsunami researcher who wants to have a handy, state of the science reference for tsunami research.

1.1 Tsunami concepts and processes

Tsunamis are usually generated when energy released by an underwater earthquake triggers an abrupt deformation of the sea floor. Because of the near incom-

The Sea, Volume 15, edited by Eddie N. Bernard and Allan R. Robinson
ISBN 978–0–674–03173–9 ©2009 by the President and Fellows of Harvard College

pressibility of seawater, this sea floor deformation immediately appears as a displacement of the sea surface. Gravity restores mean sea level by converting the energy of the local sea surface displacement into gravity waves. These gravity waves have lengths that are much longer than the ocean depth. Such long waves can propagate across ocean basins for several hours with negligible frictional loss at speeds exceeding 700 km/hour. Propagating across the continental margin, these waves amplify inversely proportional to the shoaling bottom and can reach amplitudes of several meters as they approach the coastline. At the coastline these waves can break into a turbulent mass of fluid. Tsunami can repeatedly inundate coastal regions for up to 12 hours while destroying lives and property several kilometers inland.

If the disturbance is close to the coastline, local tsunamis can demolish coastal communities within minutes. A very large disturbance can cause both local devastation and export tsunami destruction thousands of kilometers away. The word tsunami is a Japanese word, represented by two characters: tsu, meaning, "harbor", and nami meaning, "wave". Unusual wave behavior in harbors was the first sign that a tsunami was approaching, so yelling the word tsunami became the earliest warning system.

Tsunamis rank high on the scale of natural disasters. Since 1850 alone, tsunamis have been responsible for the loss of over 420,000 lives and billions of dollars damage to coastal structures and habitats. Most of these casualties were caused by local tsunamis that occur about once per year somewhere in the world. Predicting when and where the next tsunami will strike is currently not possible. Once a tsunami is generated, however, forecasting tsunami arrival and impact is possible through modeling and measurement technologies if sufficient input information is available.

In this volume, the following tsunami-related terms are used:

Generation. Tsunamis are most commonly generated by earthquakes in marine and coastal regions. Figure 1.1 shows the location of the generation sites of the 1,990 tsunamigenic events identified from 2000 B.C. until the present. Major tsunamis are produced by large (greater than 7 on the Richer scale), shallow focus (<30 km depth in the earth) earthquakes associated with the movement of oceanic and continental plates. They frequently occur in the Pacific, where dense oceanic plates slide under the lighter continental plates. When these plates fracture, the quick seafloor movement provides efficient transfer of energy from the solid earth to the ocean. When a powerful earthquake (magnitude 9.2) struck the coastal region of Indonesia in 2004, the movement of the seafloor produced a tsunami in excess of 20 meters along the adjacent coastline, killing more than 168,000 people. From this source the tsunami radiated outward and within 4 hours had claimed 60,000 additional lives in Thailand, Sri Lanka, India, and 10 other Indian Ocean nations.

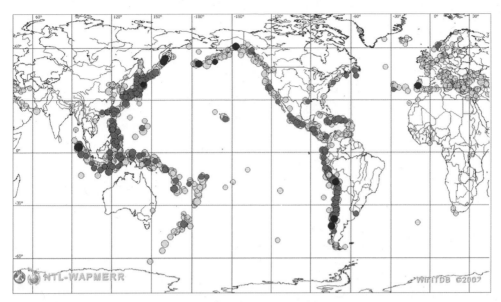

Fig. 1.1 – Locations of 1,990 global tsunamigenic events are shown for the period from 2000 BC to the present. Size of circles is proportional to event magnitude. Transoceanic events – black. Regional events with human casualties – dark gray. Other regional events – light gray. (Gusiakov – Ch.2, Fig.2, see color insert)

Underwater landslides associated with smaller earthquakes are also capable of generating destructive tsunamis. The tsunami that devastated the northwestern coast of Papua New Guinea on July 17, 1998, was generated by an earthquake that registered 7.0 on the Richter scale that apparently triggered a large underwater landslide. Three waves measuring more than 7 meters struck a 10 km stretch of coastline within 10 minutes of the earthquake/slump. Three coastal villages were swept completely clean by the deadly attack, leaving 2,200 people dead. Other large-scale disturbances of the sea surface that can generate tsunamis are explosive volcanoes and asteroid impacts. The eruption of the volcano Krakatoa in the East Indies on August 27, 1883 produced a 30-meter tsunami that killed over 36,000 people. In 1988, scientists discovered evidence of a 10km diameter asteroid that landed on the edge of the Yucatan Peninsula approximately 65 million years ago, producing a huge tsunami that swept over portions of the Gulf of Mexico and the Caribbean, leaving coarse grain sand deposits.

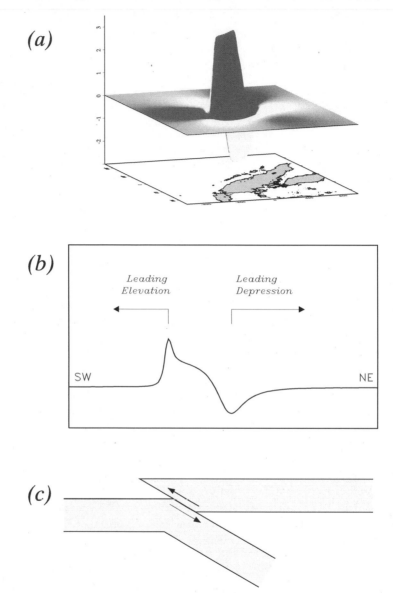

Fig. 1.2 (a) – Sea surface initial distortion along the fault strike for the 1833 South Sumatra Earthquake. (b) Cross section of (a) perpendicular to the fault. Note the asymmetry: a NE landward depression and a SW seaward elevation. (c) Schematic cross section of co-seismic displacements at a subduction zone (a,b,c not to scale). (Okal – Ch.5, Fig.5)

Propagation. Because earth movements associated with large earthquakes are thousands of square kilometers in area, any movement of the seafloor immediately changes the sea surface. The resulting tsunami propagates as a set of waves whose energy is concentrated at wavelengths corresponding to the earth movements (~100 km), at wave heights determined by vertical and horizontal displacement (~1 m), and at wave directions determined by the adjacent coastline geometry and bathymetry. Figure 1.2 illustrates the initial sea surface distortion associated with the 1833 earthquake off the coast of Sumatra (Indonesia): the depression rapidly reached the nearby Sumatran coast, causing a draw-down of water before flooding, while the oceanward elevation propagated across the basin and struck the coast of Ceylon (Sri Lanka) in a few hours. Because each tsunami is unique, every tsunami has unique wavelengths, wave heights, and directionality. Directionality coupled with the bathymetry between the source and distant coastlines can concentrate or diffuse tsunami energy. A tsunami can, therefore, flood one part of a coastline and not affect the same coastline only a few kilometers away. Basin-wide propagation of tsunami waves is illustrated in Fig. 1.3. The waves spread from the generation location in the Kuril Islands at ~45°N latitude and 150°E longitude. The tsunami that hit Crescent City, California harbor 8.5 hours after generation was steered and amplified by the submarine Mendocino Escarpment lying along 40°N latitude.

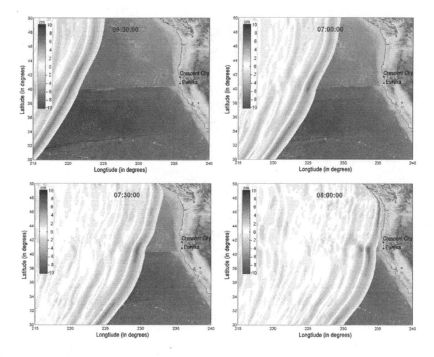

Fig. 1.3 – Snapshots of tsunami waves at t=6.5, 7.0, 7.5, 8.0 hours after the central Kuril Islands earthquake on November 15, 2006. (Liu – Ch. 9, Fig.2)

Inundation. As the tsunami approaches the coastline, the wave energy is concentrated in smaller volumes of ocean. Since these waves are not dissipated as they propagate across the ocean, their energy is conserved as they move into shallow water. A consequence of this property is that the energy amplifies the tsunami heights. If there is adequate energy, the tsunami will repeatedly flood the coastline, a process termed inundation. Since the inundation process involves multiple waves (from 3–10 waves), communities along the coastline can experience repeated inundation and draining for up to 12 hours. The inundation process can destroy structures, transforming the structures to floating debris, which, in turn, can be used as battering rams for subsequent waves to cause more destruction, as illustrated in Fig. 1.4.

Fig. 1.4 – Damage caused by a water-borne missile (a boat) in Nagappattinam India by the 2004 Indian Ocean tsunami. (Yeh – Ch.11, Fig. 8)

Forecasts. Recently developed real-time, deep ocean tsunami detectors, termed DART buoys, as illustrated schematically in Fig. 1.5 and geographical locations in Fig. 1.6, provide the data, assimilated into advanced forecast models, to make tsunami forecasts. Since 2003, DART buoys have detected 10 Pacific Ocean tsunamis, each of them adding to the National Oceanic and Atmospheric Administration's (NOAA) confidence in the accuracy of its forecast capability. NOAA scientists made experimental forecasts for tsunamis generated in the Aleutian Islands (November 2003), Kuril Islands (November 2006 and January 2007), Tonga (May 2006), Solomon Islands (April 2007), Peru (August 2007), and Chile (November 2007) for 12 U.S. coastal communities that have tide stations. When scientists compared the experimental forecasts with tide data for the seven tsunamis, they found that the forecasts were within 90 percent agreement with tide gauge recorded tsunami (Titov, Ch. 12, Table 1).

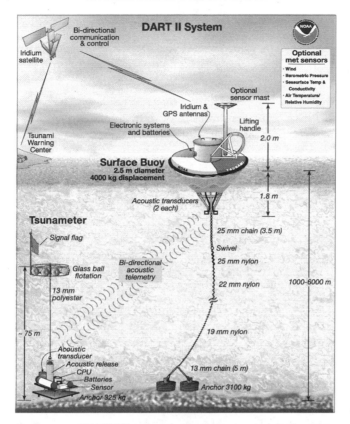

Fig. 1.5 – Schematic diagram of DART II system that measures tsunamis in the open ocean. Shown are the bottom unit (tsunameter) that measures bottom pressure and a separate surface buoy that communicates with the bottom unit using an acoustic modem and with shore installations via an Iridium satellite system for global coverage. (Mofjeld – Ch. 7, Fig. 2)

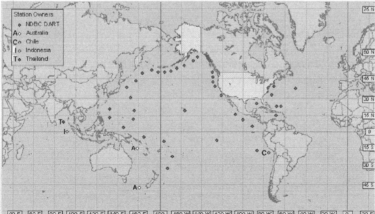

Fig. 1.6 – Geographical location of 43 deep ocean tsunami detection stations operated by Australia, Chile, Indonesia, Thailand, and the United States as of March 2008. Each DART has the same pressure sensor and communications technology, so they represent the deep ocean measurement standard for the global tsunami warning system. Real-time data from the array are available at http://www.ndbc.noaa.gov/dart.shtm.

Warning Systems. Since 1949, NOAA's tsunami warning system has provided warnings of potential tsunami danger in the Pacific basin by monitoring earthquake activity and the passage of tsunami waves at tide gauges. However, neither seismometers nor coastal tide gauges provide data that allow accurate prediction of the impact of a tsunami at a particular coastal location. Monitoring earthquakes gives a good estimate of the potential for tsunami generation, based on earthquake size and location, but gives no direct information about the tsunami itself. Tide gauges in harbors provide direct measurements of the tsunami, but the tsunami is significantly altered by local bathymetry and harbor shapes, which severely limits their use in forecasting tsunami impact at other locations. Partly because of these data limitations, 15 of 20 tsunami warnings issued since 1949 were considered false alarms because the tsunami that arrived was too weak to cause damage. Recent measurement and modeling technologies, described in detail in this volume, have provided the necessary tools to now accurately forecast tsunamis in real time.

1.2   Short summary of the Indian Ocean Tsunami

Physical characteristics (Iwan, 2006)
The total energy released by the December 26, 2004 Indian Ocean Mw 9.2 earthquake was estimated to be the equivalent of 200 billion tons of explosives. There was at least 10 m movement laterally and 4–5 m vertically along the subduction fault line. In February 2005, the Royal Navy vessel *HMS Scott* high-resolution survey of the seabed around the earthquake epicenter revealed that the earthquake had made a huge impact on the topography of the seabed. Thrust ridges, as high as 1,500 m, had collapsed, generating landslides several kilometers wide. One such landslide consisted of a single block of rock some 100 m high and 2 km long. The sudden vertical and horizontal movements of the subduction zone, massive underwater landslides, and large splay fault ruptures during the earthquake displaced enormous volumes of water, resulting in a tsunami that struck the coasts of the Indian Ocean. The total energy of the tsunami was about 2% of the energy released in the earthquake itself. In many places the waves reached as far as 3 km inland. Scientists investigating the damage in the province of Aceh, Indonesia found evidence that the wave reached a height of 24 m when coming ashore along large stretches of the coastline, rising to 37 m in some inland areas.

Because the subduction zone earthquake was in a nearly north-south orientation, the tsunami waves were mostly directed to the east and west. Bangladesh, which lies at the northern end of the Bay of Bengal, had very few casualties despite being a low-lying country near the rupture zone. Coasts that have a landmass between them and the tsunami's location of origin are usually safe; however, tsunami waves can sometimes diffract around such landmasses and reflect off steep bathymetric features. Thus, the Indian state of Kerala was hit by the tsunami despite being on the western coast of India, and the western coast of Sri Lanka also suffered substantial impacts. At Columbo, Sri Lanka, the largest tsunami recorded was the tsunami that reflected off the Maldives island chain arriving 2.5 hours after the initial tsunami wave arrival. Also, distance alone is no guarantee of safety as Somalia was hit harder than Bangladesh despite being much farther away.

The tsunami was noticed as far as Struisbaai in South Africa, some 8,500 km away, where a 1.5 m tsunami surged on shore about 16 hours after the earthquake.

The tsunami also reached Antarctica, where tidal gauges at Japan's Showa Base recorded oscillations of up to a meter, with disturbances lasting a couple of days. Some of the tsunami's energy escaped into the Pacific Ocean, where it produced small but measurable tsunamis along the western coasts of North and South America, typically around 0.2 to 0.4 m. At Manzanillo, Mexico, a 1.0 m crest-to-trough tsunami was measured. As well, the tsunami was large enough to be detected in Vancouver, British Columbia, Canada. The tsunamis measured in some parts of South America were larger than those measured in some parts of the Indian Ocean due to the mid-ocean ridges which acted as wave guides to direct tsunami energy over long propagation paths.

This tsunami has been the most studied of any tsunami in history. Surveys of the physical damage, as well as the psychological damage to humans, have produced volumes of data that will provide the research community with links to new discoveries, theories, and practices for tsunami preparedness. These data will make the 2004 tsunami the watershed event for tsunami research.

Human and Economic Impacts (Cosgrave, 2007)

The early morning earthquake of December 26, 2004 caused destruction in Banda Aceh and other parts of Aceh even before the arrival of the tsunami. No one knows how many died in the earthquake as the tsunamis overtook the rescuers. The best estimate is that the death toll from the earthquake was about 1,000 people. The December thrust earthquake led to horizontal and vertical movements of seafloor across a strip over 1,200 km long, triggered hundreds of underwater landslides, and activated hundreds of secondary faults throughout the region. The underwater land movements disturbed the ocean surface and generated a train of waves that sped across the Indian Ocean killing about 228,000 people and displacing 1.7 million across 14 countries. Indonesia, Sri Lanka, India and Thailand suffered the greatest loss of life. Entire coastal zones were destroyed, with the tsunamis causing damage up to 3 km inland in some cases. For loss of life by nationality, Germany and Sweden were the fifth and sixth worst affected, although the tsunami did not strike those countries. The impact of the tsunamis depended on location, with towering 30 m waves in Aceh and a 2 m swell in parts of the Maldives. By the time they had travelled 8,000 km to South Africa the tsunamis were barely distinguishable from the background pattern of normal waves. The tsunami was recoded by tide gauges throughout the world, making this the first global tsunami in recorded history. This difference in the severity of the tsunamis showed itself in the ratio between numbers of those who were killed and injured in different places. In Aceh the ratio of dead to injured survivors was 6:1, dropping to 1.5:1 in Sri Lanka and 0.3:1 in India. The tsunami killed more women than men: the ratio of male to female casualties ranged from 20 percent more women to more than twice as many women as men. The normal pattern with flooding is that floods kill more men than women. It is hypothesized that men are prone to take risks or to overestimate their swimming ability. However, storm surges, where the sea rushes inland like in a tsunami, typically kill more women than men. The reasons include both sex (men's greater strength allowed them to hold on to trees and fixed objects for longer), and gender (knowledge of swimming, different locations, childcare duties, clothing and others). Death rates were also higher for those under 15 and those over 50. Sometimes the elderly suffered the highest death rates and sometimes children did, but on average these groups were over twice as likely to die as adults. Adult men under 50 had the highest chance of survival, as well as the highest chance of surviving with injuries. Less than 1 percent of those who died were tourists, but these got most of the media attention in donor countries. One study found that 40 percent of western media coverage on people affected by the disaster dealt with tourists. Paradoxically, while estimates of local mortality increased up to the end of January those for tourist deaths fell from over 9,000 at the beginning of January 2005 to 3,528 at the end of January, eventually to settle at 2,218.

The assessments after the tsunami estimated losses and damage at just under $10 billion. As with all disasters this is only a very rough estimate as damage (the cost of a factory) is relatively easy to calculate; the consequent losses (being without work for a long period) are far harder to estimate. Indonesia suffered nearly half of the total economic losses and damage. The large investment in the tourist industry in Thailand and the Maldives led to significant damage to tourist facilities. Although only 108 people died in the Maldives, the estimate of economic damage and loss was nearly 80 percent of a year's gross national income. In comparison, within Indonesia the damage was almost wholly confined to the province of Aceh where the estimate of damage was equal to almost a year's economic output from the province. Industries based at the coast were the worst affected. In Aceh, ports and harbors were destroyed in addition to the fishing fleet and industries along the coastal strip. Fishing and tourism were the two worst affected sectors overall, but those with farms near the coast lost animals and saw their fields made infertile by debris and salt.

1.3   Societal Response

Following the tsunami, an immense media-fuelled, global response resulted in US$13.5 billion pledged or donated internationally for emergency relief and reconstruction, including more than US$5.5 billion from the general public in developed countries. Private donations broke many records. Donors were flexible and quite rapid in their funding. Reporting of pledges and commitments and the timeliness of official donations has been better than in other crises. The international humanitarian aid community also made a historic first effort at accountability. There was early recognition that the exceptional response to the tsunami disaster, including the amount of money given, not only demanded a high standard of accountability to a generous public but also provided an opportunity for learning by the international humanitarian community. A meeting in Geneva in February 2005 started a process that led to the formation of the Tsunami Evaluation Coalition (TEC), a group of aid agencies interested in encouraging learning from the international response to the December 2004 tsunami disaster through joint evaluation.

The TEC studies found the international response to the tsunami disaster helped the affected people and reduced their suffering. They identify many examples of good practice in emergency response, and some welcome innovations. However, overall the studies conclude that the response did not achieve the potential offered by the generous funding. The TEC Synthesis Report (2007) made four recommendations: around ownership (and accountability), capacity, quality, and funding. These are all about one central idea—that the humanitarian aid community needs to go about its business in a different way. It needs to cede ownership of the response to the affected population and become accountable to them. This change needs to be supported by more equitable and proportionate funding, the development of disaster response capacities, a greater focus on risk-reduction, and a system for controlling the quality of the work done by humanitarian agencies.

The U.S. response was generous, swift, and effective. In total, the United States government spent about $1.0 billion for a tsunami that killed 31 U.S. citizens and did not damage any U.S. coastline. The U.S. military spent about 25% of this amount by immediately dispatching an aircraft carrier and a hospital ship to aid in rescue and relief operations. The State Department was provided about 66% of the funds for reconstruction activity in India, Indonesia, Maldives, Sri Lanka, and Thailand. Of the State Department amount, about $14M (0.1%) was spent on contributing to the establishment of an Indian Ocean tsunami warning system. The Departments of Commerce (through NOAA) and Interior (through the United States Geological Survey—USGS) received $39M (0.4%) to strengthen the existing U.S. tsunami warning system. Interim tsunami warning service for the Indian Ocean, provided by NOAA, began in March, 2005. The President's Office of Science and Technology Policy (OSTP) released the "Tsunami Risk Reduction for the United States: A Framework for Action" report in December 2005 which called for, among other actions, a national strategy for tsunami research. The National Tsunami Research Plan (Bernard et al., 2007), published in April 2007, recommended six priorities for tsunami research: 1) enhance and sustain tsunami education, 2) improve tsunami warnings, 3) understand the impacts of tsunamis at the coast, 4) develop effective mitigation and recovery tools, 5) improve characterization of tsunami sources, and 6) develop a tsunami data acquisition, archival, and retrieval system.

The U.S. Congress passed the Tsunami Warning and Education Act (Public Law 109–424) as an extension of the efforts of the National Tsunami Hazard Mitigation Program (NTHMP)—a State/Federal partnership to reduce tsunami hazards to U.S. coastlines that began in 1997 (Bernard, 2005). The Act has four tsunami elements: warning, education, research, and international cooperation. Both the National Tsunami Research Plan and the Tsunami Act emphasize research that supports a community-based mitigation program. Just as the TEC found that the best approach to natural disaster recovery is to empower the local jurisdictions, the U.S. has embraced the concept of tsunami resilience, the ability of a community to quickly recover from a tsunami. The foundation of resilience is a community's ability to develop local advocates as outlined in "Preparing Your Community for Tsunamis" (Samant et al., 2007). Communities, unfortunately, will not have the technical expertise to judge the quality of state-of-the science products required to meet preparedness needs (i.e., inundation maps, building codes). The tsunami research community's challenge is to establish state of the science methods for immediate use while simultaneously conducting research on improvements in the state-of-the-science products. One of the purposes of this volume is to capture a snapshot of tsunami science-based mitigation products available today.

2. Overview of the Volume

What emerges from the 2004 Indian Ocean tsunami and the society's response is a call for research that will mitigate the effects of the next tsunami on society. The scale of the 2004 tsunami's impact, and the world's compassionate response, requires that tsunami research focus on applications that benefit society. Tsunami science will be expected to develop standards that ensure mitigation products are based on state of the science. Standards based on scientifically endorsed procedures assure the highest quality application of this science. Community educational activities will be expected to focus on preparing society for the next tsunami. This volume provides an excellent starting point for the challenges ahead, including, importantly, academic educational activities. The volume contains the technical elements of tsunami state of the science, including: three chapters (2, 3, 4) on the recorded and geologic history of tsunamis and how to assess the probability of the tsunami risk; two chapters (5, 6) on the generation of tsunamis; four chapters (7, 8, 9, 10) on measurement and modeling tsunami propagation and inundation; one chapter (11) on the impacts of tsunamis on coastlines; and two chapters (12, 13) on tsunami forecast and warnings. Together, these chapters give a technical foundation to apply tsunami science to community-based tsunami preparedness. The tsunami research community, together with relevant elements of the civic, commercial, and governmental communities, is *expected* to establish standards and procedures for tsunami education and technical products to ensure the public is receiving the highest quality product available.

A summary of each Chapter, including recommendations for future research, is provided below to prepare readers for the content and to integrate the chapters in the volume into a more coherent reference.

Chapter 2, *Tsunami History: Recorded,* by V. K. Gusiakov, covers the historical record of tsunamis widely used to evaluate the tsunami hazard of a specific coastal area. Gusiakov gives an in-depth discussion, with an extensive reference list, of the development of historical databases, problems with these databases, and an overview of historical tsunamis using the latest database containing 2130 tsunamis since 2000 B.C. He also provides a historical overview of tsunamis by their generating mechanisms, including earthquakes (see Chapter 5 for more details), slide-generated, and volcanic-generated tsunamis (see Chapter 6 for more details). One of the surprising results from his analysis of earthquake-generated tsunamis is that tsunami intensity is *not* well correlated with earthquake magnitude (Fig. 2.7). His concluding explanation is that secondary mechanisms, such as submarine slides, can play an important role in tsunami generation. He also includes two sections on other mechanisms for generating tsunami-like waves, including meteorological events, man-made explosions, and earthquake waves exciting small bodies of water. His summary includes seven conclusions, including the distribution of tsunamis by ocean basin (57% in Pacific, 25% in Mediterranean, 12% in Atlantic, and 6% in Indian), the fact that only 10% of tsunamis cause fatalities with 95% of deaths occurring within one hour propagation time from the source, and that geological evidence of tsunamis (Chapter 3) should be included in the historical databases. His overall recommendation is that data collection standards should be established and tsunami databases should be organized regionally and shared globally. A careful reading of this chapter will provide a better understanding of the range of tsunami hazards.

Chapter 3, *Geological Effects and Records of Tsunamis,* by Joanne Bourgeois, extends the historical record of tsunamis by including geological evidence of past tsunamis to better assess the hazard at specific locations. Bourgeois provides an excellent overview of the traces left by large tsunamis in the geological record of the history of our planet. She describes the study of geological records of tsunamis as "immature" because extensive publications began after 1985. By 2006, however, over 500 peer-reviewed articles had appeared, largely spurred by the high number of damaging tsunamis in the 1990s. She provides a detailed history of the evolution of the field by someone who witnessed the evolution along with an extensive reference list. In this discussion, she points out some "imaginative interpretations" that every evolving field experiences, about which a new student should be cautious. This chapter introduces us to the granddaddy of all tsunamis, the asteroid impact tsunami (see Chapters 6 and 8 for more details) and the geological traces left by these enormous tsunamis up until the Holocene geological epoch (about 9600 B.C. to present). The Holocene contains deposits from earthquake-generated tsunamis and are easily compared to contemporary tsunamis such as the 2004 Indian Ocean tsunami for validation of scientific concepts. This chapter also discusses the physics behind the deposits and techniques for dating the time of deposit using a variety of techniques and technologies. Her section on tsunami erosion provides insights to the source of the deposits. She then discusses some of the exciting research topics in this emerging geological specialty, including mapping methods for contemporary tsunamis, distinguishing deposits from storm surge vs. tsunami flooding, and quantitative methods for assessing the fluid flow associated with deposits. This interesting chapter will give the careful reader a better understanding of using deposits to infer tsunami-flooding impact.

Chapter 4, *Tsunami Probability,* by Eric L. Geist, Tom Parsons, Uri S. ten Brink, and Homa Lee, presents formal approaches for estimating tsunami probability based on probability distributions. Geist et al. choose candidate functions for the distributions that have parameters that can be fitted empirically to historic data (Chapter 2) at a coastal site or estimated via tsunami modeling (Chapters 8, 9, and 10). Since empirical data of run-up and maximum wave heights are limited, especially for large tsunamis, it is necessary to use probability distributions of tsunamigenic earthquakes (as the principle source of tsunami) to estimate tsunami probability. The use of probability distributions involves assumptions about the characteristics of tsunamis generated by earthquakes. Tsunami propagation (Chapter 9) and inundation models (Chapter 10) are then used to produce ensembles of synthetic run-up, maximum wave heights, inundation extent, and tsunami-induced currents within the coastal areas of interest. A pilot study using this approach has recently been applied to Seaside, Oregon to estimate probability of maximum tsunami flooding for 100 and 500 years. It is suggested that the reader unfamiliar with tsunami propagation and inundation models, review Chapters 9 and 10 as preparation for reading this chapter. The authors also discuss what is known about landslide sources with particular focus on the Storegga and Southern California areas. The effects of large uncertainties in the frequency and magnitude of tsunami sources are also presented, along with suggestions for further research that may reduce these uncertainties. Reducing uncertainty will require a better understanding of earthquake slip distributions, more data on landslide speed, and how uncertainty affects probability estimates. After reading this chapter, the reader will have an appreciation of the problem of determining tsunami probabilities with little historical data.

Chapter 5, *Excitation of Tsunamis by Earthquakes,* by Emile Okal, presents theories, with examples, of estimating the initial tsunami intensity from earthquake moment magnitude or normal mode theory. Okal details, with extensive references, the seismic theory behind the seismic moment as a representation of the earthquake source and then applies this theory to show how earthquake seismic moment can be converted to tsunami initial conditions. His approach is applied to several tsunamis that show the gross characteristics of the theory in the near and far field. He then details the theory behind the normal mode formalism, which treats tsunamis as free oscillations of the earth. One limit to the normal mode theory is that it cannot be applied to the near field. He then discusses the problems of relating the seismic moment to near-field tsunami inundation where success is elusive. He concludes his discussion with the "tsunami earthquake," or a slow rupturing earthquake that produces larger than expected tsunamis. Seismic moment and normal modes theories break down because of scaling laws. The time to determine the seismic moment is much longer for slow earthquakes because the process is much slower. In this discussion, Okal demonstrates that the definition of "tsunami earthquake" is vague, misleading, and misused (see his Table 1). Attempts to explain this phenomenon using seismic theory fall short and the few studies that use empirical approaches are not reliable. He concludes that for the real-time warning problem, seismic theory explains some characteristics of tsunami generation, but not reliably. He recommends, for real-time magnitude computations, research that develops different magnitude scales for different parts of the seismic spectrum. The reader who is interested in the application of seismic theory to magnitude determination is urged to carefully review this chapter.

Chapter 6, *Tsunami Generation: Other Sources,* by Galen Gisler, deals with tsunami generation by sources other than earthquakes. Gisler reports on a series of numerical experiments that explore tsunami generation by sub-aerial and submarine landslides, volcanic eruptions (pyroclastic flows and caldera collapses), and asteroid impacts (Chapter 3) using a multi-material, multi-phase, full Navier-Stokes, compressible fluid model (SAGE hydrocode). He begins by providing examples of tsunamis not generated by earthquakes and details experimental work that has successfully reproduced tsunami inundation dynamics for sub-aerial landslides. He then makes the case for using numerical models that must deal with multi-material, multi-phase processes (volcanic eruptions, landslides, and asteroid impacts) in a compressible fluid. After discussion of some of the features of the SAGE hydrocode, he then details numerical experiments for submarine volcano simulations of Kick-em Jinny volcano, for sub-aerial landslide simulations of Lituya Bay, Alaska and La Palma, Canary Islands, for sub-marine landslide simulations of hypothetical continental shelf turbidites, and for asteroid impact simulations for various size asteroids colliding at different incident angles. Validation of these experiments is only qualitative (except Lituya Bay), since tsunami data from these other sources are limited. For those readers interested in his results from a LaPalma flank collapse, the answer is local destruction, but no threat to the U.S. east coast. He recommends more experimental work with both slurries of scaled rheology and with solid blocks, field surveys of past slides, monitoring sites with potential for slides or slumps, and repeating bathymetric surveys of steeply sloping continental shelf-edge regions to accurately map the "before slump" bathymetry. The reader will find Chapter 6 both fascinating and sobering.

Chapter 7, *Tsunami Measurements,* by Harold Mofjeld, deals with the in situ instrumental measurement of tsunamis for use in tsunami warnings and research. Mofjeld provides a detailed report, with extensive references, on the history of tsunami instrumentation, an overview of present in-situ instruments and observational networks, a discussion of tsunami time series derived from these networks, and the types of data collected during a post-tsunami survey. Because observations are so fundamental to research, in situ tsunami measurements are closely linked to other chapters in this volume, including: model validation (Chapters 8, 9, and 10), forecast and warnings (Chapters 12 and 13), while post-tsunami surveys are useful for determining sources (Chapters 5 and 6), impacts (Chapter 11), contribute to the historical data base (Chapter 2), and are used as a basis for interpreting paleotsunami evidence (Chapter 3). As a result of the 2004 Indian Ocean tsunami there is an international commitment to create and maintain a global tsunami-observing network of coastal and deep-ocean instruments that will share data freely over the Internet. Deep-ocean data will be standardized for use in warnings as well as research. The tsunami research community will have its real-time, dedicated observing network to advance the understanding of tsunami dynamics. These data will also be archived in World Data Center A in Boulder, Colorado to assure access and use in future publications. The author recommends research that will sustain these observational networks by either combining tsunami networks with other observing systems and/or reducing the cost of the present system. The reader is advised to thoroughly read this important chapter, as it is the foundation of future research.

Chapter 8, *Tsunami Modeling: Development of Benchmarked Models,* by Costas Synolakis and Utku Kânoğlu, deals with the evolution and development of numerical models to be used for tsunami warnings and inundation map production. Synolakis and Kânoğlu provide an extensive report on the substantial progress achieved over the past 50 years of tsunami science, leading to the creation of numerical models to simulate tsunami dynamics. The authors take us through a fascinating history of the interplay between hydrodynamic theory and tsunami observations, through a medium called benchmark workshop, to advance our ability to forecast, in real time, tsunami behavior. Not only do the authors detail the successes and failures of this evolution, but they also give us a roadmap on how tsunami science must be accountable, through rigorous testing, and adaptable, when new observations challenge the state-of-the science. The authors are very thorough in explaining the role of analytical and experimental results to verify numerical experiments and provide an extensive reference list. They also present results from landslide- and asteroid-generated tsunami simulations that complement the model presented in Chapter 6. The history of tsunami model development has direct linkage to Chapter 7 for model validation and to Chapter 9, 10, and 12 for propagation, inundation, and forecast modeling methods. The authors recommend the continuation of benchmarking workshops, as new data from the observational systems described in Chapter 7 will challenge the present set of numerical models. They believe that the next generation of models will be higher-order approximations of the Navier-Stokes equations, such as the one described in Chapter 6. This chapter is required reading for anyone involved in tsunami research, as it integrates the various sub-disciplines of tsunami research (observation, theory, analytical/experimental/numerical modeling, and forecast) into a focused forecast framework that has value to society.

Chapter 9, *Tsunami Modeling: Propagation,* by Philip Liu, deals with the modeling of tsunami propagation. Liu covers the subject of propagation modeling thoroughly by providing an overview, formulation of numerical models, discussion of the two-dimensional governing equations, report on other applications, and an application to the 2006 Kuril Island tsunami. (This tsunami is also described in Chapter 12, Tsunami Forecasting.) He also provides an Appendix deriving the two-dimensional governing equations. This chapter is introduced in Chapter 8, an overview of tsunami modeling. For his case study, he shows how propagation models can reveal certain features of tsunami behavior that may have value in warning operations (Chapter 13). The reader will benefit from reading this chapter as an excellent example of using the simplest model to gain insights into tsunami dynamics.

Chapter 10, *Tsunami Modeling: Inundation,* by Fumihiko Imamura, deals with the modeling of tsunami inundation applied to tsunami hazard map production. Imamura gives a succinct, yet comprehensive, description of inundation models in use in 2007. He walks the reader through the details of the various numerical methods for tracking the waveform on land, for computing the advance of the tsunami, and for applying bottom friction. He elaborates on the application of friction depending on the type of land being inundated. For example, a coastal forest has more frictional resistance than a river. He then explains how to use these models for developing hazard maps. He suggests using tsunami scenarios to identify areas expected to be flooded and the arrival time and height of the initial tsunami. Once the tsunami hazard is identified for the specific community, then the public should be involved to best plan activities that would mitigate the tsunami hazard, including evacuation planning and land use activities. Inundation modeling is introduced in Chapter 8. Inundation models are also used to study impacts on structures (Chapter 11) and in tsunami forecasting (Chapter 12). The author recommends the use of inundation models to produce inundation maps for hazard identification, education, and public safety. The reader interested in the application of models to tsunami hazard identification will benefit from reading this chapter.

Chapter 11, *Tsunami Impacts on Coastlines,* by Harry Yeh, deals with the complex tsunami/shoreline interaction problem of which he has provided an excellent overview, with exceptional references. The reader interested in land use management and building code development in tsunami hazard zones will find this chapter of value. He begins with a comprehensive introduction that includes photographs of tsunami damage to illustrate the wide variations of flooding along neighboring coastlines (due to initial waveform, tsunami reflectivity, coastal topography, and local bathymetry) as well as structural damage (due to direct water forces, impact by water born missiles, i.e., fishing boats, fire spread by floating material, scour, and slope/foundation failure). He then presents processes of inundation onto the shore that takes field data and laboratory experiments and converts these results into crude mathematical models to describe initial inundation dynamics for non-breaking waves and bores. The model uses nonlinear shallow-water wave theory coupled with a run-up algorithm, derived in Appendix A, to simplify the computations. He computes many cases of run-up and drawdown to determine which flows have the greatest velocities, and hence impact. His findings are consistent with observations and experimental results. With these tools, he proceeds to apply his models to estimate forces on inshore structures in equations 28 and 30. These estimates would be a good starting point for designing tsunami-resilient structures. He acknowledges that these formulations do not consider the impact of water borne missiles, but the flow velocity could be used as a proxy for their impact speed. This chapter links to Chapters 8 and 10, a modeling overview and inundation modeling, and Chapters 2 and 3, which deal with historical data and geological deposits. The author recommends further research on the role of water-borne missiles in estimating forces on structures and investigating tsunami forces in the shallow offshore region where new structures (for example, LNG terminals and wind farms) may be built. This is an excellent chapter to read to gain insights into tsunami forces during inundation.

Chapter 12, *Tsunami Forecasting,* by Vasily Titov, deals with real-time tsunami forecasting once a tsunami has been formed. Titov has written an outstanding summary of the need for real-time forecasting, the history of the development of the measurement and modeling technologies required to produce forecasts, the stringent standards that must be met, and the ten real-time experimental forecasts that show the timeliness and skill of the forecast methodology. He starts with past problems of tsunami warnings based on seismic and tide gauge data alone, including over-warning that leads to unnecessary evacuations. He then describes the measurement (Chapter 7) and modeling (Chapters 8, 9, 10) technologies required to produce real-time forecasts. He then explains how to combine these two technologies into a real-time system capable of delivering a forecast before the tsunami arrives. Thus, accuracy, speed, and robustness must be guiding principles of the system. Accuracy is linked to Chapter 7, where deep-ocean measurements are available in real time to assimilate into numerical models, and Chapter 8, where benchmarking and testing models over a decade provide the confidence that the models are accurate. By pre-computing the propagation part of the model (based on historical sources described in Chapter 2) and selecting inundation models that can be run fast and efficiently (see Chapter 8), forecasts can be produced before the tsunami arrives at the forecast site. Robustness is defined by the reliability of the measurement system to deliver real-time data and the stability of the models to perform under a wide range of conditions. Titov demonstrates the integration of these elements into a timely, reliable forecast system by describing ten different tsunamis that were experimentally forecasted in real time. He shows the impact of these forecasts in Table 1. The November 2006 Kuril tsunami, also discussed in Chapter 9 on propagation modeling as a case study, is an excellent example of a Pacific-wide tsunami that caused no deaths, but provided tsunami data at many tide gauges, providing scientists with the perfect research tsunami. The United States will install the forecast system described in this chapter in its two warning centers in 2009. The author recommends more testing of the system with every new tsunami to identify and repair errors in the system. Every reader of this volume is encouraged to read this chapter as it provides a defined product resulting from the information contained in Chapters 2, 5, 7, 8, 9, 10, and 13.

Chapter 13, *Tsunami Warnings,* by Paul Whitmore, Director of the Alaska/West Coast Tsunami Warning System, is a comprehensive summary of tsunami warning operations and capabilities. He describes, with extensive references, the need for tsunami warning systems through a detailed history of tsunamis affecting the United States. He also describes the areas of responsibility for other tsunami warning centers throughout the world. He points out that the warning must be accompanied by the proper response of the population for the warning to be effective in saving lives and protecting property. He describes the observational networks, including real-time seismic and sea level data, required for a warning system to function. He then describes how the center processes these data to identify where the earthquake occurred, when it occurred, and what magnitude it represented. Based on earthquake location and magnitude thresholds, the warning center disseminates information in the form of information bulletins, tsunami watches, and tsunami warnings. If a warning is issued, the center will provide tsunami forecast information in the form of tsunami arrival time for predetermined locations. Then the center will process either deep-ocean or tide gauge data for confirmation that a tsunami exists. If a tsunami exists, the center continues to issue warning messages containing the latest tsunami data. It is the responsibility of local authorities to initiate evacuations. The states of Hawaii and Washington have procedures to evacuate if a warning is in effect for their state 3 hours prior to the tsunami arrival time at their coastline. Other states rely on state emergency management organizations to make these decisions. Once the recording at coastal tide gauges subsides, the center cancels the warning. A warning cancellation is issued when the center determines that the tsunami danger has passed. Other functions of the warning center include outreach and education on the products disseminated by the center. This chapter is linked to Chapter 5, earthquake generated tsunamis, Chapter 7, tsunami measurements, and Chapter 12, tsunami forecasting. The author recommends that the staff of the warning centers be properly trained and the center's operations be rigorously tested on a daily basis. The reader interested in tsunami warning operations should carefully review this chapter.

3. Concluding Remarks: Emergent Findings and New Directions

Since the watershed 2004 tsunami, new data and ideas are entering the scientific literature at a pace unprecedented in the history of tsunami science. Two of the most seminal advances in tsunami research since the Indian Ocean tsunami are: i) deep ocean tsunami measurements, and ii) their use together with advanced numerical models in accurately forecasting tsunamis after the tsunami has been generated. These two advances not only enable accurate real time tsunami forecasts, but also will lead to more insights into tsunami dynamics and help address the challenge of creating tsunami-resilient communities—our ultimate goal.

The overarching mission guiding the future direction of tsunami research is clear—to benefit society. In the case of tsunami, a community-based mitigation program, consisting of: an assessment of the tsunami hazard for each threatened community, an appropriate mitigation strategy to minimize the impact of future tsunamis, and an effective warning and education program to alert the at-risk populations, is required. One integrating principle that could be used to organize our future research is tsunami forecasting. A tsunami forecast system (Chapter 12) can now serve as an operational system that integrates the state of the science for tsunami generation, propagation, and inundation. Figure 1.7 presents the real-time forecast results for the November 15, 2006 Kuril Island tsunami. Numbered black line contours indicate the position of the advancing tsunami front in hours after the generation event. After passing the location of a DART system the data is inversely assimilated into an updated forecast for DART systems further out, and for the initialization of very high resolution inundation models at select coastal locations. Figure 1.8 shows a very good comparison between coastal tide gauge data and the inundation forecast.

If the operational forecast system had a research counterpart, then advances in generation, propagation, and inundation could be applied immediately to improve the existing forecast system. In the research mode, the forecast system could also be used to conduct hazard assessment studies by varying the location and size of potential tsunamis. The forecast approach has the advantage of having both regularly tested models (used in warning operations) and independent verification of results (tide gauges). This concept is not new to the scientific community, as many "community models" are used by researchers in a myriad of scientific fields. The reason these community models are so successful is that they reduce the research effort required by the individual scientist to stay active in the field. This characteristic is extremely relevant to tsunami science, which suffers from the cyclical nature of tsunami funding.

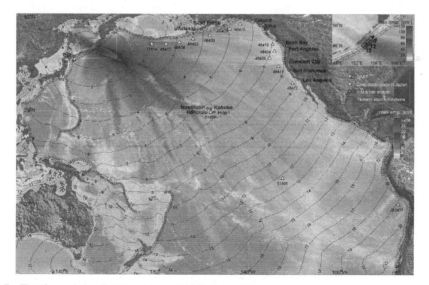

Fig. 1.7 – Test forecast for the November 15, 2006 Kuril Island tsunami. Color contours show maximum forecast tsunami amplitudes, yellow triangles show DARTs deployed at the time of the earthquake, red circles show locations of forecast sites, black rectangles indicate locations of sources for the pre-computed tsunami forecast database, star shows location of the earthquake epicenter. (Titov – Ch.12, Fig. 7, see color insert)

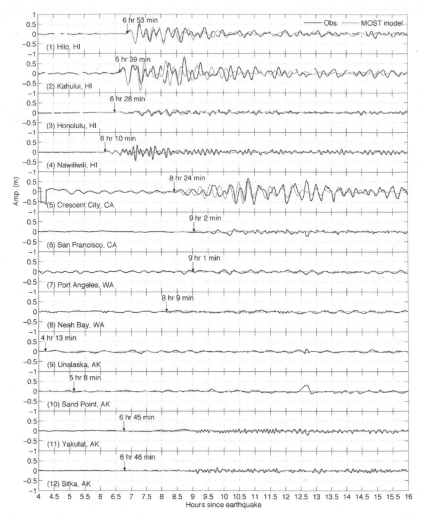

Fig. 1.8 – Comparison of high-resolution coastal forecast model results (grey lines) with de-tided gauge observations (black lines) for the November 15, 2006 Kuril Island tsunami. (Titov – Ch.12, Fig. 9)

Some important research topics identified in this volume include:

1. generation dynamics and real-time generation feature models
2. multiple generation mechanisms
3. simulations for propagation and inundation
4. effects of bathymetry for potential disaster sites
5. very high resolution simulations for inundation and destruction by debris
6. errors and uncertainty estimates for forecasts
7. numerical models for evacuation and other mitigation strategies
8. conducting offshore facility siting studies
9. estimating current velocities during inundation for construction design considerations

Many aspects of this research could be efficiently facilitated and enabled by a forecast research tool.

With some new directions identified, the challenge is to conduct tsunami research aimed at minimizing deaths from future tsunamis. One approach might be to apply the ideas of GeoHazards International (Samant et al., 2007) to use local advocates to identify community needs that can be met through research. If local advocates could interact effectively with research scientists in an ongoing workshop venue, the results could be very successful. The goal would be to conduct research to produce products that would be used immediately by the advocate community. Communication between research scientists and community advocates will be a challenge because of the lack of common technical language for both groups. For this reason, the workshops should be conducted in a way that is respectful of both groups' challenges and opportunities. For example, in Thailand, conducting community paleotsunami surveys has been an effective way to educate and inform the community of the tsunami hazard. The workshop format for scientists and advocates should encourage group exercises and minimize lecture style communication. The two groups do, however, have a common goal: to save lives from future tsunamis. This common goal is the most powerful motivator available. Hopefully, this motivation will overcome communications barriers and parochial interest for both groups and result in a powerful alliance that serves society well.

To be a successful alliance, however, both the scientist and the advocate must benefit. The advocate will benefit by a set of research products that makes a community more resilient to future tsunamis. The scientist will benefit by gaining an advocate for research, which may provide a more reliable source of research funding. The lack of sustained research is a serious problem for the tsunami scientist. The history of tsunami research funding has been a "boom" (immediately following a damaging tsunami) or "bust" (long periods of no damaging tsunamis) cycle. As a result, most tsunami scientists are part-time researchers whose funding follows the tsunami cycle. For the tsunami expert, this is a difficult way to stay abreast of tsunami science and to conduct long-term, basic research. For society, this is an inefficient way to prepare for the next tsunami because the trusted tsunami expertise required after the next tsunami may or may not be available. The consequence of not having trusted tsunami expertise is the onslaught of poorly qualified opportunists taking advantage of the disaster. Lacking trusted advice, the community then runs the risk of not applying local and donated resources wisely after the next tsunami. The scientist's expertise is compromised and dated because of the uncertainty of funding. In short, the scientist and advocate must use the alliance to survive and thrive between damaging tsunamis. A research forecasting tool may alleviate some of the effort required to stay abreast of tsunami research advances for both the advocate and scientists because the research forecasting tool would be continuously integrating new advances.

Graduate education is of utmost importance to maintain and enhance the community of tsunami scientists. Tsunami science faculty should include and encourage others to include tsunami-related topics in relevant graduate courses such as seismology, geology, oceanography, geophysical fluid dynamics, civil and ocean engineering, etc. Additionally, this volume is intended to provide the basis for one or more special topic courses on broader-based focused tsunami science as well as serving as an overview and reference source.

If the alliance is organized by nation, tsunami alliance members could serve on national committees responsible for all hazards to ensure that the tsunami hazard is properly represented. Alliance members could use every tsunami as a teaching opportunity for the public by holding press conferences or contributing to media stories on the accuracy of the tsunami forecast. The public is much attuned to the concept of forecasting, so education would be effective. Alliance members could organize tsunami awareness events to remind the public of this hazard. National workshops of the alliance could lead to products that would be immediately applied by the advocate community. With the aid of internet technology, national programs could be posted on web pages, in a common language, to promote international sharing of best practices and to coordinate research. Such research coordination, through a forecasting framework, may also allow scientists to participate in research beyond the capability of their nation. Such international research sharing is already underway in tsunami modeling as discussed in Chapters 8 and 12. An example of other international research sharing is the international survey teams, dispatched after every damaging tsunami, as described in Chapter 7. The tsunami research community, through an alliance-type mechanism, could pioneer a better way of conducting sustained research that truly benefits society.

The challenge is great, but the payoff is fewer casualties from future tsunamis.

References

Bernard, E. N. (ed.) 2005. *Developing Tsunami-Resilient Communities: The National Tsunami Hazard Mitigation Program*. Reprinted from Natural Hazards, 35(1), Springer, The Netherlands, 184 pp.

Bernard, E. N., L. Dengler, and S. Yim. 2007. National Tsunami Research Plan: report of a workshop sponsored by NSF/NOAA. NOAA Tech. Memo.OAR PMEL-133, 135 pp.

Cosgrave, J. 2007. Synthesis Report: Expanded Summary. Joint evaluation of the international response to the Indian Ocean tsunami. London: Tsunami Evaluation Coalition.

Iwan,W. D. (ed.) 2006. The Great Sumatra Earthquakes and Indian Ocean Tsunamis of 26 December 2004 and 28 March 2005 Reconnaissance Report. Earthquake Spectra Special Issue III, Volume 22, 900 pp.

Samant, L. D., L. T. Tobin, and B. Tucker. 2007. Preparing Your Community for Tsunamis: A Guidebook for Local Advocates, GeoHazards International, 54 pp.

Chapter 2.   TSUNAMI HISTORY: RECORDED

V. K. GUSIAKOV

*Tsunami Laboratory, Institute of Computational Mathematics and
Mathematical Geophysics, Siberian Division, Russian Academy of Sciences*

Contents

1. Introduction
2. Historical Tsunami Catalogs
3. Tsunami Databases
4. Problems with Historical Data
5. Tsunami Quantification
6. Geographical and temporal distribution of tsunamis
    in the historical records
7. Basic types of tsunami sources as historically recorded
8. Unusual cases of tsunami-like water disturbance
9. Conclusions
    References

1.   Introduction

Historical data on tsunami occurrence and coastal run-up are important for basic understanding of the tsunami phenomenon, its generation, propagation and run-up processes, its damaging effects. Such data are widely used for evaluating tsunami potential of coastal areas and for determining of the degree of tsunami hazard and risk for use in coastal-zone management and disaster preparedness. Also, historical data are of a critical importance for real-time evaluation of underwater earthquakes by the operational Tsunami Warning Centers, for the establishment of thresholds for issuing tsunami warnings and for design criteria for any tsunami-protective engineering construction.

In terms of documented total damage and loss of human lives, tsunamis do not come first among other natural hazards. With an estimated 700,000 fatalities, resulted from tsunamis for all historical times (Gusiakov et al., 2007), they rank fifth after earthquakes, floods, typhoons and volcanic eruptions. However, because they can affect densely populated and usually well-developed coastal areas, tsunamis can have an extremely adverse impact on the socioeconomic infrastructure of society, which is strengthened by their full suddenness, terrifying rapidity, and

The Sea, Volume 15, edited by Eddie N. Bernard and Allan R. Robinson
ISBN 978–0–674–03173–9 ©2009 by the President and Fellows of Harvard College

their potential for heavy destruction of property and high percentage of fatalities among the population exposed to their action.

There is evidence that tsunami as a catastrophic natural phenomenon has been known by humankind since antiquity. Many languages have a special word for this type of disaster coming from the sea—*tidal waves, seismic sea waves* (English), *raz de maree, vagues sismiques* (French), *flutwellen* (German), *maremoto* (Spanish), *vlogengolden* (Holland), *tsunami* (Japanese), *hai-i* (Chinese), *loka* (Fijian). The earliest tsunami event, recorded in historical chronicles, occurred in the eastern Mediterranean off the coast of Syria in 2000 B.C (Ambraseys, 1962). The second historically known case is a destructive tsunami in the Aegean Sea generated by the catastrophic eruption of the Santorini volcano on Thera Island dated to 1628 B.C. (Marinatos, 1939). For the whole B.C. period, 23 historical events are presently known. Most of them occurred in the eastern Mediterranean whose coastline was densely populated since the ancient times and was repeatedly damaged by tsunami waves. Despite a long and close relationship between humans and tsunamis, until recently they had not been considered worth of independent historical compilation, and the data on tsunamis were typically included in catalogs of other natural hazards such as earthquakes, volcanoes, hurricanes and other dangerous natural phenomena.

2. Historical Tsunami Catalogs

To the best of the author's knowledge, the first historical tsunami catalog was compiled by N. H. Heck, former Director of the US Coast and Geodetic Survey and Chairman of the American Committee of the International Commission of Raz de Maree, IUGG, who summarized the tsunami data from previous earthquake catalogs. Initially, this catalogue was published in French (Heck, 1934), the English, slightly updated version of the catalog appeared in 1947 as a journal article published in BSSA (Heck, 1947). Undoubtedly, this latter publication was inspired by the April 1, 1946 Aleutian tsunami that caused damage and losses of life on Hilo, Hawaii and further damage on certain islands in the Pacific.

The revised tsunami catalog (Heck, 1947) had a global coverage and contained 270 events spanning from 479 BC until AD 1946. The format of this catalog was just a list of tsunami observations with date, descriptive location, short account on the tsunami effect and bibliographical references. There is no specific information related to tsunami source or to coastal impact. Also, there is no quantification of tsunami intensity, only reference to the degree of associated earthquake destruction (I – moderate, II – strong, III – widespread destruction). The author notes, however, that "for older sources, the existence of record, no matter how bare or brief, indicates that the wave must have been large". Since Heck's compilation was mainly based on the available earthquake catalogs, most tsunamigenic events in his catalog are of seismic origin with few exceptions for well-known volcanic events such as the 1792 Unzen and 1883 Krakatau tsunamis.

N. H. Heck was well aware of incompleteness of his tsunami catalog and its dependence on such factors as the density and cultural level of population on the coast. He wrote, for example, that "many waves have undoubtedly occurred to the eastward of New Guinea, but the record for that region is very scant" (Heck, 1947,

p. 269). He made also an important observation that "with respect to well populated shores, absence of records may be accepted as meaning no occurrences; this conclusion may or may not be valid, however, with respect to sparsely settled coast". In a brief introduction to his catalog, he wrote that "in spite of the paucity of data for many regions and the certainty that many tidal waves are missing, there is little doubt that the list gives a good idea of the distribution of such waves and indicates where they are likely to occur" (Heck, 1947, p. 270). Finally, he indicated the immediate practical application for the catalog compiling noting that "appraisal of probability of occurrence and of the type of such waves can be made only from the historical records. The advisability of appraisal is evident from the needs of mariners and from the possible hazards of military and naval bases so placed as to be in jeopardy" (Heck, 1947, p. 270).

Heck's catalog is the only global tsunami catalog published to date. All subsequent catalogs have been compiled on the regional or national basis. Despite the briefness and incompleteness of Heck's catalog, for number of years it was the only source of information for experts studying the tsunami problem until other historical compilations began to appear in Japan, Russia, USA and other countries.

All the historical tsunami catalogs (there are more than 120 of them, published so far, a comprehensive list can be found at http://tsun.sscc.ru/tsulab/tsu_catalogs .htm) can be divided into two large groups: *descriptive* and *parametric*. Descriptive catalogs are a compilation of original descriptions of tsunami coastal effect and resulted destructions retrieved from the primary reports, scattered in different publications sometimes with a very difficult access (Heck, 1934, 1947; Imamura, 1949; Takahashi, 1951; Agostinho, 1953; Iida, 1956; Berninghausen, 1962, 1964, 1966, 1968, 1969; deLange and Healy, 1986; Zayakin and Luchinina, 1987; Murty and Rafiq, 1991; Lander et al., 2002). Quite often the compilers just repeat the original description of unusual water behavior using different styles, formats and approaches for data selection. In these catalogs, further interpretation of a described phenomenon (for example, classification, localization and quantification of its source) is left for their readers. The quantitative data in the descriptive catalogs are scattered through the text and are not easy to be retrieve and process. These limitations restrict to some extent further application of these catalogs in the tsunami research. However, they are still of great value as indicators to the degree of tsunamigenic activity and a resulting hazard for a particular region.

The second type of the tsunami catalog is parametric, where gathered information is presented in table form, listing tsunami events in chronological order and providing some set of the basic parameters on each event (Soloviev, 1978; Iida, 1984; Papadopoulos and Chalkis, 1984; Hamzan et al., 2000; Papadopoulos, 2001). These parameters varies from simple locality and magnitude of a tsunami source to very detailed sets of source parameters provided in the recent tsunami databases such as supported by NOAA's National Geophysical Data Center in Boulder, USA (http://www.ngdc.noaa.gov/seg/hazard/tsu.shtml), Tsunami Laboratory in Novosibirsk, Russia (http://tsun.sscc.ru/htdbwld) or University of Bologna, Italy (Tinti et al., 2001). A major problem with parametric catalogs is that they usually include very little original descriptive information on tsunami manifestation and thus force a reader to rely upon the interpretation made by a catalog compiler.

Another problem is that there are differences in scales used for quantification of tsunamis. It was quite typical that compilers of earlier catalogs proposed their own scales for measuring tsunami intensity, as an example we can indicate to Sieberg's (Sieberg, 1927), Iida-Imamura's (Imamura, 1942; Iida, 1958), Soloviev-Imamura's scales (Soloviev, 1972). Another problem is the reporting of maxima which might not be representative, the reporting of tsunami inundation depth instead of run-up height, or maximum amplitude instead of maximum wave height measured on mareograph records, and the potential lack of means to differentiate these values.

Some of the published catalogs have both descriptive and parametric parts, the latter presented as the tables with basic source parameters of tsunamigenic events retrieved from or estimated on the basis of historical descriptions collected in the descriptive part. As the examples of these catalogs we can indicate to Everingham, 1977, 1987; Fernandez et al., 2000; Fokaefs, Papadopoulos, 2007; Lander et al., 1993; Lander, 1996; Lockridge et al, 2002; O'Loughlin and Lander, 2003; Papadopoulos, 2000; Papadopoulos et al., 2007; Soloviev and Go, 1974, 1975; Soloviev et al., 1992; Soloviev et al., 2000; Stephenson et al., 2007; Watanabe, 1989.

3. Tsunami Databases

Even the most complete historical tsunami catalogs, having both descriptive and parametric parts, such as Soloviev and Go (1974, 1975) or Lander (1996) catalogs have somewhat limited application in the tsunami research, because the data and information from these catalogs cannot be easily retrieved and handled. Present-day information technology demands the organization of data in the form of computerized databases, where data can be kept in a constantly updated and active form and are easily accessible. The information from a database can be quickly retrieved in many different ways and formats, and can be transferred to other relational databases and to data processing and visualization programs.

Conversion of descriptive catalogs into parametric databases is not a trivial task and quite often presents a number of specific problems to be solved. One of the problems results from a fundamental feature of tsunami waves, namely, their ability to propagate over great distance from the source area. As distinct from earthquake cataloguing, an observation of unusual wave activity near a particular coast may relate to a source in quite a remote part of the same oceanic basin (or even another basin). In the database creation, this problem is usually resolved by dividing all the data into two parts—the tsunami event catalog and the tsunami run-up catalog. The event catalog contains the list of tsunamigenic sources, usually arranged in chronological order. The run-up catalog contains the list of observed and measured tsunami wave heights provided with location names or even exact geographical coordinates of observational sites. The relation between these two tables is provided through a special event identification number or, quite often, through the full date of the event (that includes also the source time) and is considered to be a unique characteristic of any record in the tsunami event table.

Initial work on the development of a computerized tsunami database was started at the International Tsunami Information Center (ITIC) in Honolulu, Hawaii (USA) in the mid-1970s. Following this initial work and in response to the recommendation of the ITSU-XI (Summary Report . . . , 1987), a standardized database format was developed, and the first tsunami database was compiled from many available sources and distributed through the ITSU National Contacts (Pararas-Caraynnis, 1991). However, at that time, the ITIC efforts were not sufficiently supported both financially and conceptually. Also, few tsunami data were available in the computer readable form. Therefore, the progress in further data collection was slow, and the proposed format did not become mandatory for data compilers and database developers.

In the middle of 1980s, the NOAA's National Geophysical Data Center (NGDC/NOAA) in Boulder, Colorado (USA) began compilation of quantitative tsunami data from all available catalogs and many special studies of tsunamis, and initiated their conversion into a computerized form. For a number of years, the NGDC/NOAA World-Wide Tsunami Database (Lockridge, Dunbar, 1995) remained the only source of tsunami information available in digital domain and was used to create tsunami databases in several research institutions and operational centers. Unfortunately, until recently their large data set, initially compiled in the 1980s, has not been subject to careful refining, checking for errors and matching to later catalogs and research publications. These limitations affect to some extent the value of a large amount of gathered and digitized information and its application in the tsunami research. However, at present the NGDC/NOAA World-Wide Tsunami Database remains the most frequently cited source of historical tsunami information.

The next step in the tsunami database development was undertaken in the beginning of the 1990s within GITEC (Genesis and Impact of Tsunamis on the European Coast) Project initiated in 1992 by the University of Bologna, Italy (GITEC, 1992). One of main outputs of this project was the development of the comprehensive historical tsunami database for the Mediterranean and other European surrounding seas which summarized data of numerous published historical catalogs for this region (Tinti et al., 2001, 2004). The present version of the European Tsunami Catalog (ETC) is being developed and maintained within the Workpackage 1 (Tsunami Catalogue) of the TRANSFER (Tsunami Risk And Strategies For the European Region) Project launched in 2005 by the European Community (TRANSFER, 2005).

Another initiative in tsunami database development was undertaken in the middle of 1990s by the Novosibirsk Tsunami Laboratory (NTL) of the Institute of Computational Mathematics and Mathematical Geophysics of the Siberian Division of Russian Academy of Sciences (ICM&MG SD RAS) under the Expert Tsunami Database (ETDB) Project (Gusiakov et al., 1997). The concept of this project is based on integration of observational data, numerical models and analytical and processing tools with the visualization and mapping tools within a single software package. Therefore, from the beginning attention was paid to development of a geographical mapping subsystem intended for easy data retrieval and visualization. Another important feature of this project is that the ETDB is intended to be a multi-entry database, which means that a particular tsunamigenic event is provided with the full set of original data and information retrieved from different sources, thus giving the user a possibility to make his/her own interpretation and judgment.

The Historical Tsunami Database for the Pacific (HTDB/PAC) Project was initiated by the IUGG Tsunami Commission (IUGG/TC) in 1995 under the leadership of the NTL/ICMMG. By that time, the NTL/ICMMG had the basic parametric tsunami catalog for the whole Pacific compiled from a variety of sources that was provided with the specialized graphic shell written in Turbo-Pascal.

At present, there are two global historical tsunami databases maintained separately by the NGDC/NOAA in Boulder, USA and the NTL/ICMMG in Novosibirsk, Russia. The NGDC/NOAA database is maintained in the Oracle RDBMS from where the data can be accessed via Web-based HTML forms and ArcIMS interactive maps at http://www.ngdc.noaa.gov/hazard/tsu.shtml as well as exported in several different formats. The NTL/ICMMG database is maintained in the MS Excel with a specialized graphic shell (PDM_TSU) providing screen forms for data editing, retrieval and listing. As a stand-alone application, the NTL/ICMMG database is distributed on a CD-ROM (ITDB/WLD, 2005). This application is provided with a specially developed GIS-type graphic shell (WinITDB) allowing the fast and efficient manipulation of maps, models and data. The web-version of this database is accessible at http://tsun.sscc.ru/htdbwld.

The content of these two databases is fairly close in terms of total number of historical events, temporal and spatial coverage and basic source parameters. However, for many historical events they differ in types of origin, number of available run-up observations, resulting fatalities and degree of validity for some older historical events. At the ITSU-XIX (Wellington, 2003) the ITIC, WDC/NOAA, and NTL/ICMMG were tasked to implement a Global Tsunami Database (GTDB) by merging the content of these two existing databases into a single unified data set. This work, being implemented under the GTDB (Global Tsunami DataBase) Project (Gusiakov, 2003), is still in progress, since it requires application to the primary sources of historical information to resolve the existing uncertainties in the parameters of many old historical events. Further analysis of historical data, given in this Chapter, is based on the content of both databases, generally referred to as the GTDB catalog, and making reference to a specific database if needed.

4. Problems with Historical Data

Anyone dealing with historical tsunami data should bear in mind several intrinsic problems closely associated with this type of information. These problems result from inaccuracy and from the fragmentary nature of available information about old or geographically remote events. Quite frequently, the information on an older event is so incomplete that it is difficult to make a reliable judgment on the nature of reported phenomenon and/or to evaluate its physical scale. Basically, there are two types of errors in tsunami catalogs. The first is confusion of tsunamis with other hazardous natural phenomena (e.g., storm surge, high tide, river flood, rogue waves). The second is errors in interpretation of available descriptions in order to retrieve the basic parameters of an event (date and time, location, intensity, type of a source).

The main reason for confusion with other phenomena is the scarcity of information and the lack of details in descriptions of reported events. Regarding this type of errors, the catalog compilers can do almost nothing but assign a low validity index 1 (very doubtful) or 2 (doubtful) to events with doubtful nature, thus alerting users to practice caution in treating said data. Sometimes it happens that additional data are found later thus allowing one to resolve the uncertainty and to increase its validity index up to 3 (probable) or 4 (definite) or alternatively, to exclude the event from the list. In practice, in databases the latter events are not excluded at all, but are kept on the list with validity 0 (false entry) to prevent re-entry because information about these events exists in archives and literature. Both NGDC/NOAA and NTL/ICMMG databases currently have about 5–6% of all entries with validity 0.

The validity index relates not only to the degree of confidence that a particular event is a tsunami but also to its relationship to the indicated source on the date given. Indeed, the correct date is one of the main problems for older historical events. The reasons include the poor event dating in the primary reports and using different calendar systems. For example, the Gregorian calendar was proclaimed in 1582, and then gradually adopted by most but not all Christian-dominated countries over the next 200 years. Another good example is the dating of the Aegean Sea tsunami resulted from the catastrophic Santorini eruption. Historical catalogs indicate very different dates for this event spanning over 270 years from 1380 BC (Soloviev et al., 2000) to 1650 BC (NGDC/NOAA database). In the NTL/ICMMG database, we have adopted 1628 BC for this event, based on (Papadopoulos, 2001) and tree-ring dating for Santorini (Baillie and Munro, 1988).

The location of a source for historical tsunamis is commonly a problem because tsunami waves can propagate over a great distance. Historical documents usually report on tsunami manifestation at some *coastal* location, but association of these reports with some earthquake, volcanic eruption or landslide that may have occurred a hundred or a thousand miles away is a complicated task. For this reason even in recently published catalogs so many old events have not been assigned the source coordinates and contain only indication to the general area of tsunami manifestation (like SW Portugal, Azores, Baleares, etc.). Such an approach, however, is unacceptable for a database compiler, because most data retrievals start from selection of a geographical area for the data search. Therefore, in the NTL/ICMMG database, for example, we try to assign source coordinates to as many events as possible, based on all sorts of available information even if the accuracy of such an assignment is poor (order of ±2°). However, in the NTL/ICMMG catalog up to 140 events still lack any source coordinates.

Another problem concerns an important group of parameters describing the "size" of a tsunami. Many of old reports do not contain any quantitative indications to run-up height, inundation depth, or a measure of in-land flooding. Quite often, available documents mention only the damage to vessels and buildings, sometimes they tell something about human fatalities. If any height estimates in old reports are given, they should be treated with care, because eyewitnesses are typically not trained observers and their reports can be greatly exaggerated or simply erroneous.

A potential problem also results from conversion of old measuring systems to the modern metric system. For instance, the old Russian fathom had several metric equivalents (from 1.62 to 2.16 m), and sometimes it is not clear which unit was used by an eyewitness. Even in contemporary reports, feet are occasionally mixed with meters and inches with centimeters. Another problem is the implied accuracy of the estimates given. For example, the report given by S. Krasheninnikov for a

maximum run-up height of 30 fathoms for the 1737 Kamchatka tsunami (Krasheninnikov, 1755) can be an approximation meaning something between 25 and 35
fathoms, and its conversion to the metric system (63 m) should not be considered
to have 1 meter accuracy. Besides, a single report, even given by a scientist
(S. Krasheninnikov was a staff naturalist of the Second Russian Research Expedition sent to Kamchatka in 1737–1741) should be treated with care, especially when
the data are not confirmed by a contemporary study of their physical or geological
traces.

Another important issue in cataloguing historical tsunamis is the nature of data
on human fatalities and on damage. Usually, these data are the very numbers that
mass media and general public are most interested in, but at the same time they
are the most difficult parameters to resolve. First of all, old historical reports rarely
give exact numbers of fatalities. Typical wording is "many people washed away" or
"all villagers drown". When digits are available, quite often they relate to the total
number of victims of both the source event (such as an earthquake) and the tsunami. For example, the widely cited 60,000 fatalities for the 1755 Lisbon tsunami
will probably never be resolved in terms of the death toll from the parent event
(earthquake) and from the resulting tsunami. (In Table 1 we adopt 30,000 as very
rough approximation of tsunami-related fatalities from the Lisbon event). The
same is true for the 1815 Tambora tsunami caused by the most catastrophic volcanic eruption of the last millennium. Death toll from this event is poorly constrained and sometime includes even those who died from starvation resulted from
the ash fall that killed vegetation over the large area (NGDC/NOAA Historical
Tsunami Database). Death tolls for a number of most destructive ancient tsunamis
like 1628 BC Santorini, AD 416 Java, AD 1452 Kuwae are not known at all. Even
for the recent events, collection of reliable data on human fatalities presents difficulties for catalog compilers. Usually, these data are taken from mass media reports and newspaper articles that rarely make reference to their information
sources. Being uncritically adopted and cited in the field reports and even peer-
reviewed articles, these numbers obtain the status of "scientific data", but they are
not. As the most recent example of, we can refer to the 2004 Indian Ocean tsunami. The estimates of the number of fatalities circulating in mass media, reported
in scientific publications and accessible through the Internet, vary from 180,000
to 300,000 (including missing people). Since in many affected countries (except
Thailand), burial of victims took place without any body identification and sometimes even without body counting, and because many people were washed to
sea, the actual numbers of fatalities will never be known. At present, the most
realistic estimates of the human toll for the 2004 Indian Ocean tsunami are available from UN Office of the Special Envoy for Tsunami Recovery (http://www
.tsunamispecialenvoy.org/country/humantoll.asp). Their site lists 186,983 persons
as killed and 42,883 as missing, giving a total of 229,866 victims. The site does not
give any supporting data for these digits (such as breakdown of losses by coastal
communities), but at least this number conforms to the published statistics of human losses in the most affected countries (Indonesia 167,736, Sri Lanka 35,322,
India 18,045, Thailand 8,212).

5.   Tsunami Quantification

As noted above, one of the main problems in cataloguing historical tsunamis is the measure of the overall "size" or "force" of an event. To compare different tsunamigenic events, we need some scale by which to measure them. The best parameter for estimating the size of different tsunamis would be their total energy. However, this value is not easy to calculate because it requires knowledge of tsunami waveforms at different locations, covering all possible propagation directions, and that is not always the case even for most recent tsunamis.

There are two types of scales for measuring the "size" of a hazardous natural phenomenon—the magnitude scales and the intensity scales. Magnitude scales relate to the source area of an event, while the intensity scales describe the resulting effects at different locations. So, one event can have a single value of its magnitude and many values of its intensity. Both scale types can be descriptive (e.g., the Mercalli scale for intensity of seismic shaking), or quantitative, based on measuring some physical parameter characterizing the source of an event (e.g., the Richter scale for the magnitude of earthquakes, the VEI scale for quantification of explosive volcanic eruptions) or combined, containing both a descriptive part and quantitative values for some measured parameter (e.g., the Saffir-Simpson hurricane scale, the Beaufort wind scale).

Historically, the first scale proposed for measuring a tsunami was the Sieberg scale (Sieberg, 1927). It was a descriptive intensity scale, based on the destructive effect of a tsunami, consisting of only four grades and not including any quantitative measures of tsunami wave height. Ambraseys (1962) slightly modified this scale, making it 6-grade, by dividing the upper grade into three additional grades. This 6-grade scale was mainly used for quantification of the Mediterranean tsunamis, most of them being old historical events with limited descriptions that often did not contain any quantitative values. Although both of these scales are typical intensity scales, based only on local tsunami effects at the coast, from the very beginning they were used for characterization of overall tsunami size (i.e., as magnitude scales) by assigning to a tsunamigenic event its maximum observed intensity at the coast. This practice is in fact still used by cataloguers of pre-instrumental historical earthquakes.

Another scale for the tsunami quantification was introduced in 1942 in Japan (Imamura, 1942). This scale was also descriptive, consisting of five grades (0 to 4), but the description of each grade contained some quantitative parameters—run-up heights and extension of the coast flooded by the waves. A. Imamura called it a *tsunami magnitude* scale, but it is a typical example of *intensity* scale, since it is based on coastal tsunami effects and does not contain any correction for distance from source.

Later K. Iida (Iida, 1963) modified this scale by adding one additional grade ($m = -1$) for characterization of weak tsunamis. He was also the first who directly connected the grade number m with a maximum observed run-up value at the coast H_{max} by the formula

$$m = \log_2 H_{max} \tag{1}$$

This so-called Imamura-Iida intensity scale was widely used in cataloguing historical Pacific tsunamis and, as a magnitude scale, for overall quantification of tsunamigenic events in the catalogs (the latter, possibly, because a large tsunami has
many reported coastal run-up observations, but only one *maximum* run-up value).
H. Watanabe (Wanatabe, 1963) was the first to point out this contradiction and
proposed to use the term *tsunami magnitude* for a value defined as

$$m = \lg H_0, \tag{2}$$

where H_0 is the wave height in the open sea at the edge (boundary) of a tsunami
source. This proposal, of course, was not implemented, because there was no practical way to measure such a parameter in the open sea.

Taking into account this difficulty (mixing of intensity and magnitude scales,
based on the same parameter), Shuto (1993) proposed to use the formula

$$i = \log_2 H, \tag{3}$$

where H is a local tsunami height in meters, for defining *tsunami intensity scale i*, to
be used for quantification of tsunami damage at the coast. His 6-grade (from 0 to
5) scale contains the description of expected damage for boats and different types
of constructions tabulated on the basis of the H value.

Another modification of this scale was made by S. Soloviev (Soloviev, 1972),
who proposed to calculate the tsunami intensity according to the formula

$$I = \tfrac{1}{2} + \log_2 H_{av}, \tag{4}$$

where H_{av} is the *average* wave height along the nearest coast. Soloviev argued that
this value is a more steady characteristic of a tsunami and closer relates to the total
tsunami energy radiated from a source. With this scale, Soloviev evaluated the
intensity for large number of Pacific tsunamis during compilation of his catalogs
(Soloviev, Go, 1974, 1975; Soloviev, 1978). The I scale is also used in the NGDC/
NOAA and NTL/ICMMG global tsunami databases as the main parameter characterizing tsunami size.

T. Hatori (1986) attempted to formalize calculation of tsunami magnitude m on
the Imamura-Iida scale by taking into account the propagation distance, that is, to
convert it from the an intensity scale to a fully magnitude scale. He proposed to
calculate m value by the following formula:

$$m = 2.7 \lg H + 2.7 \lg R - 4.3, \tag{5}$$

where H is a tsunami wave height at a coastal observation point, and R is the distance from this point to the tsunami source along the wave propagation path. Hatori's proposal was quite reasonable but did not get adopted by practitioners due
to unevenness of wave heights along the coast and to uncertainty involved in determination of the propagating distance. A typical dimension of a tsunami source is
about 100 km, which is comparable to or even greater than a distance to the

nearby coast, so that the uncertainty in the travel path length can be on the order of 100%.

A new, real magnitude-type scale M_t based on instrumental measurement of tsunami height, was introduced in 1979 by K. Abe (Abe, 1979, 1981) who proposed to calculate M_t based on a maximum amplitude of tsunami waves that were recorded by tide-gauges according to the formula

$$M_t = a \lg H + b \lg R + D, \tag{6}$$

where H is a maximum tsunami-wave amplitude (in m) measured by tide gauge, R is the epicentral distance (in km) and a, b and D are constants that are determined to make the M_t scale closely related to the earthquake M_w (moment-magnitude) scale. The M_t scale is a real magnitude scale because it is based on quantitative parameters (instrumental wave height) and includes correction for propagation distance. K. Abe determined and published the M_t value for almost 200 large tsunamis occurred in the Pacific since tide gauges came into use. However, old historical events and most of weak contemporary tsunamis do not have M_t values thus limiting to some extent usage of the M_t scale for overall size comparison of tsunamigenic events.

Another magnitude-type scale, based on the total tsunami energy and providing a wide, actually unlimited range for tsunami quantification, was proposed by T. Murty and H. Loomis (Murty and Loomis, 1980). Their ML value is defined as

$$ML = 2 (\log E - 19), \tag{7}$$

where E is tsunami energy in ergs. In their initial publication (Murty, Loomis, 1980), ML values were determined for about 25 of the largest Pacific tsunamis, and since that time almost no new determinations of ML have been made. The reason is difficulties involved in the tsunami energy calculation. In the future, however, this scale can become more applicable for quantification of contemporary tsunamis, based, for instance, on knowledge of detailed distribution of initial displacement in a tsunami source.

A new 12-grade *intensity* scale for quantification of coastal tsunami effect was proposed by Papadopoulos and Imamura (2001). A detailed description of each grade, based on (a) effects on humans, (b) effects on vessels and nature, and (c) damage to buildings, was elaborated on the basis of analysis of well-documented damaging effects of the recent destructive tsunamis in the Pacific. The new scale was made consistent with several 12-grade seismic intensity scales used in seismology for quantification of seismic shaking effects (Papadopoulos, 2003). Although no quantitative parameter is used in definition of each grade, in their original publication Papadopoulos and Imamura provided a table of a possible correlation of grade number I_{PI} on their scale with grade number i (and, therefore, with local run-up height H) on the Shuto scale as defined by (3).

In the following analysis of historical data I use mainly the tsunami intensity I on the Soloviev-Imamura scale, considering it, among other existing scales, to be most closely related to the overall size of a tsunami. This scale is incorporated in two most complete historical tsunami databases (maintained by the NGDC/ NOAA and NTL/ICMMG SD RAS), and, what is most important, the intensity I is now determined for more than 2/3 of all historical tsunamis worldwide thus allowing to rank the different tsunamigenic events by their overall size.

6. Geographical and temporal distribution of tsunamis in the historical records

The present version of the GTDB catalog on tsunamis and tsunami-like events covers the period from 2000 BC till present and currently contains nearly 2130 historical events, having the validity index equal to or more than one. Of these events, 1206 occurred in the Pacific, 263 in the Atlantic, 125 in the Indian Ocean and 545 in the Mediterranean region. The geographical distribution of sources of historical tsunamis is shown in Fig. 2.1. When analyzing this map, one should take into account that it reflects not only the level of tsunami activity, but also the regional historical and cultural conditions that strongly influence the availability of the historical data. From geographical distribution of tsunamigenic sources, we can see that most known tsunamis have been generated along subduction zones and major plate boundaries in the Pacific, Atlantic and Mediterranean regions. Very few historical events occurred in the deep ocean and central parts of marginal seas, except several cases of small tsunamis originating along middle-ocean ridges and some major transform faults.

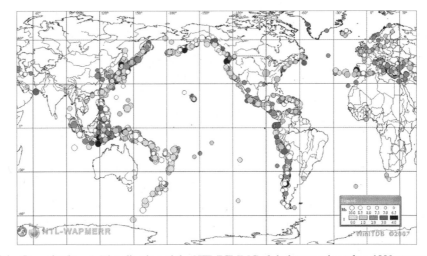

Fig. 2.1 – See color insert. Visualization of the NTL/ICMMG global tsunami catalog. 1990 tsunamigenic events with identified sources are shown for the period from 2000 BC to present time. Size of circles is proportional to event magnitude (for seismically induced tsunamis), density of gray tone represents tsunami intensity on the Soloviev-Imamura scale. White circles show the events with unknown intensity.

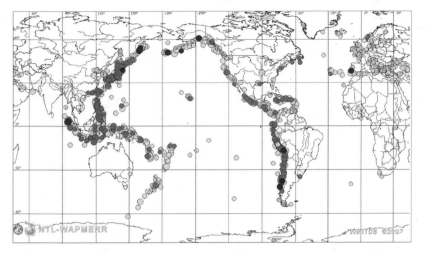

Fig. 2.2 – See color insert. The same data as in Fig.2.1. but presented with 1990 historical tsunamigenic events having identified sources divided into three groups: 1 – transoceanic tsunami (black) (see text for definition), 2 – regional tsunami resulting in human fatalities (dark gray), 3 – all other tsunamis (light gray).

In Fig. 2.2 the same data are shown in slightly different way. All the tsunamis are divided into three groups—(1) trans-oceanic, (2) resulting in human fatalities, and (3) non-fatal tsunamis. Trans-oceanic events are defined according to a formal criterion—those events whose reported run-up reaches 5 m or so at a distance exceeding 5000 km from the source, thus able to produce considerable damage on the opposite side of an oceanic basin. From the 2130 tsunamigenic events only 11 met this criterion, and all of them occurred during the last 250 years. They are shown in Fig. 2.2 in red color and listed in Table 1 with their basic source parameters. All these events were highly destructive and fatal—altogether they are re-

TABLE 1.

List of historical trans-oceanic tsunamis (see text for definition). M – magnitude (macroseismic, M_s or M_w), I – tsunami intensity on the Soloviev-Imamura scale, H_{maxNF} – maximum reported run-up in the near field in m, H_{maxFF} – maximum reported run-up in the far field (more than 5000 km) in m, FAT – number of reported fatalities due to tsunami.

Date and place	M	I	H_{maxNF}, m	H_{maxFF}, m	FAT
1 November 1755, Lisbon	8.5	4.0	30.0	7.0	30,000
7 November 1837, Chile	8.5	3.0	8.0	6.0	many
13 August 1868, Chile	9.1	3.5	15.0	5.5	612
15 June 1896, Sanriku	7.4	3.8	38.2	5.5	27122
3 February 1923, Kamchatka	8.3	3.5	8.0	6.1	3
1 April 1946, Aleutians	7.9	4.0	42.2	20.0	165
4 November 1952, Kamchatka	9.0	4.0	18.0	9.1	>10,000
9 March 1957, Aleutians	9.1	3.5	22.8	16.1	none
22 May 1960, Chile	9.5	4.0	15.2	10.7	1,260
28 March 1964, Alaska	9.2	4.5	68.0	4.9	221
26 December 2004, Sumatra	9.3	4.5	50.9	9.6	229,866

sponsible for 274,000 (39% of the total) fatalities. The remainder of documented
fatalities (426,000) occurred in 213 regional and local tsunamis. Thus, in all, tsunamis are responsible for about 700,000 human fatalities during all the historical
time. With this number, tsunamis rank fifth among other natural catastrophes,
after earthquakes, floods, hurricanes and volcanic eruptions. However, in the third
millennium, after the 26 December 2004 Indian Ocean tsunami, tsunami fatalities
rank first, and probably will, until the next big earthquake in a large metropolitan
area or a catastrophic monsoon flood in the overpopulated Bangladesh coast.

The temporal distribution of historical tsunamis is shown in Fig. 2.3 for the last
1000 years. The chart clearly demonstrates that historical data have highly non-
uniform distribution over time with three quarters of all the events reported within
the last two hundred years. In all the regions (except, possibly, Japan) there are
obvious gaps in reporting even large destructive events for the period preceding
the XIXth century. Although the total duration of the global historical catalog
exceeds 4000 years, its median date, dividing the data into two equal parts, lies
around 1885.

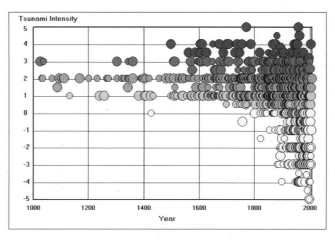

Fig. 2.3 – Recorded tsunami occurrence in the World Ocean vs time since AD 1000. Events are shown
as circles with the density of gray tone depending on tsunami intensity (on Soloviev-Imamura scale) and
size proportional to the earthquake magnitude. Systematic data on weak tsunamis appear in the catalog
only from the end of XIX century when the tide gauge network was put in operation. For the period
earlier than the XV century the data are fragmentary even for the large events.

Length and to some extent completeness of historical catalogs for different
tsunamigenic regions can be seen in Fig. 2.4. Patterns reflect, first of all, availability
of historical data which, in turn, strongly depends on population density and cultural features of a region. The longest catalog exists for the Mediterranean (4000
years, the first 2000 years not illustrated in Fig. 2.4). The second longest catalog,
exceeding 1500 years, exists for Japan and neighboring seas. For many areas, like
Kuril-Kamchatka, Aleutians, Central and South America, New Zealand, Philippines and Indonesia, the catalog starts with the first arrival of European travelers

in the area. The temporal distribution of events in the regional catalogs is very uneven, typically with a large gap (up to several centuries) between the first event in the catalog and the beginning of more or less constant coverage, at least in terms of destructive events.

Fig. 2.4 – Comparative length and completeness of regional tsunami catalogs and entire global catalogue for the last 2000 years. Each tsunamigenic event is represented by a vertical line on the horizontal time axis.

For all the tsunamigenic regions, the most complete data exist for the last 100 years, when the instrumental measurements of weak tsunamis became available and the reports on all or almost all of damaging tsunamis were carefully collected. Completeness of the catalog for the XXth century can be confirmed statistically because the position of its median date that lies around 1951. In 1901–2000, a total of 990 tsunamis were historically recorded in the World Ocean, an average rate of about ten tsunamigenic events per year. Most of these tsunamis were weak, observable only on mareograph records. About 260 tsunamis were "perceptible", defined as having a run-up height exceeding one meter. Among them, in 33 cases run-up greater than one meter was observed and recorded at a distance exceeding 1000 km from the source. During the previous century, five trans-oceanic tsunamis, all in the Pacific, occurred and all these events fall in an 18-year time interval (1946 Aleutians, 1952 Kamchatka, 1957 Aleutians, 1960 Chile, 1964 Alaska).

The histogram of tsunami occurrence in the World Ocean during the XXth century is shown in Fig. 2.5. The rate of tsunamigenesis over that century is more or less constant with only one distinct minimum of damaging events that had place in the 1980s. This minimum is the lowest in more than two last hundred years. Possibly, the low tsunami rate in this decade led to many comments on "the increased level of tsunami activity" in the 1990s. However, in the longer historical perspective, the 1990th were just a return to the normal, long-term average in tsunami occurrence.

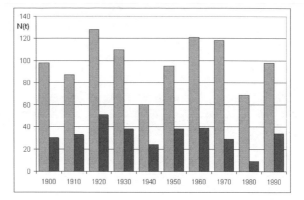

Fig. 2.5 – Histogram of global tsunami occurrence (in 10-year intervals) for the period from 1901 to 2000. Grey tone shows all tsunami, black tone – damaging tsunami with intensity $I > 1.0$.

7.   Basic types of tsunami sources as historically recorded

Analysis of distribution of historical tsunamigenic events over the types of sources is based on the content of the NTL/ICMMG database, since the NGDC/NOAA database uses slightly different classification of tsunamigenic sources. A pie-type diagram of tsunami occurrence depending on the type of sources is shown in Fig. 2.6. Most of known tsunamis (up to 75% of all historical cases) are generated by shallow-focus earthquakes capable of transferring sufficient energy to the overlying water column. Remaining cases where source is apparently known are divided between the landslide (10%), volcanic (4%) and meteorological (3%) tsunamis. Up to 8% of all the reported historical run-ups still have unidentified sources. In this case, some unusual wave activity near the coast was observed, but it was not possible to associate it with any of known potential sources (earthquake, volcano, landslide, or atmospheric event). Even for the XXth century, the NTL/ICMMG catalog contains 51 events with unidentified sources. Most of them were weak, non-damaging events, found only in tide-gauge records, however, among them there are six cases when the reported run-up was more than five meters (4 January 1923 San Felix Is., 9 August 1929 Northern Chile, 19 August 1929 Atlantic City, 14 September 1944 US East coast, October of 1954 SE Greenland, 12 September 1999 Guerrero, Mexico).

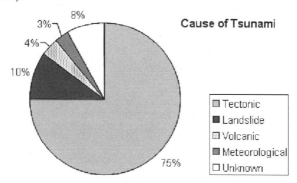

Fig. 2.6 – Pie-type diagram of historical tsunami occurrence depending on the type of tsunami source as it is classified in the NTL/ICMMG database. The NGDC/NOAA database has slightly different classification, but ratio between events of different origin is generally the same.

From the events with unidentified sources, one of the most interesting cases is 30 January 1607 when the large destructive waves hit the lowland coast of the Bristol Channel, UK, resulting in more than 2000 fatalities, possibly, the largest fatality rate from this type of a hazard in all UK history (Bryant and Haslett, 2003). The waves occurred on a fine day and surprised inhabitants. The source and nature of these waves is highly debated, with most of experts interpreting this event as an extreme high storm surge. However, a line of strong evidences, retrieved from the analysis of erosional and depositional features left by these waves along more than 570 km of coastline, may favor a long-wave such as tsunami (Haslett and Bryant, 2005). This event is not included in the NGDC and GITEC catalogs, in the NTL/ICMMG catalog it has a validity index 4 (definite), but the cause attribute is still kept as U (unknown) pending further evidence of its origin.

7.1. Seismogenic tsunamis

Seismogenic tsunamis are generated by submarine earthquakes due to the large-scale co-seismic deformation of the ocean bottom and the dynamic impulse transformed to a water column by compression waves. The size of tsunami generated by an earthquake relates to the energy release (earthquake magnitude), source mechanism, hypocentral depth, fault rupture velocity and water depth over the source region, but even in the most favorable cases, the energy transferred into tsunami waves is an order of 1% of the total energy released by an earthquake (see Chapter 5 in this volume). A list of some largest regional seismogenic tsunamis with their basic source parameters is shown in Table 2.

TABLE 2.

List of some largest regional seismogenic tsunamis in the historical catalogs.
M_s – surface wave magnitude, M_w – moment-magnitude, I – tsunami intensity on the Soloviev-Imamura scale, H_{max} – maximum reported run-up in m, CAU – cause of tsunami (T- tectonic, L – landslide), FAT – number of reported fatalities due to tsunami.

Date and Place	M_S	M_W	I	H_{max}, m	CAU	FAT
9 July 1586, Lima, Peru	8.5	-	3.5	26.0	T	many
31 January 1605, Shikoku, Japan	8.0	-	3.5	30.0	T	many
2 December 1611, Sanriku, Japan	8.1	-	4.0	25.0	T	4,783
28 October 1707, Nankaido, Japan	8.1	-	4.0	25.7	T	30,000
23 December 1854, Nankaido, Japan	8.3	-	3.0	28.0	T	5,000
15 June 1896, Sanriku, Japan	7.4	8.5	3.8	38.5	T	27,122
2 March 1933, Sanriku, Japan	8.3	8.6	3.5	29.3	T	3,064
9 July 1956, Aegean Sea	7.5	7.7	3.0	30.0	TL	4
12 December 1992, Flores Sea	7.6	7.7	2.7	26.2	TL	2,200
12 July 1993, Okushiri, Japan	7.6	7.7	3.1	31.7	T	198

One of the most important questions about seismogenic tsunamis is their possible maximum run-up value in the near-field. Available historical data, summarized in Tables 1 and 2, suggest that this value can hardly exceed a 35–50 meter even for the largest possible submarine earthquakes. Maximum run-up values of 60–70 m reported for 1771 Ishigaki, 1788 Sanah-Kodyak, and 1737 Kamchatka tsunamis are not very reliable. They are based on single anecdotal reports, which are not always confirmed by the recent geological investigation (see, for instance, Pinegina and Bourgeois, 2001), and can relate to run-up from a locally landslide–generated tsunami that could accompany the main tectonic tsunami. As it was demonstrated for the 1964 Alaska tsunami, all the major run-ups exceeding 25–30 meters were generated by slides from the fronts of the numerous deltas on the affected Alaska coast. These waves arrived at the coast almost immediately after the earthquake and were followed by the main seismically-induced tsunami that typically had a height of 12–18 meters (Lander, 1996).

Another important question about seismogenic tsunamis is the dependence of the resulting run-up height on the source magnitude of the parent event. In this analysis, I use the average run-up values at the nearest coast for calculation of the tsunami intensity I on the Soloviev-Imamura scale because average values are clearly less dependent on a particular coastal topography and thus give more stable estimates for overall tsunami intensity. The dependence tsunami intensity on earthquake magnitude is shown in Fig. 2.7 for surface-wave magnitude M_s and moment-magnitude M_w. In this analysis, I used only the historical data for the instrumental period of seismic observation (since the beginning of the XXth century). Both diagrams clearly demonstrate that there is little to no dependence of tsunami intensity on source earthquake magnitude, there is only a general tendency of increase in tsunami intensity with increase in source magnitude. This absence of direct correlation of tsunami intensity with earthquake magnitude makes operational tsunami warning, based solely on the seismic data, a difficult task. Moreover, the data scatter is considerable and exceeds six grades on the tsunami intensity scale. The

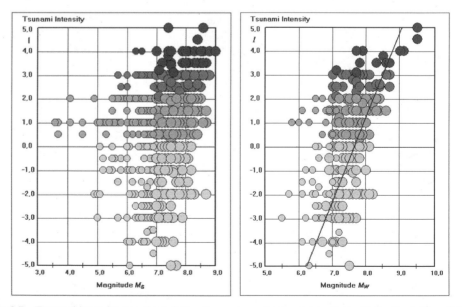

Fig. 2.7 – Tsunami intensity I on the Soloviev-Imamura scale versus magnitude M_s (on the left) and M_w (on the right) for tsunamigenic earthquakes occurring in the World Ocean from 1901 to 2005. The predicted intensity I calculated by the formula $I = 3.55 M_w - 27.1$ (Chubarov, Gusiakov, 1985) is shown as a solid line on the right diagram.

reasons for this scattering are multi-fold. First, it is a difference in the focus depth and the source mechanism. Second, there are differences in the source location (marginal seas, subduction zones, deep-water oceanic plate, etc.). Third, and possibly the most important, is the degree of involvement of secondary mechanisms (foremost being submarine slides and slumps) in the tsunami generation process (see next section).

7.2. Slide-generated tsunamis

Not as frequent as seismogenic tsunamis, but still very common world-wide, the slide-generated tsunamis result from rock and ice falling into the water, and sudden submarine landslides or slumps (see Chapter 6 in this volume). They can produce an extremely high water splash (up to 50–70 m, with the highest historical record of 525 m in Lituya Bay, Alaska in 1958) but not widely extended along the coast. In general, the energy of a landslide-generated tsunami rapidly dissipates as tsunami waves travel away from the source, but in some cases (e.g., if the landslide covers a large depth range), a long duration of slide movement can focus the tsunami energy along a narrower beam than the equivalent seismic source (Iwasaki, 1997). One of the most recent cases where the involvement of slide mechanism in tsunami generation has been confirmed is the 1998 Papua New Guinea tsunami when 15-m waves were observed after a M_w 7.0 earthquake (Synolakis et al., 2002; Tappin et al., 2002). Slide-generated water waves occur not only in the oceans and seas, but also pose a clearly recognized hazard to reservoirs, harbors, lakes and even large rivers where they may endanger lives, overtop dams, or destroy the waterside property.

In the case of large earthquakes, the accompanying landslides, locally triggered by strong shaking, can produce local waves greatly exceeding the height of the main tectonic tsunami. They are particularly dangerous as they arrive within a few minutes after the earthquake, leaving no time for an evacuation. One of the primary causes of death in the 1964 Alaska earthquake was the secondary tsunamis generated by slides from the fronts of the numerous deltas at the Alaska (Lander, 1996).

The global tsunami catalog gives many examples of historical events where involvement of subaerial and submarine landslides in tsunami generation was clearly observed and well documented. Some of these waves were destructive and resulted in considerable economic damage and loss of numerous lives. Among the best-known examples of the extreme water splash in the recent history is a well-documented 525-meter run-up in the Lituya Bay, southeastern Alaska, caused by a massive landslide occurred after the magnitude 7.8 earthquake of July 10, 1958 in the south-eastern Alaska (Miller, 1960). Lesser known cases of the extreme run-up heights in the same bay are the 1936 and 1853 events with the maximum run-up heights of 150 and 120 meters, respectively (Lander, 1996).

The first indication of involvement of the slide mechanism into tsunami generation comes from the absence of any associated seismic activity mentioned in the historical description. The global catalogs contain on average 10–15% such events. The percentage strongly varies regionally, exceeding 50% for the waters around England, the North Sea, the Norwegian Sea and the Baltic Sea. Obviously, a considerable proportion of these events could result from the meteorological and oceanographic phenomena (e.g., storm surges, rogue or freak waves). However, there are many cases where the catalog compilers specifically emphasized that an event occurred under "clear sky" and in a "calm sea".

A second indicator of slide generation is the value of the maximum run-up height for the particular event. The results of numerical modeling show that for a typical tsunamigenic earthquake in the magnitude range from 7.0 to 7.5, the co-seismic bottom displacement alone can hardly be responsible for the coastal run-ups exceeding 2–3 meters (Chubarov and Gusiakov, 1985). In fact, the instrumental data for the last decade give several examples of the shallow-depth earthquakes with magnitudes even above 7.5 when tsunami heights did not exceed several tens of centimeters (e.g., $M_w7.7$ Santa Cruz earthquake of 21 April 1997; $M_w8.1$ Balleny Islands earthquake of 25 March 1998, $M_w7.8$ Rat Island earthquake of 17 November 2003; $M_w8.1$ Macquare Island earthquake of 23 December 2004). Therefore, each case of a seismogenic tsunami with run-up heights exceeding 4–5 meters, resulted from an earthquake with magnitude below 7.5, can be considered as "suspicious" in terms of the involvement of a slide mechanism.

Table 3 lists the historical tsunamis with run-up values more than 50 meters. Five of these 13 events are known to have been generated by landslides, and in three cases the involvement of slide mechanism was confirmed by later studies. Only for four cases (1674 Indonesia, 1737 Kamchatka, 1788 Aleutians and 2004 Sumatra) the involvement of a slide mechanism has not been documented. Three of four are the old historical events with very limited data available.

TABLE 3.

List of the historical tsunamis with run-up greater than 50 m, sorted in order of their H_{max} value. M_s – surface wave magnitude, I – tsunami intensity on the Soloviev-Imamura scale, m – tsunami magnitude on the Iida scale, H_{max} – maximum reported run-up in m, CAU - cause of tsunami (T- tectonic, L – landslide, V- volcanic), FAT – number of reported fatalities due to tsunami.

Date and Place	M_s	I	m	H_{max}, m	CAU	FAT
10 July 1958 Lituya Bay, Alaska	7.9	2.5	9.1	525	TL	5
27 October 1936 Lituya Bay, Alaska	-	2.0	7.2	150	L	unknown
1853–1854 Lituya Bay, Alaska	-	2.0	6.9	120	L	unknown
6 August 1788 Sanak Is., Aleutians	8.0	4.0	6.5	88*	T	unknown
24 April 1771 Ishigaki Is, Ryukuy	7.4	3.5	6.4	85	TL	13,486
17 February 1674 Oma, Indonesia	8.0	4.0	6.3	80*	T	unknown
13 September 1936 Loen, Norway	-	2.0	6.1	70	L	4
28 March 1964 Alaska	8.5	4.5	6.1	68	TL	115
17 October 1737 Kamchatka	8.5	4.0	6.0	63*	T	unknown
7 April 1934 Tafjord, Norway	-	2.0	5.9	62	L	40
10 September 1899 Yakutat Bay, Alaska	8.6	2.5	5.9	60	TL	unknown
21 May 1792 Unzen volcano, Japan	-	2.0	5.8	55	VL	4,300
26 December 2004, Sumatra, Indonesia	8.8	4.0	5.7	51	T	229,866

*Run-up value is based on a single witness report and, therefore, is not very reliable.

The hazards of the underwater slumping range from just technical problems and economic loss due to submarine cables breaks resulted from turbidity currents (which were first documented for the 1929 Grand Bank earthquake) to the destructive effects of huge waves devastating the nearby coast and resulting in large death toll (e.g., 1998 Aitape tsunami). Mitigation and countermeasures against this hazard are quite difficult because very often the destructive water waves come without any seismic or meteorological precursors. The only way to mitigate this hazard is careful investigation of historical and pre-historical cases of waves caused by underwater slumping, identification of slump-prone zones and the evaluation of the long-term risk imposed upon the nearest populated areas by mass failures.

7.3.   Volcanic tsunamis

Though relatively infrequent, volcanically-generated tsunamis can be extremely destructive and devastating in the immediate source area and can result in numerous fatalities. Until the 2004 Indian Ocean tsunami, the 1883 Krakatau tsunami was on the top of the list of deadliest tsunamis historically known. The explosive caldera-forming eruption with 18–20 km^3 estimated volume of the eruptive material resulted in 35–40-meter tsunami waves that flooded the coast of the Sunda Strait and killed 36,416 people. Another great volcanic tsunami, devastating the northern coast of the Crete Island, was generated by a caldera-forming eruption of Santorini volcano in 1628 B.C. with the estimated volume of erupted material in 50–60 km^3 (Simkin and Siebert, 1994). The number of fatalities of this tsunami will possibly never be known, but undoubtedly it seriously affected the northern coast of the Crete Island, destroying most of the Minoan fleet. These effects, along with a general environmental downturn resulting from the giant eruption, could have led to the demise of the Minoan civilization (Baillie, 1990).

Both the 1628 BC Santorini and 1883 Krakatau eruptions were highly explosive where the following caldera collapse could have been the leading mechanism of tsunami generation. Smaller eruptions of island and coastal volcanoes can generate a significant tsunami if they are accompanied by volcano's slope failure. Active volcanoes are dynamically unstable structures, whose growth and development typically include episodes of the edifice instability and structural failure (McGuire, 2006). Part of their cones may collapse from time to time, generating giant rock slides and avalanches that can travel downslope at a speed exceeding 100 km/hour. When these mass movements enter the water, they can produce destructive tsunami waves. The latest event of this type occurred on 30 December 2002 at the Eolian Islands in the Tyrrhenian Sea, when two large slides, separated by about 7 minutes, with an estimated total volume of about 17 million cubic meters, entered the water at the NW flank of the Stromboli volcano and generated 7-meter waves that severely attacked the NW part of the island (Tinti et al., 2004).

Growing volcanoes may become unstable and experience a collapse at any scale and by different reason, some of them (heavy rainfall, erosion of the base of the volcano, deformation and tilting of volcano basement, earthquake shaking) do not include any eruption of volcano itself. A destructive collapse of the Ritter volcano (Papua New Guinea) in 1888 occurred without any sign of volcanic eruption and led to generation of 15-meter tsunami waves that killed hundred of villagers along the West New Britain coastline (Latter, 1981). In May of 1792, the southern part of the cone of the Unzen volcano in Kyushu, Japan with an estimated volume of only 0.33 km^3 collapsed, and a debris avalanche entered Ariake Bay forming a devastating tsunami with a maximum height of 55 meters that killed more than 4,300 people (Katayama, 1974)). The slope failure of the Unzen volcano occurred suddenly and was not accompanied by any major eruption which actually ended three months before.

Volcanogenic tsunamis also can be produced in lakes. On 2 January 1996, an explosive eruption of nearby Karymsky volcano (Kamchatka, Russia) generated large water waves in the 4-km diameter Karymskoye Lake (Belousov et al., 2000). The highest run-up (up to 27–29 m) occurred on the shore immediately adjacent to a tuff ring, 700 m from the center of crater. On the opposite eastern shore the run-up height was within 5–7 m. Runup heights in this case are based on physical evidence of soil erosion that was still clearly visible on the lake shore in the summer of 1996. In another case, a series of five tsunamis were generated on 28–30 September of 1965 in the Taal crater lake in the Luzon Island, Philippines. These waves were large enough to cause fatalities and to destroy several nearby villages (Hedervari, 1986).

Submarine eruptions can also generate tsunami if they occur in the water of 500–1000 m depth. The largest known event of this type occurred during the 11 April 1781 eruption near Sakura-jima, Kuyushu Island, Japan. Three boats overturned and 15 people were lost (Iida, 1984).

The list of top-ten volcanic tsunamis in the historical records is shown in Table 4. Altogether, the global catalog contains about 100 cases of volcanic tsunamis, that is about 4% of all historical tsunamigenic events. But their fatality rate is almost twice as high as seismically-induced tsunamis (almost 67,000 deaths that is about 9% of the total). However, most of them (at least, those resulting from explosive eruptions) cannot transport significant energy over long distances because even giant Santorini-class volcanic eruptions are essentially point sources and initially large waves lose their energy rapidly with distance. For example, all the fatalities of the 1883 Krakatau tsunami occurred within 1-hour propagation time. According to the NDGC/NOAA database, the largest far-field amplitude was observed in Geraldton, Western Australia and was only 1.8 m.

TABLE 4.

List of largest historical volcanic tsunamis. **VEI** – volcanic explosion index, **VOL** – total volume of eruptive material in km^3, H_{max} – maximum reported run-up, I – tsunami intensity on the Soloviev-Imamura scale, **FAT_EVE** – total number of fatalities, **FAT_TSU** – fatalities from tsunami

Date and place	VEI	VOL, km^3	H_{max}, m	I	FAT_EVE	FAT_TSU
1628 BC Santorini, Aegean Sea	6	60–70	40–90	4.0	unknown	unknown
1452 Kuwae, Vanuatu	6	32–39	10–15	3.0	unknown	unknown
31 July 1640 Komagatake, Japan	5	80–100	10	2.0	unknown	700
29 September 1650, Santorini	4	18–20	30	3.0	unknown	unknown
29 August 1741 Oshima, Japan	4	1–2	15	2.5	15,000	1,475
21 May 1792 Unzen, Japan	2		35–55	2.0	9,745	4,300
10 April 1815 Tambora, Indonesia	7		5–10	1.5	117,000	unknown
27 August 1883 Krakatau, Indonesia	6		36–41	4.0	37,000	36,416
13 March 1888 Ritter, Bismark Sea	3		12–15	3.0	3,000	500
4 August 1928, Paluweh, Flores Sea	3		5–10	2.0	226	160

7.4.   Meteorological tsunamis

All historical tsunami catalogs contain reports on unusual tsunami-like changes in water level and currents that are not directly associated with typical tsunami sources like earthquakes, volcanoes or submarine landslides. The global tsunami catalog contains about 3% of such events. These seiche-type oscillations are generated by some large-scale atmospheric disturbances like a rapidly moving atmospheric pressure front moving over a shallow sea at about the same speed as tsunami allowing the air and water to couple. They have different names in different countries or even in particular bays ("rissaga" in Spain, "abiki" and "yota" in Japan, "marrubio" in Sicily, "milghuba" in Malta). Their period varies from 2–3 min to 2–3 hours, and amplitudes can be from tens of centimeters to several meters. In the latter case, they can be destructive and sometimes result in serious damage. Typically, they occur within straits and in particular bays and harbors; however, sometimes they can happen along a straight coast. The term "meteorological tsunami" or "meteo-tsunami" has been proposed as a general name for this type of phenomena (Rabinovich and Monserrat, 1996). In some publications, another term, "atmospheric tsunami" is used, but it should not be considered as appropriate name, since this term is more applicable to infra-sonic gravitational waves in the stratified atmosphere. These waves are generated by some large volcanic explosions and can propagate far away from their source, even round the globe, like the widely known infra-sonic waves generated by the 1883 Krakatau explosion.

It is important to stress that meteorological tsunamis should be distinguished from a more common phenomenon, storm surges, which occur in many coastal areas vulnerable to hurricane and typhoon impact. A storm surge results from the hydrostatic water rise due to the low pressure zone in the cyclone center that is amplified by the dynamic surge caused by a strong wind pressure. The main difference between these phenomena is in their periods and duration. The waves observed during a storm surge have a shorter wave length but the coastal flooding can last from several hours up to several days and usually accompanied by a strong wind and a heavy rain, while meteorological tsunami lasts from 15–20 min to several hours, and quite often occurs in calm weather. However, distinguishing these two types of phenomena quite often creates a problem for catalog compilers, mainly, due to the lack of details in description of old historical events. For instance, for the Yellow Sea, Soloviev and Go (1974) catalog lists a total of nine destructive tsunamis between 1076 and 1636. Of these nine events five tsunamis were not associated with earthquakes and they are listed with remark "probably, of meteorological origin" (but no additional details on weather conditions are provided). As it is indicated in (Gusiakov, 2001), some of these cases possibly could be slide-generated tsunamis, because they occurred in the area with very high sedimentary loading, resulting from nearby mouth of the largest China's river (Huang He or Yellow River).

A typical example of a meteorological tsunami is the one that hit Daytona Beach, Florida on July 3, 1992 (Sallenger et al., 1995). The anomalously large wave, reportedly 5–6 m high, struck at least 20 km of shoreline on a clear, calm evening when the ambient waves were small. Eyewitnesses described the large wave, rapidly approaching the coast as "a wall of white water with the roar of a breaking wave". Approximately 20 vehicles, parked on the road near the beach, were lifted by incoming water and there were about 20 minor injuries, but no fatalities. The initial explanation for the cause (offshore landslide creating a tsunami) was soon replaced by a "squall-line surge" hypothesis that explained the effect as a resonant coupling with a meteorological disturbance moving along the coast with the speed of a long wave at this depth.

Among other cases of tsunami-like waves, when resonant conditions with a moving atmospheric front were observed, one occurred in the Lake Michigan at Chicago lakeshore on June 25, 1954. The case caused seven fatalities as a result of the first unexpected increase of water level, which reached 2.4 m at Mountrose Harbor, and 3.3 m at North Avenue (Ewing et al., 1954).

8.   Unusual cases of tsunami-like water disturbance

Tsunamis and tsunami-like water wave disturbances can occur not only in oceans and seas, but also in any water reservoir provided that its surface or bottom experiences some large-scale disturbance. This section describes several unusual historical cases of generation of tsunami-like waves in rivers, lakes, bays and harbors as well as long-period water waves resulted from explosions.

A rare case of tsunami-like wave in the large, ice-covered bay is described V. Semyonov (1985). In April of 1939 Semyonov was skiing across Avacha Bay, Kamchatka, covered with thick ice at that time. Being on ice, approximately 1.5 km from the coast, he saw a two-meter ice swell moving from the south-west. For several minutes the swell crossed the bay leaving behind a ridge of broken ice. A depression of up to 1.5–2.0 m preceded the swell, and then Semyonov could not see the coast; he estimated the total height of "ice wave" to be up to 3.5–4 m. Since no felt earthquake or volcanic eruption was reported for that day in Kamchatka, the only possible explanation for this large-scale disturbance was an underwater slide or a slump on the bottom of the bay, possibly from Avacha River delta or from other deltas in Avacha Bay.

Another interesting case of landslide-generated tsunami is the two-meter waves generated in Lake Coatepeque, El Salvador, soon after a 7.6 magnitude earthquake occurred on 13 January 2001 in the Pacific, 50 km off the coast. Five deaths were reported as a result of run-up of these waves at the coast of the lake (J. Borrero, 2001). It is worth to note that this earthquake did not generate any significant tsunami at the Pacific coast near the earthquake area.

Huge and destructive water waves were generated in Lake Nyos, Cameroon, on 21 August 1986, when a giant "bubble" of about 3 million cubic meters of CO_2 came from depths of the lake to the surface and generated destructive water waves that killed many people. "A wave of water was sent crashing across the lake reaching on average a height of about 25 m along the southern shore and overtopping a 75-m-high promontory" (French, 1988).

One of the most appalling phenomena accompanying the $M_s8.6$ New Madrid earthquake of 16 December 1811 was agitation of water in the Mississippi River that began almost simultaneously with seismic shaking. In a detailed report on this earthquake, Fuller (1912) cites several eyewitness accounts on this phenomenon: "The waters of the Mississippi were seen to rise up like a wall in the middle of the stream and suddenly rolling back would beat against either bank with terrible force ... The water moved inward with a front wall 15 to 20 feet high and tore boats from their mooring and carried them up a creek ... for a quarter of a mile ... Just below New Madrid, a flatboat was swamped and six men drowned ... The river was laterally covered with the wrecks of boats". The generation mechanism of these water waves in the river is not known—it could be strong seismic shaking or river bottom disturbance. Fuller says "There is no reason to doubt that fissures opened and closed beneath the water as they did on the land, giving rise to large waves by the ejection of water. These waves of great size moved upward against the current".

A similar case of large water waves in the Volga River is described by Didenku-lova and Pelinovsky (2002). The 5–6 meter waves were generated by a sudden slope failure with an estimated volume of about 150,000 m³ which occurred on a steep bank slope of the river near Nizhniy Novgorod on 18 June 1597. Waves damaged numerous vessels anchored along the river, some of them were thrown aground and left on the dry bank 40–50 m beyond the bank.

The global catalog also contains several cases of tsunamis generated by large explosions. In December of 1917, large waves were generated by the greatest man-made explosion before the nuclear era. This explosion happened in the Halifax Harbor (Nova Scotia, Canada) after collision of the munitions ship *Mont Blanc,* having 3,000 tons of TNT on board, with the relief ship *Imo.* At the coast near to the explosion site, the waves were over 10 m high, but their amplitude diminished greatly further away (Murty, 2003). Some nuclear tests, made in the 1950s and 1960s at small Pacific atolls resulted in generation of tsunami-like waves with up to 10-m height near the source area. An extensive study of water waves generated by nuclear explosions, both on and under the sea surface and up to 10 Mt yield, and also of a series of smaller-scale tests carried out in Mono Lake (California) was made Van Dorn et al. (1968) for the US Navy. The main conclusions from their study were that tsunamis from explosions have a shorter wavelength as compared to the size of the resulted cavity (a few km in diameter), in near-field their height can be very large, but rapidly decays as the waves travel outside the source area. They also noted to the effect of breaking short-length waves as they crossed the continental shelf, generating a large-scale turbulence, but leaving the coast without damaging run-up. The discovery that explosion waves or large impact-generated waves will break on the outer shelf and produce a little damage on the coast has since been known as the "Van Dorn effect", that, however, has not been confirmed by actual observations

9. Conclusions

1. The present version of the GTDB catalog on tsunami and tsunami-like events covers the period from 2000 BC till present and currently contains nearly 2100 historical events with 1206 of them occurring in the Pacific, 263 in the Atlantic, 125 in the Indian Ocean and 545 in the Mediterranean region. All together, these events are responsible for nearly 700,000 lives lost in tsunami waves during the whole historical period of available observations.

2. Of all 2100 tsunamigenic events, only 223 (10%) tsunamis resulted in any fatalities, all others were weak local events observable only in several particular areas of the nearest coast. From these 223 deadly tsunamis, 213 (95%) fall into the category of local and regional events with most damage and all fatalities limited to one-hour propagation time. In total, they are responsible for 426,000 (61%) fatalities.

3. Ten trans-oceanic tsunamis that occurred in the World Ocean during the last 250 years are responsible for 274,000 (39%) fatalities. Among them, nearly 230,000 people were killed during just one event—the December 26, 2004 Indian Ocean tsunami. All other trans-oceanic tsunamis are responsible for 34,000 deaths or 5% of all tsunami-related fatalities.

4. The study (Gusiakov et al., 2007) of the death toll for 10 most destructive trans-oceanic tsunamis occurred in the World Ocean during the last 250 years shows that although the damaging impact of large tsunamis can last up to 23–24 hours, over 84% of their fatalities occurring within the first hour of propagation time. Another 12% of fatalities occur within the second hour, with the rest of 4% occurring at the remaining time (exceeding two hours).

5. The intensity of seismically induced tsunamis is mainly controlled by an earthquake magnitude and, in general, is directly proportional to it. Detailed study of historical data for the instrumental period of available observations (since 1900) shows, however, that the actual scattering of tsunami intensity for earthquakes with the same magnitude exceeds six grades on the Soloviev-Imamura scale. That means than tsunami amplitudes can differ by a factor of 60 for earthquakes of the same magnitude that makes unreasonable an operational prediction of expected tsunami height at the coast, based solely on earthquake magnitude.

6. Both distantly and locally generated tsunamis are a typical example of "low probability – high consequence" hazard. Having long recurrence intervals (typically from 10s to 100s of years) for a particular coastal location, they can have very adverse impact on the coastal communities resulting in heavy property damage, many fatalities, and major disruption of commerce and community life.

7. There is a persistent need for any information that predates or augments historically reported and instrumentally measured past occurrence of tsunamis. Geological data on paleotsunamis therefore should be included in tsunami catalogs. Only on the basis of integrity of all data (instrumental, historical and geological) can we study the time-space patterns of large tsunamis and evaluate their long-term occurrence rates.

Finally, I would like to conclude this chapter with the words by G. Pararas-Carayannis written in his 1990 paper "Tsunamis of the 21st century". "As a starting point, we need to develop a uniform and standardized program of tsunami, seismic and geological data collection. A wealth of such data already exists, but this data is not properly organized, is not uniformly collected, and of course is not readily available. Therefore, standards must be established for the collection of such data and a tsunami database must be organized on a regional basis initially, and shared on a global scale at a late time" (Pararas-Caraynnios, 1990, p. 280). Despite since that time a lot has been done for historical tsunami data compilation and verification, the above words are still relevant today and direct us to further work in the field of tsunami cataloguing and database development.

Acknowledgements

The authors wish to thank Mrs. T. Kalashnikova for assistance in tables and figures preparation and Mrs. E. Semochkina for improvement of the initial English text of the manuscript. Two anonymous external reviewers made a number of conclusive remarks thus contributing to improvement of the quality of the paper. The author is very grateful to the internal reviewer, Jody Bourgeois, whose careful reading of the manuscript resulted in numerous remarks, corrections and refinements as well as in valuable comments on several important topics related to historical tsunami data. The work was supported by the RFBR grants 05-05-64460, 08-07-00105, SD RAS Integration Grant 2006-113 and the NTL-WAPMERR Contract TSU2005.

References

Abe, K., 1979. Size of great earthquakes of 1957–1974 inferred from tsunami data. *J. Geoph. Res.,* **84**(4), 1561–1568.

Abe, K., 1981. Physical size of tsunamigenic earthquakes of the north-western Pacific. *Phys. Earth Planet. Inter.,* **27**(3), 194–205.

Agostinho, J., 1953. Notes on some tsunamis of the Azores. *Annales de la Commission Pour l'Etude des Raz de Maree, Internat. Union Geod. Geophys.,* **5,** 21–24.

Ambraseys, N. N., 1962. Data for the investigation of the seismic sea waves in the Eastern Mediterranean. *Bull. Seis. Soc. Amer.,* **52**(4), 895–913.

Baillie, M. G. L., 1990. Irish tree-rings and an event in 1628 BC. In: *Thera and the Aegean World III Vol.3, D. A. Hardy (Ed.)* The Thera Foundation, London, 160–166.

Baillie, M. and M. Munro, 1988. Irish tree-rings, Santorini and volcanic dust-veils, *Nature,* **332,** 344–346.

Belousov, A., B. Voight, M. Belousova, and Y. Muravyov, 2000. Tsunami generated by submarine volcanic explosions: unique data from 1996 eruption in Karymskoe Lake, Kamchatka, Russia. *Landslides and Tsunami,* Edited by B.Keating, Ch.Waythomas, A.Dawson, Birkhauser, 1135–1143.

Berninghausen, W. H., 1962. Tsunamis reported from the west coast of South America, 1562–1960. *Bull. Seismological Soc. Amer.,* **52**(4), 915–921.

Berninghausen, W. H., 1964. Tsunamis and seismic seiches reported from the Eastern Atlantic ocean south of the Bay of Biscay. *Bull. Seism. Soc. Amer.,* **54**(1), 439–442.

Berninghausen, W. H., 1966. Tsunamis and seismic seiches reported from regions adjacent to the Indian Ocean. *Bull. Seismological Soc. Amer.,* **56**(1), 69–74.

Berninghausen, W. H., 1968. Tsunamis and seismic seiches reported from the Western North and South Atlantic and the coastal waters of Northwestern Europe. *U.S. Naval Oceanographic Office, Washington D.C., Informal Report.,* **6885,** 48 pp.

Berninghausen, W. H., 1969. Tsunamis and seiches of Southeast Asia. *Bull. Seism. Soc. Amer.,* **59**(1), 289–297.

Borrero, J., 2001. Message posted on Tsunami Bulletin Board on January 18, 2001.

Bryant, E. A. and S.Haslett, 2003. Was the 1607 coastal flooding event in the Severn Estuary and Bristol Channel (UK) due to a flooding event? *Archaeology in the Severn Estuary,* 14, p 163–167.

Chubarov, L .B. and V. K. Gusiakov, 1985. Tsunamis and earthquake mechanism in the island arc region. *Science of Tsunami Hazard,* **3**(1), 3–21.

deLange, W. P., and T. R. Healy, 1986. New Zealand Tsunamis, 1840–1982. *New Zealand Jour. Geology and Geophysics,* **29,** 115–134.

Didenkulova, I. I. and E. N. Pelinovsky, 2002. The 1597 tsunami in the river Volga. *Proceedings of the International Workshop "Local Tsunami Warning and Mitigation",* Moscow, 17–22.

Everingham, I. B., 1977. Preliminary Catalog of Tsunamis for the New Guinea-Solomon Islands Region, 1768–1972. *Rept. No. 180, Dept. Minerals and Energy, Bur. Miner., Resour.: Geol. and Geophysics,* Canberra, Australia, 85 pp.

Everingham, I. B., 1987. Tsunamis in Fiji. *Ministry of Lands, Energy & Mineral Resources, Mineral Resources Dept., Suva, Fiji, Report 62,* ISSN 0250-7234; Second Printing, Aug. 1988, 27 pp.

Ewing, M., F. Press, and W. L. Down, 1954. An explanation of the Lake Michigan wave of 25 June 1954, *Science,* **120,** 684–686.

Fernandez, M. A., E. Molina, J. Havskov, and K. Atakan, 2000. Tsunamis and tsunami hazards in Central America. *Nat. Hazards,* **22,** 91–116.

French, J. S., 1988. Lake Nyo Dam, *EOS,* **29**(2), p. 776.

Fuller, M. L., 1912. The New Madrid Earthquake, *US GS Bulletin,* **294,** Washington, D.C., 119 pp.

Fokaefs, A., and G. A. Papadopoulos, 2007. Tsunami hazard in the Eastern Mediterranean:

strong earthquakes and tsunamis in Cyprus and the Levantine Sea. *Nat Hazards,* **40,** 503–526.

GITEC – Genesis and Impact of Tsunamis on the European Coast, 1992. A Research Proposal Submitted to the Commission of the European Communities, January 1992.

Gusiakov, V. K., 2001. "Red", "green" and "blue" Pacific tsunamigenic earthquakes and their relation with conditions of oceanic sedimentation. *Tsunamis at the End of a Critical Decade.* G.Hebenstreit (Editor), Kluwer Academic Publishers, Dordrecht-Boston-London, 2001, 17–32.

Gusiakov, V. K., 2003. NGDC/HTDB meeting on the historical tsunami database proposal. *Tsunami Newsletter,* **XXXV**(4), 9–10.

Gusiakov, V. K., An.G. Marchuk, and A. V. Osipova, 1997. Expert tsunami database for the Pacific: motivation, design and proof-of-concept demonstration. *Perspectives on Tsunami Hazard Reduction: Observations, Theory and Planning,* G.Hebenstreit, (Editor), Kluwer Academic Publisher, Dordrecht-Boston-London, 21–34.

Gusiakov, V. K., D. G.Khidasheli, A. G.Marchuk, and T. V.Kalashnikova, 2007. Analysis of tsunami travel time maps for damaging historical tsunamis in the world ocean. Report of the NTL/ICMMG SD RAS prepared for IOC/UNESCO, Novosibirsk, Russia, 28 pp. http://tsun.sscc.ru/tsulab /TTT_rep.htm.

Hamzan, L., N. Puspito, and F. Imamura, 2000. Tsunami catalog and zones in Indonesia, *J.of Nat. Disaster Sci.,* **22**(1), 25–43.

Haslett, S. K. and E. A.Bryant, 2005. The AD 1607 coastal flood in the Bristol Channel and Severn Estuary: Historical records from Devon and Cornwall, UK. *Archaeology in the Severn Estuary,* **15,** p 81–89.

Hatori, T., 1986. Classification of tsunami magnitude scale. *Bull. Earthquake Res. Inst. Univ. Tokyo,* **61,** 503–515.

Heck, N. H., 1934. List of seismic waves. *Annales de la Commission pour l'Etude des Raz de Maree,* **4,** 20–41.

Heck, N. H., 1947. List of seismic sea waves, *Bull. Seis. Soc. Amer.,* **37**(4), 269–286.

Hedervari, P., 1986. Catalog of submarine volcanoes and hydrological phenomena associated with volcanic events January 1, 1900 to December 31, 1959, *World Data Center A for Solid Earth Geophysics, Report SE-42,* 36 pp.

Iida, K., 1956. Earthquakes accompanied by tsunami occurring under the sea of the Islands of Japan. *J. Earth Sc., Nagoya Univ.,* **4**(2).

Iida, K., 1958. Magnituda and energy of earthquakes accompanied by tsunamis, and tsunami energy. *J. Earth Sc., Nagoya Univ.,* **6**(2), 101–112.

Iida, K., 1963. Magnitude, energy and generation mechanisms of tsunamis and a catalogue of earthquakes associated with tsunamis. Proc. Tsunami Meeting, 10th Pacific Sci. Congress, 1961, *IUGG Monograph,* **24,** 7–18.

Iida, K., 1984. Catalog of Tsunamis in Japan and Its Neighboring Countries. *Special Report, Aichi Institute of Technology, Japan,* 52 pp.

Imamura, A. 1942. History of Japanese tsunamis. *Kayo-No-Kagaku (Oceanography),* **2**(2), 74–80 (in Japanese).

Imamura, A. 1949. Catalog of Japanese tsunamis. *Zisin,* Ser.2, **2**(2), 23–28 (in Japanese).

ITDB/WLD (2005) Integrated Tsunami Database for the World Ocean, Version 5.15 of July 31, 2005, CD-ROM, Tsunami Laboratory, ICMMG SD RAS, Novosibirsk, Russia

Iwasaki, S. I., 1997. The wave forms and directivity of a tsunami generated by an earthquake and a landslide. *Sci. Tsunami Hazard,* **15,** 23–40.

Katayama, N., 1974. Old records of natural phenomena concerning the Shimabara catastrophe. *Sci. Rep. Shimabara Volc. Observ, Fac. Sci. Kyushu Univ.,* **9,** 1–45.

Krasheninnikov, S. P., 1755. Description of Kamchatka Land, St.Petersburg, 1755 (in Russian). English translation: Krasheninnikov S. P. (1972) Explorations of Kamchatka North Pacific Scimitar / Transl. With Introduction and Notes by E. A. P.Crownhart-Vaughan. Portland, Oregon Historical Society, 1972, 375 pp.

Lander, J. F., 1996. Tsunamis Affecting Alaska, 1737–1996. National Geophysical Data Center (NGDC), Boulder, CO, *NGDC Key Geophys. Res., Doc. 31,* Nat. Geophys. Data Center, NOAA, Boulder, CO, 195 pp.

Lander, J. F., P. A. Lockridge, and M. J. Kozuch, 1993. Tsunamis Affecting the West Coast of the United States, 1806–1992. National Geophysical Data Center (NGDC), Boulder, CO, *NGDC Key to Geophysical Records Doc. 29,* 242 pp.

Lander J. F., L. S. Whiteside, and P. Hatori, 2002. The tsunamis history of Guam, 1849–1993, *Sci Tsunami Hazards,* **20**(3), 158–174.

Latter, J. H., 1981. Tsunamis of volcanic origin; summary of causes with particular reference to Krakatoa, 1883, *Bulletin Volcanologique,* Vol. 44, p. 467–490.

Lockridge P. A., L. S. Whiteside, and J. F. Lander, 2002. Tsunamis and tsunami-like waves of the Eastern United States, *Sci. Tsunami Hazards,* **20**(3), 120–157.

Lockridge, P. A. and P. Dunbar, 1995. World-Wide Tsunamis, 2000 B.C–1990. World Data Center A for Solid Earth Geophysics, Boulder, Colorado, 31 pp.

Marinatos, S., 1939. The volcanic destruction of Minoan Crete. *Antiquity,* **13,** 425–439.

McGuire, W. J., 2006. Lateral collapse and tsunamigenic potential of marine volcanoes, In: Trose C., De Natale G., Kilburn C. L. *Mechanism of activity and unrest of large calderas.* Geological Society London, Special Publ., **269,** 121–140.

Miller, D. J., 1960. Giant waves in Lituya Bay. *Bull. Seis. Soc. Am.,* **50**(3), 253–266

Murty, T. S., 2003. A review of some tsunamis in Canada. In: Yalciner A, Pelinovsky E, Synolakis C, Oka E (eds), *Submarine landslides and tsunamis. NATO Science Series: IV, Earth and Environmental Sciences: 21,* Kluwer Academic Publishers, Dordrecht, 173–183.

Murty,T. S. and H. G. Loomis, 1980. A new objective tsunami magnitude scale. *Geod.,* **4,** 267–282.

Murty, T. S., and M. Rafiq, 1991. Tentative List of Tsunamis in the Marginal Seas of the North Indian Ocean. *Natural Hazards,* **4**(1), 81–83.

O'Loughlin, K. F., and J. F. Lander, 2003. Caribbean Tsunamis. A 500 Year History from 1498–1998. *Advances in Natural and Technological Hazards Research*, Kluwer Academic Publishers, 263 pp.

Papadopoulos, G. A., Editor, 2000. Historical earthquakes and tsunamis in the Corinth Rift, Central Greece. National observatory of Athens, Institute of Geodynamics, Publ. **12,** 129 pp.

Papadopoulos, G. A., 2001. Tsunamis in the East Mediterranean: a catalogue for the area of Greece and adjacent seas. In *Joint IOC-IUGG International Workshop "Tsunami Risk Assessment Beyond 2000: Theory, Practice and Plans. In memory of Professor S. L. Soloviev. Moscow, Russia, June 14 to 16, 2000"*, Proceeding, Moscow, 34–43.

Papadopoulos, G. A., 2003. Quantification of Tsunamis: A Review. In: *Submarine Landslides and Tsunamis*, A. C. Yalçiner, E. N. Pelinovsky, E. Okal and C. E. Synolakis, Editors, NATO Science Series, IV. Earth and Environmental Sciences, Vol. 21, Kluwer Academic Publishers, Dordrecht – Boston-London, 285–289.

Papadopoulos, G. A., and B. J. Chalkis, 1984. Tsunamis Observed in Greece and the Surrounding Area from Antiquity up to the Present Times. *Marine Geology*, **56,** 309–317.

Papadopoulos, G. A., E. Daskalaki, A. Fokaefs, and N. Giraleas, 2007. Tsunami hazards in the Eastern Mediterranean: strong earthquakes and tsunamis in the East Hellenic Arc and Trench system. Nat. *Hazards Earth Syst. Sci.*, **7,** 57–64.

Papadopoulos, G. A. and F. Imamura, 2001. A proposal for a new tsunami intensity scale. In: *Proceedings of International Tsunami Symposium 2001*, Seattle, U.S.A., 569–577.

Pararas-Carayannis, G., 1991. ITIC Progress Report for 1989–1991, Honolulu, ITIC/NOAA, 66 pp.

Pararas-Carayannis, G., 1990. Tsunamis of the 21st century. In: *Second International Workshop on the Technical Asp[ects of Tsunami Warning Systems, Tsunami Analysis, Preparedness, Observatuion and Instrumentation, Novosibirsk, USSR, 4–5 August 1989;* IOC Workshop Reports, No.58, IOC/UNESCO, Paris, France, 277–280.

Pinegina, T. K., and J. Bourgeois, 2001. Historical and paleo-tsunami deposits on Kamchatka, Russia: long-term chronologies and long-distance correlations. *Nat. Hazards and Earth System Sciences*, **1,** 177–185.

Rabinovich, A. B., and S. Monserrat, 1996. Meteorological tsunamis near the Balearic and Kuril Islands: Descriptive and statistical analysis, *Natural Hazards*, **13**(1), 55–90.

Sallenger, Jr. A. H., J. F. List, G. Gelfenbaum, R. P.Stump, and M. Hansen, 1995. Large waves at Daytona Beach, Florida as a squall-line surge, *Journal of Coastal Research.*, **11**(4), 1383–1388.

Semyonov, V. I., 1985. Tsunami-like phenomenon in the Avacha Bay, *Voprosy Geografii Kamchatki*, **9,** 151–155 (in Russian).

Shuto, N., 1993 Tsunami intensity and disaster. In: *Tsunamis in the World*. S.Tinti, Editor, Kluwer Academic Publisher, Dordrecht, 197–216.

Sieberg, A., 1927. Geologische, physikalische und angewandte Erdbeben Kunde. *Verlag von Gustav Fischer*, Jena, 527 pp. (in German).

Simkin, T. and L. Siebert, 1994. Volcanoes of the World, Second Edition, A Regional Directory, Gazetteer, and Chronology of Volcanism During the Last 10,000 Years, Geoscience Press, Inc., Tucson, Arizona, in association with Smithsonian Institution, Global Volcanism Program, Washington, D.C., 350 pp.

Soloviev, S. L., 1972. Recurrence of earthquakes and tsunamis in the Pacific ocean. In: *Volny Tsunami, Trudy Sakhnii*, **29,** 7–47 (in Russian).

Soloviev, S. L., 1978. Basic data on tsunamis on the Pacific coast of the USSR, *Izuchenie Tsunami v Otkrytom Okeane*, Moscow, "Nauka" Publishing House, 61–135 (in Russian).

Soloviev, S. L. and Ch.N. Go, 1974. A catalogue of tsunamis on the western shore of the Pacific Ocean, Moscow, "Nauka" Publishing House, 308p. English translation: Soloviev S. L., Go Ch.N. (1984). A catalogue of tsunamis on the western shore of the Pacific ocean, Translation by Canada Institute for Scientific and Technical Information, National Research Council, Ottawa, Canada KIA OS2.

Soloviev, S. L. and Ch.N. Go, 1975. Catalog of Tsunamis on the Eastern Shore of the Pacific Ocean. Academy of Science of the USSR, Nauka Publishing House, Moscow, Translated from Russian to English by Canadian Institute for Science and Technical Information, No. 5078, National Research Council, Ottawa, Canada, 1984, 293 pp.

Soloviev, S. L., Ch.N. Go, and Kh.S. Kim, 1992. Catalog of Tsunamis in the Pacific, 1969–1982. (translated from Russian to English by Amerind Publishing Co., Pvt. Ltd., New Delhi, 1988), Academy of Sciences of the USSR, Soviet Geophysical Committee, Moscow, 208 pp.

Soloviev, S. L., O. N. Solovieva, Ch.N. Go, Kh.S. Kim and N. A. Shchetnikov, 2000. Tsunamis in the Mediterranean Sea: 2000 B.C–2000 A.D. Translation from Russian to English by Gil B. Pontecorvo and Vasiiy I. Tropin; Kluwer Academic Publishers, Dordrecht, The Netherlands, 237 pp.

Stephenson, F., A. B.Rabinovich, O. N.Solovieva, E. A.Kulikov, and O.IYakovenko, 2007. Catalogue of tsunamis, British Columbia, Canada, 1700–2007. *Institute of Oceanology RAS,* Moscow, 133 pp.

Summary Report on the XI Session of the International Coordination Group for the Tsunami Warning System in the Pacific, Bejing, People's Republic of China, 8–12 September, 1987, Paris, IOC/UNESCO, 1987.

Synolakis, C. E., J. P. Bardet, J. C. Borrero, H. L. Davies, E. A. Okal, E. A.Silver, S. Sweet, and D. R. Tappin, 2002. The slump origin of the 1998 Papua New Guinea tsunami, *Proc. Royal. Soc.,* London, **458,** 763–790.

Takahashi, E., 1951. An estimate of future tsunami damage along the Pacific coast of Japan. *Bull. Earthq. Res. Inst., Tokyo Univ.,* **29**(1).

Tappin, D. R., P. Watts, G. M. McMurtry, Y. Lafoy, and T. Matsumoto, 2002. Prediction of slump generated tsunamis: The 17 July 1998 Papua New Guinea event, *Sci. Tsunami Hazards,* **20**(4), 222–238.

Tinti, S., A. Maramai, and L. Graziani, 2001. A new version of the European tsunami catalogue: updating and revision. *Nat. Hazards Earth Syst. Sci.,* **1,** 255–262.

Tinti, S., A. Maramai, and L. Graziani, 2004. The new catalogue of the Italian tsunamis, *Natural Hazards,* **33,** 439–465.

TRANSFER - Tsunami Risk ANd Strategies For the European Region, 2005. Research Proposal presented to Sixth Framework Programme Sustainable Development, Global and Ecosystems Priority. 6.3.IV.2.2: Assessment and Reduction of Tsunami Risk in Europe, October 2005.

Van Dorn, W G., B. LeMehaute, and Li-San Hwang, 1968. Final Report : Handbook of Explosion-Generated Water Waves. Volume I - State of the Art" (October 1, 1968). *Scripps Institution of Oceanography Library.* Paper 15.

Watanabe, H., 1989. Comprehensive list of tsunamis to hit the Japanese Islands. Second Edition. *Univ. Tokyo Press,* Japan, 206 pp. (in Japanese).

Watanabe, H., 1963. Method of determination of tsunami magnitude and its application to tsunami warning. *Zisin,* **27**(4) (in Japanese).

Zayakin, Yu.A. and A. A. Luchinina, 1987. Catalog of Tsunamis in Kamchatka, Obninsk, Russia, VNIIGMI-WDC, 51 pp. (in Russian).

Chapter 3. GEOLOGIC EFFECTS AND RECORDS OF TSUNAMIS

JOANNE BOURGEOIS

University of Washington

Contents

1. Introduction

Nor should we omit to mention the havoc committed on low coasts, during earthquakes, by waves of the sea which roll in upon the land, bearing everything before them, for many miles into the interior throwing down upon the surface great heaps of sand and rock, by which the remains of drowned animals may be overwhelmed.

> Charles Lyell, 1832

Those of us working on tsunami traces in the 1980s were commonly doubted because some tsunami scientists argued that tsunamis did not leave deposits, and many geologists were skeptical. However, it is clear from the reports of several pre-1980s surveys that tsunamis eroded and deposited not only sand, but also large boulders and coral debris. Moreover, photographs of tsunamis in progress show turbidity—for example, in a well-known 1957 photo series from Oahu, turbidity clearly develops in the nearshore as the tsunami arrives from the Aleutians. Since the 1990s, and certainly since 2004, there is no doubt that tsunamis erode and deposit sediment (Figure 3.1).

The Sea, Volume 15, edited by Eddie N. Bernard and Allan R. Robinson
ISBN 978–0–674–03173–9 ©2009 by the President and Fellows of Harvard College

Fig. 3.1 – Satellite images of Jantang, Aceh (Sumatra, Indonesia) before and after the 26 December 2004 tsunami. Note widening of the river mouth by erosion, stripping of vegetation, and deposition of sand on the coastal plain (light gray color). Images from Digital Globe.

Relative to other aspects of tsunami science, the study of the geologic record of tsunamis is immature. Only since the mid- to late 1980s has extensive work been done, and 1992 is the first year when there were more than 10 papers published (Figure 3.2). The literature expanded steadily in the 1990s, largely spurred by a number of damaging tsunamis in the Pacific. The field of tsunami geology continues to expand, especially following the 26 December 2004 Indian Ocean tsunami. Our working bibliography of tsunami geology has over 500 peer-reviewed articles up through 2006, not counting the geology of tsunami sources such as papers discussing what conditions generate "tsunami earthquakes" (e.g., Pelayo and Wiens, 1992; see Chapter 5).

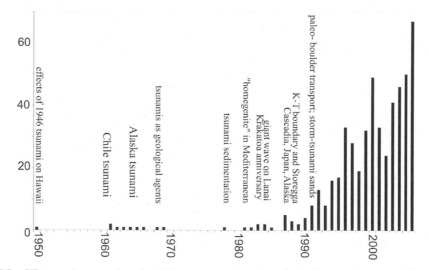

Fig. 3.2 – History of peer-reviewed articles on tsunami geology (including government publications), based primarily on GeoRef and Web-of-Science databases. Some landmarks (research triggers; pioneer papers) are noted, up through 1991. Non-English-language articles are included but probably underrepresented. Bibliography compiled by Andrew Ritchie, analyzed by the author.

For such an immature field, there have been a remarkable number of review articles on tsunami sedimentology and geomorphology. The earliest of these is Coleman (1968; also 1978). The majority of reviews have been written by Alastair Dawson—one of the modern pioneers—and co-authors (e.g., Dawson, 1994; Dawson, 1996; Dawson and Shi, 2000; Dawson and Stewart, 2007). Some syntheses (e.g., Bryant, 2001; Kelletat and Scheffers, 2003) include broad and untested speculation, some of which is pointed out in Dominey-Howes et al. (2006; also see Dominey-Howes, 2007), who review the various geological signatures of modern and paleo tsunamis. One review (Shanmugam, 2006) focuses on terminology and on the relationship between tsunamis and turbidity currents. Rhodes et al. (2006) summarize some of the observations and conclusions from a 2005 International Workshop on Tsunami Deposits. Several symposium publications and special volumes have been published with some focus on tsunami geology, the earliest being in *Marine Geology* (Einsele et al., 1996). For example, a publication spike in the year 2000 (Figure 3.2) is largely due to three special volumes that appeared that year.

In general, when there is abundant literature on a subtopic, such as the Cascadia Subduction Zone or the K-T boundary, I will cite the earliest and latest or most comprehensive publications.

2. Historical Review

More than 50% of the tsunami geology literature concerns tsunami deposition on the coastal plain, including coastal lakes, from modern and Quaternary seismogenic tsunamis. (In a few cases, the source of the tsunami is disputed or the earthquake also triggered submarine landslides.) This literature is dominated by cases from Japan, Cascadia (northern California to southern British Columbia), and the 2004 Indian Ocean tsunami. Other localities with concentrated studies include New Zealand, the Russian Far East, Alaska, Chile and Peru, the 1929 Grand Banks event, and the 1755 Lisbon earthquake and tsunami. While there are few studies of prehistoric tsunami deposits in low-latitude regions, surveys of recent tsunami effects, in addition to 2004, include Indonesia, Philippines, Papua New Guinea, Hawaii, Peru, Nicaragua, the Caribbean, and the Mediterranean.

Of the tsunami geology literature identifying tsunami sources other than earthquakes—about 150 articles to date—there is a fairly even split between landslide-generated, volcanogenic, and impact-generated tsunamis, with each category having one or two dominant cases. The landslide-generated tsunami-geology literature is dominated by the Storegga landslide and Hawaii cases, the latter currently disputed. The volcanogenic tsunami-geology literature is dominated by Santorini, with a few articles on Krakatoa. The literature concerning tsunami geology related to asteroid impacts is dominated by the Cretaceous-Tertiary (K-T) boundary case. (A recent change in the formal geologic time scale now means that this boundary is called the Cretaceous-Paleogene boundary.)

2.1   Surveys of tsunami effects

> Besides chairs, tables, bookshelves, etc . . . there were several roofs of cottages,
> which had been transported almost whole. . . . During my walk around the is-
> land, I observed that numerous fragments of rock, which, from the marine pro-
> ductions adhering to them, must recently have been lying in deep water, had
> been cast up high on the beach; one of these was six feet long, three broad, and
> two thick.
>
> Charles Darwin, 1835 (in *The Voyage of the Beagle*)

The earliest publications documenting observed effects of tsunamis are summaries
from post-tsunami surveys (e.g., Verbeek, 1886; Simons, 1888; Yamana, 1896 compiled by Unohana and Oota, 1988; Platania, 1908; Earthquake Research Institute,
1934; Shepard et al., 1950; Eaton et al., 1961; Konno et al., 1961). The earliest
known survey in Russia of the effects of a tsunami, the one produced by the great
1952 Kamchatka earthquake, was kept a military secret and is only recently being
uncovered. My review of surveyed geologic effects of modern tsunamis begins with
Shepard et al. (1950); Kajiura (1983) and Abe (2005) contain bibliographies of
many Japanese tsunami surveys before that time (see also Verbeek, 1886). Zayakin and Luchinina (1987) summarize findings of Russian surveys.

Following the destructive 1 April 1946 tsunami on Hawaii, three geologically
trained scientists conducted a post-tsunami survey (Shepard et al., 1950). Francis
Parker Shepard, a pioneering marine geologist, was on vacation in Hawaii when
the 1 April 1946 Aleutian tsunami struck. Volcanologist Gordon A. MacDonald of
the U.S. Geological Survey was mapping the geology of Hawaii in this period.
Geologist and hydrologist Doak Cox, born and raised in Hawaii, had just taken a
job with the Hawaiian Sugar Planters' Association. Their report includes 33 photographic plates, many showing tsunami erosion and sedimentation. Text sections
in the report on damage and erosion describe not only erosional but also depositional phenomena:

> Many fishpond walls of loose rock . . . were damaged or destroyed, and
> some of the ponds were partly filled with silt, sand, and rocks from the
> walls. . . . (p. 459)
>
> Where the waves rose over a high beach on a barrier of dunes and
> flooded lower lands behind, they eroded deep channels through the sand. . . .
>
> At many places sand excavated by the waves must have been carried
> seaward. Elsewhere, however, much of it was carried inland. At Haena, on
> northern Kauai, the highway was buried under 4 feet of sand. Thinner layers
> of sand covered [other] roads . . . on Kauai, Oahu and Maui. Many taro
> patches in Waipio Valley [on the big island] were completely covered by
> sand. (p. 462)
>
> . . . A great many coral heads, ranging in size up to 12 feet across, were
> torn loose and thrown up on the beaches. . . .
>
> At many places the near-shore water became muddy as a result of the
> tsunami. At most places the water cleared again within a few days. The mud-
> diness . . . may be attributable partly to the stirring up of fine terrigenous
> sediment on the shallow ocean bottom, but more largely to the result of
> washing away of soil cover on the temporarily inundated land areas. (p. 463)

Following the great 1960 Chile tsunami, several groups documented geological effects including tsunami erosion and sedimentation around the Pacific. On the south-central coast of Chile, Wright and Mella (1963) described sand covering the soil at several localities; more recent studies in Chile (e.g., Cisternas et al., 2000) describe internal characteristics of this deposit (e.g., Figure 3.3). Eaton's team in Hawaii (Eaton et al., 1961) was present during the tsunami, which arrived in the dark. Their report focuses on details of the tsunami waves and on destruction, but mentions the transport of large boulders and makes a rough calculation of the bore necessary to move them. Reports from Japan (Konno et al., 1961) include the first detailed sedimentological description of a tsunami deposit. Konno's team published cross-sections through the 1960 tsunami deposits on the coastal lowland of northeastern Japan; they documented and described thin sheets of sand and silt thickening into swales, thinning landward, and including graded layers (A. Moore, 2003).

Fig. 3.3 – Geological effects of the 1960 tsunami at Rio Lingue, central Chile. A: View from south bank near river mouth (left side of air photo); photo in 1989. B: Air photo taken of Rio Lingue in 1960 following the May earthquake, with interpretation. Dotted lines—former river channel banks before tsunami erosion. Shaded area—approximate distribution of preserved tsunami deposits as of 1989. C: Deposits of the 1960 tsunami observed in 1989. Left: proximal tsunami deposits of schist boulders (schist crops out on south bank and large rock in channel). Middle: tsunami deposit of laminated coarse sand, including intraclasts, about 1 km directly inland of the river mouth. Right: tsunami deposit of fine sand ~3 km upstream; tsunami deposit (in middle) overlies formerly farmed soil and is overlain by intertidal muds.

Other early articles that describe geological effects of tsunamis include reports on the 1953 Suva earthquake and tsunami (Houtz, 1962), the Lituya Bay landslide-generated tsunami (Miller, 1960), and the 1964 Alaska earthquake and tsunami (Reimnitz and Marshall, 1965; Plafker and Katchadoorian, 1966). Visiting the Copper River delta in Alaska shortly after the 1964 earthquake and tsunami, Reimnitz and Marshall described "extreme erosion" of tidal flats, which they attributed to "tsunami and seiches" without differentiating the two processes. They postulated that this eroded material would have been deposited in delta channel fills and would likely resemble turbidity-current deposits. The actual deposit of the 1964 tsunami was described first on Vancouver Island in British Columbia (Clague et al., 1994).

In the 1990s, post-tsunami surveys began regularly to include observations of tsunami deposits and other geologic effects of tsunamis (e.g., Abe et al., 1993; Sato et al., 1995; Shi et al., 1995; Dawson et al., 1996; Minoura et al., 1997; Bourgeois et al., 1999; Matsutami et al., 2001; Gelfenbaum and Jaffe, 2003; Rothaus et al., 2004). Already there is voluminous literature on geologic effects of the 26 December 2004 tsunami (Satake et al., 2007), and a whole new community of geoscientists has become engaged in studying the geologic effects of tsunamis. By late 2007, there were more than 35 refereed publications with a major focus on surficial physical aspects of the Indian Ocean tsunami and its aftermath (e.g., Ramasamy et al., 2006).

The December 2004 tsunami has generated a new wave of geological and related studies, many using techniques not available in the times of prior great tsunamis (Alaska, 1964, Chile, 1960, Kamchatka, 1952), and addressing questions that have arisen since then. Moreover, this tsunami affected many low-latitude coastlines, whereas the previous three great tsunamis affected primarily mid- to high-latitude coastlines. Also, because population densities were high on many coastlines affected by the 2004 Indian Ocean tsunami, environmental effects have received more attention.

Of particular interest are the first studies of tsunami effects focused on satellite imagery (e.g., Figure 3.1). Ramakrishnan et al. (2005; India and Sri Lanka) and Borrero (2005; Banda Aceh) were the first to publish analyses of post-tsunami satellite imagery; Borrero also made on-the-ground observations. Yang et al. (2007) used a new satellite dataset, FORMOSAT-2, to identify hard-hit regions of Banda Aceh and Thailand and discuss how this technology may aid future post-disaster responses. Umitsu et al. (2007) combined on-the-ground observations with interpretation of satellite images to examine local topographic effects on tsunami flow, erosion, and deposition on the coastlines of Banda Aceh, Sumatra, and Nam Khem, Thailand. The larger inundation, runup, and backflow in Banda Aceh (also see McAdoo et al., 2007) showed less geomorphic control than the studied case in Thailand, where typical runup was 4–5 m. Satellite imagery was also used in India, for example, to assess tsunami damage (Kumar et al., 2007); certainly other studies are to come.

Other field surveys that outline geological and geomorphic effects of the 2004 great tsunami include Szczucinski et al.'s studies (2005, 2007) of the environmental and geological impacts of the tsunami on the Thailand coast. Kurian et al. (2006) describe inundation and geomorphological impacts of the tsunami on the SW coast of India, documenting before-and-after beach profiles and quantifying erosion and deposition by the tsunami. Kench et al. (2006) describe geological effects of the tsunami on the Maldives, a set of low-lying, mid-ocean coral islands, where deposition dominated erosion. Lavigne et al. (2007) summarize field observations in Java, and Obura (2006) outlines impacts of the tsunami on the coast of Africa.

Whereas the onshore effects of tsunamis have received much attention, the offshore marine record of historical tsunamis has rarely been documented (e.g., van den Bergh et al., 2003), aside from speculation such as that by Shepard et al. (1950) and Reimnitz and Marshall (1965). In a number of historical cases, on the coastal plain, seaward-directed flow and evidence of seaward flow such as flopped-over plants have been observed. The drawdown phase of the tsunami is typically slower than the uprush, however, and outflow tends to be concentrated in topographic lows such as channels. Terrestrial debris from tsunami outflow has been observed and photographed in the nearshore region in many historical cases. It is likely that on the shelf a tsunami deposit looks like and might be confused with a deposit from a flooding river mouth (e.g., Wheatcroft and Borgeld, 2000), or a storm-surge return flow (e.g., Aigner and Reineck, 1982).

2.2.   Tsunamis, turbidity currents, and submarine canyons

An interesting early paper by geologists E. B. Bailey and J. Weir (1932) attributes aspects of Jurassic-age (Kimmeridgian) conglomerates along the Great Glen fault in Scotland to the action of tsunamis ("tunamis" in their spelling). Bailey was interested in sedimentation and tectonics, had mapped major faults in Scotland, and had previously described what he interpreted as submarine landslide deposits in Paleozoic rocks in Quebec. It is clear the authors had been influenced by reports from the great 1923 Kanto earthquake and tsunami in Japan. In that case, the earthquake and following firestorms generated horrific casualties; the tsunami had a runup of more than 10 m and killed hundreds of people.

Bailey and Weir's interpretation illustrates a fundamental geologic question existing around this time: What was the origin of coarse-grained sediments deposited, apparently, in quiet or deep water (Walker, 1973; Dott, 1978)? In the Kimmeridgian case, Bailey and Weir described boulder-bearing conglomerates, containing shallow-marine fossils and exhibiting some degree of grain-size sorting, interbedded with ammonite-bearing shales. The latter they interpreted to represent quiet water, offshore deposition, and they interpreted that the boulders' beds were emplaced by earthquake-triggered landslides, with the observed sorting accomplished by attendant tsunamis.

About 20 years later, in the 1950s, many coarse-grained beds (especially graded beds) interbedded with marine shales (indicating quiet water deposition) were reinterpreted as the deposits of turbidity currents. Turbidity currents themselves had been described by the late 1880s (though not by name) where rivers entered lakes or reservoirs; however, the genetic connection of these sediment-laden density currents with graded beds in the geologic record was first made around 1950 (Kuenen and Migliorini, 1950). Such beds came to be called turbidites. However, the Kimmeridgian boulder beds of Bailey and Weir (1932) were reinterpreted not as turbidites but as submarine debris-flow deposits by Pickering et al. (1984).

Early studies of and speculation about the origin of submarine canyons linked turbidity currents and tsunamis. In 1936, R. A. Daly proposed that sediment-laden undersea currents (later called density currents and turbidity currents) were responsible for the generation of submarine canyons by erosion. However, while agreeing with the basic erosional nature of submarine canyons, Bucher (1940) attributed their origin to erosion by tsunamis, arguing that the return flow would be stronger because of gravitational forces. Bucher cited effects of the 1929 Grand

Banks earthquake and tsunami (speculating, as others had and would, about the cause of the timing of submarine cable breaks); he also mentioned the 1933 Sanriku coast tsunami in Japan. Other than Bucher's mention, little attention was paid at the time to the tsunami associated with the 1929 Grand Banks earthquake, partly because it occurred during a storm surge (Piper et al., 1988).

Coleman (1968; also 1978), following Bucher (1940), speculated that the sediment-charged return flow from tsunamis might be responsible for triggering turbidity currents and by this means eroding submarine canyons. He also suggested that erosional geomorphic features associated with deltas and barrier reefs, and apparently not explainable by storms, might be attributed to tsunami action. He guessed that other unusual deposits in the geological record might be from tsunamis. In his articles, he did not consider onshore tsunami erosion or deposition.

Tsunami-triggered return (offshore) flow into deep water was invoked by Kastens and Cita (1981) for a "homogenite" in the Mediterranean Sea they attribute to the tsunami triggered by the Santorini caldera collapse c. 3500 B.P. (Cita and Aloisi, 2000). The 1981 paper was the first description since Bailey and Weir of a specific ancient deposit attributed to tsunami action. However, as noted by Shanmugam (2006), this "homogenite" as interpreted is actually a "turbidite" not a "tsunamite." (Terminology for many deposits related to earthquakes [e.g., "seismites" for liquefied beds] and tsunamis is a morass [Shanmugan, 2006]; I will use "tsunami deposit" and "tsunami-related deposit" in this article.)

The connection of earthquakes, tsunamis, and turbidites also includes literature on turbidites (triggered by land failures) triggered by earthquakes, sometimes known as "seismoturbidites" (Mutti et al., 1984). In addition to the Grand Banks case (Piper et al., 1988), such deposits have been described by Adams (1990) from cores in the Cascadia deep-sea channel off the Washington and Oregon coastline and interpreted as evidence for up to 13 prehistoric earthquakes since ~7000 B.P. Nakajima and Kanai (2000) described submarine land failures triggered by the 1983 Japan Sea earthquake as well as prehistoric cases where turbidites are inferred to be proxies for earthquakes (see also Doig, 1990; Inouchi et al., 1996; Goldfinger et al., 2003; Gutscher, 2005; McHugh et al., 2006). All the historic cases considered were also tsunamigenic, and submarine landslides produce tsunamis, but a direct genitive link between tsunamis and turbidity currents is difficult to document.

3. Tsunami Deposits

Tsunami deposits fall into the category geologists refer to as "event deposits," that is, episodic deposits of short-duration, unusual or high-energy processes relative to deposits of everyday or normal conditions, the latter sediments commonly called "background deposits" (Einsele and Seilacher, 1982; Dott, 1983; Clifton, 1988; Einsele et al., 1991). There is no precise definition of "event," and relegation of a process to that category depends partly on temporal and spatial perspective. In the marine realm, the most common such physical events are storms, turbidity currents, underwater landslides, and tsunamis. Rarer and more convulsive events with associated tsunamis would include caldera collapses such as Krakatoa 1883 (Simkin and Fiske, 1983; Carey et al., 2001) and island-sector collapses such as the prehistoric Alika 2 slide on Hawaii (J. G. Moore and G. W. Moore, 1984; J. G.

Moore et al., 1994). Catastrophic tsunamigenic events would include oceanic asteroid impacts (Bourgeois et al., 1988; Dypvik and Jansa, 2003).

Tsunamis of geologic significance, that is, ones that leave a geologic record,
include those produced by large earthquakes, large landslides, volcanic eruptions,
and asteroid impacts. The most common of these, of more than local extent, are
tsunamis from earthquakes. Tsunamis from large earthquakes (Mw 7–7.9) will
produce regional effects, and tsunamis from great earthquakes (Mw > 8, and especially > 9) can produce ocean-wide effects.

Thus far, the documented geologic record of recent (Holocene) seismogenic
and landslide-generated tsunamis is almost entirely from terrestrial settings, including lakes. There are at least two reasons for the lack of a documented offshore
record. First, little work has been done to look for offshore records of tsunamis—
in part because of expense, but also in part due to the youthfulness of the field.
Second, however, we can expect that most tsunamis will not have as great an effect
on the sea floor as storms on the continental shelf (e.g., Bourgeois et al., 1988;
Weiss and Bahlburg, 2006) and would thus be reworked. Even on the shoreface,
where tsunami effects may be large, their record is likely to be reworked or erased
by storm waves. In deeper water, excepting the case of tsunami-triggered turbidity
currents, the record is likely to be minuscule and obscure (Pickering et al., 1991).

The literature on pre-Quaternary deposits interpreted to be from tsunami action is almost exclusively about marine deposits of shelf depths and deeper; there
are at least three reasons why the onshore record is scarce in older rocks. First,
long-term terrestrial erosion removes non-marine strata wholesale. Second, the
deposits of recent tsunamis as described in most onshore coastal sites are subtle;
moreover, in the geologic record, these deposits may have been interpreted as
storm deposits (e.g., see Pratt, 2002). Interpretations of shoreface and shallow-
marine facies as tsunami deposits (e.g., Massari and D'Alessandro, 2000; Pratt,
2002) should be viewed with skepticism because storm waves are very effective in
this regime. Finally, some of the literature on ancient tsunami deposits is quite
speculative and probably wrong.

3.1   Tsunami deposits in the pre-Quaternary Record

Almost all published literature on pre-Quaternary tsunami deposits is associated
with asteroid-impact-generated megatsunamis, and about half of these articles are
about the Cretaceous-Tertiary (K-T) boundary. Deposits related to postulated
megatsunamis also have been associated geologically to several other known impact structures (Gersonde et al., 2002; Dypvik and Jansa, 2003). Most of the remainder of the literature describing ancient tsunami deposits correlates the
deposits with geologic evidence for earthquakes; deposits associated with earthquakes are commonly called "seismites."

Pre-Phanerozoic tsunami deposits have been described on a number of continents. The geologically oldest tsunami deposits described in the literature are from
Australia and are early Archaean in age, almost 3.5 billion years old (Glikson,
2004). Other Archaean deposits from Australia have been described by Hassler et
al. (2000; Hassler and Simonson, 2001) and from South Africa by Byerly et al.
(2002). All of these deposits are tied to evidence of asteroid impacts, such as impact spherules; impacts were more frequent in early Earth history than later. Pro-

terozoic deposits from India (Bhattacharya and Bandyopadhyay, 1998), North America (Pratt, 2001), and China (Du et al., 2001) have been tied to evidence for earthquake activity such as deformed beds. Pratt attributes certain layers in the Belt Supergroup to tsunami backwash. Du et al. discuss multiple hypotheses to explain what they call "earthquake event deposits" in Mesoproterozoic strata.

Literature on Paleozoic tsunami deposits is also dominated by impact cases, but includes deposits associated with evidence for earthquakes. Impact-associated cases include Devonian-aged breccias and other coarse-grained deposits known as the Alamo Breccia (Warme and Sandberg, 1995) and the Devonian Narva Breccia correlated with the Middle Devonian Kaluga impact crater (Masaitis, 2002). Speculative earthquake-associated cases include Cambrian and Ordovician strata from North America (Pratt, 2002; Pope et al., 1997).

The Mesozoic tsunami-deposit literature is dominated by K-T boundary articles, but includes a number of Triassic and Jurassic cases, including postulated land-slide-generated tsunami deposits (Brookfield et al., 2006). A well-documented impact-related tsunami deposit is associated with the Jurassic Mjolnir crater (Dyp-vik et al., 2004). There are also several deposits of Cretaceous age attributed to impact-generated or other tsunamis (e.g., Steiner and Shoemaker, 1996; Rossetti et al., 2000; Bievre and Quesne, 2004; Fujino et al., 2006; Weber and Watkins, 2007). Studying upper Cretaceous offshore deposits in Nebraska and South Dakota, USA, Weber and Watkins (2007) used redeposited nannofossils to demonstrate a resuspension event they correlate with the Manson, Iowa, impact ~74 million years ago. Several articles about Mesozoic tsunami geology invoke more than one kind of possible tsunami source (e.g., Bussert and Aberhan, 2004; Schnyder et al., 2005; Simms, 2007). Of course, impacts typically would generate earthquakes and land-slides, as can volcanoes, so multiple kinds of tsunami sources would be associated with mega-events.

With regard to the K-T tsunami deposit and associated sediments, the consensus view is that an impact of a ~10-km bolide on the edge of the (paleo-) Yucatan Peninsula generated coarse-grained deposits including tsunami deposits around the Gulf of Mexico and Caribbean (Bourgeois et al., 1988; Smit et al., 1996; Lawton et al., 2005), although there is some literature disputing a tsunami interpre-tation (e.g., Stinnesbeck and Keller, 1996). K-T deposits related to the impact have also been described in platform carbonates in Brazil (Albertão and Martins, 1996) and in deep-sea sediments of the North Atlantic (Norris et al., 2000). Norris et al. ascribe most observed K-T deposits from deep-sea cores to slope failure and asso-ciated turbidity currents; however, they suggest a tsunami may have generated erosional features on the submarine Blake Plateau and elsewhere.

Three Eocene impact structures have been identified in the subsurface record of the continental margin of eastern North America: Chesapeake Bay—35.7 Ma, Toms Canyon—35.7 Ma, Montagnais—51 Ma (Poag et al., 2002), with tsunami deposits described from the Chesapeake Bay structure (e.g., Poag et al., 1992; Poag, 1997). Other pre-Quaternary Cenozoic tsunami deposits include Miocene deposits in Japan (Shiki and Teiji, 1996) and Chile (Cantalamessa and Di Celma, 2005) and scour-and-drape features in Pliocene carbonates in Italy (De Martini et al., 2003). Some work has suggested a relationship between possible tsunami de-posits and the Pliocene Eltanin impact structure (Hartley et al., 2001; also see Paskoff, 1991).

3.2. Quaternary deposits attributed to landslide-generated tsunamis

The best-documented deposits from a prehistoric landslide-generated tsunami are from early Holocene Storegga submarine landslides in the North Sea. One of the earliest papers on tsunami deposits was the description by Dawson et al. (1988) of an unusual deposit on the eastern coast of Scotland, which they speculated was produced by tsunami runup from a Storegga event. The correlative deposit was found in Norway, best preserved in coastal lakes (Bondevik et al., 1997; Figure 3.4). Later, a combined group documented multiple landslide-generated tsunamis in this area (Bondevik et al., 2005a; also see Smith et al., 2004).

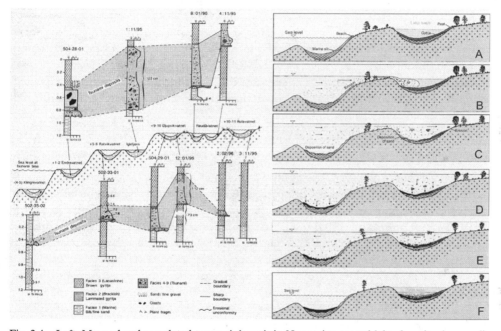

Fig. 3.4 – Left: Mapped and correlated tsunami deposit in Norwegian coastal lake deposits, from sediment cores. Right: Interpretation of the sequence of events by which this deposit was emplaced by the Storegga-slide-triggered tsunami (Bondevik et al., several publications).

Other tsunami deposits tied to earthquake-triggered landslides include one associated with the North Anatolian fault (Minoura et al., 2005), another with the 1929 Grand Banks tsunami, and another case of sublacustrine gravel ridges attributed to a prehistoric landslide-generated tsunami in Lake Tahoe, U.S. (J. G. Moore et al., 2006). Also, large coral blocks were moved by the 1771 Meiwa tsunami, interpreted by some to be landslide-enhanced but the landslide interpretation disputed by others (see Nakamura, 2006, for review). The Grand Banks tsunami, while understudied around the time of its origin, has recently been reinvestigated. Deposits from the tsunami have been described by Tuttle et al. (2004), who contrast it with an interpreted storm deposit, and A. Moore et al. (2007), who conducted detailed grain-size analysis of the deposit and describe landward fining.

Other Quaternary deposits attributed to submarine landslides are primarily cases associated with volcanic processes. Those landslides not associated directly with eruptive processes include sector collapse of volcanic edifices, of which the most studied are those in Hawaii (J. G. Moore et al., 1994). J. G. Moore and G. W. Moore (1984) mapped, described and named the Hulopoe Gravel on the island of Lanai and suggested it was deposited by a giant tsunami generated by flank collapse of a Hawaiian island. These and similar deposits on other islands had previously been interpreted as uplifted shoreline deposits (reviewed by Grigg and Jones, 1997), and their reinterpretation has remained contentious for more than 20 years. On the tsunami side, A. Moore (2000) made a quantitative argument for tsunami deposition of a similar deposit on Molokai, and McMurtry et al. (2004) described a deposit on the island of Hawaii that they interpret as the deposit of a megatsunami from a flank collapse of Mauna Loa. Most recently arguing on the other side, Felton et al. (2006), in a series of papers, make a detailed paleoecological case against the tsunami interpretation of the Hulopoe Gravel (also see Rubin et al., 2000).

On the Canary Islands, Perez Torrado et al. (2006) have described a coarse-grained deposit they also attribute to deposition from a flank-collapse, landslide-generated tsunami.

3.3   Historic and Quaternary geologic records of volcanogenic tsunamis

Explosive island volcanoes are likely to produce sudden sea-floor displacements, such as the 1883 eruption and collapse of Krakatoa, which generated a large tsunami (Simkin and Fiske, 1983; Latter, 1982). Deposits from the Krakatoa tsunamis are used by Carey et al. (2001) in a discussion of tsunami deposits from explosive volcanic eruptions. Van den Bergh et al. (2003) describe an offshore deposit they attribute to the Krakatoa tsunami.

Other historically documented examples of volcanogenic tsunamis and associated deposits include Stromboli in Italy (Tanner and Calvari, 2004), Karimsky Lake on Kamchatka (Beloussov and Beloussova, 2001), and Vesuvius from A.D. 79 (Sacchi et al., 2005). Dominey-Howes et al. (2000) describe historical and geological evidence for historical eruption of Thera (Santorini). Nishimura et al. (1999) review historic tsunami deposits in Japan from volcanogenic sources.

The most studied prehistoric volcanogenic tsunami is one generated by the Santorini caldera collapse about 3500 B.P. (Antonopoulos, 1992; McCoy and Heiken, 2000). Most of these articles describe deposits in the deeper Mediterranean, variously interpreted through the years as more and less directly related to a tsunami (Cita and Aloisi, 2000, and this group's earlier papers; Hieke and Werner, 2000). More recently, Santorini deposits from shallow water and onshore settings have been discovered and described (Minoura et al., 2000; Bruins et al., 2008). At Palaikastro in northeastern Crete, Bruins et al. (2008) reported the discovery of extensive "geoarchaeological tsunami deposits" characterized by a mixture of geological materials, including volcanic Santorini ash, and archaeological settlement debris. Identified tsunami signatures included an erosional lower contact, intraclasts and reworked building stone material in the lower part of the deposit, marine fauna, and imbrication of rounded beach pebbles, settlement debris, ceramic sherds, and even bones.

Numbers of other cases of volcanogenic tsunamis have been described, in a few cases including geologic evidence of resulting deposits. In prehistoric cases, there commonly is discussion about what kind of volcanogenic process generated the tsunami—hot pyroclastic flow, cold debris flow, or flank collapse, for example. There are both historic and prehistoric examples of volcanogenic tsunamis from Augustine volcano in Alaska (Waythomas and Neal, 1998; Waythomas et al., 2006); in a prehistoric case, a pumice-bearing tsunami deposit overlies airfall ash, leading to a reconstruction of wave travel time across Bristol Bay. Lowe and de Lange (2000) correlate tsunami deposits on New Zealand to the Taupo (c. 200 A.D.) caldera-forming eruption.

3.4. Holocene seismogenic tsunami deposits

By far the most literature on tsunami geology deals with Holocene tsunami sands deposited on coastal lowlands along seismically active continental margins. This work began in the mid-1980s, the first publications being by Atwater (1987) on Cascadia and Minoura et al. (1987) on Japan. Although there had been some descriptions of historic tsunami sediments by this time (as noted above), this work was scant, and actual tsunamis had been and would be few in the 1970s and 1980s.

Whereas Japan clearly had a historic record of locally generated seismogenic tsunamis, Cascadia did not, and Atwater's described evidence for great earthquakes on the Washington coast spurred research up and down the coast from Canada to California (e.g., Darienzo and Peterson, 1990; Clague and Bobrowsky, 1994; Atwater, 1996; Nelson et al., 1996 as early examples; comprehensive reviews by Clague et al., 2000; Atwater et al., 2005; Nelson et al., 2006; Peters et al., 2007). Moreover, tsunami deposits have been described in the interior seaway between Seattle and Vancouver, BC (Atwater and Moore, 1992; Bourgeois and Johnson, 2001; Williams et al., 2005); some of these are related to local crustal faults.

There have been many studies of Holocene tsunami deposits (and older) in Japan, the most focused of which have been on the island of Hokkaido, which has a relatively short historical record (Minoura et al., 1994; Nishimura and Miyagi, 1995). Thus research there is helping to elucidate tsunami recurrence, and also documenting that prehistoric tsunamis have been larger than historic ones (Nanayama et al., 2003, 2007; Figure 3.5; also see Minoura and Nakata, 1994). Minoura and Nakaya (1991) described sediment effects of the 1983 Japan Sea tsunami and then used that information to interpret paleotsunami deposits on northern Honshu. Kumagai (1999) documented tsunami deposits from large earthquakes along the Nankai trough in central Japan. On a beach plain in northern Japan, Yagishita (2001) described a layer of coarse gravel most probably deposited by the 1896 Meiji-Sanriku tsunami. Some of the only documented examples of recent tsunami deposits and effects found in marine and estuarine environments come from Japan (e.g., Fujiwara et al., 2000; Noda et al., 2007). However, some deposits attributed to tsunamis in Pleistocene valley fills (Takashimizu and Masuda, 2000) are probably not tsunami deposits but tidally mediated estuarine sands with associated liquefaction structures.

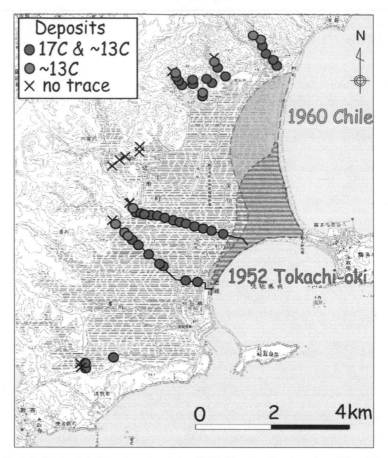

Fig. 3.5 – Historical tsunamis have inundated the Hokkaido coastline less than 2 km. However, this region has a short written history (unlike other parts of Japan), and paleotsunami deposits some centuries old have been found up to 4 km inland. Based on Nanayama et al., 2003.

There are several other regions where seismogenic paleotsunami deposits have been documented. Such deposits were first described in the Russian Far East by Melekestsev et al. (1995), work continued in particular by Pinegina (e.g., Pinegina et al., 2003). A number of studies have been done in New Zealand (e.g., Goff, 1997; Goff et al., 2001; Cochran et al., 2006; deLange and Moon, 2007), where there has been particular attention paid to the coincidence of archaeological sites with tsunami deposits (McFadgen and Goff, 2007). There is also quite a bit of work in the Mediterranean (e.g., Pirazzoli et al., 1999; Dominey-Howes, 2002; Luque et al., 2002; DeMartini et al., 2003; Reinhardt et al., 2006). Other localities with seismogenic paleotsunami studies not previously mentioned include southern California (Kuhn, 2005), Chile (Cisternas et al., 2006), and Australia (Dominey-Howes, 2007). In India, spurred by studies of the 2004 tsunami deposit, Rajendran et al. (2006) excavated evidence of two possibly comparable paleotsunami deposits, about 1000 and 1500 years old, in archaeological sites.

Because in any one region large tsunamigenic earthquakes have typical recurrence intervals of hundreds of years, few coastlines have long enough historic records to produce statistically significant recurrence probabilities (see Chapter 2). The geologic record of co-seismic deformation has been used in a number of localities to reconstruct these earthquakes (e.g., Atwater and Hemphill-Haley, 1997). Studies of long-term records of seismogenic (paleo)tsunami deposits (e.g., Minoura et al., 1994; Pinegina and Bourgeois, 2001; Kelsey et al., 2002; Nelson et al., 2004; Cisternas et al., 2005; Nanayama et al., 2007) (Figure 3.6) are becoming important to probabilistic hazard analysis not only of tsunamis but also of their parent earthquakes.

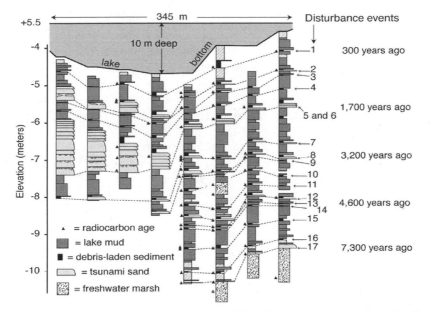

Fig. 3.6 – An example of long-term records of tsunami deposits interpreted to be from the Cascadia Subduction Zone; from Bradley Lake on the coast of southern Oregon (based on Kelsey et al., 2005).

There is at least one historical case of an inland seismogenic tsunami generated under a lakebed, with a later-studied deposit attributed to the tsunami. That event is the 1872 Owens Valley earthquake, which offset the valley by about one meter, and produced a tsunami that appears to have generated a pebbly sand deposit (Smoot et al., 2000).

3.5. Tsunami-deposit dating and correlation

Numerical age dating of tsunamis has recently been accomplished using optical luminescence dating of quartz grains (Prescott and Robertson, 1997). These grains are reset when exposed to the sun, so that wave-worked nearshore sands are typically zeroed out. Therefore, if these sands are eroded by a tsunami and rapidly buried, the age of their burial can be determined (e.g., Huntley and Clague, 1996; Banerjee et al., 2001; Ollerhead et al., 2001). Murari et al. (2007) tested this technique on the 2004 tsunami deposit in India and with their analysis concluded that most of the tsunami sediment came from near the sediment-water interface, and that a paleotsunami deposit could probably be dated to within about 50 years.

More commonly, tsunami deposits are dated by associated datable material, particularly plant material for radiocarbon dating, which works only for about the last 50,000 years. Radiocarbon dating of shell material is more complicated as local marine reservoirs of carbon must be considered, whereas plants take their carbon from the air. The association of tsunami deposits with marker tephra, that is, well-dated tephra layers (usually by radiocarbon), has proved very useful in some cases, as illustrated by aforementioned work on Kamchatka and Hokkaido.

Other techniques that have been used to date tsunami deposits are varied. Corals included in deposits have been dated by techniques including electron-spin resonance, radiocarbon, and U-series dating. Whether the coral died at the time of the tsunami or was already dead obviously affects these results. Recent tsunami deposits have been dated using included anthropogenic tracers such as PCBs and ^{137}Cs (Barra et al., 2004). Clearly archaeological context can also help date deposits (e.g., Bruins et al., 2008).

Correlation of tsunami deposits is difficult because deposits rarely have intrinsically correlative characteristics. Volcanogenic tsunamis may have associated fresh volcanic materials that aid in correlation (e.g., Waythomas and Neal, 1998; Bruins et al., 2008). On Kamchatka, while individual deposits are rarely correlated, deposits between marker tephra can be counted and thus statistical tsunami recurrence rates estimated (e.g., Pinegina et al., 2003; Bourgeois et al., 2006; Figure 3.7). Nelson et al. (2006) have attempted to correlate tsunami deposits along the Cascadia margin and thus to infer paleo-rupture lengths of the subduction zone. Switzer et al. (2006) used ground-penetrating radar to map deposit continuity in the subsurface.

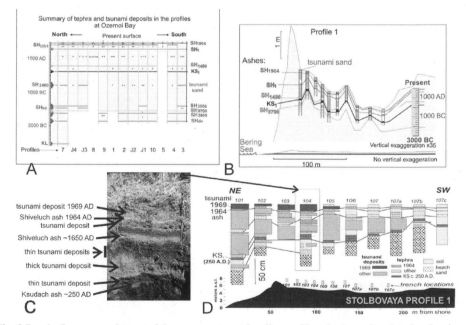

Fig. 3.7 – A: Summary of tsunami frequency at one locality on Kamchatka, using dated volcanic ash layers for correlation; lines are tephra, dots are tsunami sands. B: Profile 1 (see A) and excavation records, plotted according to true depth in profile. C: Excavation 104 in Profile 1. D: Profile 1 plotted according to field notes; note 1969 tsunami deposit (see Figure 10) (based on Bourgeois et al., 2006).

4. Other Themes in Tsunami Geology

Of themes that emerge in a literature review of tsunami geology, two that bridge disciplines are tsunami geomorphology and geoarchaeology. Many papers also consider aspects of hazard analysis using tsunami deposits (e.g., Fiedorowicz and Peterson, 2002; Szczuciński et al., 2006; Kumar et al., 2007). A few focus on field and lab methodology.

4.1 Tsunami geomorphology and tsunami erosion

Where there is tsunami deposition, there must have been erosion, but erosion and other geomorphic aspects of tsunamis are explicitly considered in less than 5% of prehistoric cases. Historical surveys and photographs, however, clearly show that tsunamis are geomorphic agents, if only temporarily. Dawson (1994) wrote a brief review of known and speculated geomorphic effects of tsunamis, including boulder deposits; Scheffers and Kelletat (2003) wrote another review, including their own speculations about chevrons, to be discussed below. Well-documented, preserved geomorphic features of tsunamis include scoured depressions through beach ridges and associated lobate accumulations landward. Vegetation and soil stripping are also common, especially in the proximal zone. Using both satellite images and on-the-ground surveys, erosion from the 2004 Indian Ocean tsunami has recently been documented in a number of cases (e.g., Srinivasalu et al., 2007 and previously cited papers; also see Fig. 3.1).

There is a very speculative literature on bedrock sculpting by tsunamis (e.g., Bryant and Young, 1996; Aalto et al., 1999), not based on observed examples. There is no fundamental basis for the argument that tsunamis are more powerful sculptors than storm waves. Bryant (2001 and earlier papers) has gone further and argued that tsunamis are important coastal geomorphic agents in Australia and elsewhere, on coastlines where tsunamis are historically rare. These interpretations have been disputed (e.g., Felton and Crook, 2003; Dawson, 2003; Dominey-Howes et al., 2006). For example, Dawson (2003) said "Such explanations are in contradiction with everything that is written and known about the Quaternary of Scotland and the author [Bryant] appears blissfully unaware that features he describes have been affected by a complex glacial and sea-level-change history. Unfortunately, there are many pages of text here that students would be well advised to avoid at all costs."

Another area of unfounded speculation, almost certainly mistaken, is the argument that along many semi-arid coastlines, certain large Holocene parabolic sand dunes, called "chevrons" by some authors, are the product of mega-tsunamis, possibly generated by asteroid impacts (e.g., Bryant, 2001; Scheffers and Kelletat, 2003). Others have interpreted these bed forms as eolian; some cases on carbonate platforms may be due to storm waves. Examples of problems with the "chevron" megatsunami speculation include the facts that very similar bed forms are present in the interiors of continents; that the bed forms do not show evidence of bathymetric and topographic steering as expected from tsunamis and not from wind; and that the bed forms are peculiar to certain climatic conditions. Moreover, the construction of large sandy bed forms requires low shear stresses (bed load transport conditions) over extended time periods (days or more); neither would be the case for impact-generated (or other) tsunamis. The chevrons are constructed principally of sand, though larger clasts have been found nearby (e.g., Kelletat and Scheffers, 2003). If those large clasts were transported at the same time as the "chevron" sand, the bed load condition for the sand would almost certainly be violated.

Let's consider just the bed load transport argument. When sediment is transported at bed load (relatively low boundary shear stress, τ_b), there will be bed forms (ripples, dunes) on the bed, leaving behind cross-stratification. When sediment is transported as suspended load (higher shear stresses), bed forms wash out, called the plane bed condition; typical deposits will be planar laminated, or massive to normally graded if deposited rapidly. Thus already, without reference to chevrons, we can infer for recent tsunami deposits that these historical sediments were transported principally as suspended load, based on their characteristic graded, planar-laminated or massive bedding (summaries in Dawson and Shi, 2000; Dominey-Howes et al., 2006). The transition from bed load to suspended load is dependent on a relationship between boundary shear stress τ_b (typically expressed as "shear velocity" $= [\tau_b/\rho]^{\frac{1}{2}}$) and grain settling velocity W_s, this relationship commonly referred to as the Rouse criterion (e.g., Vanoni, 1975). For bed load transport to take place, U_* should be less than W_s (Abbott and Francis, 1977). For sand, with a *maximum* W_s in water of about 0.1 m/sec (very coarse sand), the skin friction component of U_* should be less than 0.1 m/sec in order for bed forms to be stable on the bed.

Now let's take a *minimum conservative condition* during postulated megatsunami flow over the chevrons. Actively migrating bed forms must be inundated with water at least twice their height, and chevrons are reported to be 8–20 m high and more (and reported to be present at elevations above modern sea level of more than 100 m) (Kelletat and Scheffers, 2003). If we take a very conservative minimum flow depth of 20 m, a Froude number of 1 (a conservative choice—higher Froude numbers will give greater velocities), and a characteristic roughness of 0.5 m, the skin friction portion of U_* would be about 0.4 m/sec, giving a Rouse number of 0.6, well within the condition for suspended load. Greater flow depths, as postulated or required for these "chevrons" if they are indeed megatsunami deposits, would of course increase the shear velocity and decrease the Rouse number.

4.2. Paleontology and archaeology of tsunami deposits

Fossils in tsunami deposits have been used in a number of studies; fossils are also used extensively to document co-seismic deformation, which will not be reviewed here. Presence of marine fossils in a deposit is one piece of evidence for a tsunami origin, rather than fluvial or eolian (non-marine) processes (Hemphill-Haley, 1995; S. Dawson et al., 1996; Hutchinson et al., 1997; Williams and Hutchinson, 2000; Sawai, 2002). There has also been some discussion of the depth of origin of benthic microfossils in tsunami deposits and its possible significance. From the distribution of foraminifera and ostracodes in their samples, Hussain et al. (2006) inferred that the 26 December 2004 tsunamigenic sediments deposited on the coast of the Andaman group of islands were derived from shallow littoral to neritic depths and not from deeper bathyal territories.

A number of other types of tsunami studies have used faunal and floral elements. Hemphill-Haley (1996) showed that the landward extent of marine microfossils exceeded the landward extent of tsunami sand and silt, illustrating that a recognizable siliciclastic deposit is only a limiting minimum for inundation (Hemphill-Haley, 1996). Nanayama and Shigeno (2006) used microfossils to distinguish tsunami inflow deposits from outflow deposits. While most studies have used microfossils, Fujiwara et al. (2003) described and interpreted mixed molluscan assemblages in tsunami deposits. Hughes and Matthewes (2003) describe plant recolonization following a tsunami.

Many ancient as well as historical settlements are located on coastlines, particularly in cases of maritime societies, so it is not surprising that many coastal archaeological sites include evidence for tsunamis. For example, Veski et al. (2005) describe an anomalous deposit in early Holocene archaeological sites in Estonia that they tentatively correlate with either the Storegga-slide tsunami or an asteroid impact (with regional evidence of spherules) (also see Bondevik, 2003). McFadgen and Goff (2007) summarize several geoarchaeological investigations of mostly seismogenic tsunamis in New Zealand. As noted in the section on Santorini, Bruins et al. (2008) have reported the discovery of extensive "geoarchaeological tsunami deposits" in northeastern Crete. A number of other interdisciplinary studies mention tsunami as a possible contributor to the history of archaeological sites (e.g., Luque et al., 2002; Carson, 2004).

5.   Recent Directions in Tsunami Geology and Sedimentology

No doubt the 2004 Indian Ocean earthquake and tsunami will have a major impact on our understanding of tsunami geology, and it is difficult yet to synthesize this ongoing work (Razzhigaeva et al., 2006; Hawkes et al., 2007; Hori et al., 2007; Srinivasalu et al., 2007; and others). The level of effort and detail of work are impressive (e.g., Figure 3.8). An example of the importance of this tsunami to our understanding is its complexity—since 1960 Chile and 1964 Alaska, the 26 December 2004 tsunami is the most complex in terms of number of large waves. Unraveling this complexity via tsunami deposits is a challenge, but abundant videos and still photos, as well as eyewitness accounts, are helping sedimentologists do so in many thorough studies.

Some basics we have learned or re-learned from the Indian Ocean case, relevant to sedimentology, include the observation that the tsunami on land was rather like a river without banks. Major waves commonly had multiple waves superimposed on them, and many localities experienced withdrawals between major waves. Acceleration, deceleration, and wave reflection are shown in video images. Thus the deposits from this tsunami, and presumably others of comparable scale, are commonly complex, though still dominated by suspended load leading to massive, graded, or laminated structure.

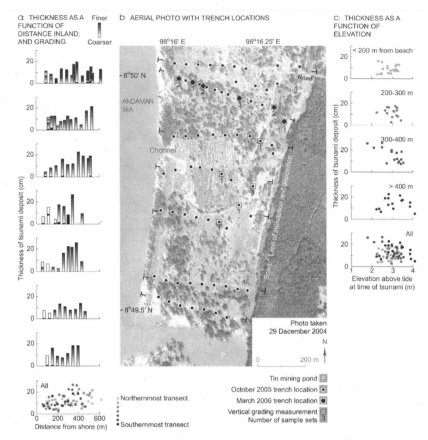

Fig. 3.8 – Distribution and analysis of the 26 December 2004 tsunami in Kao Lak Thailand (Higman figure from Alam et al., in press); used with permission.

Prompted by the 2004 tsunami, many new investigators are being trained in tsunami geology. Countries around the Indian Ocean are engaging their scientists in paleotsunami studies (e.g., Rajendran et al., 2006). Another region where such work is still in early phases is the Caribbean (Morton et al., 2006). Moya et al. (2006) review the tsunami history of Puerto Rico and describe cores with both historical and pre-historical sand layers they interpret to be tsunami deposits. The Caribbean is one of the areas where debate about whether storms (hurricanes) or tsunamis are responsible for boulder transport, gravel ridges, and other coastal features (e.g., Scheffers, 2004; Morton et al., 2006).

In a June 2005 workshop on tsunami geology, 80 scientists conferred on recent advances and new directions in the field. For example, as noted below, geologists and tsunami modelers are beginning to work together, and tsunami sedimentology is becoming more quantitative. Tsunami geology is being used for inverse modeling to earthquake sources. The web of possible interactions among tsunami geology and related fields is shown in Figure 3.9.

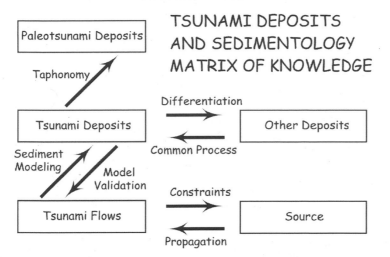

Fig. 3.9—A geologist starts with a paleotsunami deposit and works back toward an understanding of the processes by which it was deposited ("tsunami flows"). To arrive there, one must understand how the deposit has been altered since deposition ("taphonomy") and also how tsunamis transport and deposit sediment ("sediment modeling"). One must also determine that the deposit is truly from a tsunami and not from another event such as a storm ("differentiation").

This diagram also emphasizes that other processes and their deposits can help us understand tsunami deposits because many such events—such as dam-break floods, turbidity currents, and tidal bores—share "common processes."

The nature and distribution of tsunami deposits can help validate models of tsunami "flow"—propagation and runup—and that flow's geologic effects—erosion and deposition.

Ultimately, we may want to understand/invert the deposit not only to the tsunami, but also to the tsunami source. What were the rupture parameters of the earthquake? Was the tsunami from an earthquake or from a landslide? A volcanic eruption? An asteroid impact? Diagram developed by working group during NSF-sponsored Tsunami Deposits Workshop, June, 2005.

5.1 Tsunamis and neotectonics

Mapped tsunami deposits provide minimum runup heights and inundation distances, which in turn are related to earthquake rupture characteristics (see other chapters in this volume). Thus by inverse modeling paleotsunami deposits can help us reconstruct tsunami sources and the attendant tectonic character of a region. For example, Satake et al. (2005) used modern and paleotsunami records on Hokkaido to reconstruct earthquake sources on the southern Kuril trench. Bourgeois et al. (2006) reconstructed more than 2000 years of paleotsunami history on the Bering Sea coast of Kamchatka and used this information, as well as an inverse model of the local 1969 tsunami, to estimate the convergence rate on this previously understudied boundary. Nelson et al. (2006) used paleotsunami deposits and other paleoseismological evidence to examine possible segmentation of the Cascadia subduction zone. Martin et al. (2008) distinguished two earthquake sources (1969 and 1971) on Kamchatka by inverting the distribution of deposits back to the source regions (Figure 3.10).

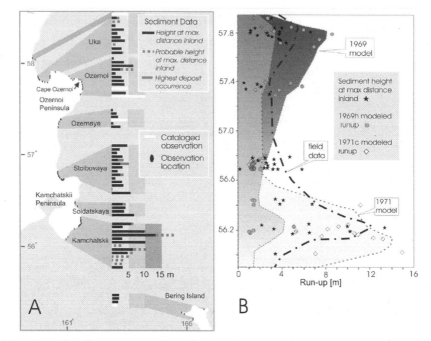

Fig. 3.10—Elevations of recent tsunami deposits along the Bering Sea coast of Kamchatka compared to runup models generated with the Method of Splitting Tsunamis (MOST) (Martin et al., 2008). Right: Distribution of maximum elevations of young tsunami deposits. Tsunami deposits can be taken only as MINIMA for tsunami elevation, runup and inundation because 1) the water must be higher than the highest deposit, but we do not know now much higher; and 2) the tsunami can outrun its deposit. Left: Comparision of runup models and tsunami sediment data. After Martin et al. (2008).

5.2.   *Storm vs. tsunami and the boulder-transport problem*

Since the beginnings of modern studies, the most pressing question has been, how does one distinguish an (onshore) tsunami deposit from an onshore storm deposit (e.g., Witter et al., 2001)? Knowledge of both is rapidly growing, and studies from the 2004 Indian Ocean tsunami and 2005 Katrina hurricane will add significantly to our body of knowledge. Several papers on this topic have taken the approach of comparing historical examples of storm and tsunami deposits (e.g., Nanayama et al., 2000; Goff et al., 2004; Tuttle et al., 2004; Morton et al., 2007). From these studies, as well as from reasoning about the differences between tsunamis and storm surges, a summary contrast is emerging (e.g., Figure 3.11). Offshore, *most* tsunamis will be less effective than storms, and their record in the nearshore may commonly be erased by storm waves (Weiss and Bahlburg, 2006). Onshore, storms are more likely to generate wedge-like, bed-load dominated units, whereas tsunamis are more likely to produce sheet-like, suspended-load dominated deposits (Figure 3.11).

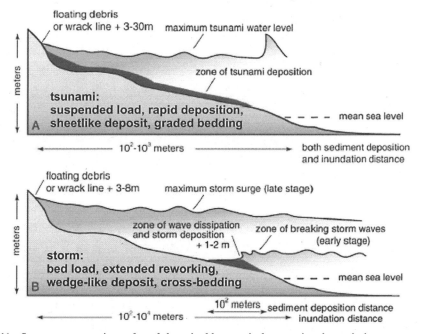

Fig. 3.11—Summary comparison of sand deposited by a typical tsunami and a typical storm surge, on a coastal profile. After Morton et al. (2007), courtesy of the authors.

In addition to discussing physical characteristics, some authors have examined whether fauna, flora, or other microscopic characters can help distinguish deposits of storms from tsunamis (e.g., Nigam and Chaturvedi, 2006). Kortekaas and Dawson (2007) found no difference in foraminifera in historical storm and tsunami deposits and concluded that only a combination of characteristics allowed distinction. Bruzzi and Prone (2000) used the surface texture of sand grains as a possible criterion for distinguishing storm and tsunami deposits.

The most controversial field of tsunami geology currently is the interpretation of certain coarse-grained deposits, particularly boulders, transported either by storms or tsunamis (if not some unimagined process). Recent large-clast transport by both storm waves and tsunamis is well documented, but even this literature must be read critically. Our bibliography includes nearly 30 refereed articles, about half favoring tsunamis and half storms. (Of course, both interpretations could be valid in different or even the same cases, such as the Caribbean.) Most of these papers consider boulders and gravel currently on the surface; for this review, pre-Quaternary deposits are excluded.

The earliest paper attributing surficial bouldery deposits to a prehistoric tsunami was the aforementioned work by J. G. Moore and G. W. Moore (1984, 1988) on the Hulopoe Gravel on Lanai. The interpretation of these deposits remains unsettled (Noormets et al., 2004). Other coastal regions where the controversy is active include Australia (Young et al., 1996; Nott, 1997; Saintilan and Rogers, 2005) and the Caribbean (Jones and Hunter, 1992; Scheffers, 2004; Morton et al., 2006). Mastronuzzi and Sanso (2000) conclude that boulders on the coast of southern Italy were tsunami-transported. Williams and Hall (2004) document accumulations of "megaclasts" on the Atlantic coast of Ireland, attribute them to large storms, and caution against the use of large transported boulders as evidence for (mega) tsunamis.

5.3. Quantitative tsunami sedimentology

Only recently have there been many substantive studies on quantitative aspects of tsunami sedimentology (e.g., Jones and Mader, 1996; Minoura et al., 1997; Hindson and Andrade, 1999; Matsutomi et al., 2001; Tonkin et al., 2003; Schlichting and Peterson, 2006; Smith et al., 2007; Gelfenbaum et al., 2007). Approaches have varied from experimental to theoretical, from forward to inverse modeling. Studies fall broadly into two categories: studies of tsunami sediment transport, and inversions of map distributions of deposits to runup and source models.

Forward models of tsunami propagation and runup (without sediment) have been benchmarked and tested, though these models are commonly limited by bathymetric and topographic data (see Chapters 8 and 9). (Clearly roughness is another major factor; for example, Minoura et al. (1996) describe how a tsunami over snow will lose less momentum than over vegetation.) Because these forward models predict the distribution of tsunami flow over a coastal area, the distribution of a tsunami deposit over this area can be inverted to a tsunami and its source. This approach has been attempted, for example, in the North Sea Storegga case (Bondevik et al., 2005b), in Japan (Satake et al., 2005), and on Kamchatka (Figure 3.10).

There are still few published quantitative studies of tsunami sediment transport, but the field is growing. In an early attempt, Eaton et al. (1961) made crude calculations of the forces necessary to move large boulders transported by the 1960 Chile tsunami on Hawaii. Bourgeois et al. (1988) calculated the shear stress necessary to move the largest clasts in a deposit at the Cretaceous-Tertiary boundary by calculating lift and drag on the clasts. Then, using Airy wave theory as an approximation, they estimated the size of the wave that could have moved these clasts in >50 m water depth. However, methods for calculating initiation of boulder transport remain unsettled.

There are two published kinds of inverse models for tsunami sediment transport, and both are works in progress, with numbers of assumptions. Also, Matsutomi et al. (2001) review a semi-empirical method of relating tsunami flow to sediment transport. The "advection model" (Soulsby et al., 2007; Moore et al., 2007; and earlier cited work) calculates a depth-velocity product that can transport a given grain size class (in suspension) from high in the flow to its farthest point inland. This model assumes no reentrainment. If neither the depth nor velocity is known, then a second method must be used to solve for these two unknowns. The "Rouse model" (my term) (Jaffe and Gelfenbaum, 2007) assumes that all the sediment at one point on the bed was deposited from the overlying water column at one time and inverts the sediment size and volume distribution of that deposit to a shear stress using the Rouse equation. This approach assumes the flow is quasi-steady and quasi-uniform.

Forward models of tsunami sediment transport have been presented at meetings but have yet to be published, except in meeting proceedings (Gelfenbaum et al., 2007; see Huntington et al., 2007). All studies of tsunami sediment transport remain in alpha testing, and this field is a very promising one for future scientists because of its importance both to evaluating tsunami hazard and to reconstructing geologic history.

Acknowledgments

I thank in particular my graduate students, especially Bretwood Higman (Hig), Bre MacInnes, Beth Martin, and Andy Ritchie, for their stimulating thoughts, and their help in many ways. Brian Atwater has been a steady and generous contributor to tsunami geology and to our group's work. Colleagues at the Center for Tsunami Research, especially Robert Weiss and Vasily Titov, have made important contributions to our understanding of tsunamis; Bruce Jaffe and Guy Gelfenbaum at the USGS, and Harry Yeh at Oregon State have contributed to advancing the field of tsunami geology. Brian Atwater helped develop the working bibliography for this chapter, and Yuki Sawai helped identify important Japanese literature. Participants in the NSF-sponsored tsunami workshops have made numerous contributions of ideas and data. Andy Moore performed a careful review of the manuscript and made important comments that improved its revision; an anonymous reviewer pointed out some errors since corrected.

References

Aalto, K. R., R. Aalto, C. E. Garrison-Laney and H. F. Abramson, 1999. Tsunami(?) sculpting of the Pebble Beach wave-cut platform, Crescent City area, California *J. Geol.*, **107,** 607–622.

Abbott, J. E. and J. R. D. Francis, 1977. Saltation and suspension trajectories of solid grains in a water stream. *Phil. Trans. Roy. Soc. London,* **284,** 225–254.

Abe, Ku., Ka. Abe, Y. Tsuji, F. Imamura, H. Katao, I. Yoshihisa, K. Satake, J. Bourgeois, E. Noguera and F. Estrada, 1993. Field survey of the Nicaragua earthquake and tsunami of September 2, 1992. *Bull. Earthquake Res. Inst.,* **68,** 23–70 [in Japanese and English]

Abe, K, 2005. Tsunami resonance curve from dominant periods observed in bays of northeastern Japan. *in* Satake, K., ed., *Tsunamis: Case Studies and Recent Developments.* Springer, 97–113.

Adams, John, 1990. Paleoseismicity of the Cascadia subduction zone; evidence from turbidites off the Oregon-Washington margin. *Tectonics,* **9,** 569–583.

Aigner, T. and H.-E. Reineck, 1982. Proximality trends in modern storm sands from the Helgoland Bight (North Sea) and their implications for basin analysis. *Senckenbergiana Marit.,* **14,** 183–215.

Alam, S., B. Higman, U. Glawe and C. Maxcia, C., 2008. Sand-sheet stratigraphy recording the 2004 Indian Ocean tsunami on a Thai coastal plain. *Sedimentology,* in press.

Albertão, G. A. and P. P. Martins Jr., 1996. A possible tsunami deposit at the Cretaceous-Tertiary boundary in Pernambuco, northeastern Brazil. *Sed. Geol.,* **104,** 189–201.

Antonopoulos, J., 1992. The great Minoan eruption of Thera volcano and the ensuing tsunami in the Greek Archipelago. *Natural Hazards,* **5,** 153–168.

Atwater, B. F., 1987. Evidence for great Holocene earthquakes along the outer coast of Washington State. *Science,* **236,** 942–944.

Atwater, B. F., 1996. Coastal evidence for great earthquakes in western Washington: *U.S. Geol. Survey Prof. Paper,* **1560,** 77–90.

Atwater, B. F. and E. Hemphill-Haley, 1997. Recurrence intervals for great earthquakes of the past 3,500 years at northeastern Willapa Bay, Washington: *U.S. Geol. Survey Prof. Paper 1576,* 108p.

Atwater, B. F. and A. L. Moore, 1992. A tsunami about 1000 years ago in Puget Sound, Washington. *Science,* **258,** 1614–1617.

Atwater, B. F., S. Musumi-Rokkaku, K. Satake, Y. Tsuji, K. Ueda and D. K. Yamaguchi, 2005. The orphan tsunami of 1700; Japanese clues to a parent earthquake in North America. *U.S. Geol. Survey Prof. Paper 1707,* 133p.

Bailey, E. B. and J. Weir, 1932. Submarine faulting in Kimmeridgian times: east Sutherland. *Trans. Royal Soc. Edinburgh,* **57,** 429–467.

Banerjee, D., A. S. Murray and I. D. L. Foster, 2001. Scilly Isles, UK; optical dating of a possible tsunami deposit from the 1755 Lisbon earthquake. *Quat. Sci. Rev.,* **20,** 715–718.

Barra, R., M. Cisternas, C. Suarez, A. Araneda, O. Pinones and P. Popp, 2004. PCBs and HCHs in a salt-marsh sediment record from south-central Chile: use of tsunami signatures and Cs-137 fallout as temporal marker. *Chemosphere,* **55,** 965–972.

Beloussov, A. and M. Beloussova, M., 2001. Eruptive process, effects and deposits of the 1996 and the ancient basaltic phreatomagmatic eruptions in Karymskoye Lake, Kamchatka, Russia. *Spec. Publ. Int. Assoc. Sedimentologists,* **30,** 35–60.

Bhattacharya, H. N. and S. Bandyopadhyay, 1998. Seismites in a Proterozoic tidal succession, Singhbhum, Bihar, India. *Sed. Geol.,* **199,** 239–252.

Bievre, G. and D. Quesne, 2004. Synsedimentary collapse on a carbonate platform margin (lower Barremian, southern Vercors, SE France). *Geodiversitas,* **26,** 169–184.

Bondevik, S., 2003. Storegga tsunami sand in peat below the Tapes beach ridge at Haroy, western Norway, and its possible relation to an early Stone Age settlement. *Boreas,* **32,** 476–483.

Bondevik, S., J. Mangerud, S. Dawson, A. Dawson, A. and O. Lohne, 2005a. Evidence for three North Sea tsunamis at the Shetland Islands between 8000 and 1500 years ago. *Quat. Sci. Rev.,* 24, 1757–1775.

Bondevik, S., F. Lovholt, C. B. Harbitz, J. Mangerud, A. Dawson and J. I. Svendsen, 2005b. The Storegga Slide tsunami; comparing field observations with numerical simulations. *Marine Petroleum Geol.,* **22**(1–2), 195–208.

Bondevik, S., J. I. Svendsen and J. Mangerud, 1997. Tsunami sedimentary facies deposited by the Storegga tsunami in shallow marine basins and coastal lakes, western Norway: *Sedimentology,* **44,** 1115–1131.

Borrero, J. C., 2005. Field data and satellite imagery of tsunami effects in Banda Aceh. Science, **308** (5728), 1596.

Bourgeois, J., T. A. Hansen, P. L. Wiberg and E. G. Kauffman, 1988. A tsunami deposit at the Cretaceous-Tertiary boundary in Texas. *Science,* **241,** 567–570.

Bourgeois, J. and S. Y. Johnson, 2001. Geologic evidence of earthquakes at the Snohomish Delta, Washington, in the past 1200 yr. *Geol. Soc. Am. Bull.,* **113,** 482–494.

Bourgeois, J., C. Petroff, H. Yeh, V. Titov, C. Synolakis, B. Benson, J. Kuroiwa, J. Lander and E. Norabuena, 1999. Geologic setting, field survey and modeling of the Chimbote, northern Peru tsunami of 21 February 1996. *Pure and Appl. Geophys.,* **154,** 513–540.

Bourgeois, J., T. K. Pinegina, V. Ponomareva and N. Zaretskaia, 2006. Holocene tsunamis in the southwestern Bering Sea, Russian Far East, and their tectonic implications. *Geol. Soc. Am. Bull.,* **118,** 449–463.

Brookfield, M. E., I. Blechschmidt, R. Hannigan, M. Coniglio, B. Simonson and G. Wilson, 2006. Sedimentology and geochemistry of extensive very coarse deepwater submarine fan sediments in the Middle Jurassic of Oman, emplaced by giant tsunami triggered by submarine mass flows. *Sed. Geol.,* **192,** 75–98.

Bruins, H. J., J. A. MacGillivray, C. E. Synolakis, C. Benjaminie, J. Keller, H. J. Kische, A. Klügelg and J. van der Plichth, 2008. Geoarchaeological tsunami deposits at Palaikastro (Crete) and the Late Minoan IA eruption of Santorini. *J. Archaeol. Sci.,* **35,** 191–212.

Bruzzi, C. and A. Prone, 2000. Une mèthode d'identification sédimentologique des depots de tempète et de tsunami; l'exoscopie des quartz, résultats préliminaires. *Quaternaire* (Paris), **11,** 167–177.

Bryant, E. A., 2001. *Tsunami. The Underrated Hazard.* Cambridge Univ. Press, 320p.

Bryant, E. A. and R. W. Young, 1996. Bedrock-sculpting by tsunami, south coast New South Wales: *J. Geol.,* **104,** 565–582.

Bucher, W. H., 1940. Submarine valleys and related geologic problems of the North Atlantic. *Geol. Soc. Am. Bull.,* **51,** 489–511.

Bussert, R. and M. Aberhan, 2004. Storms and tsunamis: evidence of event sedimentation in the Late Jurassic Tendaguru Beds of southeastern Tanzania. *J. Afr. Earth Sci.,* **39,** 549–555.

Byerly, G. R., D. R. Lowe, J. L. Wooden and X. Xie, 2002. An Archean impact layer from the Pilbara and Kaapvaal cratons. *Science,* **297,** 1325–1327.

Cantalamessa, G. and C. Di Celma, 2005. Sedimentary features of tsunami backwash deposits in a shallow marine Miocene setting, Mejillones Peninsula, northern Chile. *Sed. Geol.,* **178,** 259–273.

Carey, S., D. Morelli, H. Sigurdsson and S. Bronto, 2001. Tsunami deposits from major explosive eruptions; an example from the 1883 eruption of Krakatau. *Geology,* **29,** 347–350.

Carson, M. T., 2004. Resolving the enigma of early coastal settlement in the Hawaiian Islands: The stratigraphic sequence of the Wainiha Beach Site in Kaua'i. *Geoarchaeology,* **19,** 99–118.

Cisternas, M., I. Contreras and A. Aranada, 2000. Reconocimiento y caracterización de la facies sedimentaria depositada por el tsunami de 1960 en el estuario Maullín, Chile. *Revista Geológica de Chile,* **27**(1), 3–11.

Cisternas, M., B. F. Atwater and 13 others, 2006. Predecessors of the giant 1960 Chile earthquake. *Nature,* **437,** 404–407.

Cita, M. B. and G. Aloisi 2000. Deep-sea tsunami deposits triggered by the explosion of Santorini (3500 y BP), eastern Mediterranean. *Sed. Geol.,* **135,** 181–203.

Clague, J. J., and P. T. Bobrowsky, 1994. Tsunami deposits beneath tidal marshes on Vancouver Island, British Columbia. *Geol. Soc. Am. Bull.,* **106,** 1293–1303.

Clague, J. J., P. T. Bobrowsky and T. S. Hamilton, 1994. A sand sheet deposited by the 1964 Alaska tsunami at Port Alberni, British Columbia. *Est. Coast. Shelf Sci.,* **38,** 413–421.

Clague, J. J., P. T. Bobrowsky and I. Hutchinson, 2000. A review of geological records of large tsunamis at Vancouver Island, British Columbia, and implications for hazard. *Quat. Sci. Rev.,* **19,** 849–863.

Clague, J. J., A. Munro and T. Murty, 2003. Tsunami hazard and risk in Canada; An assessment of natural hazards and disasters in Canada. *Natural Hazards,* **28,** 433–461.

Clifton, H. E., ed., 1988. Sedimentologic Consequences of Convulsive Geologic Events. *Geol. Soc. Am. Special Paper 229,* 157p.

Coleman, P. J., 1968. Tsunamis as geological agents. *J. Geol. Soc. Australia,* **15,** 267–273.

Coleman, P. J., 1978. Tsunami sedimentation, *in* Fairbridge, R. W. and J. Bourgeois, eds., *Encyclopedia of Sedimentology:* Stroudsburg, PA, Dowden, Hutchinson & Ross, 828–832.

Daly, R. A., 1936. Origin of submarine canyons. *Am. J. Sci.,* **31,** 401–420.

Darienzo, M. E. and C. D. Peterson, 1990. Episodic tectonic subsidence of late Holocene salt marshes, northern Oregon central Cascadia margin. *Tectonics,* **9,** 1–22.

Darwin, Charles, 1845. *The Voyage of the Beagle.*

Dawson, A. G., 1994. Geomorphological effects of tsunami run-up and backwash. *Geomorphology,* **10,** 83–94.

Dawson, A. G., 1996. The geological significance of tsunamis. *Zeitschrift fur Geomorphologie, N.F., Suppl.-Bd.,* **102,** 199–210.

Dawson, A. G., 2003. Book Review: Tsunami the Underrated Hazard. *J. Quaternary Sci.,* **18,** 581–582.

Dawson, A. G., D. Long and D. E. Smith, 1988. The Storegga slides: Evidence from eastern Scotland for a possible tsunami: *Mar. Geol.,* **82,** 271–276.

Dawson, A. G., S. Shi, S. Dawson, T. Takahashi and N. Shuto, 1996. Coastal sedimentation associated with the June 2nd and 3rd, 1994 tsunami in Rajegwesi, Jav. *Quat. Sci. Rev.,* **15,** 901–912.

Dawson, A. G. and S. Shi, 2000. Tsunami deposits. *Pure and Appl. Geophys.*, **157**, 875–897.

Dawson, A. G. and I. Stewart, 2007. Tsunami deposits in the geological record. *Sed. Geol.*, **200**, 166–183.

Dawson, S., D. E. Smith, A. Ruffman and S. Shi, 1996. The diatom biostratigraphy of tsunami sediments: Examples from recent and middle Holocene events. *Phys. Chem. of the Earth*, **21**, 87–92.

de Lange, P. J. and V. G. Moon, 2007. Tsunami washover deposits, Tawharanui, New Zealand: *Sed. Geol.*, **200**, 232–247.

De Martini, P. M., P. Burrato, D. Pantosti, A. Maramai, L. Graziani and H. Abramson, 2003. Identification of tsunami deposits and liquefaction features in the Gargano area (Italy): paleoseismological implication. *Annals Geophys.*, **46**, 883–902.

Doig, R., 1990. 2300 yr history of seismicity from silting events in Lake Tadoussac, Charlevoix, Quebec. *Geology*, **18**, 820–823.

Dominey-Howes, D., 2002. Documentary and geological records of tsunamis in the Aegean Sea region of Greece and their potential value to risk assessment and disaster management. *Natural Hazards*, **25**, 195–224.

Dominey-Howes, D., 2007. Geological and historical records of tsunami in Australia. *Mar. Geol.*, **239**(1–2), 99–123.

Dominey-Howes, D. T. M., G. A. Papadopoulos and A. G. Dawson, 2000. Geological and historical investigation of the 1650 Mt. Columbo (Thera Island) eruption and tsunami, Aegean Sea, Greece. *Natural Hazards*, **21**, 83–96.

Dominey-Howes, D. T. M., G. S. Humphreys and P. P. Hesse, 2006. Tsunami and palaeotsunami depositional signatures and their potential value in understanding the late-Holocene tsunami record. *The Holocene*, **16**, 1095–1107.

Dott, R. H., Jr., 1978. Tectonics and sedimentation a century later. *Earth-Sci. Rev.*, **14**, 1–34.

Dott, R. H., Jr. 1983. Episodic sedimentation—How normal is average? How rare is rare? Does it matter? *J. Sed. Petrology*, **53**, 5–23.

Dott, R. H., Jr, 1996. Episodic event deposits versus stratigraphic sequences; shall the twain never meet? *Sed. Geol.*, **104**, 243–247.

Du, Y. S., C. H. Zhang, X. Han, S. Z. Gu and W. J. Lin, 2001. Earthquake event deposits in Mesoproterozoic Kunyang Group in central Yunnan Province and its geological implications. *Science in China Series D-Earth Sciences*, **44**(7), 600–608.

Dypvik, H., P. T. Sandbakken, G. Postma and A. Mork, 2004. Early post-impact sedimentation around the central high of the Mjolnir impact crater (Barents Sea, Late Jurassic). *Sed. Geol.*, **168**, 227–247.

Dypvik, H. and L. F. Jansa, 2003. Sedimentary signatures and processes during marine bolide impacts; a review. *Sed. Geol.*, **161**, 309–337.

Earthquake Research Institute, 1934. Papers and reports on the tsunami of 1933 on the Sanriku coast, Japan. *Bulletin of the Earthquake Research Institute, Tokyo Imperial University, Supplement 1.* [in Japanese; as cited in Abe, 2005]

Eaton, J. P., D. H. Richter and W. U. Ault, 1961. The tsunami of May 23, 1960, on the Island of Hawaii: *Seis. Soc. Am. Bull.*, **51**, 135–157.

Einsele, G., W. Ricken and A. Seilacher, eds., 1991. *Cycles and Events in Stratigraphy.* Springer, 956p.

Einsele, G. and A. Seilacher, eds., 1982. *Cyclic and Event Stratification.* Springer, 536p.

Einsele, G., S. K. Chough and T. Shiki, 1996. Depositional events and their records—an introduction. *Sed. Geol.*, **104**, 1–9.

Felton, E. A. and K. A. W. Crook, 2003. Evaluating the impacts of huge waves on rocky shorelines: an essay review of the book 'Tsunami—The Underrated Hazard.' *Mar. Geol.*, **197**, 1–12.

Felton, E. A., K. A. W. Crook, B. H. Keating and E. A. Kay, 2006. Sedimentology of rocky shorelines: 4. Coarse gravel lithofacies, molluscan biofacies, and the stratigraphic and eustatic records in the type area of the Pleistocene Hulopoe Gravel, Lanai, Hawaii. *Sed. Geol.,* **184,** 1–76.

Fiedorowicz, B. K. and C. D. Peterson, 2002. Tsunami deposit mapping at Seaside, Oregon, USA, *in* Bobrowsky, P.T., ed., *Geoenvironmental Mapping; Methods, Theory and Practice.* Netherlands: A. A. Balkema, 629–648.

Fujino, S., F. Masuda, S. Tagomori and D. Matsumoto, 2006. Structure and depositional processes of a gravelly tsunami deposit in a shallow marine setting: Lower Cretaceous Miyako Group, Japan. *Sed. Geol.,* **187,** 127–138.

Fujiwara,O., F. Masuda, T. Sakai, T. Irizuki and K. Fuse, 2000. Tsunami deposits in Holocene bay mud in southern Kanto region, Pacific coast of central Japan. *Sed. Geol.,* **135,** 219–230.

Fujiwara, O., T. Kamataki and K. Fuse, 2003. Genesis of mixed molluscan assemblages in the tsunami deposits distributed in Holocene drowned valleys on the southern Kanto region, east Japan. *Daiyonki-Kenkyu = Quaternary Research,* **42,** 389–412. [in Japanese]

Gelfenbaum, G. and B. Jaffe, 2003. Erosion and sedimentation from the 17 July, 1998 Papua New Guinea tsunami. *Pure and Appl. Geophys.,* **160,** 1969–1999.

Gelfenbaum, G., D. Vatvani, B. Jaffe and F. Dekker, 2007. Tsunami inundation and sediment transport in vicinity of coastal mangrove forest. *Coastal Sediments '07, Proc. 6th Int. Symp. On Coastal Eng. & Science of Coastal Sediment Processes Vol. 2,* Eds. N. C. Kraus and J. D. Rosati, Am. Soc. Civ. Eng., 1117–1128.

Gersonde, R., A. Deutsch and B. A. Ivanov, eds., 2002. Ocean Impacts: Mechanisms and Environmental Perturbations. *Deep Sea Research Part II: Topical Studies in Oceanography,* **49,** 951–1169.

Glikson, A. Y., 2004. Early Precambrian asteroid impact-triggered tsunami: Excavated seabed, debris flows, exotic boulders, and turbulence features associated with 3.47-2.47 Ga-old asteroid impact fall-out units, Pilbara Craton, Western Australia. *Astrobiology,* **4,** 19–50.

Goff, J., C. Chagué-Goff and S. Nichol, 2001. Palaeotsunami deposits; a New Zealand perspective. *Sed. Geol.,* **143,** 1–6.

Goff, J., B. G. McFadgen and C. Chagué-Goff, 2004. Sedimentary differences between the 2002 Easter storm and the 15th-century Okoropunga tsunami, southeastern North Island, New Zealand: *Mar. Geol.,* **204,** 235–250.

Goff, J. R., 1997. A chronology of natural and anthropogenic influences on coastal sedimentation, New Zealand. *Mar. Geol.,* **138,** 105–117.

Goldfinger, C., C. H. Nelson and J. E. Johnson, 2003. Deep-water turbidites as Holocene earthquake proxies: the Cascadia subduction zone and northern San Andreas fault systems. *Ann. Geophys.,* **46,** 1169–1194.

Gutscher, M., 2005. Destruction of Atlantis by a great earthquake and tsunami? A geological analysis of the Spartel Bank hypothesis. *Geology,* **33,** 685–688.

Hartley, A., J. Howell, A. E. Mather and G. Chong, 2001. A possible Plio-Pleistocene tsunami deposit, Hornitos, northern Chile. *Revista Geologica de Chile,* **28,** 117–125.

Hassler, S. W., H. F. Robey and B. M. Simonson, 2000. Bedforms produced by impact-generated tsunami, approximately 2.6 Ga Hamersley Basin, western Australia: *Sed. Geol.,* **135,** 283–294.

Hassler, S. W. and B. M. Simonson, 2001. The sedimentary record of extraterrestrial impacts in deep-shelf environments: Evidence from the early Precambrian. *J. Geol.,* **109,** 1–19.

Hemphill-Haley, E., 1996. Diatoms as an aid in identifying late-Holocene tsunami deposits. *The Holocene,* **6,** 439–448.

Hemphill-Haley, E., 1995. Diatom evidence for earthquake-induced subsidence and tsunami 300 yr ago in southern coastal Washington. *Geol. Soc. Am Bull.,* **107,** 367–378.

Hieke, W. and F. Werner, 2000, The Augias megaturbidite in the central Ionian Sea (central Mediterranean) and its relation to the Holocene Santorini event. *Sed. Geol.,* **135,** 205–218.

Hindson, R. A. and C. Andrade, 1999. Sedimentation and hydrodynamic processes associated with the tsunami generated by the 1755 Lisbon earthquake. *Quat. Intern.,* **56,** 27–38.

Houtz, R. E., 1962. The 1953 Suva earthquake and tsunami. *Bull. Seis. Soc. Am.,* **52,** 1–12.

Hughes, J. F. and R. W. Mathewes, 2003. A modern analogue for plant colonization of palaeotsunami sands in Cascadia, British Columbia, Canada. *The Holocene,* **13,** 877–886.

Huntington, K., J. Bourgeois, G. Gelfenbaum, P. Lynett, B. Jaffe, H. Yeh and R. Weiss, 2007. Sandy signs of tsunami onshore depth and speed. *EOS, Trans. Am. Geophys. Union,* **88**(52), 577–578.

Huntley, D. J. and J. J. Clague, 1996. Optical dating of tsunami-laid sands. *Quat. Res.,* **46,** 127–140.

Hutchinson, I., J. J. Clague and R. W. Mathewes, 1997. Reconstructing the tsunami record on an emerging coast; a case study of Kanim Lake, Vancouver Island, British Columbia, Canada. *J. Coastal Res.,* **13,** 545–553.

Inouchi, Y., Y. Kinoshita, F. Kumon, S. Nakano, S. Yasumatsu and T. Shiki, 1996. Turbidites as records of intense palaeoearthquakes in Lake Biwa, Japan. *Sed. Geol.,* **104,** 117–125.

Jones, A. T. and C. L. Mader, 1996. Wave erosion on the southeastern coast of Australia; tsunami propagation modeling. *Austral. J. Earth Sci.,* **43,** 479–483.

Jones, B. and I. G. Hunter, 1992. Very large boulders on the coast of Grand Cayman; the effects of giant waves on rocky coastlines. *J. Coastal Res.,* **8,** 763–774.

Kajiura, K., 1983. Some statistics related to observed tsunami heights along the coast of Japan: *in* Iida, K. and T. Iwasaki, eds., *Tsunamis: Their Science and Engineering.* Tokyo: Terra Scientific Publishing Company, 131–145.

Kastens, K. A. and M. B. Cita, 1981. Tsunami-induced sediment transport in the abyssal Mediterranean Sea. *Geol. Soc. Am. Bull. Part I,* **92,** 845–857.

Kelsey, H. M., A. R. Nelson, E. Hemphill-Haley and R. C. Witter, 2005. Tsunami history of an Oregon coastal lake reveals a 4600 yr record of great earthquakes on the Cascadia subduction zone. *Geol. Soc. Am. Bull.,* **117,** 1009–1032.

Kelsey, H. M., R. C. Witter and E. Hemphill-Haley, 2002, Plate-boundary earthquakes and tsunamis of the past 5500 yr, Sixes River estuary, southern Oregon. *Geol. Soc. Am. Bull.,* **114,** 298–314.

Kelsey, H. M., R. C. Witter and E. Hemphill-Haley, 1998. Response of a small Oregon estuary to coseismic subsidence and postseismic uplift in the past 300 years. *Geology,* **26,** 231–234.

Kench, P. S., R. F. McLean, R. W. Brander, S. L. Nichol, S. G. Smithers, M. R. Ford, K. E. Parnell and M. Aslam, 2006. Geological effects of tsunami on mid-ocean atoll islands: The Maldives before and after the Sumatran tsunami. *Geology,* **34,** 177–180.

Konno, E., J. Iwai, N. Kitamura, T. Kotaka, H. Mii, H. Nakagawa, Y. Onuki, T. Shibata and Y. Takayanagi, 1961. Geological observations of the Sanriku coastal region damaged by the tsunami due to the Chile earthquake in 1960. *Contr. Inst. Geol. Paleont., Tohoku Univ.,* **52,** 40p.

Kortekaas, S. and A. G Dawson, 2007. Distinguishing tsunami and storm deposits: an example from Martinhal, SW Portugal. *Sed. Geol.,* **200,** 208–221.

Kuenen, P. H. and Migliorini, C. I., 1950. Turbidity currents as a cause of graded bedding. *J. Geol.,* **58,** 91–127.

Kuhn, G. G., 2005. Paleoseismic features as indicators of earthquake hazards in North Coastal, San Diego County, California, USA. *Engin. Geol.,* **80,** 115–150.

Kumar, A., R. K. Chingkhei and T. Dolendro, 2007. Tsunami damage assessment: a case study in Car Nicobar Island, India. *Internat. J. Remote Sensing,* **28,** 2937–2959.

Kurian N. P., P. Abilash P. Pillai, K. Rajith, B. T. Murali Krishnan and P. Kalaiarasan, 2006. Inundation characteristics and geomorphological impacts of December 2004 tsunami on Kerala coast. *Current Science,* **90**(2), 240–249.

Latter, J. H., 1982. Tsunamis of volcanic origin; summary of causes, with particular reference to Krakatoa, 1883. *Bull. Volcanol.,* **44,** 467–490.

Lavigne, F., C. Gomez, M. Giffo, P. Wassmer, C. Hoebreck, D. Mardiatno, J. Prioyono and R. Paris, R., 2007. Field Observations of the 17 July 2006 tsunami in Java.Nat. Hazards Earth Syst. Sci., 7, 177–183.

Lawton, T. F., K. W. Shipley, J. L. Aschoff, K. A. Giles and F. J.Vega, 2005. Basinward transport of Chicxulub ejecta by tsunami-induced backflow, La Popa basin, northeastern Mexico, and its implications for distribution of impact-related deposits flanking the Gulf of Mexico. *Geology, 33,* 81–84.

Lowe, D. J. and W. P. de Lange, 2000. Volcano-meteorological tsunamis, the c. AD 200 Taupo eruption (New Zealand) and the possibility of a global tsunami. *The Holocene, 10,* 401–407.

Luque, L., J. Lario, J. Civis, P. G. Silva, C. Zazo, J. L. Goy and C. J. Dabrio, 2002. Sedimentary record of a tsunami during Roman times, Bay of Cadiz, Spain; Sea-level changes and neotectonics. *J. Quat. Sci., 17,* 623–631.

Lyell, Charles, 1832. *Principles of Geology, Volume 2.* London: John Murray, 330p.

Martin, M. E., R. Weiss, J. Bourgeois, T. Pinegina, H. Houston and V. Titov, 2008. Combining constraints from tsunami modeling and sedimentology to untangle the 1969 Ozernoi and 1971 Kamchatskii tsunamis. Geophys. Res. Lett., 35, L01610.

Masaitis, V. L., 2002. The middle Devonian Kaluga impact crater (Russia): new interpretation of marine setting. *Deep Sea Res. Part II: Topical Stud. Oceanog., 49,* 1157–1169.

Massari, F. and A. D'Alessandro, 2000. Tsunami-related scour-and-drape undulations in Middle Pliocene restricted-bay carbonate deposits (Salento, south Italy). *Sed. Geol., 135,* 265–281.

Matsutomi, H., N. Shuto, F. Imamura and T. Takahashi 2001. Field Survey of the 1996 Irian Jaya earthquake tsunami in Biak Island. *Natural Hazards, 24,* 199–212.

McAdoo, B. G., N. Richardson and J. Borrero, 2007. Inundation distances and run-up measurements from ASTER, QuickBird and SRTM data, Aceh coast, Indonesia. *Internat. J. Remote Sensing, 28,* 2961–2975.

McCoy, F. W. and G. Heiken, 2000. Tsunami generated by the late Bronze Age eruption of Thera (Santorini), Greece. *Pure Appl. Geophys., 157,* 1227–1256.

McFadgen, B. G. and J. R. Goff, 2007. Tsunamis in the New Zealand archaeological records: *Sed. Geol., 200,* 263–274.

McHugh, C. M. G., L. Seeber, M. H. Cormier, J. Dutton, N. Cagatay, A. Polonia, W. B. F. Ryan and N. Gorur, 2006. Submarine earthquake geology along the North Anatolia Fault in the Marmara Sea, Turkey: A model for transform basin sedimentation. *Earth Planet. Sci. Lett., 248,* 661–684.

McMurtry, G. M., G. J. Fryer, D. R. Tappin, I. P. Wilkinson, M. Williams, J. Fietzke, D. Garbe-Schoenberg and P. Watts, 2004. Megatsunami deposits on Kohala Volcano, Hawaii, from flank collapse of Mauna Loa. *Geology, 32,* 741–744.

Melekestsev, I. V., A. V. Kurbatov, M. M. Pevzner and L. D. Sulerzhitskiy, 1995. Prehistoric tsunamis and large earthquakes on the Kamchatskiy Peninsula, Kamchatka, based on tephrochronological data. *Volcanology and Seismology, 16,* 449–459.

Miller, D., 1960. Giant waves in Lituya Bay Alaska. *U.S. Geol. Survey Prof. Paper 354,* 51–86.

Minoura, K., V. G. Gusiakov, A. Kurbatov, S. Takeuti, J. I. Svendsen, S. Bondevik and T. Oda, 1996. Tsunami sedimentation associated with the 1923 Kamchatka earthquake. *Sed. Geol., 106,* 145–154.

Minoura, K., F. Imamura, U. Kuran, T. Nakamura, G. A. Papadopoulos, D. Sugawara, T. Takahashi and A. C. Yalciner, 2005. A tsunami generated by a possible submarine slide: Evidence for slope failure triggered by the North Anatolian Fault movement. *Natural Hazards, 36,* 297–306.

Minoura, K., F. Imamura, U. Kuran, T. Nakamura, G. A. Papadopoulos, T. Takahashi and A. C. Yalciner, 2000. Discovery of Minoan tsunami deposits. *Geology, 28,* 59–62.

Minoura, K., F. Imamura, T. Takahashi and N. Shuto, 1997. Sequence of sedimentation processes caused by the 1992 Flores tsunami; evidence from Babi Island. *Geology, 25,* 523–526.

Minoura, K. and T. Nakata, 1994. Discovery of an ancient tsunami deposit in coastal sequences of Southwest Japan; verification of a large historic tsunami. *The Island Arc, 3,* 66–72.

Minoura, K., S. Nakaya and M. Uchida, 1994. Tsunami deposits in a lacustrine sequence of the Sanriku coast, Northeast Japan. *Sed. Geol.,* **89,** 25–31.

Minoura, K. and S. Nakaya, 1991. Traces of tsunami preserved in inter-tidal lacustrine and marsh deposits; some examples from Northeast Japan. *J. Geol.,* **99,** 265–287.

Minoura, K., S. Nakaya and H. Sato, 1987. Traces of tsunamis recorded in lake deposits; an example from Jusan, Shiura-mura, Aomori: *Zisin = Jishin,* **40,** 183–196. [in Japanese]

Moore, A. L., 2000. Landward fining in onshore gravel as evidence for a late Pleistocene tsunami on Molokai, Hawaii. *Geology,* **28,** 247–250.

Moore, Andrew, 2003. Tsunami deposits. *in* Middleton, G.V., ed., *Encyclopedia of Sedimentology and Sedimentary Rocks.* Dordrecht: Kluwer Academic, 755–757.

Moore, A. L., B. G. McAdoo and A. Ruffman, 2007. Landward fining from multiple sources in a sand sheet deposited by the 1929 Grand Banks tsunami, Newfoundland. *Sed. Geol.,* **200,** 336–346.

Moore, J. G. and G. W. Moore, 1984. Deposit from a giant wave on the island of Lanai, Hawaii. *Science,* **226,** 1312–1315.

Moore, G. W. and J. G. Moore, 1988. Large-scale bedforms in boulder gravel produced by giant waves in Hawaii. *Geol. Soc. Am. Special Paper* **229,** 101–110.

Moore, J. G., W. R. Normark and R. T. Normark, 1994. Giant Hawaiian landslides. *Ann. Rev. Earth Planet. Sci.,* **22,** 119–144.

Moore, J. G., R. A. Schweickert, J. E. Robinson, M. M. Lahren and C. A. Kitts, 2006. Tsunami-generated boulder ridges in Lake Tahoe, California-Nevada. *Geology,* **34,** 965–968.

Morton, R. A., B. M. Richmond, B. E. Jaffe and G. Gelfenbaum, 2006. Reconnaissance investigation of Caribbean extreme wave deposits—Preliminary observations, interpretations, and research directions. *U.S. Geol. Survey Open-File Report* 2006–1293; in review, *J. Sed. Res.*

Morton, R. A., G. Gelfenbaum and B. E. Jaffe, 2007. Physical criteria for distinguishing sandy tsunami and storm deposits using modern examples. *Sed. Geol.,* **200,** 184–207.

Moya, J. C. and A. Mercado, A., 2006. Geomorphologic and stratigraphic investigations on historic and pre-historic tsunami in northwestern Puerto Rico: Implications for long term coastal evolution. In A. Mercado-Irizarry and P. Liu, P., eds., *Caribbean Tsunami Hazard.* World Scientific Publishing, 149–177.

Murari, M. K., H. Achyuthan and A. K. Singhvi, 2007. Luminescence studies on the sediments laid down by the December 2004 tsunami event: Prospects for the dating of palaeo tsunamis and for the estimation of sediment fluxes. *Current Science,* **92,** 367–371.

Nakajima, T. and Y. Kanai, 2000. Sedimentary features of seismoturbidites triggered by the 1983 and older historical earthquakes in the eastern margin of the Japan Sea. *Sed. Geol.,* **135,** 1–19.

Nakamura, M., 2006. Source fault model of the 1771 Yaeyama tsunami, southern Ryukyu

Islands, Japan, inferred from numerical simulation. *Pure Appl. Geophys.,* **163,** 41–54.

Nanayama, F., K. Satake, R. Furukawa, K. Shimokawa, B. F. Atwater, K. Shigeno and S. Yamaki, 2003. Unusually large earthquakes inferred from tsunami deposits along the Kuril Trench. *Nature,* **424,** 660–663.

Nanayama, F., R. Furukawa, K. Shigeno, A. Makino, Y. Soeda and Y. Igarashi, 2007. Nine unusually large tsunami deposits from the past 4000 years at Kiritappu march along the southern Kuril trench. *Sed. Geol.,* **200,** 275–294.

Nanayama, F., K. Shigeno, K. Satake, K. Shimokawa, S. Koitabashi, S. Miyasaka and M. Ishii, 2000. Sedimentary differences between the 1993 Hokkaido-nansei-oki tsunami and the 1959 Miyakojima typhoon at Taisei, southwestern Hokkaido, northern Japan. *Sed. Geol.,* **135,** 255–264.

Nelson, A. R., A. C. Asquith and W. C. Grant, 2004. Great earthquakes and tsunamis of the past 2000 years at the Salmon River estuary, central Oregon coast, USA. *Bull. Seis. Soc. Am.* **94,** 1276–1292.

Nelson, A. R., H. M. Kelsey and , R. C. Witter, 2006. Great earthquakes of variable magnitude at the Cascadia subduction zone. *Quat. Res.,* **65,** 354–365.

Nelson, A. R., I. Shennan and A. J. Long, 1996. Identifying coseismic subsidence in tidal-wetland stratigraphic sequences at the Cascadia subduction zone of western North America. *J. Geophys. Res.,* **101**(B3), 6115–6135.

Nigam, R and S. K. Chaturvedi, 2006. Do inverted depositional sequences and allochthonous foraminifers in sediments along the Coast of Kachchh, NW India, indicate palaeostorm and/or tsunami effects? *Geo-Marine Lett.,* **26,** 42–50.

Nishimura, Y., N. Miyaji and M. Suzuki, 1999. Behavior of historic tsunamis of volcanic origin as revealed by onshore tsunami deposits. *Physics and Chemistry of the Earth .Part A: Solid Earth and Geodesy,* **24,** 985–988.

Nishimura, Y. and N. Miyaji, 1995. Tsunami deposits from the 1993 southwest Hokkaido earthquake and the 1640 Hokkaido Komagatake eruption, northern Japan; Tsunamis; 1992–1994, their generation, dynamics, and hazard. *Pure Appl. Geophys.,* **144,** 719–733.

Noormets, R., K. A. W. Crook and E. A. Felton, 2004. Sedimentology of rocky shorelines: 3.: Hydrodynamics of megaclast emplacement and transport on a shore platform, Oahu, Hawaii. *Sed. Geol.,* **172,** 41–65.

Norris, R. D., J. Firth, J. S. Blusztajn and G. Ravizza, 2000. Mass failure of the North Atlantic margin triggered by the Cretaceous-Paleogene bolide impact. *Geology,* **28,** 1119–1122.

Nott, J., 1997. Extremely high-energy wave deposits inside the Great Barrier Reef, Australia; determining the cause, tsunami or tropical cyclone. *Mar. Geol.,* **141,** 193–207.

Obura, D., 2006. Impacts of the 26 December 2004 tsunami in Eastern Africa. *Ocean & Coastal Management,* **49,** 873–888.

Paskoff, R., 1991. Likely occurrence of a mega-tsunami in the middle Pleistocene, near Coquimbo, Chile. *Revista Geologica de Chile,* **18,** 87–91.

Pelayo, A. M. and Wiens, D. A., 1992. Tsunami earthquakes: Slow thrust-faulting events in the accretionary wedge. *J. Geophys. Res.,* **97,** 15,321–15,337.

Perez Torrado, F. J., R. Paris, M. C. Cabrera, J. Schneider, P. Wassmer, J. Carracedo, A. Rodriguez Santana and F. Santana, 2006. Tsunami deposits related to flank collapse in oceanic volcanoes; the Agaete Valley evidence, Gran Canaria, Canary Islands. *Mar. Geol.,* **227,** 135–149.

Peters, R., B. Jaffe and G. Gelfenbaum, 2007. Distribution and sedimentary characteristics of tsunami deposits along the Cascadia margin of western North America. *Sed. Geol.,* **200,** 372–386.

Pickering, K. T., 1984. The Upper Jurassic "boulder beds" and related deposits; a fault-controlled submarine slope, NE Scotland: *J. Geol. Soc. London,* **141**(2), 357–374.

Pickering, K. T., W. Soh and A. Taira, 1991. Scale of tsunami-generated sedimentary structures in deepwater: *J. Geol. Soc.,* **148,** 211–214.

Pinegina, T. K. and J. Bourgeois, 2001. Historical and paleo-tsunami deposits on Kamchatka, Russia; long-term chronologies and long-distance correlations. *Nat. Haz. and Earth Syst. Sci.,* **1,** 177–185.

Pinegina, T. K., J. Bourgeois, L. I. Bazanova, I. V. Melekestsev and O. A. Braitseva, 2003. A millennial-scale record of Holocene tsunamis on the Kronotskiy Bay coast, Kamchatka, Russia. *Quat. Res.,* **59,** 36–47.

Piper, D. J. W., A. N. Short and J. E. H. Clark, 1988. The 1929 "Grand Banks" earthquake, slump and turbidity current. *Geol. Soc. Am. Special Paper* **229,** 77–92.

Pirazzoli, P A, S. C. Stiros, M. Arnold, J. Laborel and Laborel-Deguen, 1999. Late Holocene coseismic vertical displacements and tsunami deposits near Kynos, Gulf of Euboea, central Greece. *Phys. Chem. Earth Pt A, Solid Earth and Geodesy,* **24,** 361–367.

Plafker, G. and R. Kachadoorian, 1966. Geologic effects of the March 1964 earthquake and associated seismic sea waves on Kodiak and nearby islands, Alaska. *U.S. Geol. Survey Prof. Paper,* **543d,** 1–46.

Platania, G., 1909. Il maremoto dello Stretto di Messina del 28 Dicembre 1908: *Bollettino della Societa Sismologica Italiana,* **13,** 369–458.

Poag, C. W., 1997. The Chesapeake Bay bolide impact; a convulsive event in Atlantic Coastal Plain evolution. *Sed. Geol.,* **108,** 45–90.

Poag, C. W., D. S. Powars, L. J. Poppe, R. B. Mixon, L. E. Edwards, D. W. Folger, and S. Bruce, 1992. Deep Sea Drilling Project Site 612 bolide event; new evidence of a late Eocene impact-wave deposit and a possible impact site, U.S. east coast. *Geology,* **20,** 771–774.

Poag, C. W., J. B. Plescia and P. C. Molzer, 2002. Ancient impact structures on modern continental shelves: The Chesapeake Bay, Montagnais, and Toms Canyon craters, Atlantic margin of North America. *Deep Sea Res. Part II: Topical Studies in Oceanography,* **49,** 1081–1102.

Pope, M. C., J. F. Read, R. Bambach and H. J. Hofmann, 1997. Late Middle to Late Ordovician seismites of Kentucky, southwest Ohio and Virginia: Sedimentary recorders of earthquakes in the Appalachian basin. *Geol. Soc. Am. Bull.,* **109,** 489–503.

Pratt, B. R., 2001. Oceanography, bathymetry and syndepositional tectonics of a Precambrian intracratonic basin: integrating sediments, storms, earthquakes and tsunamis in the Belt Supergroup (Helena Formation, ca. 1.45 Ga), western North America. *Sed. Geol.,* **141–142,** 371–394.

Pratt, B. R., 2002. Storms vs. tsunamis: Dynamic interplay of sedimentary, diagenetic and tectonic processes in the Cambrian of Montana. *Geology,* **30,** 423–426.

Prescott, J. R. and G. B. Robertson, 1997. Sediment dating by luminescence: A review: *Radiation Measurements,* **27,** 893–922.

Ramakrishnan, D., S. K. Ghosh, V. K. M. Raja, R. V. Chandran, and A. Jeyram, 2005. Trails of the killer tsunami: A preliminary assessment using satellite remote sensing technique. *Current Science,* **88,** 709–711.

Ramasamy, S. M., C. J. Kumanan, J. Saravanavel and R. Selvakumar, 2006. Geosystem responses to December 26, 2004 tsunami and mitigation strategies for Cuddalore-Nagapattinam coast, Tamil Nadu. *J. Geol. Soc. India,* **68,** 967–983.

Reimnitz, E. and N. F. Marshall, 1965. Effects of the Alaska earthquake and tsunami on recent deltaic sediments. *J. Geophys. Res.,* **70,** 2363–2376.

Reinhardt, E. G., B. N. Goodman, J. I. Boyce, G. Lopez, P. van Hengsturn, W. J. Rink, Y. Mart, and A. Raban, 2006. The tsunami of 13 December A.D. 115 and the destruction of Herod the Great's harbor at Caesarea Maritima, Israel. *Geology,* **34,** 1061–1064.

Rhodes, B., M. Tuttle, B. P. Horton, L. Doner, H. Kelsey, A. Nelson and M. Cisternas, 2006. Paleotsunami research. *Eos, Trans. Am. Geophys. Union,* **87,** 205, 209.

Rossetti, D. D., A. M. Goes, W. Truckenbrodt and J. Anaisse, 2000. Tsunami-induced large-scale scour-and-fill structures in Late Albian to Cenomanian deposits of the Grajau Basin, northern Brazil. *Sedimentology,* **47,** 309–323.

Rothaus, R. M., E. Reinhardt and J. Noller, 2004. Regional considerations of coastline change, tsunami damage and recovery along the southern coast of the Bay of Izmit (the Kocaeli (Turkey) earthquake of 17 August 1999). *Natural Hazards,* **31,** 233–252.

Rubin, K .H., C. H. Fletcher, III and C. Sherman, 2000. Fossiliferous Lana'i deposits formed by multiple events rather than a single giant tsunami. *Nature,* **408,** 675–681.

Sacchi, M., D. Insinga, A. Milia, F. Molisso, A. Raspini, M. M Torrente and A. Conforti, 2005. Stratigraphic signature of the Vesuvius 79 AD event off the Sarno prodelta system, Naples Bay; Mediterranean prodelta systems. *Mar. Geol.,* **222–223,** 443–469.

Saintilan, N. and K. Rogers, 2005. Recent storm boulder deposits on the Beecroft Peninsula, New South Wales, Australia. *Geog. Res.,* **43,** 429–432.

Satake, K., F. Nanayama, S. Yamaki, Y. Tanioka and K. Hirata, K., 2005 . Variability among tsunami sources in the 17th-21st centuries along the southern Kuril Trench. *Adv. in Natural and Tech. Haz. Res.,* **23,** 157–170.

Satake, K., E. A. Okal and J. C. Borrero, 2007. Introduction to Tsunami and its Hazard in the Indian and Pacific Oceans. *Pure Appl. Geophys.,* **164,** 249–259.

Sato, H., T. Shimamoto, A. Tsutsumi and E. Kawamoto, 1995. Onshore tsunami deposits caused by the 1993 southwest Hokkaido and 1983 Japan-Sea earthquakes. *Pure Appl. Geophys.*, **144**, 693–717.

Sawai, Y., 2002. Evidence for 17th-century tsunamis generated on the Kuril-Kamchatka subduction zone, Lake Tokotan, Hokkaido, Japan. *J. Asian Earth Sci.*, **20**, 903–911.

Scheffers, A., 2004, Tsunami imprints on the Leeward Netherlands Antilles (Aruba, Curacao, Bonaire) and their relation to other coastal problems. *Quat. Internat.*, **120**, 163–172.

Scheffers, A. and D. H. Kelletat, 2003. Sedimentologic and geomorphologic tsunami imprints world-wide; a review. *Earth-Science Reviews*, **63**, 83–92.

Schlichting, R. B. and C. D. Peterson, 2006. Mapped overland distance of paleotsunami high-velocity inundation in back-barrier wetlands of the central Cascadia margin, U.S.A. *J. Geol.*, **114**, 577–592.

Schnyder, J., F. Baudin and J. F. Deconinck, 2005. A possible tsunami deposit around the Jurassic-Cretaceous boundary in the Boulonnais area (northern France). *Sed. Geol.*, **177**, 209–227.

Shanmugam, G., 2006. The tsunamite problem: *J. Sed. Res.*, **76**, 718–730.

Shepard, F. P., G. A. Macdonald and D. C. Cox, 1950. The tsunami of April 1, 1946 [Hawaii]. *Calif. Univ., Scripps Inst. Oceanography Bull.*, **5**, 391–528.

Shi, S. Z., A. G. Dawson and D. E. Smith, 1995. Coastal sedimentation associated with the December 12th, 1992 tsunami in Flores, Indonesia. *Pure Appl. Geophys.*, **144**, 525–536.

Shiki, T. and T. Yamazaki, 1996. Tsunami-induced conglomerates in Miocene upper bathyal deposits, Chita Peninsula, central Japan; Marine sedimentary events and their records: *Sed. Geol.*, **104**, 175–188.

Simkin, T., and R. S. Fiske, 1983. *Krakatau 1883; The Volcanic Eruption and its Effects*. Washington, DC: Smithsonian Inst. Press, 464p.

Simms, M. J., 2007. Uniquely extensive soft-sediment deformation in the Rhaetian of the UK: Evidence for earthquake or impact? *Palaeogeog., Palaeoclimat., Palaeoecol.*, **244**, 407–423.

Smit, J., T. B. Roep, W. Alvarez, A. Montanari, P. Claeys, J. M. Grajales-Nishimura and J. Bermudez, 1996. Coarse-grained, clastic sandstone complex at the K/T boundary around the Gulf of Mexico; deposition by tsunami waves induced by the Chicxulub impact? *Geol. Soc. Am. Special Paper*, **307**, 151–182.

Smith, D. E., I. D. L. Foster, D. Long and S. Shi, 2007. Reconstructing the pattern and depth of flow onshore in a palaeotsunami from associated deposits. *Sed. Geol.*, **200**, 362–371.

Smith, D. E., S. Shi, R. A. Cullingford, A. G. Dawson, S. Dawson, C. R. Firth, I. D. L. Foster, P. T. Fretwell, B. A. Haggart, L. K. Holloway and D. Long, 2004. The Holocene Storegga Slide tsunami in the United Kingdom. *Quat. Sci. Rev.*, **23**, 2291–2321.

Smoot, J. P., R. J. Litwin, J. L. Bischoff, and S. J. Lund, 2000, Sedimentary record of the 1872 earth-quake and "tsunami" at Owens Lake, Southeast California. *Sed. Geol.*, **135**, 241–254.

Soulsby, R. L., D. E. Smith and A. Ruffman, 2007. Reconstructing tsunami run-up from sedimentary characteristics—a simple mathematical model. *Coastal Sediments '07, Proc. 6th Int. Symp. On Coastal Eng. & Science of Coastal Sediment Processes Vol. 2*, Eds. N. C. Kraus and J. D. Rosati, Am. Soc. Civ. Eng., 1075–1088.

Srinivasalu, S., N. Thangadurai, A. D. Switzer, V. Ram Mohan and T. Ayyamperumal, 2007. Erosion and sedimentation in Kalpakkam (N Tamil Nadu, India) from the 26th December 2004 tsunami. *Mar. Geol.*, **240**, 65–75.

Steiner, M. B. and E. M. Shoemaker, 1996. A hypothesized Manson impact tsunami; paleomagnetic and stratigraphic evidence in the Crow Creek Member, Pierre Shale. *Geol. Soc. Am. Spec. Paper*, **302**, 419–432.

Switzer, A. D., C. S. Bristow and B. G. Jones, 2006, Investigation of large-scale washover of a small barrier system on the southeast Australian coast using ground penetrating radar. *Sed. Geol.*, **183**, 145–156.

Symons, G. J., ed., 1888. *The Eruption of Krakatoa and Subsequent Phenomena.* Rept. Krakatoa Committee Royal Soc. London, 494p.

Szczuciński, W., N. Chaimanee, P. Niedzielski, G. Rachlewicz, D. Saisuttichai, T. Tepsuwan, S. Lorenc and J. Siepak, 2006. Environmental and geological impacts of the 26 December 2004 tsunami in coastal zone of Thailand—Overview of short and long-term effects. *Polish J. Env. Stud.,* **15,** 793–810.

Szczuciński, W., P. Niedzielski, G. Rachlewicz, T. Sobczyński, A. Zioła, A. Kowalski, S. Lorenc and J. Siepak, 2005. Contamination of tsunami sediments in a coastal zone inundated by the 26 December 2004 tsunami in Thailand. *Environ. Geol.,* **49,** 321–331.

Szczuciński, W., P. Niedzielski, L. Kozak, M. Frankowski, A. Zioła and S. Lorenc, 2007. Effects of rainy season on mobilization of contaminants from tsunami deposits left in a coastal zone of Thailand by the 26 December 2004 tsunami. *Environ. Geol.,* 53, 253–264.

Takashimizu, Y. and F. Masuda, 2000, Depositional facies and sedimentary successions of earthquake-induced tsunami deposits in Upper Pleistocene incised valley fills, central Japan. *Sed. Geol.,* **135,** 231–239.

Tanner, L. H. and S. Calvari, 2004. Unusual sedimentary deposits on the SE side of Stromboli volcano, Italy; products of a tsunami caused by the ca. 5000 years BP Sciara del Fuoco collapse? *J. Volcanol. Geotherm. Res.,* **137,** 329–340.

Tonkin, S., H. Yeh, F. Kato and S. Sato, 2003. Tsunami scour around a cylinder. *J. Fluid Mech.,* **496,** 165–192.

Tuttle, M. P., A. Ruffman, T. Anderson and H. Jeter, 2004. Distinguishing tsunami from storm deposits in eastern North America; the 1929 Grand Banks tsunami versus the 1991 Halloween storm. *Seis. Res. Lett.,* **75,** 117–131.

van den Bergh, G. D., W. Boer, H. de Haas, T. C. E. van Weering and R. van Wijhe, 2003. Shallow marine tsunami deposits in Teluk Banten (NW Java, Indonesia), generated by the 1883 Krakatau eruption. *Mar. Geol.,* **197,** 13–34.

Umitsu, M., C. Tanavud and B. Patanakanog, 2007. Effects of landforms on tsunami flow in the plains of Banda Aceh, Indonesia, and Nam Khem, Thailand. *Mar. Geol.,* **242,** 141–153.

Unohana, M. and T. Oota, 1988. Disaster records of [1896] Meiji Sanriku tsunami by Soshin Yamana. *Faculty of Engineering, Tohoku University, Tsunami Engineering Report* **5,** 57–379. [in Japanese; as cited by Abe, 2005].

Vanoni, V., ed., 1975. *Sedimentation Engineering.* NY: Am. Soc. Civ. Eng., 745p.

Verbeek, R. D. M., 1886. *Krakatau.* Batavia, Indonesia [Dutch Jakarta], 567p. [in French]

Veski, S., A. Heinsalu, V. Klassen, A. Kriiska, L. Lougas, A. Poska and U. Saluaar, 2005. Early Holocene coastal settlements and palaeoenvironment on the shore of the Baltic Sea at Parnu, southwestern Estonia. *Quat. Internat.,* **130,** 75–85.

Walker, R. G., 1973. Mopping up the turbidite mess. *in* Ginsburg, R. N., ed., *Evolving Concepts in Sedimentology.* Baltimore: John Hopkins Univ. Press, 1–37.

Warme, J. and C. Sandberg, 1995. The catastrophic Alamo Breccia of southern Nevada: record of a late Devonian extraterrestrial impact. *Cour. Forsch.-Inst. Senckenberg,* **188,** 31–57.

Waythomas, C. F. and C. A. Neal, 1998. Tsunami generation by pyroclastic flow during the 3500-year B.P. caldera-forming eruption of Aniakchak Volcano, Alaska. *Bull. Volcanol.,* **60,** 110–124.

Waythomas, C. F., P. Watts, P. and J. S. Walder, 2006. Numerical simulation of tsunami generation by cold volcanic mass flows at Augustine Volcano, Alaska. *Nat. Hazards and Earth Syst. Sci.,* **6,** 671–685.

Weber, R. D. and D. K. Watkins, 2007. Evidence from the Crow Creek Member (Pierre Shale) for an impact-induced resuspension event in the late Cretaceous Western Interior Seaway. *Geology,* **35,** 1119–1122.

Weiss, R. and H. Bahlburg, 2006. A note on the preservation of offshore tsunami deposits. *J. Sed. Res.,* **76,** 1267–1273.

Wheatcroft, R. A. and J. C. Borgeld, 2000. Oceanic flood deposits on the northern California shelf: large-scale distribution and small-scale physical properties. *Cont. Shelf Res.,* **20,** 2163–2190.

Williams, D. M. and A. M. Hall, 2004. Cliff-top megaclast deposits of Ireland, a record of extreme waves in the North Atlantic; storms or tsunamis? *Mar. Geol.,* **206,** 101–117.

Williams, H. and I. Hutchinson, 2000. Stratigraphic and microfossil evidence for late Holocene tsunamis at Swantown Marsh, Whidbey Island, Washington. *Quat. Res.,* **54,** 218–227.

Williams, H. F. L., I. Hutchinson and A. R. Nelson, 2005. Multiple sources for late-Holocene tsunamis at Discovery Bay, Washington State, USA. *The Holocene,* **15,** 60–73.

Witter, R.C., H. M. Kelsey and E. Hemphill-Haley, 2001. Pacific storms, El Nino and tsunamis; competing mechanisms for sand deposition in a coastal marsh, Euchre Creek, Oregon. *J. Coastal Res.,* **17,** 563–583.

Wright, C. and A. Mella, 1963. Modifications to the soil pattern of south-central Chile resulting from seismic and associated phenomena during the period May to August 1960. *Bull. Seis. Soc. Am.,* **53,** 1367–1402.

Yagishita, K., 2001. Transportation and sedimentation of beach and sea-floor gravels caused by tsunami; an example at Attari Coast, Iwate Prefecture. *Chigaku Zasshi = Journal of Geography,* **110,** 689–697.

Yang, M. D., T. C. Su, C. H. Hsu, K. C. Chang and A. M. Wu, 2007. Mapping of the 26 December 2004 tsunami disaster by using FORMOSAT-2 images. *Internat. J. Remote Sensing,* **28,** 3071–3091.

Zayakin, Y. A. and A. A. Luchinina, 1987. *Catalogue of Tsunamis on Kamchatka:* Vniigmi-Mtsd, Obninsk, Russia [in Russian], 50p.

Chapter 4.   Tsunami Probability

ERIC L. GEIST, TOM PARSONS, URI S. TEN BRINK, AND HOMA J. LEE

U.S. Geological Survey

Contents

1. Introduction
2. Empirical Tsunami Probabilities
3. Computational Tsunami Probabilities
4. Uncertainties
5. Summary
References

1.   Introduction

Evaluating the probability of tsunami occurrence is a crucial step in the assessment of tsunamis hazards. Deterministic tsunami hazard studies involve hydrodynamic modeling of tsunami propagation, runup, and inundation from a particular source, usually defined as the maximum credible earthquake, landslide, or another tsunami trigger. Scenario-based modeling such as this is useful in emergency planning, but transferring the modeling results to other applications, such as estimating risk, is difficult. Risk assessment relies heavily on determining the probability that a tsunami of a certain size will occur within a given time frame.

A tsunami hazard curve that plots tsunami size against probability for a given exposure time (T) is a central concept in such analyses (Fig. 4.1). There are two ways in which a tsunami hazard curve can be used. The most common way is specifying a particular probability and exposure time of interest, and then determining the magnitude that a hazard variable (e.g., runup R) that will be met or exceeded (Fig. 4.1a). An example of such an approach is flood hazard analysis for insurance applications in which the wave height and extent are determined for annualized probabilities (T=1 yr.) P=0.01 and P=0.002 (Houston and Garcia, 1978). Alternatively, engineering applications may specify a risk tolerance value for a particular structure and use the hazard curve to determine the probability that that value will be met or exceeded during the exposure time (Fig. 4.1b). In addition to tsunami hazard curves for a particular site, probabilistic-based hazard assessment tools commonly include regional assessments (Rikitake and Aida, 1988; Geist and Parsons, 2006) and probabilistic inundation maps (Tsunami Pilot Study Working Group, 2006).

The Sea, Volume 15, edited by Eddie N. Bernard and Allan R. Robinson
ISBN 978–0–674–03173–9 ©2009 by the President and Fellows of Harvard College

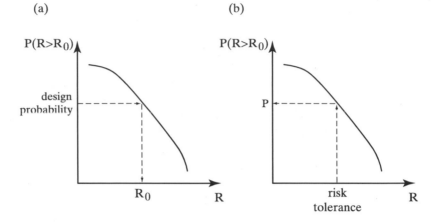

Fig. 4.1 – Schematic tsunami hazard curve showing different applications: (a) exceedance runup determined from design probability; (b) probability determined from risk tolerance.

In this paper we review techniques to determine tsunami probability using both empirical and computational approaches (Sections 2 and 3, respectively). Two distributions that form the basis for determining the probability of tsunamis (empirical) and their sources (computational) are the frequency-size distribution and the occurrence distribution of events in time. The former is nominally a power-law distribution, whereas the latter is often assumed to be an exponential distribution associated with a Poisson process. For the empirical analysis of tsunami observations at a particular coastal site, these assumptions greatly simplify the probability calculation: the probability (P) that a tsunami of a certain size (R_0) or greater will occur within time T is given by $P = 1 - e^{-\lambda T}$. λ is the long-term rate at which tsunamis of this size occur, which is given directly by the power-law size distribution $\log(\lambda) \sim -\beta \log(R_0)$, where β is the slope of the size distribution. One consequence of these assumptions is that for small values of $\lambda T \ll 1$, $P \approx \lambda$. There are, however, important deviations from these fundamental distributions and significant sources of uncertainty that are discussed in Section 2.

For most coastal locations, there is insufficient data to determine tsunami probability empirically. In these cases, probability is calculated using a combination of source specifications and numerical propagation models (Section 3). Aggregation of tsunami propagation results from all relevant sources, both near- and far-field, yield a tsunami hazard curve (Fig. 4.1) that should be equivalent to the empirical hazard curve, if there were sufficient historic data available. In addition to tsunami generation parameters, an important part of the source specification is defining the source probability: i.e., the inter-event distribution and the distribution of source sizes (e.g., seismic moment for earthquakes and volume for landslides). As part of the aggregation step of computing probabilities, various sources of uncertainty are also importantly included as we discuss in Section 4. To determine the probability of the largest tsunamis, methods to account for uncertainty become increasingly important. We briefly review methods to determine the probability of extreme events in Section 4, focusing on estimation techniques that define the tail of the size distribution for tsunamis (empirical approach) and their sources (computational approach).

2. Empirical Tsunami Probabilities

Tsunamis can be considered a stochastic process based on two essential characteristics: (1) once generated, tsunamis can propagate long distances such that many different sources in different tectonic and geologic environments can influence the tsunami hazard at a particular coastal location and (2) each type of source that generates tsunamis is itself characterized by a high degree of complexity and nonlinear interactions. As a result, tsunami probabilities can be defined by the frequency distribution of sizes and the distribution of inter-event or waiting times. Nominally, the frequency-size distribution follows a power-law relationship and the inter-event times are that of a Poisson process, typical of many natural hazards (Daley and Vere-Jones, 2003; Sornette, 2004). Furthermore, these two fundamental distributions are inter-related in that the scaling of mean frequency with size is linked to the scaling of inter-event times (Corral, 2005b). Each of the fundamental distributions is discussed further below, including how the parameters that define the specific probability distributions can be empirically obtained.

2.1 Frequency-Size Distribution

It is important to first define the variable that defines the size of a tsunami. Although runup is the measurement most often associated with tsunamis, because it is defined as the wave height with respect to ambient sea level at the maximum inundation distance, runup will occur at different geographic locations for different tsunamis. Tide gauges, on the other hand, record wave amplitude at a fixed location (see Chapter 7). For most probability problems, comparisons are made over broad geographic regions that may include both runup and wave amplitude measurements. An exception is the development of probabilistic inundation maps at a given location (Tsunami Pilot Study Working Group, 2006). Throughout this study, we will refer to runup as the tsunami size or hazard variable, although this may include other amplitude measurements of tsunamis as well.

Like many other natural hazards, the frequency-runup distribution for tsunamis at a particular location tends to follow a power-law relationship (Burroughs and Tebbens, 2005):

$$\log\left[\dot{N}(R)\right] = \alpha - \beta \log(R),\tag{1}$$

where $\dot{N}(R)$ is the annual frequency of tsunami runup R or larger and the empirical constants α and β can be thought of as activity and scale parameters, respectively, that are determined from tsunami catalog data. The power-law nature of tsunamis indicates that there is no characteristic size and stems from the fundamental physics of the tsunami source (e.g., earthquakes, landslides) as explained by Sornette (2004). For empirical analysis, because catalogs will undersample below some size, we need to include a catalog-completeness threshold, R_t. This results in a Pareto distribution, for which the probability density function (pdf) is

$$\phi(R) = \frac{\beta(R_t)^{\beta}}{R^{(1+\beta)}}, \text{ for } R_t \le R\tag{2}$$

and the complementary cumulative distribution function (ccdf) or survivor function is

$$\Phi(R) = \left(\frac{R_t}{R}\right)^{\beta}, \text{ for } R_t \leq R \tag{3}$$

(cf., Kagan, 2002a). Kagan (2002a) and Vere-Jones et al. (2001) also provide other modified Pareto distributions that have a soft taper for the roll-off parameterized by a corner runup R_c, which can be estimated using maximum likelihood techniques.

Moreover, because tsunami runup is size limited due to size limitations of the source, non-linear propagation, and wave-breaking effects near shore (e.g., Korycansky and Lynett, 2005), the Pareto distribution must be limited at large runup values. However, the limit and shape of the distribution tail are generally unclear. Burroughs and Tebbens (2001; 2005) suggest a truncated power-law relationship based on the value of the largest event (i.e, truncation of the pdf at R_x) that is equivalent to the truncated Gutenberg-Richter (G-R) size distribution for earthquakes (Kagan, 2002a):

$$\phi(R) = \frac{\beta R_t^{\beta} R_x^{\beta}}{\left(R_x^{\beta} - R_t^{\beta}\right) R^{(1+\beta)}}, \text{ for } R_t \leq R \leq R_x \tag{4}$$

and

$$\Phi(R) = \frac{(R_t / R)^{\beta} - (R_t / R_x)^{\beta}}{1 - (R_t / R_x)^{\beta}}, \text{ for } R_t \leq R \leq R_x. \tag{5}$$

The observed roll-off for an empirically-determined power-law size distribution may be due to either undersampling or the physical controls on the size of the largest event (Burroughs and Tebbens, 2001). Various statistical techniques developed for earthquake observations can also be applied to constrain the tail of the tsunami size distribution as discussed in Section 4 below.

Undersampling of tsunamis in catalogs occurs both through censoring of small events and having a catalog of insufficient duration to capture the rate of large events. For global catalogs prior to the mid-20th century, the locations of tide gage stations are sparse in comparison to the locations of earthquakes that generate measurable small tsunamis (spatial or geographic censoring). In addition, many routine catalogs of tide gage stations sampled water level on an hourly basis until the installation of the 6-minute sample period Analog-to-Digital-Recording (ADR) tide gages starting in the late 1960s and early 1970s (temporal or instrumental censoring). Because the average tsunami period is typically smaller than 1 hour and tsunami amplitudes less than ~10–20 cm are difficult to identify in the presence of ambient noise, smaller events (in amplitude and wavelength) tend to be temporally censored at tide gage stations. Archived analog records of tsunami events can be digitized at a much smaller sampling rate for future probability studies. Finally, in many locations, the tsunami catalog covers only 100–300 years, which is insufficient to accurately determine the rate of occurrence for the largest tsunamis.

To demonstrate the effects of censoring and catalog completeness on empirical power-law frequency-size distributions, we compare the 275-year tsunami catalog at Acapulco, Mexico with a computationally derived frequency-size distribution described by Geist and Parsons (2006). The tsunami catalog consists of both eye-witness observations of runup height and tide-gage measurements starting in 1950. While the entire catalog includes large runup events prior to the installation of the tide gage station that match the computational curve, censoring is evident in the divergence of the two curves for small event sizes (Fig. 4.2a). Using the tide-gage sub-catalog only, we observe a good correspondence for the small events, but incomplete data to constrain the empirical curve for large events (Fig. 4.2b). The roll-off in the computational curve is caused by a physical limitation of earthquake sizes along the offshore subduction zone (Geist and Parsons, 2006).

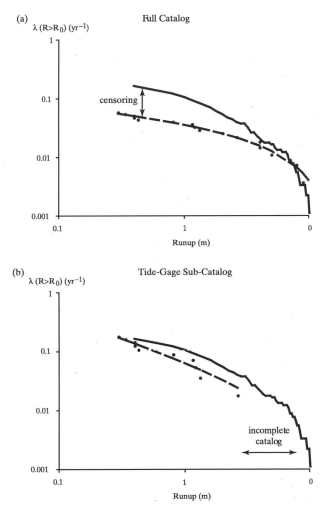

Fig. 4.2 – Observations (solid circles) and best-fit empirical curve (dashed line) representing frequency-size distribution for local tsunamis at Acapulco, Mexico. Computationally-derived distribution (solid line) is shown for comparison. (a) Empirical curve from merged catalog including eye-witness observations and tide-gage measurements (1732–2006). (b) Empirical curve from tide-gage sub-catalog (1950–2006). Modified from Geist and Parsons (2006).

The Acapulco tsunami catalog is an example of a break in an empirical power-law relationship to the left is caused by instrumental censoring (insufficient sampling rate or dynamic range). The completeness threshold for tsunami amplitudes recorded by tide gages as represented in standard tsunami catalogs is approximately 0.1 m. Properly processed, digital tide gages can record tsunamis < 0.1 m. However, many small events are not retained in a permanent archive. Modern deep-sea pressure-sensors that can record micro-tsunamis (Hino et al., 2001; Hirata et al., 2003) will help obviate censoring of small events in the future. In contrast, for a tsunami catalog of sufficient duration that depends on the rate of tsunami activity, a roll-off in an empirical power-law relationship to the right may be an indication of physical limitations to event size (Burroughs and Tebbens, 2001; 2005).

Distribution of Inter-Event Times

Because sources of tsunamis are generally uncorrelated, tsunami inter-event times can be assumed to be independent, identically distributed (iid) random variables such that the number of events in a particular time increment is independent of the number of events in any other increment. The occurrence of tsunamis under this assumption would accordingly be that of a Poisson process. The probability that n events will occur within a particular time t is given by the Poisson distribution:

$$P_n(t) = \frac{(\lambda t)^n e^{-\lambda t}}{n!}$$

(6)

where λ is the intensity or rate parameter. We consider here only homogeneous or stationary Poisson processes in which $\lambda \neq f(t)$. (An example of a non-stationary Poisson process is the arrival times of earthquakes in an aftershock sequence following a main shock that is discussed below.) The single rate parameter λ at a given location is derived from the aggregation of many sources each with a different source rate parameter as explained in Section 3. The cumulative distribution function (cdf) for n or more events occurring in time t, or equivalently, for the nth event time less than t, is given by

$$F_{N(t) \geq n}(t) = F_{T_n \leq t}(t) = 1 - \sum_{j=0}^{n-1} \frac{(\lambda t)^j e^{-\lambda t}}{j!}.$$

(7)

In hazard analysis, the commonly used special case is of one or more events occurring in time t resulting in the cdf

$$F_{N(t) \geq 1}(t) = 1 - e^{-\lambda t}.$$

(8)

The pdf of event times T_n can be derived from equation (7) (Kempthorne and Folks, 1971), resulting in the Erlang distribution:

$$f_{T_n}(t) = \frac{\lambda(\lambda t)^{n-1}}{(n-1)!} e^{-\lambda t} \qquad (9)$$

The pdf of the first event time $T_1 = \tau$, otherwise known as the inter-event or first waiting time, for a Poisson process is an exponential distribution:

$$f(\tau) = \lambda e^{-\lambda \tau}, \qquad \text{for } \tau > 0. \qquad (10)$$

Empirical estimates of tsunami inter-event times determined by Geist and Parsons (2008) indicate a more complex distribution than would be expected from a stationary Poisson process, but similar to what is observed for earthquake inter-event time distributions. As with empirically derived frequency-size distributions, it is necessary to have a large catalog of events, especially for small amplitude tsunamis that will dominate the left-hand side of the inter-event time distribution. We use, as an example, a global compilation of tide gage tsunami measurements from the National Geophysical Data Center (NGDC) based on a number of original catalogs (http://www.ngdc.noaa.gov/seg/hazard/tsu_db.shtml). In this case, the event time is defined as the origin time for the tsunami source. An examination of the cumulative number of tsunamis since the start of the 20th century indicates that the rate of reported events gradually becomes constant soon after the pivotal 1946 Aleutian tsunami (Fig. 4.3). This time approximately coincides with a sharp increase in the number of sea-level recording station around the world (Caldwell and Merrifield, 2006). With this in mind, we therefore use the portion of the NGDC catalog from 1952–2006. Maximum runup and inter-event time series for the 20th century catalog are shown in Fig. 4.4.

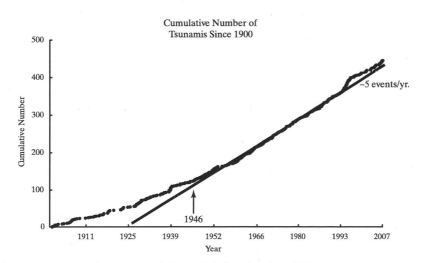

Fig. 4.3 – Cumulative number of tsunamis in the global catalog since 1900.

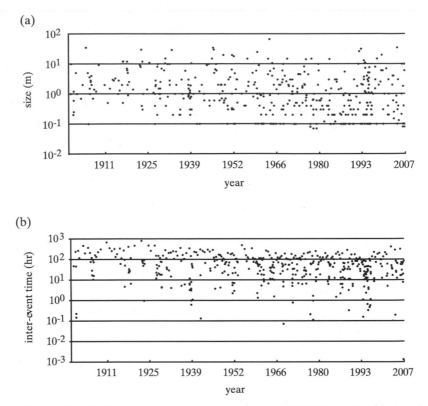

Fig. 4.4 – Time series of global tsunami catalog showing from 1900–2007 showing (a) size of events recorded as run-up heights and (b) inter-event time. Event times are defined as the origin time of the tsunami source.

To determine the empirical pdf, inter-event times were calculated and binned by Geist and Parsons (2008) according to c^i $i = 1, 2, 3 \dots$, where the binning constant c was chosen such that the range of inter-event times encompassed as many bins with non-zero entries as possible (Corral, 2005a). The results of the empirical analysis are shown in Fig. 4.5 in comparison with an exponential inter-event time pdf (equation 10). Here, the rate parameter $\lambda = N_{cat}/T_{cat}$ is determined from the number of events (N_{cat}) over the duration of the catalog (T_{cat}) (including the open interval since the last event). From the global catalog, the mean rate is one measured tsunami every 1600 hours, or approximately 5 tsunamis per year. If the frequency-size distribution is also known and can be approximated by a power-law distribution, then the rate parameter is simply 10^{α_t}, where α_t is relative to the completeness threshold of the tsunami inter-event time catalog (equation 1) (cf., Kagan, 2002a; Ward, 2002).

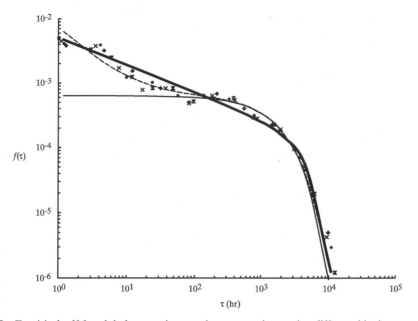

Fig. 4.5 – Empirical pdf for global tsunami source inter-event times using different binning parameters (c = 2.0, 2.2, 2.4, 2.6 hrs.; star, x, circle, diamond, respectively). Theoretical inter-event time distributions also shown for an exponential distribution (light solid line), gamma distribution where $\gamma = 0.6$ (heavy solid line), and Omori-type aftershock distribution where T_a = 9.6 hrs. (dashed line).

The discrepancy between the empirical and Poisson distribution shown in Fig. 4.5 can be investigated by considering a universal scaling law and an aftershock decay law for earthquake inter-event times. In general, the inter-event pdf is described by the functional form:

$$f(\tau) = \lambda g(\lambda \tau). \tag{11}$$

The universal scaling law proposed by Corral (2004; 2005a) indicates that $g(\theta)$, where $\theta = \lambda \tau$ is the dimensionless inter-event time, can be expressed as a generalized gamma distribution that captures many stationary and non-stationary aspects of observed seismicity:

$$g(\theta) = \frac{C|\delta|}{a\Gamma(\gamma/\delta)} \left(\frac{a}{\theta}\right)^{1-\gamma} e^{-(\theta/a)^{\delta}}, \tag{12}$$

where Γ is the complete gamma function, C and a are normalization and scale constants, respectively, and λ and δ are shape parameters. This distribution spans a range of temporal characteristics where $\gamma = \delta = 1$ is the exponential distribution (equation 10). The information gain (G) of the standard gamma distribution over a Poisson process with rate parameter λ is given by

$$G = \lambda \left[1 - \log \lambda + \int_0^\infty f(\tau) \log f(\tau) d\tau \right]. \tag{13}$$

(Daley and Vere-Jones, 2004; Harte and Vere-Jones, 2005). The information gain is highest for clustering ($\gamma < 1$), but is also significant for $\gamma > 1$ (i.e., that of a quasiperiodic process) (Harte and Vere-Jones, 2005).

When aftershocks are removed from an earthquake catalog, as in the case of southern California seismicity ($M \leq 2.5$) analyzed by Corral (2005b), $\gamma \approx 0.7$ indicating that seismicity exhibits weak, longer-term correlations. The physical mechanism of this long-term correlation, which can extend outside the classic aftershock zone, is likely triggering of secondary earthquake from either changes in static stress after each earthquake or dynamic effects from the passage of seismic waves (Parsons, 2002). This distribution also corresponds to world-wide seismicity ($M \geq 5$), suggesting a universal scaling law for earthquake inter-event times, though γ may vary depending on the presence of nonstationary aftershock sequences (Bak et al., 2002; Corral, 2004; Davidsen and Goltz, 2004; Altmann and Kantz, 2005; Molchan, 2005).

Alternatively, an aftershock decay distribution can be used to fit the observed inter-event pdf. This is based on a simplified version of Omori's law. In this case, we fit the observed tsunami inter-event times with an exponential distribution modified with a short-term Omori-law component as

$$f(\tau) = C\lambda \left(e^{-\lambda \tau} + \frac{T_a}{\tau} \right), \tag{14}$$

where T_a is an aftershock duration time constant. The best-fit gamma and aftershock-decay distribution for the NGDC tsunami compilation are determined by χ^2-minimization. This results in estimates of $\gamma = 0.6$ and $T_a = 9.6$ hrs. (Fig. 4.5). Other methods of probability model fitting are discussed by Vere-Jones and Ogata (2003). In addition, the Kullback-Leibler, Kolmogorov-Smirnov, and Anderson-Darling statistics can be used to provide an estimate of the goodness of fit (Conover, 1971; Finkelstein and Schafer, 1971; Stephens, 1974; Kotz and Nadarajah, 2000; Parsons, 2002; Daley and Vere-Jones, 2004).

Geist and Parsons (2008) demonstrate that a similar analysis can also be performed for tsunami arrivals at a particular location that has a long record of smaller tsunamis: for example, the Hilo, Hawaii tide gage station (Fig. 4.6). In contrast to the global catalog, the event time in this case is the arrival time of the tsunami at Hilo. The tide gage catalog spans a 60-year range from 1946 to 2007; however, it has only been since 1976 that the tide gage records have been systematically sampled every 6 minutes, rather than every hour. (In most cases, undigitized analog records are also available.) The best fit gamma distribution for these data has a shape parameter $\gamma = 0.8$, and a value of $T_a = 1.2$ days for the tsunami aftershock-decay distribution. An examination of the tsunami record indicates several source doublets (2 earthquakes of similar magnitude that occur close in space and time) in the Kurile and Solomon Islands that both triggered tsunamis recorded in Hilo. There is a significant degree of uncertainty in the Hilo distribution, owing to a less than optimal number of data points (N=63) required to establish reliable inter-event time statistics.

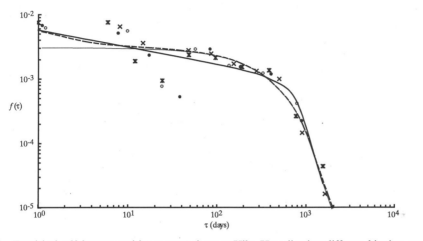

Fig. 4.6 – Empirical pdf for tsunami inter-event times at Hilo, Hawaii using different binning parameters (c = 1.8, 2.0, 2.2, 2.4 days; star, x, filled circle, open circle, respectively). Event times are defined as the absolute arrival time of the tsunami at Hilo. Theoretical inter-event time distributions also shown for an exponential distribution (light solid line), gamma distribution where $\gamma = 0.8$ (solid line), and Omori-type aftershock distribution where T_a = 1.2 days (dashed line).

In locations where there are sparse runup data, spatial binning of runup observations and Monte Carlo techniques can be used to estimate the mean rate λ (Geist and Parsons, 2006) (e.g., Fig. 4.7). The single-parameter Poisson distribution should be used for the sparse data case, rather than other distributions such as the gamma distribution above, that require estimation of two or more parameters to define the inter-event distribution. The effect of open intervals (i.e., the time before the first event and the time since the last event) on λ can be estimated by randomly drawing multiple sets of event times from a range of possible λ. By keeping track of which distributions fit the catalog data (N events over a catalog duration T_{cat}) and open intervals, we can estimate the uncertainty in λ (Parsons, 2008). The example shown in Fig. 4.7b shows the range of probability for a 30-year exposure time resulting from Monte Carlo uncertainty analysis of the catalog data. A similar analysis is performed for the U.S. West Coast by Geist and Parsons (2006). Further discussion of uncertainty related to empirical probabilities is given in Section 3.2 below in the case of empirical earthquake observations.

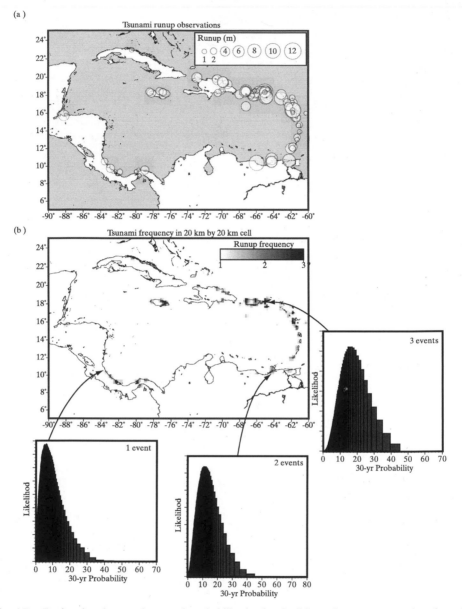

Fig. 4.7 – Regional estimate of tsunami probability in the Caribbean from sparse catalog data. (a) Spatial distribution of tsunami runup observations. Diameter of circle proportional to runup in meters. (b) Frequency of events binned into 20km-by-20km cells. Histograms represent results of Monte Carlo analysis of possible 30-yr. probabilities that fit the catalog data taken from an exponential distribution at 3 representative locations.

3. Computational Tsunami Probabilities

In many cases, tsunami probabilities cannot be determined empirically from existing tsunami records. Often this is because there is an insufficient catalog of events for the risk tolerance or design probability of interest. For such a situation, a computational approach to determine tsunami probabilities can be undertaken. In this section, we review the framework of Probabilistic Tsunami Hazard Analysis (PTHA), focusing in particular on how tsunami source probabilities are determined. Even where there is a long catalog of tsunami records, computational PTHA is often a useful technique to test the extent that censoring and catalog completeness affect empirical probabilities as discussed in the previous section and by Geist and Parsons (2006).

3.1   Structure of PTHA

PTHA is derived from Probabilistic Seismic Hazard Analysis (PSHA) developed by Cornell (1968) and others and fully described in SSHAC (1997). Like PSHA, PTHA consists of three basic steps: (1) define source parameters, including source probabilities, for all relevant sources; (2) calculate wave heights and other hydrodynamic parameters from a numerical propagation and inundation model for each source; and (3) aggregate the results to determine either the tsunami hazard curve for a particular coastal site or the probabilistic inundation map for a particular coastal region. Lin and Tung (1982) first applied Cornell's (1968) PSHA technique to seismogenic tsunamis, by using simplifying assumptions for the earthquake source and propagation parameters (e.g., constant water depth, etc.). More recent forms of PTHA (e.g., Geist and Parsons, 2006; Thio et al., 2007) involve the use of numerical tsunami propagation models for the second step. Indeed, the processing of this step is one of the primary differences between PTHA and PSHA. As a result of the complexity of wave propagation in the solid earth, standard forms of PSHA are dependent on empirical attenuation relationships (and their attendant uncertainty). PTHA, on the other hand, can take advantage of recent advances in numerical modeling of tsunami propagation and the availability of high-quality bathymetry in most of the world's oceans. The other primary difference is that, in addition to regional and local sources, a comprehensive PTHA must include far-field sources not included as part of PSHA. Depending on the design probabilities and region of interest for PTHA, other sources for tsunamis such as submarine landslides and volcanic sources may also have to be included.

Defining the size parameter linked to source probability is central in PTHA calculations. For earthquakes, this parameter is seismic moment defined as $M = \mu \overline{D} A$, where μ is the shear modulus or rigidity and $\overline{D}$ is the average slip over the area A of the fault that ruptured during the earthquake. The moment magnitude (M_w) is related to seismic moment according to $M_w = \frac{2}{3}[\log(M) - 9.05]$ (Hanks and Kanamori, 1979).

For seismogenic sources, the vertical component of seafloor displacement dominates tsunami generation. The horizontal component provides an additional small effect on tsunami generation in regions with steep bathymetry over the source region (Tanioka and Satake, 1996). Fault rupture is modeled as an elastic dislocation either using uniform slip (termed a Volterra dislocation) or more generally using distributed slip (cf., Geist and Dmowska, 1999). From this description, co-seismic displacement can be computed using analytic expressions for a homogeneous earth structure (e.g., Okada, 1985) and numerical techniques for an inhomogeneous structure (e.g., Yoshioka et al., 1989). For dislocation modeling, most parameters such as average slip and rupture area approximately scale with seismic moment. Other parameters such as fault dip and elastic rock properties are determined from analyses of past earthquakes, controlled-source geophysical surveys, and laboratory tests. For a more complete description of tsunami generation by earthquakes, see Chapter 5 and review papers by Kajiura (1981), Geist (1999), and Satake (2002).

For landslides, the primary size parameter linked to source probability is volume. Recent studies have indicated that submarine landslides may follow a power-law frequency-volume distribution, similar to their counterparts on land (ten Brink et al., 2006a). Unlike earthquakes in which a single parameter, seismic moment, is the principal parameter influencing tsunami generation, landslide tsunami generation is also heavily influenced by landslide speed (or more specifically, time history of landslide movement). During propagation, recent modeling suggests that landslide tsunamis dissipate more quickly than earthquake tsunamis (Gisler et al., 2006). Also unlike earthquakes, there is not a single constitutive relation that describes tsunami generation from landslides. Tsunami generation depends on the type of failure that occurs—e.g., rotational and translational slides, rock falls, lateral spreads, etc. (Varnes, 1978)—which in turn relates to the mechanical properties of the failed material and the bathymetric slope. Examples of different types of landslide tsunami models developed include mudflows (Jiang and Leblond, 1994), translational slides (Ward, 2001), and granular slides (Heinrich et al., 2001). Chapter 6 reviews tsunami generation by landslides in detail.

The second step of PTHA involves computing tsunami wave heights, runup values, and inundation distances at a particular coastal location for each relevant source. In the far-field, seismogenic tsunami amplitudes closely scale with seismic moment (Okal, 1988; Pelayo and Wiens, 1992). Abe (1995) has developed empirical relationships in which wave height at a particular coastal site can be estimated from seismic moment and distance to the earthquake. However, these expressions do not account for the beaming patterns from long ruptures and focusing or defocusing from propagation path effects (Ben-Menahem and Rosenman, 1972; Okal, 1988; Satake, 2002; Geist et al., 2007). In most cases, tsunami propagation is calculated using various forms of the shallow water wave equation and numerical approximates such as finite-difference or finite-element methods. Initial conditions are primarily provided by the vertical displacement from the earthquake or landslide movement. At a coastal location, if near-shore and overland flow are to be computed, non-linear terms of the shallow water equations, bottom friction, and moving boundary conditions must be included. For a review of these methods, please see Chapters 9 and 10 and Shuto (1991).

For the third step in PTHA, the general equation for aggregating probabilities from different sources can be directly adapted from PSHA methodology (e.g., Senior Seismic Hazard Analysis Committee (SSHAC), 1997) as follows:

$$\lambda(R > R_0) = \sum_{\text{source type}=i} v_i \iint P(R > R_0 \mid \psi_i, r) f(\psi_i) f(r \mid \psi_i) dr d\psi_i, \tag{15}$$

where $\lambda(R > R_0)$ is the mean rate of tsunamis at a coastal location with runup greater than R_0, v_i is the mean rate for source type i (e.g., $i = 1 \Rightarrow$ earthquakes, $i = 2 \Rightarrow$ landslides, etc.) $P(R > R_0 \mid \psi_i, r)$ is the conditional probability that runup

R will exceed a value R_0, given a distance to the source r and source parameter(s) ψ [1]. In other words, the exceedance probability of R_0 depends on the specified values of r and ψ. In addition, $f(\psi_i)$ and $f(r|\psi_i)$ are pdf's for ψ_i and r, respectively. For example, to calculate the mean rate at which a particular runup R_0 is exceeded for earthquakes randomly distributed in space, Abe's (1995) empirical relationships could be used to determine $P(R > R_0 | M, r)$, $f(\psi_1 = M)$ could be determined from the modified Gutenberg-Richter (G-R) relationship (Kagan, 2002a) and $f_R(r|M)$ could be dependent on the particular earthquake zonation scheme used (Cornell, 1968; Working Group on California Earthquake Probabilities, 1995; Wesson et al., 1999; Frankel et al., 2002). For earthquakes, a standard zonation scheme such as the Flinn-Engdahl zones can be used (Flinn et al., 1974).

However, in contrast to seismic wave propagation in the solid earth, which is essentially unobstructed, obstruction and scattering from landmasses during tsunami propagation indicates that a distribution based on scalar distance $f_R(r|m)$ is not practical. In addition, although the approach described by equation (15) may be applicable for asteroid-generated tsunamis randomly distributed throughout the ocean, earthquakes and landslides of tsunamigenic size often occur in distinct source zones. Therefore, an alternative formulation for these sources is

$$\lambda(R > R_0) = \sum_{\text{type}=i} \sum_{\text{zone}=j} v_{ij} \int P(R > R_0 | \psi_{ij}) f_\psi(\psi_{ij}) d\psi . \tag{16}$$

In this case, $P(R > R_0 | \psi_{ij})$ is determined for each source type (i) and source zone (j) from numerical propagation modeling that explicitly includes distance attenuation and propagation path effects, as well as uncertainty in location within the source zone. For source parameters that correlate strongly with a source parameter linked to v_{ij} (e.g., average slip scaling with scalar seismic moment in the earthquake case and runout linked to volume in the landslide case), uncertainty can be included in the $P(R > R_0 | \psi_{ij})$ term, particularly when the parameters are normally distributed. Details are described in Section 4 on uncertainties. As with empirical tsunami probabilities, the rate term λ is used in the inter-event time distribution to determine the exceedance probability (e.g., equation 8 for the Poisson case). In Sections 3.2 and 3.3 below, we describe how source probability distributions are determined for both earthquakes and landslides.

3.2   Source Probabilities: Earthquakes

Because the primary source parameter linked to tsunamigenesis for a particular source zone j is the scalar seismic moment ($\psi_{1j} = M_j$), we review approaches used to establish frequency-moment relationships for earthquakes. Time-independent and time-dependent probabilities that are described below have frequently been used in the past in relation to earthquake hazards. Examples include probability specifications used in seismic hazard mapping for the U.S. (Wesson et al., 1999; Frankel et al., 2002).

[1] If there is no uncertainty in source location or generation parameters, $P(R > R_0 | \psi_i, r)$ is simply 1-H, where H is the Heaviside step function.

Time Independent Probabilities

Time-independent probabilities assume that earthquakes follow a Poisson process in which the rate term is determined from the G-R power-law relationship. Because earthquakes, like most other natural hazards are size limited, the power-law relationship is modified based on an estimate of maximum moment or *corner moment*. This results in a modified G-R relationship in which the tail of the distribution falls off faster than the power-law exponent (β). Earthquakes of tsunamigenic magnitude occur near the tail of this distribution where earthquakes catalogs may be incomplete. Therefore, different forms of the frequency-magnitude distribution tail (Fig. 4.8) need to be understood.

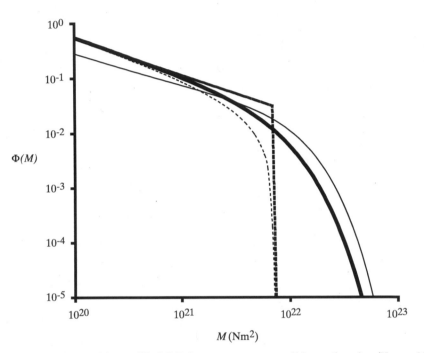

Fig. 4.8 – Different forms of the modified G-R frequency-moment ccdf for earthquakes (Kagan, 2002a). Two pairs of distributions are indicated: characteristic (heavy dashed line) and truncated (light dashed line) distributions have a finite limit. Tapered G-R (heavy solid line) and gamma (light solid line) distributions have a soft taper. For each distribution, the limiting moment and corner moment are identical: 6.92×10^{21} Nm^2 ($M_w = 8.5$).

In addition to the modified G-R distribution, a different distribution called a characteristic earthquake distribution has been proposed. Because the characteristic earthquake distribution is so widely referred to in the literature, it is important to describe the differences between the two basic distribution types. Characteristic earthquakes are usually defined from the largest recorded earthquake or geometrical differences along a fault zone defining an earthquake segment (Kagan, 1993; Wesnousky, 1994). In the case of tsunamigenic earthquakes, for example, morphological features of the downgoing plate in subduction zone can define segment boundaries and/or coincide with the extent of historic ruptures (McCann et al.,

' 1979; Nishenko, 1991). The characteristic earthquake distribution considered by Wesnousky (1994) is for the case where there is a gap in earthquake magnitudes (in the discrete form) between the characteristic magnitude M_{max} and the largest aftershock specified according to Båth's law. The identification of characteristic earthquake distributions is prone to undersampling. Several authors (Howell, 1985; Kagan, 1993; Stein and Newman, 2004) have demonstrated that if a G-R distribution is undersampled it tends toward an apparent characteristic distribution. In addition, physical models of rupture over many earthquake cycles indicate that rupture can appear to follow an characteristic mode for a number of earthquake cycles and then revert back to a G-R mode (Shaw and Rice, 2000; Shaw, 2004). For uniform fault properties, the rupture mode may be persistently characteristic (Ben-Zion, 1996; Lu and Vere-Jones, 2001; Zeng et al., 2005). However, many faults appear to be characterized by a combination of three factors: self-affine complexity caused by material heterogeneity (e.g., Perfettini et al., 2001; Shaw, 2004); static changes in stress on the fault from neighboring earthquakes (e.g., Marsan, 2005; Parsons, 2005; 2006); and the dynamics of earthquakes themselves (e.g., Cochard and Madariaga, 1996; Ben-Zion and Rice, 1997; Shaw and Rice, 2000). This complexity is typically expressed as a modified G-R distribution of earthquake magnitudes.

As summarized by Kagan (2002a), there are four basic forms of the modified G-R distributions: two truncated forms (i.e., hard corner) and two tapered forms (i.e., soft corner) (Fig. 4.8). Kagan (2002a) notes that most driven dissipative systems, such as earthquakes, exhibit a smooth transition toward the extreme value of the distribution defined by a corner moment M_c. For this reason, tapered distributions have been developed for both the ccdf (tapered G-R distribution with corner moment M_{cm})

$$\Phi(M) = \left(\frac{M_t}{M}\right)^{\beta} \exp\left(\frac{M_t - M}{M_{cm}}\right), \text{ for } M_t \leq M \qquad (17)$$

and for the pdf (gamma distribution with corner moment M_{cg})

$$\phi(M) = \frac{\beta}{CM}\left(\frac{M_t}{M}\right)^{\beta} \exp\left(\frac{M_t - M}{M_{cg}}\right), \text{ for } M_t \leq M, \qquad (18)$$

where C is a normalizing coefficient (Kagan, 2002a). (See Kagan 2002a for the pdf and ccdf of the tapered and gamma distributions, respectively.) The parameters that define the earthquake distribution (β, M_c) have been estimated for different types of plate boundaries and a number of different seismic zonation schemes using primarily maximum likelihood methods (Kagan, 1997; 1999; 2002a; b; Bird and Kagan, 2004). The commonly used b-value for earthquakes is based on M_w, whereas β is based on seismic moment M such that $\beta = \frac{2}{3}b$ according to the definition of moment magnitude.

Because large earthquakes are rare, the tails of earthquake distributions are difficult to determine from historical seismicity alone. For this reason, a seismic moment balance argument has previously been used to determine the expected earthquake distribution. Moment balance requires knowledge of the long-term fault slip rate, seismogenic thickness, and the seismic efficiency or coupling constant. For plate boundary faults, recent global studies yield information on long-term fault slip rate (Bird, 2003; Kreemer et al., 2003). The rate parameter (v_{1j}, cf., equation 16) for seismic moments greater than or equal to M_0 can be linked to the parameters to the modified G-R distributions expressed in equations 17 and 18 as follows (Kagan, 2002b):

$$v(M \geq M_0) = \xi_m^{-1} \left[\frac{1-\beta}{\Gamma(2-\beta)} \right] \frac{\dot{M}_s}{M_0^\beta M_{cm}^{1-\beta}} \qquad (19)$$

and

$$v(M \geq M_0) = \xi_g^{-1} \left[\frac{1-\beta}{\Gamma(2-\beta)} \right] \frac{\dot{M}_s}{\beta M_0^\beta M_{cg}^{1-\beta}} , \qquad (20)$$

respectively, where ξ_m and ξ_g are correction coefficients given in Kagan (2002b). Ward (1994) develops a similar expression for the truncated G-R distribution. In equations 19 and 20, $\dot{M}_s$ is the rate of seismic moment release that ideally can be determined from earthquake catalogs. Using the seismic moment balance approach, however, the maximum seismic moment rate at seismogenic depths can be constrained by the fault slip rate if one assumes an efficiency or coupling constant $\chi = 1$ (i.e., no aseismic slip at depths where earthquakes typically occur):

$$\dot{M}_s = \mu L H_s \dot{s}_{tect} , \qquad (21)$$

where μ is the shear modulus, L fault length, H_s effective seismogenic thickness (which includes the parameter χ), and $\dot{s}_{tect}$ the long-term (tectonic) fault-slip rate (Ward, 1994).

Note that the rate of large earthquakes along subduction zones is primarily linked to relative convergence rates and that factors such as age of subducted lithospehere, time since the last event, etc. do not strongly correlate with seismic activity (Bird and Kagan, 2004). The seismic coupling constant (χ) that appears in the definition of H_s above remains an elusive parameter. However, statistical analysis suggests that there may be little variation among plate boundary faults, such that $\chi \geq 0.5$ in most cases (Kagan, 2002b) and that $\chi \rightarrow 1$ for shallow faults (Kagan, 1999). A value of $\chi = 1$ provides an upper limit of activity rate for the seismic moment-balanced size distribution of earthquakes.

Time-Dependent Probabilities
In the past, large tsunamigenic earthquakes along subduction zones have considered to follow a quasiperiodic process specifically conditioned on the preceding event, in which the probability of the next earthquake is dependent on the time since the last earthquake. Such time-dependent probabilities are most often applied to characteristic earthquakes as defined above. Three basic forms are commonly used for earthquake probabilities in which the occurrence of one or more

earthquakes in the time interval T is conditional upon the time since the last earthquake τ: $F_{N(T)\geq1}(T \mid \tau)$ (Utsu, 1984). The corresponding pdf's for the inter-event time are given for the (1) Weibull distribution (e.g., Rikitake, 1999)

$$f(\tau) = \alpha v (\tau)^{\alpha-1} \exp\left(- v\tau^\alpha\right) \tag{22}$$

(2) Log-normal distribution (e.g., Nishenko and Buland, 1987)

$$f(\tau) = \frac{1}{\alpha\tau\sqrt{2\pi}} \exp\left[-\frac{(\ln(v\tau))^2}{2\alpha^2}\right] \tag{23}$$

and (3) Brownian-Passage time (BPT) distribution (Matthews et al., 2002)

$$f(\tau) = \left(\frac{1}{2\pi v \alpha^2 \tau^3}\right)^{1/2} \exp\left[-\frac{v(\tau-1/v)^2}{2\alpha^2\tau}\right]. \tag{24}$$

All three forms are similar in that three parameters are needed to define the distributions: the mean inter-event time $(1/v)$ as in the Poisson distribution (equation 8), the time since the last earthquake (τ), and the shape or aperiodicity parameter α. (In equations 23 and 24, α is similar to standard deviation and coefficient of variation, respectively; although strictly speaking these terms refer to sample statistics.) Parameter estimation for the Weibull and log-normal distributions using the method of moments and maximum likelihood are given by Utsu (1984).

As noted by Matthews et al. (2002), there are important differences in the hazard rate function $h(\tau)$ among these distributions:

$$h(\tau) \equiv f(\tau)/[1 - F(\tau)], \tag{25}$$

where $F(\tau)$ is the cdf corresponding to $f(\tau)$. The hazard rate function can also be thought of as the instantaneous failure rate or the failure probability conditional upon surviving up to point τ. For equivalent distribution parameters, the BPT distribution is characterized by a nearly constant hazard rate at long waiting times, whereas the hazard rate decreases for the log-normal distribution and increases for the Weibull distribution for long waiting times. Matthews et al. (2002) indicates that the gamma distribution for a quasiperiodic process is characterized by an increasing hazard rate (though not as sharply as the Weibull distribution) for $\gamma > 1$ (cf., equation 12). As shown in Fig. 4.9, the gamma distribution discussed in connection with tsunami inter-event times in Section 2 characterized by $\gamma < 1$ is, in contrast, associated with decreasing hazard rate function for short waiting times and nearly constant hazard rate at long waiting times (Corral, 2005c). Thus at $\gamma = 1$, the change in exponent represents a fundamental difference between a quasiperiodic process $\gamma > 1$ and a clustering process $\gamma < 1$ (Utsu, 1984), and a corresponding change from a increasing hazard rate to a decreasing hazard rate, respectively.

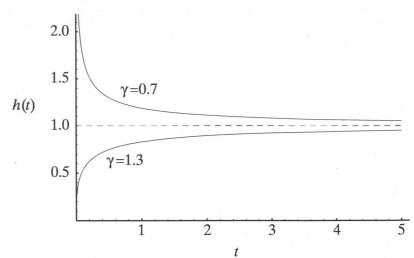

Fig. 4.9 – Hazard rate as a function of non-dimensional time $h(t)$ for two gamma distributions with different shape parameters: increasing hazard rate ($\gamma > 1$) with time associated with quasiperiodic models (e.g., Ogata, 1999); decreasing hazard rate ($\gamma < 1$) associated with temporal clustering of events (e.g., Corral, 2004). Constant hazard rate for exponential distribution (dashed line) shown for comparison.

For multiple time-dependent tsunamigenic earthquakes, a different method of aggregation must be used instead of equation 16. The probability of a tsunami at a particular coastal location generated by the jth characteristic earthquake and within an exposure time T is given by $P(R \geq R_0 | \psi_{i=1,j}, T)$. The parameter set $\psi_{i=1,j}$ includes the parameters that define the time-dependent probability distribution (t_0, υ, α) as well as the source parameters needed to compute tsunami generation. The aggregate tsunami probability of N characteristic earthquakes $P(R \geq R_0 | T)$ is given by Rikitake and Aida (1988):

$$P(R \geq R_0 | T) = 1 - \prod_{j=1}^{N} \left[1 - P(R \geq R_0 | \psi_{i=1,j}, T) \right]. \qquad (26)$$

For subduction zones, McCann et al. (1979) and Nishenko (1991) presented seismic hazard maps based on characteristic earthquakes (described previously) using the log-normal event distribution (equation 23) that could be used in the aggregation equation (26). The seismic hazard was expressed as the probability of occurrence during exposure times of 5, 10, and 20 years for a characteristic earthquake occurring along a segment defined primarily by historic earthquake rupture lengths. This is generally known as the *seismic gap hypothesis*. The Nishenko/ McCann time-dependent probabilities have been evaluated using different statistical tests in a number of papers (Lomnitz and Nava, 1983; Kagan and Jackson, 1991; 1995; Rong et al., 2003). The general conclusion of these papers is that in comparison to actual earthquake occurrence, the time-dependent probability estimates in most cases did not perform as well as the probabilities calculated from a time-independent, Poisson null or reference model. Along fault zones where there is a significant difference between the seismic moment rate ($\dot{M}_s$) determined from an earthquake catalog and that determined from tectonic parameters (equation 21), the deficit may be resolved by future large earthquake(s), depending on the uncertainty in the coupling constant χ. However, according to the statistical tests, no segment is more or less likely to fail based on the time since the last event (Bird and Kagan, 2004).

The other problematic issue related to time-dependent probabilities is the estimation of the three parameters that define the distributions (τ can be uncertain if the preceding earthquake is prehistoric.) The assumption of a generic aperiodicity parameter ($\alpha = 0.21$) for all faults, termed the Nishenko-Buland hypothesis, is shown not to be valid by Savage (1991). To determine the mean inter-event time in the absence of numerous paleoseismic records of past events, a time-predictable model (Shimazaki and Nakata, 1980) is often assumed. This is termed the *direct method* for determining earthquake recurrence in which the mean slip for an earthquake segment and the long-term slip rate can be used to determine mean inter-event time. Each of these parameters, however, has large uncertainties. Savage (1991) demonstrates that the corresponding uncertainty in the probability distribution can be so large such that the probability estimate is not very different from an informationless system. Recent observational studies indicate that neither the time-predictable nor slip-predictable models are valid starting assumptions for determining inter-event time or magnitude, respectively (Murray and Segall, 2002; Weldon et al., 2004; Weldon et al., 2005).

For faults where there are many paleoseismically-identified event horizons, it may be possible to avoid the direct method and time-predictable assumptions indicated above. For tsunamigenic earthquakes generated along subduction zones where there is no near-surface exposure of the fault, evidence of coastal subsidence in the geologic record is used to identify pre-historic earthquakes (see Chapter 3 and Atwater et al., 2004). However, in these cases a purely empirical determination of mean inter-event time and aperiodicity remains problematic. This is the classic case of how probability varies as a function of the number of Bernoulli trials—that is, the use of additional observations to improve a prior probability distribution through Bayes' theorem. For earthquake observations, Savage (1994) indicates that the uncertainty in probability decreases slowly with the number of observed paleoseismic events. Suppose there are m out of n recorded inter-event times less than exposure time T, then the probability density $P(p \mid m,n)$ that the next event will occur within time T since the most recent earthquake follows the distribution:

$$P(p \mid m,n) = \left[\frac{(n+1)!}{m!(n-m)!} \right] p^m (1-p)^{n-m},$$ (27)

with expected value

$$\langle p \rangle = \frac{m+1}{n+2}$$ (28)

and variance

$$\sigma^2 = \frac{\langle p \rangle (1 - \langle p \rangle)}{n+3}. \tag{29}$$

Without using assumptions about the underlying event distribution, determination of empirical probabilities using this method is based on a uniform distribution in the absence of any observations, also known as the principle of indifference. For a summary of arguments and interpretations of this technique, the reader is referred to Howson and Urbach (1993) and Jaynes (2003).

As an example, we consider the uncertainty in the probability of future tsunamigenic earthquakes along the Cascadia subduction zone, using the identified horizons indicated top left corner of Fig. 4.10a . Measuring inter-event times from the center of the age range for each horizon, the mean inter-event time is 517 yrs, with $m=4$ out of $n=6$ inter-event times being less than this value. The expected probability of having another earthquake within 517 years after the last event (1700) is therefore 0.62. However, the 95% confidence interval for this estimate is 0.34-0.87, indicating a high degree of uncertainty. For 10 or fewer paleoseismic horizons, Savage (1994) indicates that the probability cannot be determined better than ±0.2.

Next, we consider uncertainty associated with age-dating of each event horizon that results in a complex probability distribution for each event (Bronk Ramsey, 1998; Ogata, 1999). In light of this and the open intervals on either end of a paleoseismic sequence (i.e., before the earliest identified event and after the last event) (Davis et al., 1989), Ogata (1999) estimates the uncertainty in the time-dependent distribution parameters (α and v). Ogata (1999) assumes a uniform pdf for the possible age of a given paleoseismic horizon, and proposes an inverse probability (Bayesian) method to improve these estimates. Alternatively, Parsons (2005) uses a Monte Carlo method to determine the range of possible values of α and v for a given paleoseismic sequence. An example of the uncertainty in probability for the events described in the above example using the BPT distribution (equation 24) is shown in Fig. 4.10. In this case, the results from Monte Carlo analysis indicate that $F_{N(\tau) \geq 1}(T = 30 yr., t_0 = 1700)$ for a time window starting in 2005 ranges between 0.01 and 0.15 (Fig. 4.10b). The conditional probability for an exposure time window that does not begin with the preceding event is given by (Parsons, 2005)

$$F_{N(\tau) \geq 1}(t_0 < \tau \leq t_0 + T) = \int_{t_0}^{t_0 + T} f(t) dt. \tag{30}$$

Plots such as shown in Fig. 4.10 are useful for determining how well the paleoseismic data constrain a particular probability distribution. In general, such analysis shows that while a quasiperiodic model of earthquake recurrence is an intuitive representation for an individual fault, in practice it is often difficult to determine the parameters of a particular distribution for subduction zone earthquakes with any degree of confidence.

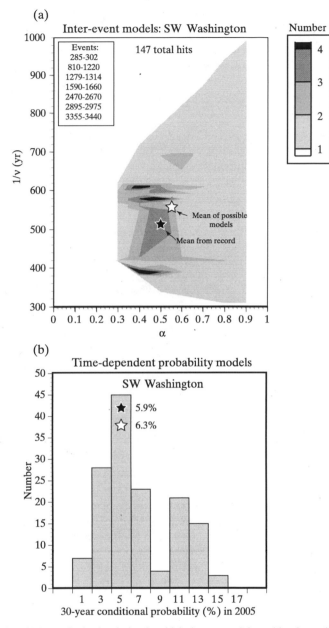

Fig. 4.10 – Results of Monte Carlo simulation in which the successful combinations of parameters ν and α for the BPT distribution fit the observed paleoseismic record of Cascadia subduction zone earthquakes. (a) number of successful hits for the two distribution parameters. (Mean inter-event time represented by $1/v$.) (b) resulting histogram of 30-year conditional probabilities for the distributions that fit the data.

Another particular type of time-dependent probability is clustering of earthquakes in time, caused by foreshocks, aftershocks, and triggering of earthquakes outside the classic aftershock zone through coseismic changes in the static-stress field and through dynamic triggering. In addition to the gamma distribution proposed by Corral (2004) discussed earlier, Kagan and Jackson (2000) model this as a short-term forecasting problem using a negative binomial distribution. Whereas foreshocks are rare, aftershocks are a common phenomena in which the rate of occurrence v is described by Omori's law:

$$v = c/(p+t)^n , \qquad (31)$$

where c, p, and n are constants and t is time since the main shock (Parsons, 2002). In addition, changes in the loading stress on a fault from the occurrence of a nearby large earthquake(s), termed the Coulomb failure stress, can significantly affect the short term probabilities (Parsons, 2004; 2005; 2006). Postseismic viscous relaxation of the lithosphere gradually decreases this probability effect over time (Michael, 2005). For tsunami applications, Geist and Parsons (2005) demonstrate that in some cases, large strike-slip earthquakes may trigger tsunamigenic dip-slip aftershocks within days after the mainshock.

3.3   Source Probabilities: Landslides

The primary source parameter linked to tsunamigensis for landslides is volume ($\psi_{2,j} = V_j$); however, landslide speed is also an important controlling variable linked to tsunami generation efficiency. Below we discuss recent work to define landslide volume distributions and how uncertainty in landslide speed can be included in PTHA calculations. In general, landslide source probabilities are particularly difficult to determine, as a result of the lack of age dates for most of the world's submarine slides. In addition, it is difficult to determine a landslide-equivalent quantity such as long-term fault slip rate to estimate the overall activity of landslide occurrence along a particular margin. Nonetheless, recent research suggests the existence of a power-law size distribution for landslides (equivalent to the standard G-R relationship for earthquakes) in specific areas. Also, some geologic analysis has been conducted that may lead to frequency-size distributions and other parameters needed to determine probabilities for landslide tsunami sources.

ten Brink et al. (2006a) demonstrated that the distribution of submarine landslides north of Puerto Rico follow a power-law relationship with an exponent (β) similar to that found for rock falls onland (Stark and Hovius, 2001; Guzzetti et al., 2002; Dussauge et al., 2003; Malamud et al., 2004). Including small landslide sizes on land, the most descriptive distribution for the entire range of sizes is in fact the double Pareto distribution described by Stark and Hovious (2001). An empirical analysis of the submarine Storegga landslide complex initially suggested that the number-size distribution follows a logarithmic distribution (Issler et al., 2005). However, when the largest distinct landslides are included, ten Brink et al. (2006a) confirms that the distribution follows a power-law relationship. Unlike earthquakes, the value of β varies significantly for landslides (e.g., comparison of the Storegga and Puerto Rico landslide regions: $\beta = 0.44$ and 0.64, respectively), indicating that the failure process significantly affects scaling (Malamud et al., 2004).

In examining the physcial mechanisms that give rise to this power law relationship, Hergarten and Neugebauer (1998) indicate that a state variable in addition to slope gradient is necessary for landslides to follow a power-law size distribution.

This is generally termed a time-weakening effect (Densmore et al., 1998; Hergarten, 2003) and is similar to a quasiperiodic process in that the probability of failure increases with waiting time after the last event at a particular source location. Examples of time-weakening effects include strain softening, creep, and redistribution of pore pressures following, for example, earthquakes (Biscontin et al., 2004; Biscontin and Pestana, 2006). Dugan and Flemings (2000) also describe a process of lateral pressure equilibration over time for submarine fans, with a gradual increase in the likelihood for failure.

The power-law exponent β for landslide volumes can be determined from either cumulative distributions or rank-order distributions. For the latter, observations are sorted from largest (rank $n=1$) to smallest (rank $n=N$). An advantage of rank-ordering is that it tends to avoid bias introduced by correlations of statistical fluctuations that can be present in cumulative distributions (Sornette, 2004). As an example, the rank-order distributions of landslides offshore northern Puerto Rico (ten Brink et al., 2006a) are presented in Fig. 4.11. The power-law portion of the distribution given by

$$V_n \propto n^{-1/\beta}, \ 1 << n \le N \tag{32}$$

where n is the rank order and V_n is the volume associated with the nth rank order (Sornette, 2004). In this case, $\beta = 0.62\pm0.03$ determined from the rank-order distribution is similar to $\beta = 0.64$ determined from the cumulative statistics of landslide volumes presented by ten Brink et al. (2006a). This is smaller than the exponent found for clay-rich, less cohesive landslides, but similar to that found for sub-aerial rockfalls, indicating the effect that mechanical properties has on landslide statistics (Densmore et al., 1998; Dussauge et al., 2003; ten Brink et al., 2006a)

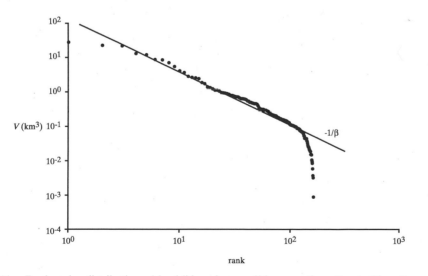

Fig. 4.11 – Rank order distribution of landslide volumes offshore northern Puerto Rico. Power-law exponent ($\beta=0.62\pm0.03$) determined from this distribution is similar to that determined from the cumulative distribution of volumes (ten Brink et al., 2006a).

Even if landslides are assumed to follow a Poisson process, in most locations there is insufficient information with which to determine the rate parameter (v_{2j}, equation 16) or equivalently the activity constant α in the frequency-size power-law relationship (equation 1). There is some indication from modeling and empirical results that the rate of landslide occurrence is non-stationary, depending on long-term global sea level fluctuations (Hutton and Syvitski, 2004). Determination of both α and β rely on the correct identification of individual slide events (in terms of the tsunami they generate) and age dating. In the past, low resolution imaging of landslide features led in some cases to the misidentification of individual landslide events. Large complexes and amphitheaters that spanned a considerable geologic age range in geomorphic development were often considered one event. With the advent of high-resolution sea floor imaging techniques such as multibeam bathymetry, it has been easier to identify individual landslide events (Lee, 2005; ten Brink et al., 2006b).

Age-dating of individual events necessitates identification of geologic horizons that span the age of failure. For example, Normark et al. (2004) recognized a debris flow unit in a piston core that appeared to correspond to a part of a large landslide complex, the Palos Verdes debris avalanche near Los Angeles identified using multibeam imagery. By obtaining C-14 ages of microfauna in units above and below the debris flow in the piston core, they were able to estimate the age of the landslide component as 7500 yr. In some locations, multiple failures at the same location are observed using seismic-reflection profiling. For example, for the Goleta landslide complex in Santa Barbara Channel, Fisher et al. (2005) identify 7 different failure events beneath one of the three major lobes. By following acoustic reflectors to the location of a nearby ODP boring, the authors determined that three of the failures occurred in the last 160 ka. Lee et al. (2004) assembled available age-date information for submarine landslides in southern California and estimated that large failures ($V > 0.5$ km^3) recur with a time interval in the range of 5,000 to 10,000 years.

Seven large pre-Holocene landslides have occurred at the location of the massive Storegga Slide complex off Norway (Solheim et al., 2005a). These were identified and dated using seismic reflection profiling, borings and core samples. The investigations show that at least one large landslide apparently occurs during every 100 ky, following glacial-interglacial cyclicity, with the most recent occurring 8,150 years ago. Accordingly, investigators have been able to conclude that the slope stability environment at the site of the Storegga complex will take another glacial-interglacial cycle to form a situation that could lead to another major tsunami-generating slide (Solheim et al., 2005b). The site is considered to be safe for development of a large gas field for the foreseeable future. Few other locations worldwide have received the kind of attention directed to southern California and Norwegian submarine landslides so recurrence interval information available for these environments is generally lacking. To identify these sediments requires either coring the debris field multiple times in the hope of collecting cores with sediment and the underlying debris, or conducting high-resolution seismic imaging to direct the coring. Even with these techniques, the age of some landslides may not be dated precisely, because of a very low sedimentation rate or lack of datable material.

If the rate of landslide recurrence can be determined through accurate event identification and age dating, it is still necessary for tsunami computations to determine the time history of landslide failure. As indicated by several authors (e.g., Ward, 2001; Todorovska et al., 2002; Trifunac et al., 2002), tsunami generation efficiency is dependent on landslide speed. Landslide dynamics is a complex field of research, and there are different measurements of landslide speed, the most applicable for the outgoing tsunami being the spreading velocity (Todorovska et al., 2002; Trifunac et al., 2002; 2003). In the near-field, other parameters such as slide shape and submergence relative to slide height are important factors in determining runup (Liu et al., 2005). Some source parameters such as runout may scale with volume and may not be mutually independent. In addition, examination of a global tsunami catalog indicates that most tsunamigenic landslides are associated with earthquakes, much like subaerial landslides in seismically active regions (Keefer, 1994). The origin of the triggering mechanism is thought to be direct loading from the earthquake and changes in the pore pressure from successive seismic loading cycles (Biscontin et al., 2004; Biscontin and Pestana, 2006). For non-seismically triggered landslides, very low tidal excursions is a common triggering mechanism in which the slide looses its hydraulic support and does not dewater rapidly. More research is needed to determine the inter-relationship among landslide source parameters that affect tsunami generation and the mobility of submarine landslides in general (Locat and Lee, 2002; Locat et al., 2004). In the meantime, equation 16 can be used to determine landslide probabilities using characteristic parameters (and their attendant uncertainty) for a given region, such that normally distributed uncertainty is included in the $P(R > R_0 | \psi_{i,j})$ term.

4. Uncertainties

The determination of uncertainty in the seismic component of the computational PTHA is highly dependent on the underlying assumptions of earthquake physics. For example, if a characteristic, time-dependent rupture model is assumed, there is little uncertainty in the magnitude of the characteristic event, whereas there is large uncertainty in the parameters that define the earthquake recurrence distribution (Savage, 1991; 1992). Similarly, if a slip-predictable model is assumed, then there is little uncertainty in the magnitude of the next event, but large uncertainty in the waiting time of the next event (Shimazaki and Nakata, 1980). Recent research in earthquake physics has indicated that the earthquake rupture process is sufficiently complex over multiple earthquake cycles and considering multiple faults. For the purposes of probabilistic analysis, tsunamigenesis can be considered a stochastic process, with random variables described in the aggregation equations. In this section we will briefly describe the ergodic assumption used in several aspects of PTHA and indicate how epistemic and aleatory uncertainties are incorporated in PTHA. We will also examine the special case of estimating extreme values for tsunami runup.

4.1. The Ergodic Assumption

In estimating certain source parameters or their uncertainty, it is often necessary to assume that the physical process (landslides, earthquakes, etc.) is ergodic.

Although the ergodic theorem originating from statistical physics is complex and multifaceted (Anosov, 2001), one important application of the theory is that the time average of a process (x) at a particular geographic point is equal to the average at a particular time (t_0) over an ensemble of points x_k (Beichelt and Fatti, 2002):

$$\lim_{T \to \infty} \frac{1}{2T} \int_{-T}^{T} x(t)dt = \lim_{N \to \infty} \frac{1}{N} \sum_{k=1}^{N} x_k(t_0) \qquad (33)$$

For natural hazards, this allows replacing an estimate of the source or hazard statistics at a particular location where there is limited knowledge throughout time with the statistics of an ensemble of known source or hazard variables over a broad region (or even globally).

An example of where the ergodic assumption is used is estimation of corner moment for a particular fault. Because earthquake catalogs are very limited at large magnitudes for a particular fault zone or fault segment throughout time, it is necessary to analyze the statistics of corner moment for a number of faults around the world, as done by Bird and Kagan (2004). However, Bird and Kagan (2004) note that different types of faults (oceanic transform faults, subduction zones) are separated because of differences in tectonic environment (stress, thermal structure, etc.) (see also Pisarenko and Sornette, 2003). Grouping all subduction zones together, Bird and Kagan (2004) were able to estimate a corner moment magnitude of 9.58. Even with the expanded catalog of global subduction zone earthquakes, however, uncertainty is still difficult to estimate. They indicate 95% confidence limits of $9.58^{?}_{-0.46}$ (upper confidence limit not found) using a merged 20th century earthquake catalog and $9.58^{+0.48}_{-0.46}$ using the seismic moment conservation argument (Section 3).

The same type of analysis could be performed to estimate submarine landslide recurrence, though it is unclear what geologic factors are key to defining the ergodic ensemble. If there are multiple dates in a given region where the offshore sediment composition, tectonics, ground shaking, etc. are similar such as southern California, the age dates of multiple landslides within in the region can be grouped together to estimate the recurrence of landslides (Lee et al., 2004). For a global ensemble, however, one has to take into account differences in sediment (clastic vs. carbonate), tectonic movement (passive margin vs, active movement as in Puerto Rico), glacial activity, and peak ground acceleration. Too large of an ensemble can result in ergodicity breaking where the assumption no longer applies. This is discussed in the context of estimating uncertainty in the seismic attenuation relationship for earthquake ground motion studies by Anderson and Brune (1999). Lutz (2004) also indicates that physical systems that follow Lévy Law distributions (discussed below in terms of slip distributions) may also exhibit ergodicity breaking. While rigorously proving the ergodic assumption for complex systems is difficult, the assumption should at minimum be closely examined for specific situations to determine its domain of applicability.

4.2   Incorporating Uncertainties into PTHA

Similar to PSHA, it is convenient to classify PTHA uncertainty as being epistemic or aleatory (Senior Seismic Hazard Analysis Committee (SSHAC), 1997; Toro et al., 1997; National Research Council (NRC), 2000). Epistemic uncertainty is also referred to as knowledge uncertainty that can be reduced by the collection of new data. Such uncertainty is often incorporated into probabilistic calculations through logic trees and computation of multiple hazard curves. From these, a mean or percentile hazard curve is determined, depending on the particular application (Abrahamson and Bommer, 2005; McGuire et al., 2005). An example of epistemic uncertainty is the mode of earthquake occurrence along the Aleutian-Alaskan subduction zone described by Wesson et al. (1999). In this case, two different segmentation models are considered involving both a G-R distribution of magnitudes and a characteristic mode of rupture at the site of the 1964 Alaska earthquake. These are associated with two different branches of a logic tree, each with specified weights, that are incorporated into computation of ground acceleration in the case of Wesson et al. (1999), and for tsunamis originating from the Aleutian-Alaskan subduction zone in the case of the Seaside, Oregon pilot study (Tsunami Pilot Study Working Group, 2006).

Aleatory uncertainty relates to the natural or stochastic uncertainty inherent in the physical system. Aleatory uncertainty is incorporated into probabilistic analysis through direct integration in the $P(R > R_0 | \psi_{ij})$ term (equation 16). Recall that ψ_{ij} is the parameter(s) that is directly linked to the source frequency-size distribution. Other source parameters that result in a range of runup for a particular value of ψ_{ij} can be considered as random variables y_k, $k=1,2,3...$ For a single random variable y where $R(y)$ is normally distributed and has expected value (μ_y), and variance (σ_y^2),

$$P(R(y) > R_0 | y) = \int_{R_0}^{\infty} \frac{1}{\sigma_y \sqrt{2\pi}} \exp\left(-\frac{1}{2}\frac{(R(y)-\mu_y)^2}{\sigma_y^2}\right) dR . \quad (34)$$

Multiple independent random variables (y_k), each resulting in a normally distributed $R(y_k)$, will combine such that the aggregate pdf $p(R|\psi_{i,j})$ is normally distributed according to the central limit theorem and under the conditions for which that theorem applies.

An example of how aleatory uncertainty is incorporated into PTHA calculations is described by Mofjeld et al. (2007) for the case of tsunami arrival time relative to tidal stage. This is a non-trivial situation, since the tsunami wave train can extend over at least an entire tidal cycle. In this application, the pdf for the maximum tsunami wave height (η) is a normal distribution that depends on the initial amplitude of the incident wave (A) and a set of tidal constants (κ):

$$p(\eta | A, \kappa) = \frac{1}{\sigma_\eta \sqrt{2\pi}} \exp\left(-\frac{1}{2}\frac{(\eta-\mu_\eta)^2}{\sigma_\eta^2}\right), \quad (35)$$

where

$$\mu_\eta = A + \text{MSL} + C(\text{MHHW} - \text{MSL})\exp\left[-\alpha(A/\sigma_0)^\beta\right], \tag{36}$$

and

$$\sigma_\eta = \sigma_0 - C'\sigma_0 \exp\left[-\alpha'(A/\sigma_0)^\beta\right], \tag{37}$$

MSL and MHHW are mean sea level and mean higher high water, respectively, σ_0^2 is the variance of the predicted tide, and $C, C', \alpha, \alpha', \beta, \beta' \in \kappa$, are all site specific constants (Mofjeld et al., 2007). The cumulative probability that the wave height will exceed a particular value (η_0) over the entire duration of the tsunami is given by

$$P(\eta > \eta_0 \mid A, \kappa) = \int_{\eta_0}^\infty \frac{1}{\sigma_\eta \sqrt{2\pi}} \exp\left(-\frac{1}{2}\frac{(\eta - \mu_\eta)^2}{\sigma_\eta^2}\right) d\eta \tag{38}$$

Another example of aleatory uncertainty is variation in slip distribution patterns for an earthquake of a given magnitude. In this case, slip on a fault plane inclined below the surface of the earth $u(\xi, y)$ (cf., Geist and Dmowska, 1999) is determined by specifying the slip spectrum in the radial wavenumber domain that can be directly linked to seismic observations (Hanks, 1979; Andrews, 1980; Frankel, 1991; Herrero and Bernard, 1994; Tsai, 1997; Somerville et al., 1999; Hisada, 2000; 2001; Mai and Beroza, 2002). Phase space is then randomized to yield a suite of $u(\xi, y)$ distributions that correspond to the same scalar seismic moment and seismic source spectrum. The suite of $u(\xi, y)$ can be used to estimate the uncertainty in local tsunami amplitude for a given seismic moment (Geist, 2002; 2005) as shown in Fig. 4.12 for a M~9 earthquake in the U.S. Pacific Northwest. In this case, the hazard variable is peak nearshore amplitude (A) rather than runup. The distribution of tsunami incident amplitudes $P(A > A_0 \mid u(\xi, y))$ arising from different slip distribution patterns is approximately a normal distribution, although there can be site-specific deviations owing to propagation path effects (Geist, 2005; Geist and Parsons, 2006).

For descriptive purposes, we can assign a random variable y_φ that defines the phase space for the slip distribution $u(\xi, y)$. In most formulations of stochastic slip distributions (including Geist, 2002; 2005), y_φ is assumed to be normally distributed. Recent research, however, indicates that the standard self-affine slip model described above may not accurately encompass possible large fluctuations in slip (Lavallée and Archuleta, 2003; Lavallée et al., 2006). The modification proposed by Lavallée et al. (2006) is to use the more general Lévy law distributions to describe y_φ. (Lavallée et al., 2006, describes 1D and 2D stochastic modeling in more detail than the simple parameterization by y_φ described here.) The Lévy law describes a class of stable distributions characterized by several parameters including the Lévy exponent $\mu : 0 < \mu < 2$ (Sornette, 2004; Lavallée et al., 2006). Specific

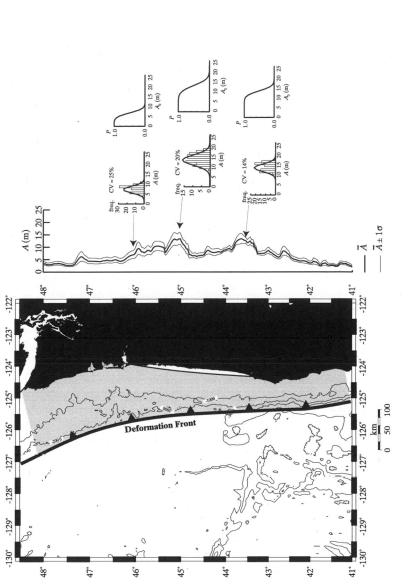

Fig. 4.12 – Example of how aleatory uncertainty in the earthquake source is integrated into the P term in equation 16. In this case for the U.S. Pacific Northwest, uncertainty of peak near shore tsunami amplitude (A) from variations of slip distribution patterns for a local $M\sim9$ earthquake (shaded region) is determined from numerical tsunami propagation models. Left column of plots indicate histograms of A values and a normal distribution approximation (solid line). CV: coefficient of variation. Right column of plots indicates the function $P(A > A_0 \mid u(\xi, y))$ using the cumulative normal distribution function (equation 34) for different threshold values, A_0. Details given by Geist and Parsons (2006).

analytic cases of symmetric Lévy distributions include the Cauchy-Lorentz distribution ($\mu = 1$) and the Lévy distribution ($\mu = 1/2$). Each of these are heavy tail pdf's in comparison to the normal distribution ($\mu = 2$). Slip distributions determined from tsunami data collected during the 2004 Indian Ocean event (Fujii and Satake, 2007) suggest that this was a case of high slip fluctuation (Geist et al., 2007) and indicating that Lévy law distributions may need to be considered when estimating uncertainty of tsunami runup from distributed slip. Note that for the two examples of aleatory uncertainty described above, $p(\eta \mid A, \kappa)$ and $p(A \mid u(\xi, y))$ are obviously not independent, in that the distribution of combined tsunami and tidal wave heights are functionally dependent on A (Mofjeld et al., 2007).

As demonstrated by Geist and Parsons (2006), Monte Carlo methods are particularly useful for incorporating multiple sources of uncertainty. In this case, slip distribution, hypocenter (within the bounds specified by the seismogenic zone), and magnitude (taken from a sample of a seismic moment-balanced G-R distribution) all were randomized to yield a tsunami hazard curve at a particular coastal site. For the case of Acapulco, the computational curve compares well with the empirical curve determined from the available tsunami catalog data, when the limitations of the catalog are taken into account (Fig. 4.2).

4.3   Probabilities of Extreme Tsunamis

In some applications, it is necessary to determine the severity of an extreme event located at the tail of the tsunami frequency-size distribution. Issues such as knowing total insurance risk, economic impact, and long-term hazard for critical facilities necessitate estimating extreme values in natural hazards. However, there is a high level of uncertainty associated with these estimates, stemming from a lack of historical data and knowledge of what the maximum possible event may be.

Empirical Approach
A standard approach to such empirical problems is to use an asymptotic model of extremes, in which the cumulative distribution of the largest events approaches one of a member of the class of distributions called the General Extreme Value Distributions (GEVD) (Castillo et al., 2005). The basic members of the GEVD class include the Gumbel distribution (tail tapering off faster than a power law), Fréchet (tail tapering off as a power law), and Weibull (tail with finite right end point) (Kotz and Nadarajah, 2000; Sornette, 2004). Of the three, the Gumbel distribution (Gumbel, 1958) has been most often applied to natural hazards, since it has an infinite right tail but corresponds in general to a tapered power-law distribution. In particular, Hogben (1990) uses GEVD to determine extreme storm wave heights and Kulikov et al. (2005) presents a probabilistic analysis of tsunami hazards in Peru and northern Chile based on the Gumbel distribution.

The main problem with using the asymptotic models of extremes is that they do not make use of the full dataset or underlying pdf (Knopoff and Kagan, 1977; Sornette, 2004). Instead, rank-ordering statistics of an observed dataset (as in the Peaks-Over-Threshold modeling in extreme value theory) can be used to define the power-law exponent at extreme values (Knopoff and Kagan, 1977; Sornette et al., 1996). In addition, the generalized Pareto distribution (GPD) as described for earthquakes, floods, and extreme tides (Pugh, 1987; Stedinger et al., 1993; Pisarenko and Sornette, 2003) can be used to establish the extreme value distribution. The GPD is as follows:

$$G_{\xi,s}(x) = \begin{cases} 1-(1+\xi x/s)^{-1/\xi}, & \text{for } \xi \neq 0 \\ 1-\exp(-x/s), & \text{for } \xi = 0 \end{cases}, \tag{39}$$

where ξ is a shape parameter that is dependent on the pdf of the observations and s is a scale parameter which depends on the threshold of observations. Kijko (2004) also provides techniques to estimate the hard maximum limit R_x, whether or not the underlying distribution is known.

Computational Approach

For the computational approach, one can follow the general method described in Section 4, paying careful attention to the underlying source-size distribution function and the attendant uncertainties. Of course, the analysis of extreme tsunamis greatly depends on the assumed mode of occurrence for the sources. For example, under the seismic gap hypothesis, the maximum size of a tsunami from a given fault segment is limited by the characteristic earthquake magnitude. This greatly reduces the uncertainty associated with determination of the maximum tsunamis, as does the assumption of a universal value for the aperiodicity parameter (α). However, a more realistic view of earthquake occurrence based on a soft-taper in the power-law distribution of sizes necessitates a more difficult analysis of uncertainties as described below.

In establishing the distribution of extreme source sizes, one can again use the GPD as described above or establish the corner seismic moment (earthquakes) or corner volume (landslides) using a modified power-law distribution. For the latter, parameter estimation of both M_c and β from global seismicity use the maximum likelihood method (e.g., Bird and Kagan, 2004). In this case the log-likelihood function (ℓ) for the tapered Pareto distribution (i.e., modified G-R distribution) is given by

$$\ell = \sum_{i=1}^{n} \log\left(\frac{\beta}{M_i}+\frac{1}{M_c}\right) + \beta n \log(M_t) - \beta \sum_{i=1}^{n} \log(M_i) + \frac{nM_t}{M_c} - \frac{1}{M_c}\sum_{i=1}^{n} M_i , \tag{40}$$

where M_i are $i=1,2,3...n$ iid observations of seismic moment (cf., equations 17 and 18). In addition, Kagan and Schoenberg (2001) describe several other approaches to estimate the corner moment, including average likelihood estimates and a method of moments estimator. Pisarenko and Sornette (2003) argue that the shape parameter ξ of the GPD in equation 39 (estimate constrained by the Kolmogorov distance) is more tightly constrained than the corner moment.

Care must also be taken in how epistemic uncertainty in determining tsunami amplitudes from a particular extreme source is propagated through the calculations. One particular concern is determining which quantile curve is most representative of the hazard where there is significant epistemic uncertainty. This has been discussed recently for PSHA studies in comparing mean, median and fractile hazard curves by Abrahamson and Bommer (2005) derived from different branches of an epistemic logic tree. They indicate that the mean hazard curve diverges significantly from the median hazard curve to higher hazard levels for very low hazard rates (10^{-7}-10^{-8} yr^{-1}). However McGuire et al. (2005) suggests that this is more of a problem with how the logic tree is formulated and weights assigned (cf., Senior Seismic Hazard Analysis Committee (SSHAC), 1997) and that the mean hazard curve is more in line with current understanding of the meaning of probability. Ideally, one would look at the entire probability distribution at a particular hazard level to determine how much information is contained in the probability estimate (Savage, 1991).

5.   Summary

In this chapter, we have outlined empirical and computational approaches to estimate tsunami probability. Both approaches are centered around determining the frequency-size and inter-event time distributions. In the case of the empirical approach, these distributions apply to the tsunamis themselves, whereas in the computational approach, the distributions apply to the tsunami sources. Previous studies (e.g., Burroughs and Tebbens, 2005) have indicated that tsunami sizes are nominally distributed according to a power-law that is modified to include a taper or roll-off at large sizes. It is in fact uncertain what the limiting size of a tsunami is, since there are very low probability geologic processes that can theoretically produce much larger tsunamis than discussed here (e.g., catastrophic volcanic flank failures and asteroid impacts). In addition, there is likely a hydrodynamic limit for tsunami size, influenced by nonlinear shoaling effects and offshore wave breaking (Korycansky and Lynett, 2005). This chapter shows that for the inter-event time distribution, global tsunamis are characterized by a deviation from the exponential distribution associated with a Poisson process, especially at short inter-event times.

For the computational approach, the size distribution of earthquakes is better constrained than that for landslides and tsunamis themselves. Substantial progress has been made in determining the power-law exponent (β) and corner moment for the earthquake size distribution. Moreover, recent studies (ten Brink et al., 2006a) have indicated that offshore landslides may also follow a modified power-law distribution of sizes, with β dependent on mechanical properties of the slide material. For inter-event time distributions, both time-independent (Poissonian) and time-dependent distributions are described for earthquakes. Aside from universal distributions recently proposed (Corral, 2004; Davidsen and Goltz, 2004; Molchan, 2005), fault-specific parameters for time-dependent probability models are in practice difficult to determine. This is a result of questionable underlying assumptions of earthquake rupture (e.g., characteristic, time-predictable) and the necessary number and precision of historic and pre-historic observations. The situation for offshore landslides is even more uncertain, as there is a general lack of age dates for individual events. While we await more data to constrain the inter-event distributions, the null hypothesis of an exponential distribution is likely our best model.

Reducing sources of epistemic uncertainty is key to developing more accurate tsunami probability estimates for the computational approach in the future. While most tsunami source parameters are approximately normally distributed, there are some parameters such as landslide speed where there is insufficient data to make this assumption. In addition, other parameters such as slip distribution appear to exhibit stronger fluctuations than expected from a normal distribution (Lavallée et al., 2006). Further research is needed to better quantify uncertainty for source parameters that scale with source size. For any tsunami probability study, it is important to understand how uncertainty affects the probability estimate, either through a determination of information content (Savage, 1991; 1992) or through Monte Carlo techniques (Parsons, 2004; 2005).

Acknowledgements

The authors gratefully acknowledge constructive reviews of this chapter by David Divoky, Hal Mofjeld, Hong Kie Thio, and Annabel Kelly. James Savage also provided helpful comments related to empirical probability.

References

Abe, K., 1995. Estimate of tsunami run-up heights from earthquake magnitudes. In: Y. Tsuchiya and N. Shuto (Editors), Tsunami: Progress in Prediction, Disaster Prevention and Warning, v. 4. Kluwer Academic Publishers, Dordrecht, pp. 21–35.

Abrahamson, N. A. and Bommer, J. J., 2005. Probability and uncertainty in seismic hazard analysis. Earthquake Spectra, 21, 603–607.

Altmann, E. G. and Kantz, H., 2005. Recurrence time analysis, long-term correlations, and extreme events. Physical Review E, 71, doi: 10.1103/PhysRevE.71.056106.

Anderson, J. G. and Brune, J. N., 1999. Probabilistic seismic hazard analysis without the ergodic assumption. Seismological Research Letters, 70, 19–28.

Andrews, D. J., 1980. A stochastic fault model 1. Static case. Journal of Geophysical Research, 85, 3867–3877.

Anosov, D. V., 2001. Ergodic Theory. In: M. Hazewinkel (Editor), Encyclopaedia of Mathematics. Springer-Verlag, Berlin, http://eom.springer.de/e/e036150.htm.

Atwater, B. F., Tuttle, M. P., Schweig, E. S., Rubin, C. M., Yamaguchi, D. K. and Hemphill-Haley, E., 2004. Earthquake recurrence inferred from paleoseismology. Developments in Quaternary Science, 1, 331–350.

Bak, P., Christensen, K., Danon, L. and Scanlon, T., 2002. Unified scaling law for earthquakes. Physical Review Letters, 88, doi: 10.1103/PhysRevLett.88.178501.

Beichelt, F. E. and Fatti, L. P., 2002. Stochastic processes and their applications. Taylor & Francis, London, 326 pp.

Ben-Menahem, A. and Rosenman, M., 1972. Amplitude patterns of tsunami waves from submarine earthquakes. Journal of Geophysical Research, 77, 3097–3128.

Ben-Zion, Y., 1996. Stress, slip, and earthquakes in models of complex single-fault systems incorporating brittle and creep deformations. Journal of Geophysical Research, 101, 5677–5706.

Ben-Zion, Y. and Rice, J. R., 1997. Dynamic simulations of slip on a smooth fault in an elastic solid. Journal of Geophysical Research, 102, 17,771–17,784.

Bird, P., 2003. An updated digital model of plate boundaries. Geochemistry, Geophysics, Geosystems, 4, doi:10.1029/2001GC000252.

Bird, P. and Kagan, Y. Y., 2004. Plate-tectonic analysis of shallow seismicity: apparent boundary width, beta-value, corner magnitude, coupled lithosphere thickness, and coupling in 7 tectonic settings. Bulletin of the Seismological Society of America, 94, 2380–2399.

Biscontin, G. and Pestana, J. M., 2006. Factors affecting seismic response of submarine slopes. Natural Hazards and Earth System Sciences, 6, 97–107.

Biscontin, G., Pestana, J. M. and Nadim, F., 2004. Seismic triggering of submarine slides in soft cohesive soil deposits. Marine Geology, 203, 341–354.

Bronk Ramsey, C., 1998. Probability and dating. Radiocarbon, 40, 461–474.

Burroughs, S. M. and Tebbens, S. F., 2001. Upper-truncated power laws in natural systems. Pure and Applied Geophysics, 158, 741–757.

Burroughs, S. M. and Tebbens, S. F., 2005. Power law scaling and probabilistic forecasting of tsunami runup heights. Pure and Applied Geophysics, 162, 331–342.

Caldwell, P. and Merrifield, M. A., 2006. Joint archive for sea level data report: March 2006. JIMAR Contribution No. 06-360, Data Report No. 19.

Castillo, E., Hadi, A. S., Balakrishnan, N. and Sarabia, J. M., 2005. Extreme Value and Related Models with Applications in Engineering and Science. Wiley Series in Probability and Statistics. John Wiley & Sons, Inc., Hoboken, NJ, 362 pp.

Cochard, A. and Madariaga, R., 1996. Complexity of seismicity due to highly rate dependent friction. Journal of Geophysical Research, 101, 25,321–25,336.

Conover, W. J., 1971. Practical Nonparametric Statistics. John Wiley and Sons Inc., New York, 462 pp.

Cornell, C. A., 1968. Engineering seismic risk analysis. Bulletin of the Seismological Society of America, 58, 1583–1606.

Corral, A., 2004. Long-term clustering, scaling, and universality in the temporal occurrence of earthquakes. Physical Review Letters, 92, doi: 10.1103/PhysRevLett.92.108501.

Corral, A., 2005a. Mixing of rescaled data and Bayesian inference for earthquake recurrence times. Nonlinear Processes in Geophysics, 12, 89–100.

Corral, A., 2005b. Statistical features of earthquake temporal occurrence. In: B. K. Chakrabarti and P. Bhattacharyya (Editors), International Workshop on Models of Earthquakes: Physics Approaches, Calcutta, India.

Corral, A., 2005c. Time-decreasing hazard and increasing time until the next earthquake. Physical Review E, 71, doi: 10.1103/PhysRevE.71.017101.

Daley, D. J. and Vere-Jones, D., 2003. An Introduction to the Theory of Point Processes, 1. Springer, New York, 469 pp.

Daley, D. J. and Vere-Jones, D., 2004. Scoring probability forecasts for point processes: The entropy score and information gain. Journal of Applied Probability, 41A, 297–312.

Davidsen, J. and Goltz, C., 2004. Are seismic waiting time distributions universal? Geophysical Research Letters, 31, doi: 10.1029/2004GL020892.

Davis, P. M., Jackson, D. D. and Kagan, Y. Y., 1989. The longer it has been since the last earthquake, the longer the expected time till the next? Bulletin of the Seismological Society of America, 79, 1439–1456.

Densmore, A. L., Ellis, M. A. and Anderson, R. S., 1998. Landsliding and the evolution of normal-fault-bounded mountains. Journal of Geophysical Research, 103, 15,203–15,219.

Dugan, B. and Flemings, P. B., 2000. Overpressure and fluid flow in the New Jersey continental slope: Implications for slope failure and cold seeps. Science, 289, 288–291.

Dussauge, C., Grasso, J. R. and Helmstetter, A., 2003. Statistical analysis of rockfall volume distributions: Implications for rockfall dynamics. Journal of Geophysical Research, 108, doi:10.1029/2001JB000650.

Finkelstein, J. M. and Schafer, R. E., 1971. Improved goodness-of-fit tests. Biometrika, 58, 641–645.

Fisher, M. A., Normark, W. R., Greene, H. G., Lee, H. J. and Sliter, R. W., 2005. Geology and tsunami-genic potential of submarine landslides in Santa Barbara Channel, southern California. Marine Geology, 224, 1–22.

Flinn, E. A., Engdahl, E. R. and Hill, A. R., 1974. Seismic and geographical regionalization. Bulletin of the Seismological Society of America, 64, 771–992.

Frankel, A. D., 1991. High-frequency spectral falloff of earthquakes, fractal dimension of complex rupture, b value, and the scaling of strength on faults. Journal of Geophysical Research, 96, 6291–6302.

Frankel, A. D., Petersen, M. D., Mueller, C. S., Haller, K. M., Wheeler, R. L., Leyendecker, E. V., Wesson, R. L., Harmsen, S. C., Cramer, C. H., Perkins, D. M. and Rukstales, K. S., 2002. Documentation for the 2002 Update of the National Seismic Hazard Maps. Open-File Report 02-420, U.S. Geological Survey.

Fujii, Y. and Satake, K., 2007. Tsunami source of the 2004 Sumatra-Andaman earthquake inferred from tide gauge and satellite data. Bulletin of the Seismological Society of America, 97, S192-S207.

Geist, E. L., 1999. Local tsunamis and earthquake source parameters. Advances in Geophysics, 39, 117–209.

Geist, E. L., 2002. Complex earthquake rupture and local tsunamis. Journal of Geophysical Research, 107, doi:10.1029/2000JB000139.

Geist, E. L., 2005. Local Tsunami Hazards in the Pacific Northwest from Cascadia Subduction Zone Earthquakes. U.S. Geological Survey Professional Paper 1661B.

Geist, E. L. and Dmowska, R., 1999. Local tsunamis and distributed slip at the source. Pure and Applied Geophysics, 154, 485–512.

Geist, E. L. and Parsons, T., 2005. Triggering of Tsunamigenic Aftershocks from Large Strike-Slip Earthquakes: Analysis of the November 2000 New Ireland Earthquake Sequence. Geochemistry, Geophysics, Geosystems, 6(Q10005), doi:10.1029/2005GC000935.

Geist, E. L. and Parsons, T., 2006. Probabilistic analysis of tsunami hazards. Natural Hazards, 37, 277–314.

Geist, E. L., and Parsons, T., 2008. Distribution of tsunami inter-event times, Geophys. Res. Lett., 35, L02612, doi:101029/102007GL032690.

Geist, E. L., Titov, V. V., Arcas, D., Pollitz, F. F. and Bilek, S. L., 2007. Implications of the December 26, 2004 Sumatra-Andaman earthquake on tsunami forecast and assessment models for great subduction zone earthquakes. Bulletin of the Seismological Society of America, 97, S249-S270.

Gisler, G., Weaver, R. and Gittings, M. L., 2006. SAGE calculations of the tsunami threat from La Palma. Science of Tsunami Hazards, 24, 288–301.

Gumbel, E. J., 1958. Statistics of Extremes. Columbia University Press, New York, 375 pp.

Guzzetti, F., Malamud, B. D., Turcotte, D. L. and Reichenbach, P., 2002. Power-law correlations of landslide areas in central Italy. Earth and Planetary Science Letters, 195, 169–183.

Hanks, T. C., 1979. b values and ω^7 seismic source models: Implications for tectonic stress variations along active crustal fault zones and the estimation of high-frequency strong ground motion. Journal of Geophysical Research, 84, 2235–2242.

Hanks, T. C. and Kanamori, H., 1979. A moment magnitude scale. Journal of Geophysical Research, 84, 2348–2350.

Harte, D. and Vere-Jones, D., 2005. The entropy score and its uses in earthquake forecasting. Pure and Applied Geophysics, 162, 1229–1253.

Heinrich, P., Piatanesi, A. and Hébert, H., 2001. Numerical modelling of tsunami generation and propagation from submarine slumps: the 1998 Papua New Guinea event. Geophysical Journal International, 145, 97–111.

Hergarten, S., 2003. Landslides, sandpiles, and self-organized criticality. Natural Hazards and Earth System Sciences, 3, 505–514.

Hergarten, S. and Neugebauer, H. J., 1998. Self-organized criticality in a landslide model. Geophysical Research Letters, 25, 801–804.

Herrero, A. and Bernard, P., 1994. A kinematic self-similar rupture process for earthquakes. Bulletin of the Seismological Society of America, 84, 1216–1228.

Hino, R., Tanioka, Y., Kanazawa, T., Sakai, S., Nishino, M. and Suyehiro, K., 2001. Micro-tsunami from a local interplate earthquake detected by cabled offshore tsunami observation in northeastern Japan. Geophysical Research Letters, 28, 3533–3536.

Hirata, K., Takahashi, H., Geist, E. L., Satake, K., Tanioka, Y., Sugioka, H. and Mikada, H., 2003. Source depth dependence of micro-tsunamis recorded with ocean-bottom pressure gauges; the January 28, 2000 Mw 6.8 earthquake off Nemuro Peninsula, Japan. Earth and Planetary Science Letters, 208(3–4), 305–318.

Hisada, Y., 2000. A theoretical omega-square model considering the spatial variation in slip and rupture velocity. Bulletin of the Seismological Society of America, 90, 387–400.

Hisada, Y., 2001. A theoretical omega-square model considering the spatial variation in slip and rupture velocity. Part 2: Case for a two-dimensional source model. Bulletin of the Seismological Society of America, 91, 651–666.

Hogben, N., 1990. Long term wave statistics. In: B. Le Méhauté and D. M. Hanes (Editors), Ocean Engineering Science. The Sea, v. 9. John Wiley & Sones, New York, pp. 293–333.

Houston, J. R. and Garcia, A. W., 1978. Type 16 Flood Insurance Study: Tsunami Predictions for the West Coast of the Continental United States. Technical Report H-78-26, U.S. Army Engineer Waterways Experiment Station, Vicksburg, MS.

Howell, B. F., 1985. On the effect of too small a data base on earthquake frequency diagrams. Bulletin of the Seismological Society of America, 75, 1205–1207.

Howson, C. and Urbach, P., 1993. Scientific Reasoning: The Bayesian Approach. Open Court Publishing Company, Chicago, Illinois, 470 pp.

Hutton, E. W. H. and Syvitski, J. P. M., 2004. Advances in the numerical modeling of sediment failure during the development of a continental margin. Marine Geology, 203, 367–380.

Issler, D., De Blasio, F. V., Elverhøi, A., Bryn, P. and Lien, R., 2005. Scaling behaviour of clay-rich submarine debris flows. Marine and Petroleum Geology, 22, 187–194.

Jaynes, E. T., 2003. Probability Theory: The Logic of Science. Cambridge University Press, Cambridge, UK, 727 pp.

Jiang, L. and Leblond, P. H., 1994. Three-dimensional modeling of tsunami generation due to a submarine mudslide. Journal of Physical Oceanography, 24, 559–572.

Kagan, Y. Y., 1993. Statistics of characteristic earthquakes. Bulletin of the Seismological Society of America, 83, 7–24.

Kagan, Y. Y., 1997. Seismic moment-frequency relation for shallow earthquakes: Regional comparison. Journal of Geophysical Research, 102, 2835–2852.

Kagan, Y. Y., 1999. Universality of the seismic-moment-frequency relation. Pure and Applied Geophysics, 155, 537–573.

Kagan, Y. Y., 2002a. Seismic moment distribution revisited: I. Statistical Results. Geophysical Journal International, 148, 520–541.

Kagan, Y. Y., 2002b. Seismic moment distribution revisited: II. Moment conservation principle. Geophysical Journal International, 149, 731–754.

Kagan, Y. Y. and Jackson, D. D., 1991. Seismic gap hypothesis: Ten years after. Journal of Geophysical Research, 96, 21,419–21,431.

Kagan, Y. Y. and Jackson, D. D., 1995. New seismic gap hypothesis: Five years after. Journal of Geophysical Research, 100, 3943–3959.

Kagan, Y. Y. and Jackson, D. D., 2000. Probabilistic forecasting of earthquakes. Geophysical Journal International, 143, 438–453.

Kagan, Y. Y. and Schoenberg, F., 2001. Estimation of the upper cutoff parameter for the tapered Pareto distribution. Journal of Applied Probability, 38A, 158–175.

Kajiura, K., 1981. Tsunami energy in relation to parameters of the earthquake fault model. Bulletin of the Earthquake Research Institute, 56, 415–440.

Keefer, D. K., 1994. The importance of earthquake-induced landslides to long-term slope erosion and slope-failure hazards in seismically active regions. Geomorphology, 10, 265–284.

Kempthorne, O. and Folks, L., 1971. Probability, Statistics, and Data Analysis. Iowa State University Press, Ames, Iowa, 555 pp.

Kijko, A., 2004. Estimation of the maximum earthquake magnitude, m_{max}. Pure and Applied Geophysics, 161, 1655–1681.

Knopoff, L. and Kagan, Y. Y., 1977. Analysis of the theory of extremes as applied to earthquake problems. Journal of Geophysical Research, 82, 5647–5657.

Korycansky, D. G. and Lynett, P. J., 2005. Offshore breaking of impact tsunami: The Van Dorn effect revisited. Geophysical Research Letters, 32, doi:10.1029/2004GL021918.

Kotz, S. and Nadarajah, S., 2000. Extreme Value Distributions: Theory and Applications. Imperial College Press, London, 187 pp.

Kreemer, C., Holt, W. E. and Haines, A. J., 2003. An integrated global model of present-day plate motions and plate boundary deformation. Geophysical Journal International, 154, 8–34.

Kulikov, E. A., Rabinovich, A. B. and Thomson, R., 2005. Estimation of tsunami risk for the coasts of Peru and northern Chile. Natural Hazards, 35, 185–209.

Lavallée, D. and Archuleta, R. J., 2003. Stochastic modeling of slip spatial complexities for the 1979 Imperial Valley, California, earthquake. Geophysical Research Letters, 30, doi:10.1029 /2002GL015839.

Lavallée, D., Liu, P. and Archuleta, R. J., 2006. Stochastic model of heterogeneity in earthquake slip spatial distributions. Geophysical Journal International, 165, 622–640.

Lee, H., Normark, W. R., Fisher, M. A., Greene, H. G., Edwards, B. D. and Locat, J., 2004. Timing and extent of submarine landslides in southern California, Offshore Technology Conference, Houston, Texas, U.S.A., OTC Paper Number 16744.

Lee, H. J., 2005. Undersea landslides: extent and significance in the Pacific Ocean, an update. Natural Hazards and Earth System Sciences, 5, 877–892.

Lin, I. and Tung, C. C., 1982. A preliminary investigation of tsunami hazard. Bulletin of the Seismological Society of America, 72, 2323–2337.

Liu, P. L. F., Wu, T.-R., Raichlen, F., Synolakis, C. E. and Borrero, J. C., 2005. Runup and rundown generated by three-dimensional sliding masses. Journal of Fluid Mechanics, 536, 107–144.

Locat, J. and Lee, H. J., 2002. Submarine landslides: advances and challenges. Canadian Geotechnical Journal, 39, 193–212.

Locat, J., Lee, H. J., Locat, P. and Imran, J., 2004. Numerical analysis of the mobility of the Palos Verdes debris avalanche, California, and its implication for the generation of tsunmis. Marine Geology, 203, 269–280.

Lomnitz, C. and Nava, F. A., 1983. The predictive value of seismic gaps. Bulletin of the Seismological Society of America, 73, 1815–1824.

Lu, C. and Vere-Jones, D., 2001. Statistical analysis of synthetic earthquake catalogs generated by models with various levels of fault zone disorder. Journal of Geophysical Research, 106, 11,115–11,125.

Lutz, E., 2004. Power-law tail distributions and nonergodicity. Physical Review Letters, 93, doi:10.1103/PhysRevLett.93.190602.

Mai, P. M. and Beroza, G. C., 2002. A spatial random field model to characterize complexity in earthquake slip. Journal of Geophysical Research, 107, doi:10.1029/2001JB000588.

Malamud, B. D., Turcotte, D. L., Guzzetti, F. and Reichenbach, P., 2004. Landslide inventories and their statistical properties. Earth Surface Processes and Landforms, 29, 687-7111.

Marsan, D., 2005. The role of small earthquakes in redistributing crustal elastic stress. Geophysical Journal International, 163, 141–151.

Matthews, M. V., Ellsworth, W. L. and Reasenberg, P. A., 2002. A Brownian model for recurrent earthquakes. Bulletin of the Seismological Society of America, 92, 2233–2250.

McCann, W. R., Nishenko, S. P., Sykes, L. R. and Krause, J., 1979. Seismic gaps and plate tectonics: seismic potential for major boundaries. Pure and Applied Geophysics, 117, 1082–1147.

McGuire, R. K., Cornell, C. A. and Toro, G. R., 2005. The case for using mean seismic hazard. Earthquake Spectra, 21, 879–886.

Michael, A. J., 2005. Viscoelasticity, postseismic slip, fault interactions, and the recurrence of large earthquakes. Bulletin of the Seismological Society of America, 95, 1594–1603.

Mofjeld, H. O., González, F. I., Titov, V. V., Venturato, A. J. and Newman, A. V., 2007. Effects of tides on maximum tsunami wave heights: Probability distributions. Journal of Atmospheric and Oceanic Technology, 24, 117–123.

Molchan, G., 2005. Interevent time distribution in seismicity: A theoretical approach. Pure and Applied Geophysics, 162, 1135–1150.

Murray, J. and Segall, P., 2002. Testing time-predictable earthquake recurrence by direct measure ment of strain accumulation and release. Nature, 419, 287–291.

National Research Council (NRC), 2000. Risk Analysis and Uncertainty in Flood Damage Reduction Studies. National Academy Press, Washington, DC, 202 pp.

Nishenko, S. P., 1991. Circum-Pacific seismic potential: 1989–1999. Pure and Applied Geophysics, 135, 169–259.

Nishenko, S. P. and Buland, R., 1987. A generic recurrence interval distribution for earthquake forecasting. Bulletin of the Seismological Society of America, 77, 1382–1399.

Normark, W. R., McGann, M. and Sliter, R. W., 2004. Age of Palos Verdes submarine debris avalanche, southern California. Marine Geology, 203, 247–259.

Ogata, Y., 1999. Estimating the hazard of rupture using uncertain occurrence times of paleoearthquakes. Journal of Geophysical Research, 104, 17,995–18,014.

Okada, Y., 1985. Surface deformation due to shear and tensile faults in a half-space. Bulletin of the Seismological Society of America, 75, 1135–1154.

Okal, E. A., 1988. Seismic parameters controlling far-field tsunami amplitudes: A review. Natural Hazards, 1, 67–96.

Parsons, T., 2002. Global Omori law decay of triggered earthquakes: Large aftershocks outside the classical aftershock zone. Journal of Geophysical Research, 107, 2199, doi:10.1029/2001JB000646.

Parsons, T., 2004. Recalculated probability of M ≥ 7 earthquakes beneath the Sea of Marmara, Turkey. Journal of Geophysical Research, 109, B05304.

Parsons, T., 2005. Significance of stress transfer in time-dependent earthquake probability calculations. Journal of Geophysical Research, 110, doi:10.1029/2004JB003190.

Parsons, T., 2006. M ≥ 7.0 earthquake recurrence on the San Andreas fault from a stress renewal model. Journal of Geophysical Research, 111, doi:10.1029/2006JB004415.

Parsons, T. 2008. Monte Carlo method for determining earthquake recurrence parameters from short paleoseismic catalogs: Example calculations for California, J. Geophys. Res., 112, doi:10.1029/2007JB004998.

Pelayo, A. M. and Wiens, D. A., 1992. Tsunami earthquakes: Slow thrust-faulting events in the accretionary wedge. Journal of Geophysical Research, 97, 15,321–15,337.

Perfettini, H., Schmittbuhl, J. and Vilotte, J. P., 2001. Slip correlations on a creeping fault. Geophysical Research Letters, 28, 2137–2140.

Pisarenko, V. F. and Sornette, D., 2003. Characterization of the frequency of extreme earthquake events by the Generalized Pareto Distribution. Pure and Applied Geophysics, 160, 2343–2364.

Pugh, D. T., 1987. Tides, Surges and Mean Sea-Level. John Wiley & Sons, Chichester, UK, 472 pp.

Rikitake, T., 1999. Probability of a great earthquake to recur in the Tokai district, Japan: reevaluation based on newly-developed paleoseismology, plate tectonics, tsunami study, micro-seismicity and geodetic measurements. Earth Planets Space, 51, 147–157.

Rikitake, T. and Aida, I., 1988. Tsunami hazard probability in Japan. Bulletin of the Seismological Society of America, 78, 1268–1278.

Rong, Y., Jackson, D. D. and Kagan, Y. Y., 2003. Seismic gaps and earthquakes. Journal of Geophysical Research, 108, ESE 6-1 - 6-14.

Satake, K., 2002. Tsunamis. In: W. H. K. Lee, H. Kanamori, P. C. Jennings and C. Kisslinger (Editors), International Handbook of Earthquake and Engineering Seismology, Part A. Academic Press, San Diego, pp. 437–451.

Savage, J. C., 1991. Criticism of some forecasts of the National Earthquake Prediction Evaluation Council. Bulletin of the Seismological Society of America, 81, 862–881.

Savage, J. C., 1992. The uncertainty in earthquake conditional probabilities. Geophysical Research Letters, 19, 709–712.

Savage, J. C., 1994. Empirical earthquake probabilities from observed recurrence intervals. Bulletin of the Seismological Society of America, 84, 219–221.

Senior Seismic Hazard Analysis Committee (SSHAC), 1997. Recommendations for Probabilistic Seismic Hazard Analysis: Guidance on Uncertainty and Use of Experts. Main Report NUREG/CR-6372 UCRL-ID-122160 Vol. 1, U.S. Nuclear Regulatory Commission.

Shaw, B. E., 2004. Dynamic heterogeneities versus fixed heterogeneities in earthquake models. Geophysical Journal International, 156, 275–286.

Shaw, B. E. and Rice, J. R., 2000. Existence of continuum complexity in the elastodynamics of repeated fault ruptures. Journal of Geophysical Research, 105, 23,791–23,810.

Shimazaki, K. and Nakata, T., 1980. Time-predictable recurrence model for large earthquakes. Geophysical Research Letters, 7, 279–282.

Shuto, N., 1991. Numerical simulation of tsunamis—Its present and near future. Natural Hazards, 4, 171–191.

Solheim, A., Berg, K., Forsber, C. F. and Bryn, P., 2005a. The Storegga Slide complex: repetitive large scale sliding with similar cause and development. Marine and Petroleum Geology, 22, 97–107.

Solheim, A., Bryn, P., Sejrup, H. P., Mienert, J. and Berg, K., 2005b. Ormen Lange—an intergrated study for the safe development of a deep-water gas field within the Storegga Slide Complex, NE Atlantic continental margin: executive summary. Marine and Petroleum Geology, 22, 1–9.

Somerville, P., Irikura, K., Graves, R., Sawada, S., Wald, D., Abrahamson, N. A., Iwasaki, Y., Kagawa, T., Smith, N. and Kowada, A., 1999. Characterizing crustal earthquake slip models for the prediction of strong ground motion. Seismological Research Letters, 70, 59–80.

Sornette, D., 2004. Critical Phenomena in Natural Sciences. Springer Series in Synergetics. Springer-Verlag, Berlin, 528 pp.

Sornette, D., Knopoff, L., Kagan, Y. Y. and Vanneste, C., 1996. Rank-ordering statistic of extreme events: Application to the distribution of large earthquakes. Journal of Geophysical Research, 101, 13,883–13,893.

Stark, C. P. and Hovius, N., 2001. The characterization of landslide size distributions. Geophysical Research Letters, 28, 1091–1094.

Stedinger, J. R., Vogel, R. M. and Foufoula-Georgiou, E., 1993. Frequency analysis of extreme events. In: D. R. Maidment (Editor), Handbook of Hydrology. McGraw Hill, Inc., New York, pp. 18.1–18.66.

Stein, S. and Newman, A., 2004. Characteristic and uncharacteristic earthquakes as possible artifacts: Applications to the New Madrid and Wabash seismic zones. Seismological Research Letters, 75, 173–187.

Stephens, M. A., 1974. EDF statistics for goodness of fit and some comparisons. Journal of the American Statistical Association, 69, 730–737.

Tanioka, Y. and Satake, K., 1996. Tsunami generation by horizontal displacement of ocean bottom. Geophysical Research Letters, 23, 861–865.

ten Brink, U. S., Geist, E. L. and Andrews, B. D., 2006a. Size distribution of submarine landslides and its implication to tsunami hazard in Puerto Rico. Geophysical Research Letters, 33, doi:10.1029/2006GL026125.

ten Brink, U. S., Geist, E. L., Lynett, P. J. and Andrews, B. D., 2006b. Submarine slides north of Puerto Rico and their tsunami potential. In: A. Mercado and P. Liu (Editors), Caribbean Tsunami Hazard. World Scientific, Singapore, pp. 67–90.

Thio, H. K., Ichinose, G., Polet, J. and Somerville, P., 2007. Application of probabilistic tsunami hazard analysis to ports and harbors, Ports 2007. American Society of Civil Engineers, San Diego.

Todorovska, M. I., Hayir, A. and Trifunac, M. D., 2002. A note on tsunami amplitudes above submarine slides and slumps. Soil Dynamics and Earthquake Engineering, 22, 129–141.

Toro, G. R., Abrahamson, N. A. and Schneider, J. F., 1997. Model of strong ground motions from earthquakes in central and eastern North America: Best estimates and uncertainties. Seismological Research Letters, 68, 41–57.

Trifunac, M. D., Hayir, A. and Todorovska, M. I., 2002. A note on the effects of nonuniform spreading velocity of submarine slumps and slides on the near-field tsunami amplitudes. Soil Dynamics and Earthquake Engineering, 22, 167–180.

Trifunac, M. D., Hayir, A. and Todorovska, M. I., 2003. A note on tsunami caused by submarine slides and slumps spreading in one dimension with nonuniform displacement amplitudes. Soil Dynamics and Earthquake Engineering, 23, 223–234.

Tsai, C. P., 1997. Slip, stress drop and ground motion of earthquakes: A view from the perspective of fractional Brownian motion. Pure and Applied Geophysics, 149, 689–706.

Tsunami Pilot Study Working Group, 2006. Seaside, Oregon Tsunami Pilot Study—Modernization of FEMA Flood Hazard Maps. Open-File Report 2006-1234, U. S. Geological Survey, http://pubs.usgs .gov/of/2006/1234/.

Utsu, T., 1984. Estimation of parameters for recurrence models of earthquakes. Bulletin of the Earthquake Research Institute, 59, 53–66.

Varnes, D. J., 1978. Slope movement types and processes. In: R. L. Schuster and R. J. Krizek (Editors), Landslides: Analysis and Control. National Academy of Sciences, Washington, D.C., pp. 11–33.

Vere-Jones, D. and Ogata, Y., 2003. Statistical principles for seismologists. In: W. H. K. Lee, H. Kanamori, P. C. Jennings and C. Kisslinger (Editors), International Handbook of Earthquake & Engineering Seismology, Part B. Academic Press, pp. 1573–1586.

Vere-Jones, D., Robinson, R. and Yang, W., 2001. Remarks on the accelerated moment release model: problems of model formulation, simulation and estimation. Geophysical Journal International, 144, 517–531.

Ward, S. N., 1994. A multidisciplinary approach to seismic hazard in southern California. Bulletin of the Seismological Society of America, 84, 1293–1309.

Ward, S. N., 2001. Landslide tsunami. Journal of Geophysical Research, 106, 11,201–11,215.

Ward, S. N., 2002. Tsunamis. In: R. A. Meyers (Editor), The Encyclopedia of Physical Science and Technology, v. 17. Academic Press, pp. 175–191.

Weldon, R. J., Fumal, T. E., Biasi, G. P. and Scharer, K. M., 2005. Past and future earthquakes on the San Andreas fault. Science, 308, 966–967.

Weldon, R. J., Scharer, K. M., Fumal, T. E. and Biasi, G. P., 2004. Wrightwood and the earthquake cylce: What a long recurrence record tells us about how faults work. GSA Today, 14, doi:10.1130/1052-5173(2004)014.

Wesnousky, S. G., 1994. The Gutenberg-Richter or characteristic earthquake distribution, which is it? Bulletin of the Seismological Society of America, 84, 1940–1959.

Wesson, R. L., Frankel, A. D., Mueller, C. S. and Harmsen, S. C., 1999. Probabilistic Seismic Hazard Maps of Alaska. Open-File Report 99-36, U.S. Geological Survey.

Working Group on California Earthquake Probabilities, 1995. Seismic hazards in southern California: Probable earthquakes, 1994–2024. Bulletin of the Seismological Society of America, 85, 379–439.

Yoshioka, S., Hashimoto, M. and Hirahara, K., 1989. Displacement fields due to the 1946 Nankaido earthquake in a laterally inhomogeneous structure with the subducting Philippine Sea plate—a three-dimensional finite element approach. Tectonophysics, 159, 121–136.

Zeng, J. L., Heaton, T. H. and DiCaprio, C., 2005. The effect of slip variability on earthquake slip-length scaling. Geophysical Journal International, 162, 841–849.

Chapter 5. EXCITATION OF TSUNAMIS BY EARTHQUAKES

EMILE A. OKAL

Northwestern University

Contents

"Comparisons suggest that each tsunami is unique and unpredictable in some respects and predictable in others."

[Fraser et al., 1959]

1. Introduction: Earthquakes

The subject of this chapter is to describe both qualitatively and quantitatively the excitation of tsunamis by earthquake sources. Indeed, most large tsunamis, and especially those destructive in the far field, are generated by earthquakes, and thus an adequate understanding of the excitation of tsunamis by earthquake sources is critical to advances in the fields of tsunami mitigation and warning. In this context, our emphasis will be the nature of the physical process of excitation of the tsunami, and the identification of general properties of earthquake-generated tsunamis which are robust invariants with respect to variations in the details of the seismic source.

1.1. The earthquake process

An earthquake consists of the release of stresses constantly accumulated under the tectonic forces applied within the Earth, when they become greater than the strength of the relevant rocks. To be considered an earthquake, this process must

The Sea, Volume 15, edited by Eddie N. Bernard and Allan R. Robinson
ISBN 978–0–674–03173–9 ©2009 by the President and Fellows of Harvard College

occur in the brittle regime, in which the failing material rebounds to its undeformed state, thereby releasing the elastic energy accumulated during the interseismic deformation, in the form of seismic waves. The failure takes place
through the development of a cut of finite dimensions, or *dislocation,* in an otherwise elastic medium (the country rock). It is along this usually planar structure that
the seismic slip takes place (Fig. 5-01).

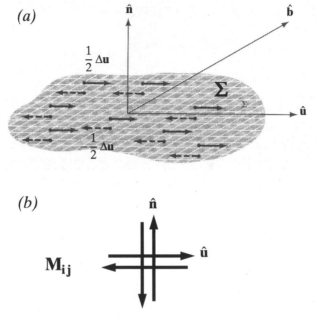

Fig. 5.01 – *(a)* Seismic slip along a fault (represented as the hatched area). The total slip Δu is the
difference between the displacements of the top wall (shown as solid arrows) and of the bottom wall
(dashed arrows). This dislocation, of area Σ, cut into an elastic medium, is equivalent to the *double-
couple* shown in *(b)*, with components in the directions of the slip vector ($\hat{\mathbf{u}}$) and of the normal to the
fault plane ($\hat{\mathbf{n}}$), acting in an unruptured medium. After *Aki and Richards* [1980].

A fundamental characteristic of the earthquake source is that the extent of the
dislocation is bounded in space, which amounts to saying that even though slip
takes place along the fault during the earthquake, the elastic material containing
the source keeps its overall cohesion. In particular, this means that it is possible to
find a continuous path through unruptured material from one side of the dislocation to the other. For example and in lay terms, following a large earthquake on
the San Andreas Fault, one can always drive far enough along the fault to eventually cross it *on an uncracked road,* beyond the ending point of rupture, and return
to the other side uneventfully. This is of course in contrast to a *landslide* or *slump*
during which the mass set in motion is permanently and totally detached from its
foot wall, with no possible connection between the two sides through unperturbed
material. This apparently trivial remark is at the core of the physical differences
between earthquake and landslide sources, and has far-reaching consequences on
their respective properties, including their potential for tsunami generation.

Because the earthquake fault is finite, and the elastic material extends continuously beyond the end of the fault, the seismic slip Δu is limited by the length L of

the fault. Within a geometrical factor of order one, their ratio is governed by the maximum strain allowed by the strength of the rock, ε_{max}. For a given class of rocks, this dimensionless parameter is a fundamental invariant of the seismic source, on the order of 2×10^{-4}. As a result, the earthquake process can be described as moving considerable amounts of matter over relatively short distances: the maximum documented value of seismic slip Δu is about 15 to 20 m, during gigantic earthquakes such as Chile (1960) or Sumatra (2004), even though these events had fault ruptures extending 800 and 1200 km, respectively [Plafker and Savage, 1970; Banerjee et al., 2007]. Furthermore, because the walls of the fault move a relatively short distance Δu during a seismic event, the duration of this motion, called *rise time* τ, is always very short, usually a few seconds, and at most a few tens of seconds for the largest events. The total source duration of an earthquake is then controlled by the *propagation* of the rupture along the potentially very long fault zone, which can last at most $t_R = 10$ mn for gigantic events such as Sumatra (2004).

This is in contrast to landslides, which displace relatively small amounts of material (rarely exceeding 10 km in size), over distances comparable to, or greater than, their own dimensions, and which can last as much as several hours if they develop into underwater turbidity currents [Schwarz, 1982].

1.2. Representation of the earthquake source; the seismic moment M_0.

During an earthquake, the Earth as a global system is deformed, but remains isolated in space, and thus receives no linear or angular momentum. This remark explains the unsuccessful attempts, prior to the 1950s, to represent the earthquake source with a single force or a couple of forces exerting a torque on the planet. Rather, the theoretical description of a seismic source uses a system of forces called a double-couple, a concept introduced originally by *Vvedenskaya* [1956], formalized by *Knopoff and Gilbert* [1959], and first applied to the quantification of an earthquake by *Aki* [1966]. In this framework, the earthquake is modeled using a second-order symmetric moment tensor **M,** whose only non-zero components are those indexed along the direction of seismic slip, **û,** and the direction normal to the fault plane, **n̂**:

$$\mathbf{M} = M_0 \, (\hat{\mathbf{u}}\hat{\mathbf{n}} + \hat{\mathbf{n}}\hat{\mathbf{u}}) \tag{1}$$

where the scalar M_0, known as the *seismic moment* of the earthquake, takes the value

$$M_0 = \mu \, \Sigma \, \Delta u \tag{2}$$

μ being the rigidity of the source rocks, Σ the area of faulting, and Δu the amplitude of the seismic slip[1]. The representation theorem forming the basis of quantita-

---

[1] Here, the notation, **a b** means the column vector **a** multiplied on its *right* by the row vector **b**^{T}, which is the transposed of the column vector **b**. The result is thus a 3×3 tensor, as opposed to the case of a *left* multiplication, which would yield the 1×1 classic scalar product $\mathbf{b}^T\mathbf{a} = \mathbf{b} \cdot \mathbf{a}.$

tive seismology then states that the response of the Earth to the slip on the bounded dislocation cut into the elastic medium (as shown in Fig. 5-01a) is equivalent to that of the unfractured medium to the system of forces represented by (1) and shown on Fig. 5-01b. The latter can be used as a forcing term in the equations of dynamics to model the excitation and the propagation of all seismic waves, including, as we shall see, tsunamis.

The derivation of the representation theorem is given in all advanced seismology textbooks [*e.g., Aki and Richards,* 1980], and will not be repeated. It is important however to note that it involves integration by parts, using the familiar argument of a surface of integration sufficiently distant from the source that the fully integrated term vanishes. As noted by *Dahlen* [1993], this requires that the unperturbed medium be continuous around the source, and thus cannot be applied to the case of a landslide or slump, in which it leads to additional source terms in the representation theorem. The latter express mathematically the fundamental difference in material properties between the two kinds of sources.

It can be verified from (1) that the description of a seismic moment tensor **M** requires 4 real numbers. The latter can be expressed as the three components of the seismic slip vector [$\mu \Sigma \Delta \mathbf{u}$], plus the orientation in space of the fault normal $\hat{\mathbf{n}}$, or alternatively, the scalar moment M_0 and three angles orienting the process in space. The latter representation emphasizes the common physical nature of earthquake sources, individual events differing only through an Euler solid rotation of the source, with the three familiar angles (strike ϕ and dip δ of fault, and rake λ of the slip on the fault plane) merely describing the orientation of the process with respect to a practical system of coordinates (vertical, North, East) at the epicenter. This remark explains that the seismic moment M_0 is the primary descriptor of the "size" of an earthquake, and consequently of the excitation and generation of all waves, seismic and others, including tsunamis, whose amplitudes are only moderately affected by the other parameters, namely the three angles describing the geometry of rupture. Conversely, M_0 (and more generally the full seismic moment tensor **M**) can be inverted from extensive datasets of seismic waves, a procedure pioneered in the pre-digital era by *Gilbert and Dziewonski* [1975] and now used routinely by automated algorithms operating in quasi-real time at various data centers worldwide [*e.g., Dziewonski et al.,* 1981].

The seismic moment M_0 defined by (2) is often expressed as a so-called moment magnitude, $M_w = (\log_{10} M_0 - 16.1)/1.5$, introduced by *Kanamori* [1977] to coincide with conventional magnitudes in a range of earthquake sizes where the latter have started to be affected by source finiteness, without however reaching full saturation [*Geller,* 1976].

1.3. Scaling laws

The size of an earthquake will grow with the extensive parameters Σ and Δu in (2). As discussed above, earthquake rupture occurs when the strain accumulated in the source region reaches its critical value ε_{max}, leading to an expected proportionality between Δu and the fault length L. Furthermore, it is generally observed (for example from the study of aftershock distributions) that the aspect ratio of the fault zone remains contained, and thus Σ grows like L^2, resulting in an overall proportionality between M_0 and L^3, which is generally confirmed on observational datasets [*Kanamori and Anderson*, 1975; *Geller*, 1976; *Wells and Coppersmith*, 1994], the ratio M_0/L^3 (related within a geometrical constant to the stress drop $\Delta\sigma$ of the event) being about 50 bars. The following practical formula was proposed by *Geller* [1976]

$$M_0 = 1.45 \times 10^{20} \cdot L^3 \cdot \Delta\sigma \tag{3}$$

where M_0 is in dyn*cm, $\Delta\sigma$ in bars, and L in km. Similarly, both rise time τ and rupture time t_R are expected to scale with Δu or L, and hence as $M_0^{1/3}$. Such scaling laws further simplify the interpretation of seismic sources by implying that estimates of parameters such as slip Δu and fault length L (which do not lend themselves to an easy measurement, especially under the operational constraints of real-time tsunami warning) can be obtained from the knowledge of M_0, now generally available within one hour of the occurrence of major earthquakes worldwide.

There exist cases, however, where scaling laws will be violated, under conditions of anomalous material properties in the source region, leading to low (or conversely, high) stress drops, slow rupture velocities, or unusual fault zone geometries [*Romanowicz*, 1992; *Polet and Kanamori*, 2000]. Such earthquakes can have irregular tsunami excitation, and consequently constitute a challenge for the warning community. A particular class of them, known as "tsunami earthquakes", will be discussed in Section 6.

2. Generation of a tsunami by an earthquake: The classical model

2.1. A simple model

The framework of the generation of a tsunami by an earthquake source is presented in very simple terms on Fig. 5-02. In this model, a fraction of the ocean floor is uplifted during the seismic event, displacing a volume of seawater. If this deformation occurs sufficiently fast, and neglecting the compressibility of seawater, one predicts an immediate and identical hump on the ocean surface (Frame *(b)*). However, because the ocean is fluid, this situation is of course unstable, and the hump flows sideways *(c)*: this flow constitutes the tsunami *(d)*. Eventually, the entire mass of water in the hump returns to the ocean surface, expected to be unchanged in the final state of equilibrium *(e)*, the cross-section S of the hump being in all cases negligible compared with the total area of the ocean basin. Thus, the energy of the tsunami, expressed in detail in the Appendix, consists of the excess of potential energy of the displaced water in the hump on Frame *(b)* over its value at equilibrium in Frame *(e)*. This excess, and hence the occurrence of the tsunami, results from the transfer of the mass of water from the seabed to the hump occurring out of equilibrium, *i.e.*, as a thermodynamically irreversible process. This will be the case only if the deformation takes place rapidly, in practice faster than it takes the hump to propagate away from the epicentral area as a tsunami wave. The critical parameter in this respect is the ratio of the velocities characteristic of the propagation of seismic rupture along the fault, V_R, and of the tsunami wave, C. The former are typically around 3.5 km/s, and even for anomalously slow earthquakes, at least 1 km/s; the latter are typically 200 m/s, and never exceed 340 m/s, even in the deepest oceanic trenches. The bottom line is that seismic rupture is always hypersonic with respect to the tsunami, which justifies the arguably naive model of Fig. 5-02b and its use as an initial condition in most numerical simulations. By contrast, the slow tectonic deformations accumulated during the interseismic cycle involve velocities at most in the range of a few cm/yr, leaving ample time for the ocean surface to equilibrate, and thus qualifying as reversible processes generating no energy glut and no tsunami.

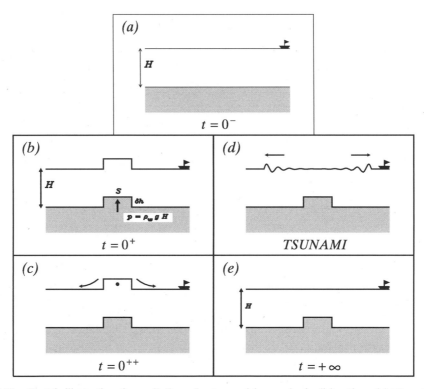

Fig. 5.02 – Sketch illustrating the excitation of a tsunami by a seismic dislocation. *(a)* Unperturbed oceanic column before the earthquake. *(b)* During the earthquake, a hump is generated on the ocean floor, resulting in an immediate and identical deformation of the ocean surface. *(c)* Because the ocean is fluid, the hump is unstable and starts flowing sideways, creating the tsunami wave *(d)*. *(e)* Eventually, at equilibrium, the ocean surface has returned to its pre-earthquake level, due to the much larger area of the ocean basin, as compared to the cross section of the hump, *S*. After *Okal* [2003].

2.2. From earthquake seismic moment to tsunami initial conditions

In the general framework of Fig. 5-02, it becomes possible to derive the excitation of a tsunami by computing the static field of vertical deformation u_z (x, y) of the ocean floor, and interpreting it as the field of initial displacements of the ocean surface, η (x, y).

At any point of an infinite homogeneous elastic medium, the three-dimensional static deformation incurred from a point source double-couple is obtained from the classical Somigliana tensor [*e.g., Aki and Richards,* 1980], and for an infinite half-space, there exists a formulation of the static displacement on the boundary of the medium [*Steketee,* 1958]. In the case of slip on a rectangular fault of finite size buried in a homogeneous half-space, the static solution on the surface can be worked out analytically, with independent but equivalent expressions published by *Mansinha and Smylie* [1971] and *Okada* [1985].

Such computations require the knowledge of the geometry of rupture, and of the fault parameters L, W, and Δu. For deferred simulations of the tsunamis of well-studied events, these are often available from investigations in source tomography [*e.g., Ishii et al.,* 2005] or from the combined use of the spatial extent of aftershocks and of the scalar moment M_0. In real time, under operational conditions, scaling laws such as (3) can be used to obtain estimates of L, W and Δu from an inverted value of M_0, and an estimated focal geometry can be derived from our understanding of regional plate tectonics.

As an example, Fig. 5-03 illustrates the computation of the vertical component of the static displacement, u_z , in the scenario of a repeat of the great 1833 earthquake in Southern Sumatra [*Synolakis and Okal,* 2006], based on parameters obtained from the modeling of emerged coral structures by *Zachariasen et al.* [1999]. A numerical simulation can then be launched by using the field $u_z(x, y)$ as the initial condition for $\eta(x, y)$ at time $t = 0^+$, for all points not located on initially dry land. This is complemented by setting to zero the initial field of the depth-averaged horizontal velocities of the fluid: $v_x(x, y) = v_y(x, y) = 0$. Fig. 5-04 presents the results of the simulation, using the MOST finite difference code, introduced by *Titov and Synolakis* [1998], which solves the non-linear equations of hydrodynamics in the shallow-water approximation using the method of splitting integration steps. The final product is a map of the maximum amplitude η expected during a possible repeat of the 1833 event. We stress that it is valid only on the high seas, and does not include the localized effects of shoaling and run-up at the beaches. Note the strong directivity of the wave in the far field, at right angles from the coast of Sumatra, discussed more in detail in Section 5, as well as the effect of shallow bathymetric features acting as optical lenses to focus the energy of the tsunami in the far field [*Woods and Okal,* 1987; *Satake,* 1988].

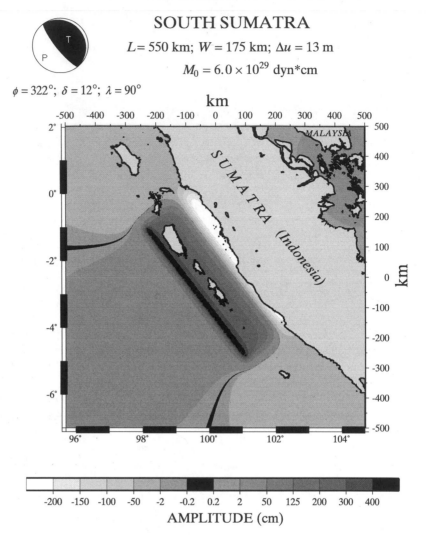

SOUTH SUMATRA

$L = 550$ km; $W = 175$ km; $\Delta u = 13$ m

$M_0 = 6.0 \times 10^{29}$ dyn*cm

$\phi = 322°$; $\delta = 12°$; $\lambda = 90°$

AMPLITUDE (cm)

Fig. 5.03 – Field of vertical static displacement u_z computed using *Mansinha and Smylie*'s [1971] algorithm in the scenario of the possible repeat of the 1833 mega-thrust earthquake along the coast of Southern Sumatra. The shallow-dipping thrust fault mechanism is sketched at upper left, and relevant fault parameters listed at the top. The two-dimensional field of u_z values is then interpreted as the initial condition $\eta(t = 0^+)$ of the hydrodynamic simulation shown on Fig. 5-04.

1833 SIMULATION (Excluding Runup)

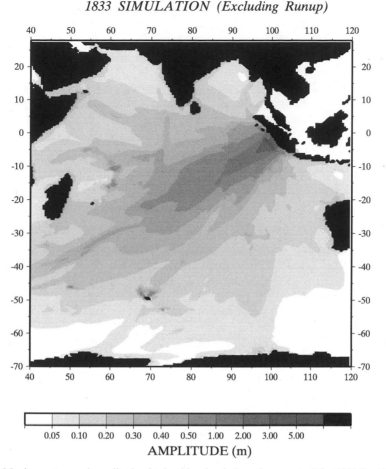

AMPLITUDE (m)

Fig. 5.04 – Maximum tsunami amplitude obtained by simulation of a repeat of the 1833 South Sumatra earthquake, based on the source model of *Zachariasen et al.* [1999]. Note the strong lobe of directivity at right angles to the strike of the fault, and the local focusing effect of shallow bathymetric features such as the Southwest Indian Ocean Ridge or the Kerguelen Plateau. After *Synolakis and Okal* [2006].

An interesting aspect of the field of static displacements presented on Fig. 5-03 is its asymmetry: along a cross-section perpendicular to the fault strike (Fig. 5-05), it features a trough near the coast, and a large hump oceanwards. This results from a combination of factors, namely the thrusting mechanism of the earthquake, the shallow dip angle δ of the fault (taken here as 12°), and the position of the centroid of rupture seawards of the shoreline. These parameters are remarkably robust for the overwhelming majority of the large interplate thrust earthquakes generating most damaging tsunamis, as they simply express the geometry of rupture in the context of the unifying theory of plate tectonics. Since the static displacement field of the rupture constitutes the initial condition of the hydrodynamic problem, *Tadepalli and Synolakis* [1994, 1996] predicted theoretically that the initial waveform of the tsunami should be negative (*i.e.,* a down-draw, or so-called "leading depression") in the direction of the beach, and positive (*i.e.,* a "leading elevation") out at

sea. Thus, the tsunami phenomenon on a nearby coastline should start with a recess of the sea. This was first confirmed during the 1995 tsunami in Manzanillo,
Mexico [*Borrero et al.,* 1997], and as summarized by *Synolakis et al.* [2007], widely
reported during the Sumatra-Andaman tsunami at shorelines located downdip of
the epicenter (Aceh Province, Thailand, Malaysia). By contrast, at distant shores
located across the ocean basin from the epicenter (*e.g.,* Sri Lanka), the first oscillation of the tsunami was reported as a leading elevation, as was the first impulse
along the JASON satellite profile [*Scharroo et al.,* 2005].

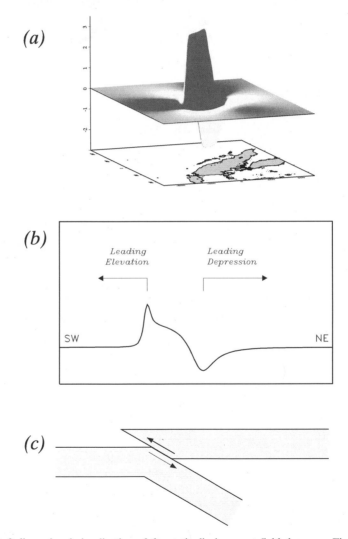

Fig. 5.05 – *(a):* 3-dimensional visualization of the static displacement field shown on Fig. 5-03, looking
from East-Southeast. Note the landwards trough generating a leading-depression wave towards Sumatra, and the oceanward hump. *(b):* Cross-section of the static displacement u_z shown on Fig. 5-03, in the
azimuth N52°E, *i.e.,* perpendicular to the fault strike. Note the asymmetry of the profile, generating a
leading depression towards the shore and a leading elevation towards the high sea. *(c):* Cross-section of
co-seismic displacements at a subduction zone, in the plate tectonic framework. Figures not drawn to
scale.

The onset of a tsunami as a leading depression wave in the most common geometries involving shorelines at subduction zones provides a natural warning, as the phenomenon starts with its most benign part, namely the recess of the sea, which can serve as a harbinger of the catastrophic elevation to follow. Indeed, this is recognized as a form of warning warranting self-evacuation, in the ancestral traditions of indigenous populations regularly exposed to such cataclysms. However, it should be borne in mind that not all tsunamis are generated by interplate thrust events, and that subduction zones can host large intraplate earthquakes, most often featuring normal faulting (a geometry exactly opposite that of thrust faulting), as the result of the internal rupture of the subducting lithospheric plate. Among them, the events of 1977 in Sumbawa, Indonesia [*Kato and Tsuji*, 1995], and especially 1933 in Sanriku, Japan [*Kanamori*, 1971] generated catastrophic tsunamis, the latter being the second most lethal tsunami in the 20th century, with upwards of 3000 deaths. Because of their normal faulting mechanism, such events should generate tsunamis approaching nearby coastlines as leading elevations, and thus providing no advance notice of inundation. However, observations during the 1977 Sumbawa earthquake reveal a consistent pattern of leading depressions as reported by witnesses [*Kato and Tsuji*, 1995]. This is most readily explained by these authors' simulations, which predict a leading elevation, but of small enough amplitude as to go unnoticed by the local population. This interpretation is upheld by *Abe*'s [1978] detailed study of tidal gauge records of the 1933 Sanriku earthquake, where he shows in particular that the exact waveform of the leading elevation can be very sensitive to the dip angle of the fault. A similar pattern was also observed during the 2004 Sumatra tsunami, this time in the farfield as the earthquake featured a classic underthrusting mechanism.

2.3.   The possible contribution of a horizontal displacement

In the context described above, a legitimate question can be the possible role, in the generation of the tsunami, of the horizontal motion of the boundaries of the ocean basin during the earthquake. Such motion could be important in the presence of a sloping seafloor above the fault zone. This problem was examined both theoretically and in the laboratory by *Iwasaki* [1982] and *Synolakis* [1986], who concluded that this contribution will generate a sustainable tsunami only for steep slopes, in practice greater than 1/3, a value much larger than typically found on the ocean floor, where even bathymetry described as "steep" rarely exceeds a slope of 10°. However, the limiting case of the horizontal displacement of a vertical wall (which for example would describe a strike-slip fault cutting through an island shore) could lead to a significant local wave at the beach, given a Froude number (*i.e.*, the ratio of fluid particle velocity to the phase velocity of the wave) approaching 1. This could explain the large tsunami observed in the near field at Mindoro, Philippines on 14 November 1994 following a strike-slip earthquake whose fault ran across the island shoreline [*Imamura et al.*, 1995].

In the wake of the 1994 Java "tsunami earthquake", *Tanioka and Satake* [1996a] examined the possible contribution to the vertical displacement of water of the horizontal motion of a sloping seafloor, which they suggested could be relatively steep in that area. They concluded that the contribution can affect slightly the shape and amplitude of the tsunami, but not to the extent of changing its order of magnitude.

3.   Tsunamis as free oscillations of the Earth: the normal mode formalism

The simulation algorithm described above suffers from a number of intrinsic limitations due to the simplifying assumptions used in its two steps. On the one hand, the computation of the static deformation ignores the presence of the water layer; on the other hand, the numerical simulation of the tsunami in the water considers a perfectly rigid (hence undeformable) bottom as a boundary condition. It is clear that these two assumptions are mutually exclusive, and thus their separate use in subsequent parts of the modeling is less than satisfactory, although it can be lent some support from *Tinti and Armigliato*'s [1998] demonstration that the static deformation of an elastic half-space is not affected when it is overlain by a liquid half-space. In addition, the model of a homogeneous elastic half-space neglects the possible influence of topography and structural layering on the response of the ocean floor.

In a series of landmark papers, *Ward* [1980, 1981, 1982] has proposed to interpret a tsunami wave as a particular case of the free oscillations of the Earth. Because the spheroidal modes of the planet involve changes in the distribution of mass in its interior, it has been realized since *Love* [1911] that gravity forms part of the restoring force in their oscillations, and that its contribution must be included in their computational algorithms [e.g., *Pekeris and Jarosch*, 1958; *Saito*, 1967; *Wiggins*, 1976]. As a result, the latter handle tsunami modes seamlessly when an appropriate combination of wave number and frequency is targeted, and given a sufficient sampling of the eigenfunction inside the water column.

This approach allows both the consideration of the finite elasticity of the oceanic column (which turns out to be negligible), and the use of a layered structure for the solid Earth, which can be more representative of the source properties of large earthquakes. The eigenfunction of the tsunami mode is no longer limited to the water column, but rather prolonged into the solid Earth, as described on Fig. 5-06. In the ocean, and at a typical period of 1000 s, the energy of the mode remains 99.9% gravitational and only 0.1% elastic, whereas it is mostly elastic (77%) in the solid Earth. The coupling at the ocean bottom results from the [positive] excess pressure caused by an [upwards] deformation η of the sea surface, to which a solid with a finite rigidity μ responds through a [downwards] vertical displacement y_1. In the limit of a low phase velocity C (characteristic of tsunamis), and for a Poisson solid, *Okal* [1988, 1991, 2003] has shown that

$$\frac{y_1}{\eta} = -\frac{3}{4} \cdot \frac{\rho_w g}{\mu k} \tag{4}$$

where ρ_w is the density of water, g the acceleration of gravity, and k the wavenumber of the tsunami. The ratio (4) remains on the order of -0.01 for typical values of μ and k. Even though the ocean bottom is not a welded interface, this vertical deformation y_1 is accompanied by a horizontal component of the eigenfunction in the solid, which is required in order to propagate a viable inhomogeneous wave through the solid medium with the phase velocity imposed by the tsunami, its amplitude at the interface being $ly_3 = -(1/3) \, y_1$ in the notation of *Saito* [1967]. Both y_1 and ly_3 decay with depth in the solid, with a skin depth approximately equal to the wavelength $\Lambda = 2\pi /k$ of the tsunami [*Okal*, 2003].

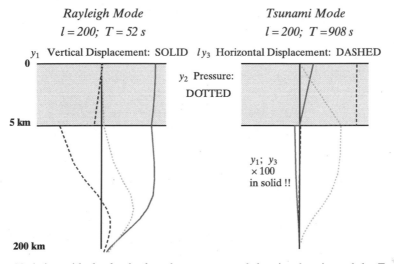

Rayleigh Mode
$l = 200; \ T = 52 \, s$

Tsunami Mode
$l = 200; \ T = 908 \, s$

y_1 Vertical Displacement: SOLID ly_3 Horizontal Displacement: DASHED

y_2 Pressure: DOTTED

0

5 km

$y_1; \, y_3$
$\times 100$
in solid !!

200 km

Fig. 5.06. – Variation with depth of selected components of the eigenfunctions of the Earth's free oscillations in the case of a classical Rayleigh mode *(left)* and of a tsunami mode *(right)*. In both cases, the angular order is $l = 200$, corresponding to a wavelength $\Lambda \approx 200$ km. The vertical displacement is plotted as the solid line, the horizontal component as the dashed line (on the same scale), and the overpressure as the dotted one (on an unrelated scale). In the case of the tsunami, the scale for displacements has been multiplied by 100 in the solid Earth to make them visible. Vertical scales in the water layer and solid Earth are also different. Note the weak penetration of the tsunami eigenfunction into the solid Earth, and its slow decay with depth.

One of the most seminal aspects of the normal mode formalism is that it simplifies greatly the investigation of tsunami excitation by seismic sources. We recall that *Gilbert* [1970] has shown that the excitation of a free oscillation of the Earth by a point source double-couple **M** buried at the location **r**ₛ inside the planet is proportional to the full scalar product $\langle \mathbf{M} : \varepsilon(\mathbf{r}_s) \rangle$ of the moment tensor by the eigenstrain of the mode at the source. In this framework, excitation coefficients expressed in a more familiar geometry related to the physical angles ϕ (strike), δ (dip) and λ (strike) were derived for seismic modes by *Kanamori and Cipar* [1974], and are computed routinely as part of the numerical solution of the free oscillation problem. Such calculations are immediately applicable, with only cosmetic changes to the codes, to the case of tsunami modes [*Ward*, 1980; *Okal*, 1988].

Before detailing the most important results regarding tsunami excitation by seismic sources, we must stress some limitations of the normal mode formalism. First and foremost, normal mode theory considers a laterally homogeneous Earth, or at most can handle a slightly heterogeneous structure [*Madariaga*, 1972; *Woodhouse and Dahlen*, 1978]. Thus, it will not be applicable to the simulation of tsunamis in the near field, where the interaction with the receiving beach is of primordial importance. In addition, seismic normal mode theory has been developed only as a linear algorithm, and thus will be applicable to tsunamis only when wave amplitudes η are small compared to water depth H and to wavelength Λ. On the other hand, given sufficient vertical sampling of the water column, it can be applied effortlessly outside the shallow-water approximation, in conditions where traditional methods become computationally prohibitive in the far field.

In the general framework of normal mode theory, we now discuss several properties of tsunami excitation; we re-emphasize that they apply only to tsunamis observed in the far field, traditionally defined as extending more than a few wavelengths away from the source, which in the case of large tsunamis, would translate into distances greater than 1500 km.

3.1.   Excitation is always small

Most importantly, and as shown on Fig. 5-06, the eigenfunction penetrates the solid Earth, but only with very small amplitude. Therefore, and because the excitation is proportional to the eigenstrain at the depth of the source, it remains relatively small. Just like a violinist does not make music by raking the bow near the bridge or the nut where the string cannot move, most earthquakes are fundamentally inefficient at exciting tsunamis, because they do so at what amounts to a node of the eigenfunction. It requires a truly gigantic source to generate a tsunami capable of inundation and destruction, especially in the far field. Synthetic maregrams generated by summation of normal modes suggest only millimetric peak-to-peak amplitudes in the far field on the high seas for a reference moment of 10^{27} dyn*cm [*Ward*, 1982; *Okal*, 1988].

3.2.   Excitation is proportional to M_0

In the absence of interference effects due to source finiteness and directivity (which will be discussed in Section 5), *Gilbert*'s [1970] theory predicts an excitation growing linearly with seismic moment, like that of all seismic spectral amplitudes unaffected by source finiteness. This is valid only for amplitudes η on the high seas, excluding the response of coastlines and harbors, and as such was for a long time difficult to verify. This is now becoming possible through the direct recording of tsunamis at sea, using the so-called "tsunameters" [*González et al.*, 2005]. In the past, observations at selected harbors known to involve few if any non-linear resonant effects have generally upheld this linearity [*e.g., Talandier and Okal*, 1989].

3.3.   Thresholds; when and where

The combination of the above two remarks leads to an estimate of $M_0 \geq 5 \times 10^{28}$ dyn*cm for the minimum seismic moment required to generate a transoceanic tsunami inflicting damage in the far field, as a zero-to-peak amplitude of ~10 cm on the high seas is capable of running up to several meters on land. This threshold estimate is generally borne out by observations, and incidentally forms the backbone of decision algorithms at warning centers. In principle, frequency-magnitude relationships, first proposed by *Gutenberg and Richter* [1941] and justified theoretically by *Rundle* [1989], suggest that there should be approximately one such event every five years, but their extrapolation to earthquakes of very great size suffers from both theoretical limitations [*Okal and Romanowicz,* 1994] and under-sampling. In practice, there have been 11 such earthquakes since 1940, listed in Table 1, of which ten were at sea. All tsunamis having inflicted casualties in the far field during that period were generated by six of those ten events, which confirms the validity of the above threshold as a necessary, if not sufficient, condition. We note further that eight of the 11 events in Table 1 occurred within a 19-yr interval between 1946 and 1965, which expresses the well-known fluctuations in seismic moment release over the past few decades.

TABLE 1.
Earthquakes with $M_0 \geq 5 \times 10^{28}$ dyn*cm, 1940–2007

Region	Date D M (J) Y	M_0 (10^{27} dyn*cxm)	Reference	Tsunami death toll in near and far field
Aleutian	01 APR (091) 1946	85	a	Near field sparsely populated: 5 deaths Far field: 161 deaths
Assam	15 AUG (227) 1950	140	b	Continental; no tsunami
Kamchatka	04 NOV (309) 1952	230	b,c,d	Near field: 2000–14000 deaths Far field: Damage strong in Hawaii; no deaths
Aleutian Is.	09 MAR (068) 1957	> 200	b,e,f	No fatalities reported despite large run-up in both near and far field
Chile	22 MAY (143) 1960	5000	g	Near field: 1000+ deaths Far field (Hawaii, Japan): 200+ deaths
Kurile Is.	13 OCT (286) 1963	75	h,i	No reported deaths or far-field damage
Alaska	28 MAR (088) 1964	820	j	Near field: 106 deaths Far field (Western US): 17 deaths
Rat Island	04 FEB (035) 1965	125	k,l	No deaths in near or far field
Sumatra-Andaman	26 DEC (261) 2004	1150	m	Near field (Aceh): 220,000 Far field: 60,000
Nias-Simeulue	28 MAR (087) 2005	105	n	No fatalities due to tsunami
Bengkulu, Sumatra	12 SEP (255) 2007	55	o	No fatalities due to tsunami

References. a: *López and Okal* [2006]; b: *Okal* [1992]; c: *Kaistrenko and Sedaeva* [2001]; d: *Macdonald and Wentworth* [1954]; e: *Johnson et al.* [1994]; f: *Fraser et al.* [1959]; g: *Cifuentes and Silver* [1989]; h: *Kanamori* [1970a]; i: *Solov'ev* [1965]; j: *Kanamori* [1970b]; k: *Wu and Kanamori* [1973]; l: *Hwang et al.* [1972]; m: *Tsai et al.* [2005]; n: *McAdoo et al.* [2006]; o: *Borrero et al.* [2007].

Until 2004, it was generally thought, following *Ruff and Kanamori* [1980], that the maximum expectable earthquake at a subduction zone could be predicted on the basis of two simple parameters (age of subducting lithosphere and rate of convergence), leading to the concept of "safe" subduction zones, from which trans-oceanic tsunamis would not be generated. Unfortunately, the 2004 Sumatra-Andaman earthquake violated this paradigm, and the re-examination of an updated Ruff-Kanamori dataset has led *Stein and Okal* [2007] to the precautionary conclusion that all sufficiently long subduction zones ($L > 500$ km) must be considered potential sources of mega-thrust earthquakes.

In the near-field, it is more difficult to define a universal moment threshold for a damaging tsunami, which may depend on local conditions. Shallow events ($h \leq 70$ km) with $M_0 \geq 10^{27}$ dyn*cm occurring in the vicinity of coastlines will generally result in tsunamis with run-ups of metric amplitude, but smaller events can occasionally be tsunamigenic. Also, the inflicted damage reflects the density of local population and infrastructure. For example, the considerable local run-up (reaching 21 m on Matua) from the earthquakes of 15 November 2006 and possibly 13 January 2007 ($M_0 = 3.4$ and 1.7×10^{28} dyn*cm, respectively) was not documented before a surveying team landed on the uninhabited Central Kuril Islands in August 2007 [Bourgeois et al., 2007]. On the average, worldwide historical records show 2 tsunamis per year inflicting damage in the near field.

3.4   Influence of source depth

The relatively slow decay of the eigenfunction with depth, illustrated on Fig. 5-06, predicts that source depth should not be a major parameter in the excitation of far-field tsunamis by earthquakes, at least in the range 10–100 km. Okal [1988] has given a detailed justification of this arguably paradoxical remark, using both frameworks—normal modes and classical gravity theory. In this respect, several examples deserve mentioning, such as the 1977 Tonga earthquake, whose tsunami amplitude (12 cm peak-to-peak at Papeete) was not deficient given its moment of 1.4 to 2.0×10^{28} dyn*cm [Talandier and Okal, 1979, 1989], despite a centroid depth later estimated at 100 km [Lundgren and Okal, 1988], and the recent event on 03 May 2006, also in Tonga (1.1×10^{28} dyn*cm; centroid depth 68 km), whose tsunami reached 25 cm peak-to-peak in Papeete [O. Hyvernaud, pers. comm., 2006] and 1.2 cm on a tsunameter off the coast of Oahu [Tang et al., 2006].

3.5.   Crustal structure

On the other hand, crustal stratification, shown by Okal [1982] to have little effect on the dispersion of tsunamis, can affect their excitation when the source is located, even partially, in a "sedimentary" layer featuring low rigidity [Okal, 1988]. This situation could contribute to the enhanced tsunami excitation for events occurring on splay faults rupturing through an accretionary prism, such as the "tsunami earthquakes" (see below) of 1975 in the Kuriles or 1896 off the coast of Sanriku [Fukao, 1979; Tanioka and Satake, 1996b].

3.6.   Focal geometry

The question of the dependence of tsunami excitation on the focal geometry of the earthquake has long been controversial. Common wisdom would suggest that strike-slip earthquakes produce little if any vertical deformation of the ocean floor, and as such should not be efficient tsunami generators. This is contradicted by the structure of the eigenfunction plotted on Fig. 5-06, which predicts a ratio as large as 1/3 between the horizontal and vertical components of displacement at the top of the solid medium and, in turn, comparable values for the excitation coefficients characteristic of strike-slip and 45° faults ($l^2 K_2/K_0 \approx 1$ in the notation of Kanamori *and Cipar* [1974]) [*Okal,* 2003]. The origin of this paradox lies in the need, under the classical gravity wave formalism (Fig. 5-03), to integrate contributions of the vertical displacement over the whole source area (formally over the entire ocean floor), which forbids any quick intuitive conclusion regarding the far-field tsunami. In particular, for any source of finite dimension, algorithms such as *Mansinha and Smylie*'s [1971] or *Okada*'s [1985] do predict poles of accumulation of vertical displacements localized in the vicinity of the tips of a strike-slip fault. They express the strong perturbation of the field of released strain in those areas, which is necessary to ensure the material's cohesion along a continuous path from one side of the fault to the other, around the fault tip. In this respect, it is worth noting that the large strike-slip event of 23 December 2004 near Macquarie Island generated a tsunami recorded in the far field by a seismic station located on the Ross Ice shelf along the coast of Antarctica, in a geometry approaching that of a tsunameter detection of the wave on the high seas [*Okal and MacAyeal,* 2006].

3.7. *Energy of a tsunami generated by an earthquake*

As detailed in *Okal* [2003] and summarized in the Appendix, the energy transferred from the seismic source to the tsunami can be computed from normal mode theory, and reconciled with its more classical derivation [*Kajiura,* 1981], the latter based essentially on the sketch on Fig. 5-02. The most remarkable aspect of these results is that the energy of the tsunami grows like $M_0^{4/3}$, *i.e.,* faster than the seismic moment, and hence than the total energy released at the source, expressed by the product $M_0 \cdot \varepsilon_{max}$. This is to say that larger earthquakes transfer an increasingly large fraction of their energy to the tsunami. However, this fraction remains in all cases very small (at most 1.3% in the case of the 1960 Chilean earthquake, the largest one ever measured instrumentally), and would reach clearly irrational values of 1 or greater only for a fault length equivalent to the Earth's circumference, a scale on which scaling laws have long ceased to be applicable. Another interesting aspect of the formulæ given in the Appendix is that the total energy fed into the tsunami, E_T, does not depend on the depth H of the water column. This is implicit from the sketch on Fig. 5-02. However, the actual amplitude of the wave in the far field can indeed be depth-dependent, as it follows the repartition with frequency of the total energy computed by *Kajiura* [1981] or *Okal* [2003]. In particular, once a tsunami is generated with a given total energy, and propagates over irregular bathymetry, the resulting amplitude in the far field will be affected by the details of the bathymetry over its source and path. This explains why tsunamis occurring in deeper, rather than shallower, water are eventually more damaging in the far field, as recently verified numerically by *Synolakis and Okal* [2006]. It also accounts partially for the deceptively low far-field tsunami generated by the "second" Sumatra event, the Nias-Simeulue earthquake of 28 March 2005, despite its large moment $M_0 = 1.05 \times 10^{29}$ dyn*cm, as its epicentral area involved much shallower bathymetry than that of the 2004 event. In addition, and as pointed out by *Synolakis and Arcas* [quoted by *Kerr,* 2005], the presence of large islands above the fault zone significantly reduced the amount of water available to be displaced and generate the tsunami.

3.8. A tsunami magnitude scale: M_{TSU}

The possibility of assigning a magnitude to an earthquake based on the amplitude of its tsunami was investigated by *Abe* [1983, 1989], who proposed a scale M_t using instrumentally measured tsunami amplitudes. This approach, derived mainly for the near field, remains empirical, and M_t has no theoretically derived relation to M_0. In addition, until recently, the only instruments capable of measuring the amplitude of tsunamis were tidal gauges; the interpretation of their records is hampered by the complexity of the response of coastlines, bays and harbors, where most of these devices are located, and by their intrinsic mechanical non-linearity.

On the other hand, the linearity derived from normal mode theory between the amplitude of a tsunami mode (and hence of the whole wave) and the seismic moment of the parent earthquake opens the possibility of developing a tsunami magnitude directly related to M_0, and fully justified from the theoretical standpoint. With the advent of tsunameters directly measuring the tsunami by recording its overpressure on the ocean bottom [*González et al.*, 1991; 2005], and to a lesser extent of direct measurements of tsunamis using satellite altimetry [*Okal et al.*, 1999; *Scharroo et al.*, 2005; *Kulikov et al.*, 2005], it is now possible to eliminate non-linear difficulties at the shorelines, and to measure the tsunami directly on the high seas. In this framework, *Okal and Titov* [2007] have developed a tsunami magnitude M_{TSU} , inspired from the mantle magnitude M_m introduced for long-period Rayleigh waves by *Okal and Talandier* [1989], and directly related to M_0 through

$$M_{TSU} = \log_{10} M_0 - 20 = \log_{10} X(\omega) + C_S + C_D + C_0 \qquad (5)$$

where $X(\omega)$ is the spectral amplitude of the tsunami at angular frequency ω, C_D a distance correction expressing geometrical spreading on the spherical Earth, C_S a source correction depending only on ω and derived form normal mode theory, and C_0 a locking constant depending on the nature of the observable X (amplitude η at the surface or overpressure p at the ocean floor), and also fully justifiable on theoretical grounds. All details can be found in *Okal and Titov* [2007].

This concept was successfully applied to five tsunamis recorded by the growing network of tsunameters. As none were deployed in the Indian Ocean at the time of the 2004 Sumatra-Andaman tsunami, *Okal and Titov* [2007] processed the trace from the JASON satellite altimeter (allowing for its specific character of being neither a time nor a space series). Fig. 5-07 shows that the M_{TSU} algorithm recovers the seismic moment satisfactorily, even in conditions where the concept should not be applicable, such as the great circle path crossing a continental mass. In addition, *Okal* [2007a] has shown that the concept can be applied to seismic records of tsunami waves obtained at island or continental stations.

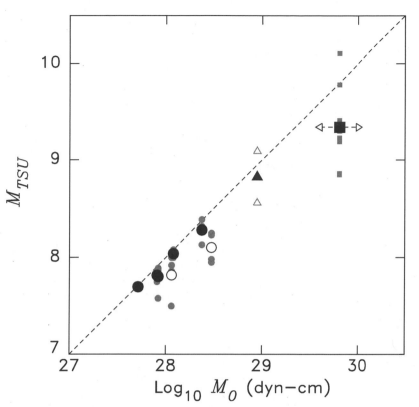

Fig. 5.07 – Performance of the magnitude M_{TSU} studied for recent tsunamis, as a function of the published moment M_0. For each earthquake, the small gray dots are the individual measurements at each tsunameter, averaged over usable frequencies. The large symbol is the value averaged over all tsunameters. The two open symbols refer to events for which the great circle paths intersect either a continent or the Aleutian arc. The square symbols are for the JASON satellite profile of the 2004 Sumatra-Andaman tsunami, with the horizontal arrows indicating the range of published moments. The triangles refer to the 1946 Aleutian event and are virtual measurements made on synthetic maregrams computed by *Okal and Hébert* [2007]. See *Okal and Titov* [2007] for details.

3.9.   The high-frequency components of the tsunami wave

Because it works seamlessly without special assumptions outside the domain of validity of the shallow-water approximation ($k \cdot H \ll 1$), normal mode formalism is particularly well suited to investigate the higher-frequency components of tsunamis, typically in the 2–10 mHz frequency band. Since they are strongly dispersed according to the classical formula

$$\omega^2 = g\,k \cdot \tanh\,(k \cdot H) \tag{6}$$

such components arrive both later and with a smaller amplitude than the lower-frequency (and conventional) part of the tsunami spectrum (typically 0.3–1 mHz). However, during surveys of the 2004 Sumatra tsunami in Réunion and Madagascar, *Okal et al.* [2006*a,b*] have reported occurrences of large vessels breaking their moorings in harbors several hours after the arrival of the tsunami; these phenomena were interpreted by the resonance of harbor basins under the delayed high-frequency components of the tsunami [*Pančošková et al.*, 2006]. As they carry obvious implications in terms of tsunami warning and civil defense, it is imperative to understand to which extent such effects can be predicted and quantified. In this respect, *Okal et al.* [2007] have used records of the 2004 tsunami on hydrophones of the Comprehensive Nuclear-Test Ban Treaty Organization to show that its high-frequency spectral amplitudes can be successfully quantified from the seismic moment of the earthquake, using normal mode theory.

3.10. Tsunamis in the atmosphere and above

The argument used to justify the continuation of the tsunami eigenfunction into the solid Earth is also applicable at the other boundary of the oceanic column, namely the sea surface. In all above computations, the latter is taken as a "free" boundary where pressure must vanish, but it is really overlain by an atmosphere "only" 1000 times less dense than water. Accordingly, the tsunami mode is actually continued upwards in the form of a gravitational oscillation of the atmosphere. This idea of treating various scenarios of coupling between the solid, liquid and gaseous layers of the Earth as the propagation of a single wave through a combined structure was pioneered by *Haskell* [1951] and developed by *Harkrider et al.* [1974] for the conjugate problem of the generation of seismic Rayleigh waves by atmospheric explosions. While the energy density of the mode continued into the atmosphere is expected to decrease with height, the concurrent rarefaction of the material density $\rho_{atm.}$ results in an actual *increase* of the particle displacement with altitude, the amplification of sea-surface disturbances reaching up to 5 orders of magnitude at the base of the ionosphere. This concept, observed for Rayleigh waves by *Yuen et al.* [1969] was predicted for tsunamis by *Peltier and Hines* [1976], and first observed by *Artru et al.* [2005] through the disruption of GPS signals over Japan during the 2001 Peruvian tsunami; it was modeled quantitatively in the case of the 2004 Sumatra tsunami by *Occhipinti et al.* [2006]. It could bear some potential for real-time detection of tsunamis over non-instrumented portions of the ocean basins, using various space-based technologies.

4. Earthquake tsunamis in the near field: Scaling laws and invariants

In general, the run-up of a tsunami on a local beach following an earthquake will be controlled by a very large number of parameters, comprising those describing the earthquake source (seismic moment and geometry), its physical environment (earthquake depth, bathymetry of the source area), and the properties of the receiving shore line (above- and below-sea level topography of the beach, presence of bays, harbors, estuaries, etc.). Successful simulations can be run only on a case-by-case basis taking into account all specific parameters. Nevertheless, it is desirable to understand which of them can define trends, and if possible robust invariants.

In this general framework, one notes the linearity of the Somigliana tensor, which is carried into the algorithm of *Mansinha and Smylie* [1971] described in Section 2, suggesting that, everything else being equal, the excitation of the tsunami should grow linearly with the slip on the fault plane, Δu, and so should the final run-up ζ on the shore, in the simplified model of an ocean-beach-shore system offering translational symmetry, and for moderate amplitudes precluding the development of non-linearities. Similarly, the lateral extent of the inundation by the tsunami should be expected to grow like the dimension L of the fault over which the rupture takes place. These simple arguments then suggest that the aspect ratio of the distribution of run-up along the beach should express the ratio $\Delta u / L$, and hence the strain ε released by the dislocation, known to be, at least in principle, an invariant of the seismic source.

Motivated by this remark, *Okal and Synolakis* [2004] have considered the idealized model of an earthquake source occurring offshore of a perfectly linear beach (Fig. 5-08a), and studied the field of run-up ζ along the direction y of the waterfront. Their simulations used the MOST code, for a very large number of scenarios obtained by varying the scalar moment M_0 of the earthquake, the three angles ϕ, δ, λ of its focal geometry, the distance D of the earthquake from the shore, its hypocentral depth h, the thickness H of the oceanic column at the source, and the slope β of beach. For each case under study, they fit the distribution of run-up with a bell-shaped curve of the form

$$\zeta(y) = \frac{b}{\left(\dfrac{y-c}{a}\right)^2 + 1} \tag{7}$$

and focus on the two dimensionless parameters $I_1 = b/\Delta u$ and $I_2 = b/a$. As detailed in *Okal and Synolakis* [2004], both I_1 and I_2 are found to be robust, and thus play the role of invariants in this study. Most importantly, and as shown on Fig. 5-08, they feature upper bounds (respectively 1.35 and 7×10^{-5} in the numerical experiments), clearly controlled by the properties of the seismic sources and its own invariants. By contrast, a similar study using representative models of submarine landslides produced much steeper distributions of run-up along coastlines, with aspect ratios I_2 always remaining greater than 10^{-4} (Fig. 5-08c).

These theoretical predictions were upheld by fitting profiles of the form (7) to run-up datasets documented during field surveys following the major tsunamis of 1985–2001 [*Synolakis and Okal*, 2005] (allowing for a slight increase of I_1 to a maximum value of 2, in view of our imprecise experimental knowledge of Δu). Among all earthquakes tested, the only two clearly violating the pattern are the 1998 Papua New Guinea and 1946 Aleutian events, whose local tsunamis are explained through the triggering of major underwater landslides.

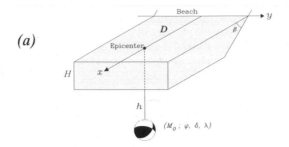

(a)

(b)   Simulate Tsunami Propagation to Beach and Run-up

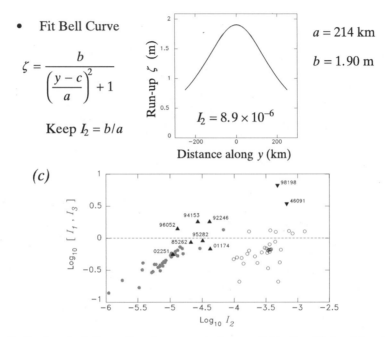

- Fit Bell Curve

$$\zeta = \frac{b}{\left(\dfrac{y-c}{a}\right)^2 + 1}$$

Keep $I_2 = b/a$

$a = 214$ km

$b = 1.90$ m

$I_2 = 8.9 \times 10^{-6}$

(c)

Fig. 5.08 – Scaling invariants for earthquake sources in the near field, after *Okal and Synolakis* [2004]. *(a)* Sketch of the geometry considered, with parameters varied in numerical experiments. *(b)* Run-up ζ at the beach plotted as a function of coordinate y, fitted with a bell-shaped curve of aspect ratio I_2. *(c)* Logarithmic plot of $I_1 = b/\Delta u$ as a function of I_2 for a large population of dislocation sources (solid gray dots). Note the bounds on both quantities. By contrast, the open symbols refer to dipolar sources modeling landslides, and the ordinate is then $\log_{10} I_3$, the ratio of b to the depth of the trough in the initial dipole. Note that the two populations are separated at the $I_2 = 10^{-4}$ threshold. The triangles denote the aspect ratios of the distributions of runup surveyed after recent tsunamis (identified by Julian date). The only clear violators of the bounds $I_1 < 2$; $I_2 < 10^{-4}$ are the 1998 Papua New Guinea and 1946 Aleutian earthquakes, whose local tsunamis were due to landslides.

In the case of the 2004 Sumatra-Andaman earthquake, postdating *Okal and Synolakis'* [2004] study, the parameter I_2 cannot be adequately studied, as a continuous coastline does not parallel the full extent of the fault rupture. On the other hand, regarding I_1, we note a maximum surveyed run-up of 31 m at Lhoknga (discounting splashes on vertical cliffs) [*Borrero,* 2005]. As large as this figure can seem, it remains within the bounds of I_1, given estimates of Δu reaching for example up to 19 m on specific segments of the fault plane, in the model of *Banerjee et al.* [2007]. Thus, near-field run-up during the 2004 Sumatra-Andaman tsunami is accountable within the framework of seismic scaling laws, with a regular, if of course exceptionally large, seismic source, and does not require ancillary phenomena such as catastrophic underwater landslides, or the activation of substantial splay faults.

Finally, *Okal and Synolakis'* [2004] results underscore the value of field survey data acquired in the aftermath of tsunamis, which can give insight into the mechanism of generation of the tsunami through the application of robust discriminants. In turn, the "Plafker rule of thumb (*"Run up on a straight beach* [unaffected by river valleys, bays or harbors] *shall not exceed twice the amount of slip on the fault"*) can be applied as a gross, zeroth-order estimate of expectable local inundation during future tsunamis.

5. Earthquake tsunamis in the far field: Directivity

It has long been known that transoceanic tsunamis feature very strong *directivity, i.e.,* that the amplitude of their waves is a strong function of the azimuth of their path from the parent earthquake. For example, the 2004 Sumatra-Andaman tsunami was particularly destructive in Sri Lanka [*Goffe et al.,* 2006] and Somalia [*Fritz and Borrero,* 2006], but relatively benign at similar or shorter distances in Western Australia. The origin of this effect, lies in the spatial extent of the source, which results, in the far field, in a destructive interference of the waves generated by its individual components, in all azimuths except in the direction perpendicular to the strike of the fault.[2] It was investigated for standard seismic surface waves by *Ben-Menahem* [1961], and applied to tsunamis by *Ben-Menahem and Rosenman* [1972]. For a source with a fault length L rupturing at velocity V_R, spectral amplitudes at angular frequency ω must be multiplied by a directivity function

$$DIR = \left| \frac{\sin X}{X} \right| \quad \text{with} \quad X = \frac{\omega L}{2C} \cdot \left(\frac{C}{V_R} - \cos\psi \right) \tag{8}$$

where C is the phase velocity of the wave, and ψ the azimuth of the path measured from the fault strike. Note that $DIR = 1$ only for $X = 0$; otherwise, $DIR < 1$, expressing destructive interference. Directivity patterns are expected to be sharply different for seismic waves and tsunamis. For the former, the rupture velocity V_R (typically 3.5 km/s) is only slightly less than the typical phase velocity of a Rayleigh wave ($C = 4$ km/s), and X will be smallest (and consequently DIR largest, approaching 1) for $\psi = 0$, in the direction of propagation of rupture. On the other hand, for tsunamis, the phase velocity C is much less than V_R (typically 1/15, but still 1/5 in the case of the slow rupture of a "tsunami earthquake"), and X vanishes for an angle ψ close to $\pi/2$, meaning that the directivity lobe is directed at right angles from the fault. The physical interpretation of this result is that the faulting

---

[2] Directivity is superimposed on the radiation pattern expressing the azimuthal variation of excitation from a point-source moment tensor as a function of its focal geometry which, for tsunamis, is usually dominated by an isotropic term in the case of dipping fault planes [*Okal,* 1988, 1991].

takes place instantaneously compared to the propagation times of the tsunami, and therefore that the interference between waves coming from the various elements of the source can be constructive only if the distances traveled are stationary, which is the case in the far field for $\psi = \pi / 2$. It also follows from (8) that directivity lobes are narrower for tsunamis than seismic surface waves, and that their width decreases with increasing earthquake size [*Okal and Talandier,* 1991]. Theoretical examples of tsunami directivity functions are presented in Fig. 5-09*a,* and the case of the 2004 Sumatra-Andaman event is illustrated conceptually on Fig. 5-09*b* with a numerical simulation given on Fig. 5-10. It explains for example the relatively moderate amplitudes reported in Madagascar and the Mascarene Islands [*Okal et al.,* 2006*a,b*], as opposed to Somalia [*Fritz and Borrero,* 2006].

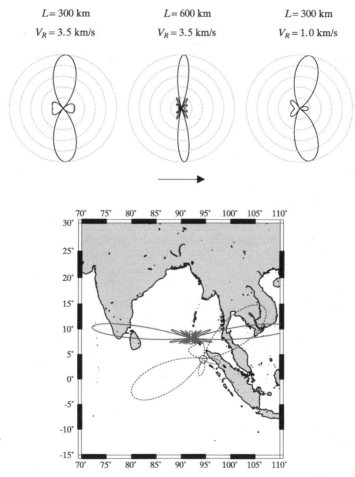

Fig. 5.09 – *Top:* Theoretical azimuthal directivity functions (8) of tsunami waves computed for typical earthquake source scenarios. In all cases, the fault strike is taken along the *x* axis, as indicated by the arrow. The dashed lines are drawn from *DIR* = 0.2 (innermost) to *DIR* = 1(outermost) in steps of 0.2 units. Note the narrowing of the lobe with increasing fault length, and its direction essentially at right angles to the fault. The plot at right represents a slow-rupturing "tsunami earthquake". *Bottom:* Application to the case of the 2004 Sumatra-Andaman earthquake. The solid line represents the directivity function for a 1000–km fault rupturing to the North and centered at 7°N, the dotted one for the model proposed initially, extending only 350 km at an azimuth of 327°, off the coast of Sumatra.

Sumatra: Composite Solution (Excluding Runup)

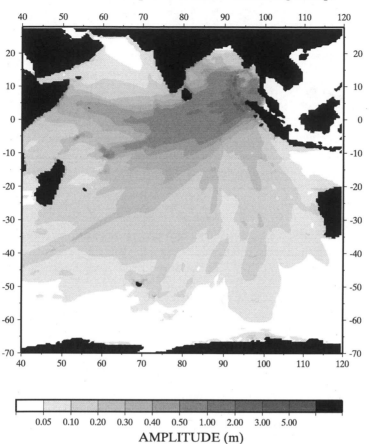

AMPLITUDE (m)

Fig. 5.10 – Same as Fig. 5-04 for the 2004 Sumatra-Andaman tsunami, modeled using the composite seismic source of *Tsai et al.* [2005], extending along the curved arc from Aceh to the Northern Andaman Islands. Note the directivity lobe at right angles to the average direction of rupture. This diagram explains the enhanced amplitudes in Sri Lanka and Somalia, as compared to Western Australia and Madagascar. After *Synolakis and Okal* [2006].

The concept of directivity leads to simple and robust results regarding the relative vulnerability of distant shores to transoceanic tsunamis. For example, as a gross approximation, the Pacific Basin can be taken as circular with plate boundaries tangent to its circumference, and Hawaii at its center. This suffices to explain the repeated disastrous amplitudes in Hawaii from sources as diverse as Chile (1960, which also inflicted death and damage to Japan but largely spared Mexico or California), Unimak (1946, which spared California and South America), or Kamchatka (1952, which spared Japan or California). The exception would be the great 1964 earthquake in Alaska, where the plate boundary striking N66°E strongly departs from this simple model. As a result, the 1964 tsunami spared Hawaii, but was directed at Northern California, inflicting heavy damage and 11 deaths in the community of Crescent City. An important aspect of the directivity concept is that the azimuth of long faults is well known in the context of plate tectonics, and thus the azimuths of the so-called "swaths of death" featuring maximum tsunami risk in the far field can be predicted in advance.

In addition to source directivity, the amplitude of a tsunami in the far field can also be affected by propagation over irregular bathymetry. In particular, because the velocity of a tsunami, $C = \sqrt{gH}$ under the shallow water approximation, varies with the depth H of the water column, a zone of reduced bathymetry (ridge, plateau) can act as an optical lens and focus the energy of the wave [*Woods and Okal,* 1987; *Satake,* 1988]. This effect, identifiable on Fig. 5-04, is not related to the excitation of the tsunami by the earthquake, and thus transcends the scope of this Chapter. Because it does not grow with source size (as opposed to the focusing by source directivity; *Okal and Talandier* [1991]), it remains of secondary importance in the case of catastrophic transoceanic tsunamis.

Finally, note that the presence of strong lobes of directivity, *i.e.,* a strong variation with azimuth of the function $DIR(\psi)$ in (8) requires both a large value of X (and thus a duration of rupture L/V_R at least comparable to the period of the wave $2\pi/\omega$), and a substantial azimuthal variation of the term in brackets, $(C/V_R - \cos\psi)$. This excludes cases where $C \gg V_R$, corresponding to exceedingly slow ruptures for which the interference is destructive in all azimuths ψ. In practice, this is the case for submarine landslides, and the existence of a strong lobe of directivity in the far field can be used as a source discriminant in this respect, for example in the case of the 1946 Aleutian tsunami [*Okal et al.,* 2002; *Okal,* 2003].

6. The case of "tsunami earthquakes"

This special class of earthquakes was defined by *Kanamori* [1972] as events whose tsunamis are greater than would be expected from their magnitudes, especially conventional ones. In this context, it is worth repeating that *"tsunami earthquakes"* are not to be confused with earthquakes merely generating a tsunami, which are simply referred to as "tsunamigenic". "Tsunami earthquakes" are obvious violators of scaling laws and we now understand that they generally involve a slower than expected rupture, which results in strongly destructive interference patterns for the higher-frequency seismic waves on which conventional magnitudes are measured: 1–s body waves for m_b and 20–s surface waves for the "Prague" magnitude M_s. In the case of particularly slow events, even seismic moments obtained in their immediate aftermath from automated centroid moment tensor inversions can be underestimated as they are derived from mantle surface waves typically in the 100–200 s range. By contrast, the excitation of tsunami waves, typically in the 1000–s period range, involves the full seismic moment, integrated over a much longer time.

"Tsunami earthquakes" can be particularly tragic in the near field where little time is available for a centralized warning and where efficient mitigation relies on self-evacuation of the coastal populations upon their feeling the earthquake. An event lacking high frequencies will produce minimal accelerations resulting in little if any damage to structures, and may even go unnoticed by humans. Such was the case of the great Sanriku, Japan earthquake of 15 June 1896, which was felt along the coast at a weak intensity not exceeding MM IV, but was followed by a devastating tsunami with run-up of 30 m, killing more than 27,000 people [*Kanamori,*

1972; *Solov'ev and Go*, 1984]. On a smaller scale, similar scenarios of tsunami devastation in the near field along sections of coast line where the earthquake *had not even been felt* were reported during the recent "tsunami earthquakes" in Nicaragua (02 September 1992), Java (02 June 1994 and 17 July 2006) and Chimbote, Peru (21 February 1996) [*Abe et al.*, 1993; *Tsuji et al.*, 1995a; *Bourgeois et al.*, 1999; *Fritz et al.*, 2006a]. It is clear that such events pose a special challenge to the warning community, as it is imperative to recognize their anomalous character in real-time, in order to issue a warning to populations who may not suspect danger. Table 2 gives a list of documented "tsunami earthquakes". The two earthquakes listed as "probable" are aftershocks of major interplate thrust events, which generated locally catastrophic tsunamis featuring in both instances greater run-up heights than those of the main shocks, despite clearly lower seismic magnitudes. However, in the absence of detailed seismological studies of these ancient earthquakes, it is difficult to formally establish the properties of their sources.

TABLE 2.
Parameters of "tsunami earthquakes", 1896–2006

Origin	Date D M (J) Y	m_b	M_s†	Moment M_0 (10^{27} dyn*cm)	Θ	Tsunami Fatalities	Reference
		Documented					
Sanriku, Japan	15 JUN (167) 1896		7.2	12		27,000	a,b
Unimak, Aleutian Is.	01 APR (091) 1946		7.4	85	−7.0	166	a,c
Northern Peru	20 NOV (325) 1960		6.75	2.7	−6.13	13	d
Kuriles	20 OCT (293) 1963		6.8	7.5	−6.42		e,f
Kuriles	10 JUN (161) 1975	5.8	7.0	0.8	−6.43		e,f
Tonga	19 DEC (353) 1982	5.9	7.7	2.0	−5.76		g,h
Nicaragua	02 SEP (246) 1992	5.3	7.2	3.4	−6.30	160	d,i
East Java	02 JUN (153) 1994	5.7	7.2	5.1	−6.03	223	d,i
Chimbote, Peru	21 FEB (052) 1996	5.8	6.6	2.2	−6.00	12	d,i
Sumatra-Andaman	26 DEC (361) 2004	7.2	8.1	1100	−5.89	280,000	j,k
Central Java	17 JUL (198) 2006	6.1	7.7	4.6	−6.13	668	l
		Probable					
Kamchatka	13 APR (103) 1923		7.2			23	m,n
Cuyutlan, Mexico	22 JUN (174) 1932		6.9			75	o

†Note that magnitudes M_s are not systematically catalogued before 1968. Earlier values are estimates from individual researchers.

References. a: *Kanamori* [1972]; b: *Tanioka and Satake* [1996b]; c: *López and Okal* [2006]; d: *Okal and Newman* [2001]; e: *Okal et al.* [2003]; f: *Fukao* [1979]; g: *Okal and Talandier* [1989]; h: *Newman and Okal* [1998]; i: *Polet and Kanamori* [2000]; j: *Stein and Okal* [2005]; k: *Choy and Boatwright* [2007]; l: *Reymond and Okal* [2006]; m: *Menyaïlov* [1946]; n: *Solov'ev and Ferchev* [1961]; o: *Farreras and Sanchez* [1991].

While most regular earthquake sources feature ruptures propagating along the length of the fault at V_R = 3 to 3.5 km/s, detailed seismological studies of "tsunami earthquakes" have revealed smaller values, reaching as low as V_R = 1 km/s [*Polet and Kanamori*, 2000; *López and Okal*, 2006]. This inherent slowness may be detected by making simple, yet robust and rapid measurements targeting widely different properties of the seismic source. One such approach is the computation of radiated seismic energy, obtained by integrating the squared ground velocity as recorded in teleseismic generalized-P wavetrains. Following the method of *Boatwright and Choy* [1986], *Newman and Okal* [1998] have developed an algorithm providing an estimate E^E of the radiated energy, while ignoring source details unavailable in real time, such as exact focal mechanism and source depth. They then compare it with the seismic moment M_0 of the source, through a "slowness" parameter

$$\Theta = \log_{10} \frac{E^E}{M_0} \qquad (9)$$

Scaling laws would predict a constant value of −4.90 for Θ, which is indeed close to the worldwide average computed on a population of regular earthquakes. On the other hand, and as shown on Fig. 5-11, "tsunami earthquakes" feature significantly depleted values of Θ, deficient by typically 1 to 2 units, the use of a logarithmic scale giving particular robustness to the concept. The algorithm has been implemented as part of the real-time assessment of teleseismic events at the Pacific Tsunami Warning Center [*Weinstein and Okal,* 2005]. Incidentally, higher-than-normal values of Θ are characteristic of "snappy" earthquakes, featuring greater stress drops, and hence higher accelerations and producing more damage than

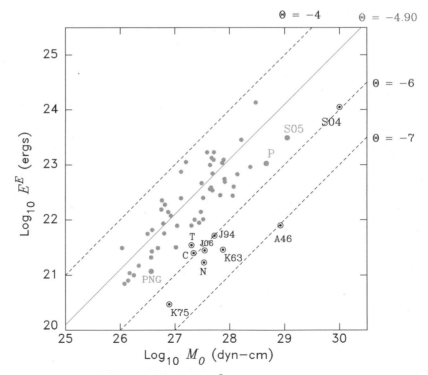

Fig. 5.11 – Logarithmic plot of Estimated Energy E^E computed for 54 recent earthquakes, *vs.* Harvard CMT seismic moment M_0. Dashed lines are drawn for constant Θ with the value predicted by scaling laws (−4.90) shown solid. Dark bull's eye symbols identify "tsunami earthquakes", listed in Table 2. Other events shown include Peru 2001 (P), Papua New Guinea 1998 (PNG) and Nias-Simeulue 2005 (S05). Updated from *Newman and Okal* [1999].

expected from seismic moments. Examples include such intraplate earthquakes as the 1939 Chillan, Chile and 1999 Oaxaca, Mexico earthquakes [*Okal and Kirby,* 2002], and the recent Central Kuriles earthquake of 13 January 2007. Finally, note that the concept of Θ shares its philosophy of comparing two measurements made at different periods with the time-honored m_b : M_s discriminant, introduced to identify large explosions [*e.g., Marshall and Basham,* 1972].

Along similar lines, *Okal et al.* [2003] have documented that "tsunami earthquakes" are also noticeably deficient in the generation of hydroacoustic T phases, and quantified this property through a parameter Γ expressing the ratio of T–phase energy flux to seismic moment. They show deficiencies of as much as two orders of magnitude for "tsunami earthquakes", as well as an excellent correlation between Γ and Θ [*Okal,* 2007*b*]. However, because T phases travel slower than seismic waves, they bear less promise towards real-time tsunami warning.

Finally, following a remark by *Ni et al.* [2005], *Reymond and Okal* [2006] have shown that source duration can be meaningfully extracted from the generalized P wave of teleseismic records by band-pass filtering in the 2–4 Hz high-frequency range, thus eliminating later phases whose contributions may affect the more conventional properties of P waves for long source durations. These authors concluded that this procedure recovers the exceptional duration of mega earthquakes such as the 2004 Sumatra-Andaman event, and of some (but unfortunately not all) "tsunami earthquakes". Thus, while this approach holds promise for the early detection of "tsunami earthquakes", more research is clearly needed in this area.

Any progress regarding the mitigation of the risk from "tsunami earthquakes" will obviously require an understanding of the conditions leading to their exceptional source slowness, and of whether such behavior can be realistically predicted in recognizable tectonic environments. In this framework, *Okal and Newman* [2001] conducted a detailed analysis of the parameter Θ in the three subduction zones where "tsunami earthquakes" occurred during the 1990s (Central America, Java and Northern Peru), and concluded that regional background seismicity at lower magnitudes did not share anomalously low values of Θ, which suggests that slowness, at least in those subduction zones, would require a minimum size of earthquake, in turn hinting at source complexity as a potential requirement for slow earthquakes. In addition, they confirmed the 1960 event in Northern Peru, about 350 km NW of Chimbote, as a "tsunami earthquake", with a parameter Θ comparable to that of the 1996 event. This suggests that common conditions along a subduction zone may give it a uniform potential for "tsunami earthquakes". Posterior to their study, the recent occurrence of the 2006 "tsunami earthquake" in Java [*Reymond and Okal,* 2006], 600 km to the West of the 1994 epicenter, lends considerable support to this suggestion. Finally, *Okal and Newman* [2001] noted that no very large interplate thrust earthquakes ($M_0 \geq 3 \times 10^{28}$ dyn*cm) were documented in the instrumental (and available historical) seismic record of any of the three relevant subduction zones, leading to the legitimate question of whether the occurrence of "tsunami earthquakes" can exclude that of mega-thrust earthquakes in the same region. In the framework of our newly revised concepts regarding the maximum size of earthquakes at subduction zones [*Stein and Okal,* 2007], and given that the 1946 Aleutian "tsunami earthquake" is itself very large [*López and Okal,* 2006], this question must be considered wide open, but is of course of considerable interest, for example regarding tsunami risk in Australia from earthquakes in the Java trench.

Regarding the nature of potential tectonic environments for the generation of "tsunami earthquakes", two classes of models have been proposed. The first one consists of events rupturing through shallow material featuring poor mechanical properties, such as accretionary prisms, as exemplified by the 1896 Sanriku event [*Tanioka and Satake*, 1996b], and the 1963 (20 October) and 1975 Kuriles earthquakes [*Fukao*, 1979]. Note that the latter two can be regarded as aftershocks, occurring on nearby splay faults, of larger shocks with conventional rupture properties, namely the great 1963 (13 October) Kuriles earthquake and the 1973 Nemuro-Oki event. The lesser rigidity of the source region of the "tsunami earthquake" would explain the greater excitation of tsunami waves, following the results of *Okal* [1988] discussed above, as well as the reduced rupture velocity V_R leading to exceptional source slowness and destructive interference for higher-frequency waves. The probable "tsunami earthquakes" of 13 April 1923 in Kamchatka (following the great interplate thrust earthquake of 03 February 1923) and 22 June 1932 in Mexico (following the Colima event of 03 June 1932) may also fit this pattern.

By contrast, a second class of models for "tsunami earthquakes" considers a sediment-starved environment leading to a partially decoupled interface between the downgoing slab and the overriding plate [*Tanioka et al.*, 1997]. The absence of a sedimentary plug may allow the upward propagation of rupture rather than the creep generally prevailing along the shallowest portion of the interface [*Byrne et al.*, 1988]. In the presence of irregular bathymetry on the subducting ocean floor, the interface is expected to be jagged and the upward rupture will be irregular and, on the average, slow. This model was expanded by *Polet and Kanamori* [2000] and applied to the 1992 Nicaragua, 1994 Java and 1996 Chimbote, Peru "tsunami earthquakes". It does hold a promise for interpreting this class of "tsunami earthquakes" in a tectonic context, giving them some level of geographic predictability.

We include the 2004 Sumatra-Andaman earthquake in the list of "tsunami earthquakes" in Table 2, based on several lines of evidence. It has generally been very difficult to obtain the source parameters of this event, as routine measuring algorithms were simply not adapted to its exceptional size; examples would include the target frequency of moment tensor inversions from surface waves, or at the other end of the spectrum, the duration of integration windows of generalized P waves for the computation of radiated energy. As a result, there remains some uncertainty, and even controversy, regarding the final values of moment M_0 and energy E^E. We use here the moment of 1.1×10^{30} dyn*cm derived by *Stein and Okal* [2005] from the analysis of the Earth's gravest modes, in excellent agreement with the composite moment tensor inversion of *Tsai et al.* [2005] (1.16×10^{30} dyn*cm), and the radiated energy estimated at 1.4×10^{24} ergs by *Choy and Boatwright* [2007], who used a special algorithm eliminating the influence of later phases. This translates into a slowness parameter $\Theta = -5.89$, roughly one unit less than predicted by scaling laws, and essentially equivalent to that of the 1996 Chimbote, Peru earthquake, thus qualifying the 2004 Sumatra-Andaman earthquake as a "tsunami earthquake". In addition, detailed studies in source tomography based on traditional seismic waves [*Ishii et al.*, 2005], the generation of T phases [*de-Groot-Hedlin*, 2005; *Tolstoy and Bohnenstiehl*, 2005; *Guilbert et al.*, 2005] or the composite inversion by *Tsai et al.* [2005] all lead to an average rupture velocity of 2.7 km/s, which is significantly less than the value of 3.5 km/s predicted under scaling laws. Furthermore, all these models suggest a slowing down of the rupture towards the North, to values of ~ 2.0 km/s or less, also in agreement with the conclusions of *Choy and Boatwright* [2007], further suggesting that the 2004 earthquake may have featured increasing slowness in the later part of its rupture, as also discussed by *Seno and Hirata* [2007]. Finally, from a qualitative standpoint, it is interesting to note that many pictures and videos of the inundation of Banda Aceh by the tsunami show that most buildings were standing unaffected by the earthquake before they were taken down by the tsunami, confirming a deficiency of high-frequency components in the source. While not reaching the record values of slowness of the 1946 Aleutian earthquake, or even the 1992 Nicaragua event, the 2004 Sumatra shock clearly features a source spectrum which cannot be explained by simply extrapolating universal scaling laws to its exceptional size.

We wish to stress, however, that the qualification of the 2004 Sumatra event as a "tsunami earthquake" may be controversial, depending on the particular definition of such events. We note in particular that a definition such as *"an event whose tsunami has significantly larger amplitude than expected from estimates of its size obtained from seismic waves"*, will depend on the particular nature of such estimates. Earthquakes whose source violates scaling laws will feature moment estimates crucially dependent on the period of the measurement; in their aftermath, scientists will make increasingly sophisticated measurements, in practice at increasingly long periods. Eventually, the seismic models obtained will provide adequate representations of the source of the tsunami. This was the case, for example, with the 1946 Aleutian earthquake: its moment estimates using 100–s waves [*Brune and Engen,* 1969; *Kanamori,* 1972] or *a fortiori* 20–s M_s measurements [*Gutenberg and Richter,* 1954] led to deficiencies in tsunami excitation, but the far-field tsunami could be successfully modeled from *López and Okal*'s [2006] very-low frequency seismic model, obtained at 250 s [*Okal and Hébert,* 2007]. The situation was remarkably similar in the case of the Sumatra event: early estimates of its seismic moment (*e.g.,* original Harvard CMT solution using 300–s waves) are clearly incompatible with observed run-up values under the Plafker rule of thumb, while the tsunami can be successfully modeled using the more definitive moment values obtained later by investigators working at much longer periods [*Stein and Okal,* 2005; *Tsai et al.,* 2005]. Indeed, one could make the provocative statement that the characterization of an event as a "tsunami earthquake" under the above definition is transient in time, and that given enough time, seismologists will come up with a model explaining the generation of both seismic and tsunami waves. A better definition of a "tsunami earthquake" should focus on the disparity of its properties in terms of its tsunami and its [most conventional] seismic waves, which expresses a violation of the paradigm commonly used to relate them, in other words of seismic scaling laws. It is in this respect that the 2004 Sumatra earthquake can qualify as a "tsunami earthquake".

Among mega-thrust earthquakes, the 1960 Chilean event (total moment $M_0 = 5 \times 10^{30}$ dyn*cm) was also determined to have involved an ultra-slow component to its source, in that case as a precursor [*Kanamori and Cipar,* 1974; *Kanamori and Anderson,* 1975; *Cifuentes and Silver,* 1989], although the absence of adequate digital records makes it impossible to formally compute a parameter Θ. On the other hand, the 2005 Nias-Simeulue earthquake features no more than a trend towards slowness ($\Theta = -5.54$), common to many other earthquakes in subduction zones. Thus excessive source slowness may not be a necessary feature of all mega-thrust events.

Finally, we want to stress that we do not categorize as genuine "tsunami earth-quakes" those events for which exceptionally large local tsunamis resulted from a phenomenon ancillary to the seismic rupture, namely the triggering of an underwa-ter landslide, the most dramatic example being the Papua New Guinea event of 17 July 1998 [*Synolakis et al.*, 2002]. In particular, we reemphasize that all large earth-quakes can trigger landslide failures in precarious materials, which in turn can generate tsunamis with run-up values locally much larger than those of the parent earthquake. For example, during the Flores, Indonesia earthquake of 02 Decem-ber 1992, which featured no clear anomaly in its seismic source properties, run-up reached a catastrophic 26 m at Riang-Kroko Village, as opposed to an average of 4–5 m at other locations along the Northern Coast of Flores Island [*Tsuji et al.*, 1995*b*]. Underwater surveys [*Plafker*, 1997] later identified submarine landslides triggered locally by the earthquake, as the probable source of such inconsistent run-up values. This interpretation is also likely for the exceptional run-up of 21 m observed at Nusa Kambangan during the 2006 Java tsunami [*Fritz et al.*, 2006*a*].

7. Conclusion

In conclusion, this Chapter has reviewed the theoretical bases of our present un-derstanding of the excitation of tsunamis by earthquake sources. The exceptional character of tsunamis as natural hazards results from their ability to export death and destruction across the largest oceanic basins. In this respect, a very large seis-mic moment is clearly a pre-requisite for fatalities in the far field. However, the dataset in Table 1 clearly shows that other factors can affect the impact of a trans-oceanic tsunami. Among them, directivity is the most important one, explaining the patterns of loss of life in the far field in 1946, 1960, 1964 and 2004. By contrast, the 1963 Kuriles and 1965 Rat Island earthquakes had directivity lobes at azimuths N133°E and 200°E, respectively 33° and 61°E away from the direction of Hawaii. There are no high (volcanic) islands lying in the lobes of directivity for their tsu-namis, only the sparsely populated atolls of Micronesia and the Marshall Islands. Such structures featuring small islands with steep underwater slopes, disfavor shoaling and run-up by long waves, as observed in the Maldives during the 2004 Sumatra-Andaman tsunami [*Fritz et al.*, 2006*b*]. This explains the lack of damage and fatalities from the tsunamis of 1963 and 1965 in the far field.

The case of the 1952 Kamchatka and 1957 Aleutian Islands events is more puz-zling. These were truly huge earthquakes, even though some uncertainty remains regarding the seismic moment of the latter [*Johnson et al.*, 1994]. Documents re-leased during the 1990s indicate that the 1952 tsunami totally destroyed the city of Severo-Kuril'sk and the adjacent Soviet Navy base, with a death toll estimated as high as 14,000 [*Kaistrenko and Sedaeva*, 2001]. As expected from its directivity lobe oriented N122°E, only 5° away from the azimuth of Hawaii, the tsunami im-pacted the Hawaiian Islands, resulting in considerable flooding and damage, as did the 1957 Aleutian earthquake, whose directivity lobe is within a few degrees of the azimuth to Hawaii. Miraculously, neither of the two events caused any deaths in

Hawaii; by contrast, 61 lives were lost in Hawaii during the 1960 Chilean tsunami. *Fraser et al.*'s [1959] account of the 1957 event reports what amounts to a conservatively cautious warning which, most importantly, was well heeded by the population, whereas the 1960 warning was poorly understood by the residents, and actually called off by officials after the first wavetrains [*Eaton et al.,* 1961].

As for the 2005 Nias event, in addition to the small far field tsunami reflecting the shallow bathymetry and presence of islands at the source, the absence of tsunami casualties in the near field can be attributed to the combination of a population educated by oral tradition, a strong coseismic uplift on Simeulue Island which helped reduce the impact of the wave, and the fact that many residents were living in refugee camps inland as a result of the December 2004 event [*McAdoo et al.,* 2006].

As detailed in this chapter, we now understand many facets of tsunami generation by earthquakes. Yet, as Table 1 shows, even the largest events can generate tsunamis of widely varying impact, both in the far field and, perhaps more surprisingly, in the near field as well. After fifty years, *Fraser et al.*'s [1959] remark quoted in the epigraph and written in the wake of the 1957 Aleutian tsunami, still prevails. Only through more research on past and future tsunamis, and through the education of populations at risk, will tragedies comparable to the 2004 Sumatra-Andaman tsunami be avoided in the future.

Appendix

We compute the energy transferred to the tsunami in the framework of Fig. 5-02. Assuming that a hump is created on the sea surface above the deformed section of the ocean bottom, the work W necessary to lift a mass $\rho_w\, S\, \delta h$ of water over the height H of the ocean column is

$$W = \rho_w\, S\, \delta h \cdot g \cdot H \tag{A1}$$

which is also the work of the pressure forces at the sea bottom, $\rho_w\, gH \cdot S$, displacing the ocean floor an amount δh. This work will be applied in part towards changing the potential energy E_p of the relevant mass of water between its average heights above undeformed sea floor, $\delta h/2$ in the original state *(a)* and H in the final state

$$\Delta E_p = \rho_w\, S\, \delta h \cdot g \cdot \left(H - \frac{\delta h}{2} \right) \tag{A2}$$

The difference $W - \Delta E_p$ is the energy of the tsunami:

$$E_T = \frac{1}{2} \rho_w\, g\, S\, (\delta h)^2 \tag{A3}$$

It can also be interpreted as the release of the potential energy of the unstable hump *(b)* as it collapses on the ocean surface *(e)*. Note in particular that this energy, growing like $(\delta h)^2$ is always positive, regardless of the sign of δh.

If the deformation were to proceed slowly and reversibly, without the creation of the hump, the pressure at the bottom would vary over the course of the deformation, adjusting from $\rho_w g H$ to $\rho_w g (H - \delta h)$, leading to W exactly equal to ΔE_p, as given by (A2), with no energy available to generate a tsunami.

Under earthquake scaling laws, all linear dimensions in (A3) are expected to grow like seismic slip Δu or equivalently fault length L, leading to E_T proportional to L^4, or $M_0^{4/3}$. This dependence was formally derived by *Kajiura* [1981], who proposed the expression

$$\left(E_T\right)_{Kaj.} = \frac{F_{Kaj.}}{2^{4/3}} \cdot \frac{\rho_w g}{\mu^{4/3}} \, \varepsilon_{\max}^{2/3} \cdot M_0^{4/3} \qquad (A4)$$

where $F_{Kaj.}$ is the square of the dimensionless ratio of δh to the slip Δu, averaged over focal geometries and hypocentral depths. *Okal* [2003] has derived a very similar expression

$$\left(E_T\right)_{modes} = 0.22 \cdot \frac{\rho_w g}{\mu^{4/3}} \, \varepsilon_{\max}^{2/3} \cdot M_0^{4/3} \qquad (A5)$$

by integrating over frequency the power spectrum density of the individual tsunami modes, weighted by their expected excitation and taking into account the source spectrum and its seismic corner frequencies. Note that (A3), (A4) and (A5) do not depend on the depth H of the oceanic column.

Acknowledgments

I am indebted to Costas Synolakis for many years of friendship and collaboration. A special word of appreciation goes to George Plafker for instilling in my mind the concept of the "Plafker rule of thumb" while roaming the fog-shrouded, bear-infested beaches of Unimak Island in 2001. I also thank many other scientists, too numerous to list individually, for enlightening my vision of tsunamis. The preparation of this chapter was supported by the National Science Foundation, under Grant CMS-03-01054.

References

Abe, Ka., A new scale of tsunami magnitude, M_t, in: *Tsunamis — their science and engineering*, ed. by K. Iida and T. Iwasaki, pp. 91–101, TERRA- PUB, Tokyo, 1983.

Abe, Ka., Quantification of tsunamigenic earthquakes by the M_t scale, *Tectonophysics*, **166,** 27-34, 1989.

Abe, Ku., A dislocation model of the 1933 Sanriku earthquake consistent with the tsunami waves, *J. Phys. Earth*, **26,** 381–396, 1978.

Abe, Ku., Ka. Abe, Y. Tsuji, F. Imamura, H. Katao, I. Yohihisa, K. Satake, J. Bourgeois, E. Noguera, and F. Estrada, Field survey of the Nicaragua earthquake and tsunami of September 2, 1992, *Bull. Earthq. Res. Inst. Tokyo Univ.*, **68,** 23–70, 1993.

Aki, K., Generation and propagation of *G*-waves from the Niigata earthquake of June 16, 1964. Part 2. Estimation of earthquake moment, released energy, and stress-strain drop from the *G*-wave spectrum, *Bull. Earthq. Res. Inst. Tokyo Univ.*, **44,** 73–88, 1966.

Aki, K., and P. G. Richards, *Quantitative Seismology,* W. H. Freeman, San Francisco, 932 pp., 1980.

Artru, J., V. Dučić, H. Kanamori, P. Lognonné, and M. Murakami, Ionospheric detection of gravity waves induced by tsunamis, *Geophys. J. Intl.,* **160,** 840–848, 2005.

Banerjee, P., F. Pollitz, B. Nagarajan, and R. Bürgmann, Coseismic slip distributions of the 26 December 2004 Sumatra-Andaman and 28 March 2005 Nias earthquakes from GPS static offsets, *Bull. Seismol. Soc. Amer.,* **87,** S86–S102, 2007.

Ben-Menahem, A., Radiation of seismic surface waves from finite moving sources, *Bull. Seismol. Soc. Amer.,* **51,** 401–435, 1961.

Ben-Menahem, A., and M. Rosenman, Amplitude patterns of tsunami waves from submarine earthquakes, *J. Geophys. Res.,* **77,** 3097–3128, 1972.

Boatwright, J., and G. L. Choy, Teleseismic estimates of the energy radiated by shallow earthquakes , *J. Geophys. Res.,* **91,** 2095–2112, 1986.

Borrero, J. C., Field data and satellite imagery of tsunami effects in Banda Aceh, *Science,* **308,** 1596, 2005.

Borrero, J. C., M. Ortiz, V. V. Titov, and C. E. Synolakis, Field survey of Mexican tsunami produced new data, unusual photos, *Eos, Trans. Amer. Geophys. Un.,* **78,** 85 and 87–88, 1997.

Borrero, J. C., R Hidayat, Suranto, C. Bosserelle, and E. A. Okal, Field survey and preliminary modeling of the near-field tsunami from the Bengkulu earthquake of 12 September 2007, *Eos, Trans. Amer. Geophys. Un.,* **88,** *(52),* U54A-04, 2007 [abstract].

Bourgeois, J., C. Petroff, H. Yeh, V. V. Titov, C. E. Synolakis, B. Benson, J. Kuroiwa, J. Lander, and E. Norabuena, Geologic setting, field survey and modeling of the Chimbote, northern Peru tsunami of 21 February 1996, *Pure Appl. Geophys.,* **154,** 513–540, 1999.

Bourgeois, J., T. Pinegina, N. Razhegaeva, V. Kaistrenko, B. Levin, B. MacInnes, and E. Kravchunovskaya, Tsunami runup in the Middle Kuril Islands from the great earthquake of 15 November 2006, *Eos, Trans. Amer. Geophys. Un.,* **88,** *(52),* S51C-02, 2007 [abstract].

Brune, J. N., and G. R. Engen, Excitation of mantle Love waves and definition of mantle wave magnitude, *Bull. Seismol. Soc. Amer.,* **59,** 923–933, 1969.

Byrne, D. E., D. M. Davis, and L. R. Sykes, Loci and maximum size of thrust earthquakes and the mechanisms of the shallow region of subduction zones, *Tectonics,* **7,** 833–857, 1988.

Choy, G. L., and J. Boatwright, The energy radiated by the 26 December 2004 Sumatra-Andaman earthquake estimated from 10-minute *P*-wave windows, *Bull. Seismol. Soc. Amer.,* **97,** S18–S24, 2007.

Cifuentes, I. L., and P. G. Silver, Low-frequency source characteristics of the great 1960 Chilean earthquake, *J. Geophys. Res.,* **94,** 643–663, 1989.

Dahlen, F. A., Single force representation of shallow landslide sources, *Bull. Seismol. Soc. Amer.,* **83,** 130–143, 1993.

de Groot-Hedlin C. D., Estimation of the rupture length and velocity of the great Sumatra earthquake of Dec. 26, 2004 using hydroacoustic signals, *Geophys. Res. Letts.,* **32,** *(11),* L11303, 4 pp., 2005.

Dziewonski, A. M., A.-T. Chou, and J. H. Woodhouse, Determination of earthquake source parameters from waveform data for studies of global and regional seismicity, *J. Geophys. Res.,* **86,** 2825–2852, 1981.

Eaton, J. P., D. H. Richter, and W. U. Ault, The tsunami of May 23, 1960 on the Island of Hawaii, *Bull. Seismol. Soc. Amer.,* **51,** 135–157, 1961.

Farreras, S. F., and A. J. Sanchez, The tsunami threat on the Mexican West coast: A historical analysis and recommendations for hazard mitigation, *Natur. Haz.,* **4,** 301–316, 1991.

Fraser, G. D., J. P. Eaton, and C. K. Wentworth, The tsunami of March 9, 1957 on the Island of Hawaii, *Bull. Seismol. Soc. Amer.,* **49,** 79–90, 1959.

Fritz, H. M., and J. C. Borrero, Somalia field survey after the December 2004 Indian Ocean tsunami, *Earthquake Spectra,* **22,** S219–S233, 2006.

Fritz, H., J. Goff, C. Harbitz, B. McAdoo, A. Moore, H. Latief, N. Kalligeris, W. Kodjo, B. Uslu, V. Titov, and C. Synolakis, Survey of the July 17, 2006 Central Java tsunami reveals 21 m runup heights, *Eos, Trans. Amer. Geophys. Un.,* **87,** *(52),* S14A-06, 2006*a* [abstract].

Fritz, H. M., C. E. Synolakis, and B. G. McAdoo, Maldives field survey after the December 2004 Indian Ocean tsunami, *Earthquake Spectra,* **22,** S137-S154, 2006*b.*

Fukao, Y., Tsunami earthquakes and subduction processes near deep-sea trenches, *J. Geophys. Res.,* **84,** 2303–2314, 1979.

Geller, R. J., Scaling relations for earthquake source parameters and magnitudes, *Bull. Seismol. Soc. Amer.,* **66,** 1501–1523, 1976.

Gilbert, F., Excitation of the normal modes of the Earth by earthquake sources, *Geophys. J. Roy. astron. Soc.,* **22,** 223–226, 1970.

Gilbert, F, and A. M. Dziewonski, An application of normal mode theory to the retrieval of structural parameters and source mechanisms from seismic spectra, *Phil. Trans. Roy. Soc. London, Ser. A,* **278,** 187–269, 1975.

Goffe, J., P. L.-F. Liu, B. Higman, R. Morton, B. E. Jaffe, H. Fernando, P. Lynett, H. Fritz, C. Synolakis, and S. Fernando, Sri Lanka field survey after the December 2004 Indian Ocean tsunami, *Earthquake Spectra,* **22,** S173–S186, 2006.

González, F. I., C. L. Mader, M. Eble, and E. N. Bernard, The 1987–88 Alaskan Bight Tsunamis: Deep ocean data and model comparisons, *Nat. Hazards,* **4,** 119–139, 1991.

González, F. I., E. N. Bernard, C. Meinig, M. C. Eble, H. O. Mofjeld, and S. Stalin, The NTHMP tsunameter network, *Nat. Hazards,* **35,** 25–39, 2005.

Guilbert, J., J. Vergoz, E. Schisselé, A. Roueff, and Y. Cansi, Use of hydroacoustic and seismic arrays to observe rupture propagation and source extent of the M_w = 9.0 Sumatra earthquake, *Geophys. Res. Letts.,* **32,** *(15),* L15310, 5 pp., 2005.

Gutenberg, B., and C. F. Richter, Seismicity of the Earth, *Geol. Soc. Amer. Spec. Pap.,* **34,** 125 pp., 1941.

Gutenberg, B., and C. F. Richter, *Seismicity of the Earth and associated phenomena,* Princeton Univ. Press, Princeton, N.J., 308 pp., 1954.

Harkrider, D. G., C. A. Newton, and E. A. Flinn, Theoretical effect of yield and burst height of atmospheric explosions on Rayleigh wave amplitudes, *Geophys. J. Roy. astr. Soc.,* **36,** 191–225, 1974.

Haskell, N. A., A note on air-coupled surface waves, *Bull. Seismol. Soc. Amer.,* **41,** 295–300, 1951.

Hwang, L.-S., H. L. Butler, and D. J. Divoky, Tsunami mode: generation and open sea characteristics, *Bull. Seismol. Soc. Amer.,* **62,** 1579–1596, 1972.

Imamura, F., C. E. Synolakis, E. Gica, V. Titov, E. Listanco, and H. J. Lee, Field survey of the 1994 Mindoro Island, Philippines, tsunami, *Pure Appl. Geophys.,* **144,** 875–890, 1995.

Ishii, M., P. Shearer, H. Houston, and J. E. Vidale, Extent, duration and speed of the 2004 Sumatra-Andaman earthquake imaged by Hi-Net array, *Nature,* **435,** 933–936, 2005.

Iwasaki, S. I., Experimental study of a tsunami generated by a horizontal motion of a sloping bottom, *Bull. Earthq. Res. Inst. Tokyo Univ.,* **57,** 239–262, 1982.

Johnson, J. M., Y. Tanioka, L. J. Ruff, K. Satake, H. Kanamori, and L. R. Sykes, The 1957 great Aleutian earthquake, *Pure Appl. Geophys.,* **142,** 3–28, 1994.

Kaistrenko, V. M., and V. M. Sedaeva, 1952 North Kuril tsunami: new data from archives, **in:** *Tsunami research at the end of a critical decade,* ed. by G. T. Hebenstreit, pp. 91–102, Kluwer, Netherlands, 2001.

Kajiura, K., Tsunami energy in relation to parameters of the earthquake fault model, *Bull. Earthq. Res. Inst. Tokyo Univ.,* **56,** 415–440, 1981.

Kanamori, H., Synthesis of long-period surface waves and its application to earthquake source studies—Kurile Islands earthquake of October 13, 1963, *J. Geophys. Res.,* **75,** 5011–5027, 1970*a.*

Kanamori, H., The Alaska earthquake of 1964: Radiation of long-period surface waves and source mechanism, *J. Geophys. Res.*, **75**, 5029–5040, 1970b.

Kanamori, H., Seismological evidence for a lithospheric normal faulting—The Sanriku earthquake of 1933, *Phys. Earth Planet. Inter.*, **4**, 289–300, 1971.

Kanamori, H., Mechanism of tsunami earthquakes, *Phys. Earth Planet. Inter.*, **6**, 346–359, 1972.

Kanamori, H., The energy release in great earthquakes, *J. Geophys. Res.*, **82**, 2981–2987, 1977.

Kanamori, H., and D. L. Anderson, Theoretical basis of some empirical relations in seismology, *Bull. Seismol. Soc. Amer.*, **65**, 1073–1095, 1975.

Kanamori, H., and J. J. Cipar, Focal processes of the Great Chilean earthquake May 22, 1960, *Phys. Earth Planet. Inter.*, **9**, 128–136, 1974.

Kato, K., and Y. Tsuji, Tsunami of the Sumbawa earthquake of August 19, 1977, *J. Natur. Disaster Sci.*, **17**, 87–100, 1995.

Kerr, R. A., Model shows islands muted tsunami after latest Indonesian quake, *Science*, **308**, 341, 2005.

Knopoff, L., and F. Gilbert, Radiation from a strike-slip earthquake, *Bull. Seismol. Soc. Amer.*, **49**, 163–178, 1959.

Kulikov, E. A., P. P. Medvedev, and S. S. Lappo. Satellite recording of the Indian Ocean tsunami on December 26, 2004, *Doklady Earth Sci.*, **401**, 444–448, 2005.

López, A. M., and E. A. Okal, A seismological reassessment of the source of the 1946 Aleutian "tsunami" earthquake, *Geophys. J. Intl.*, **165**, 835–849, 2006.

Love, A. E. H., *Some problems in geodynamics,* Cambridge Univ. Press, 1911.

Lundgren, P. R., and E. A. Okal, Slab decoupling in the Tonga arc: the June 22, 1977 earthquake, *J. Geophys. Res.*, **93**, 13355–13366, 1988.

Macdonald, G. A., and C. K. Wentworth, The tsunami of November 4, 1952 on the Island of Hawaii, *Bull. Seismol. Soc. Amer.*, **44**, 463–469, 1954.

Madariaga, R. I., Toroidal free oscillations of the laterally heterogeneous Earth, *Geophys. J. Roy. astr. Soc.*, **27**, 81–100, 1972.

Mansinha, L., and D. E. Smylie, The displacement fields of inclined faults, *Bull. Seismol. Soc. Amer.*, **61**, 1433–1440, 1971.

Marshall, P. D., and P. W. Basham, Discrimination between earthquakes and underground explosions using an improved M_s scale, *Geophys. J. Roy. astr. Soc.*, **28**, 431–458, 1972.

McAdoo, B. G., L. Dengler, G. Prasetya, and V. Titov, *Smong:* How an oral history saved thousands on Indonesia's Simeulue Island during the December 2004 and March 2005 tsunamis, *Earthquake Spectra*, **22**, S661–S669, 2006.

Menyaïlov, A. A., Tsunami v Ust'-Kamchatskom raïone, *Byull. Vulkan. Stantsii na Kamchat., Akad. Nauk SSSR*, **12**, 9–13, 1946 [in Russian].

Newman, A. V., and E. A. Okal, Teleseismic estimates of radiated seismic energy: The E/M_0 discriminant for tsunami earthquakes, *J. Geophys. Res.*, **103**, 26885–26898, 1998.

Ni, S., H. Kanamori, and D. V. Helmberger, D., High-frequency radiation from the 2004 Great Sumatra-Andaman earthquake *Nature*, **434**, 582, 2005.

Occhipinti, G., P. Lognonné, E. Alam Kherani, and H. Hébert, 3-Dimensional waveform modeling of ionospheric signature induced by the 2004 Sumatra tsunami, *Geophys. Res. Letts.*, **33**, *(20)*, L20104, 5 pp., 2006.

Okada, Y., Surface deformation due to shear and tensile faults in a half-space, *Bull. Seismol. Soc. Amer.*, **75**, 1135–1154, 1985.

Okal, E. A., Mode-wave equivalence and other asymptotic problems in tsunami theory, *Phys. Earth Planet. Inter.*, **30**, 1–11, 1982.

Okal, E. A., Seismic parameters controlling far-field tsunami amplitudes: A review, *Natural Hazards*, **1**, 67–96, 1988.

Okal, E. A., Erratum [to "Seismic parameters controlling far-field tsunami amplitudes: A review"], *Natural Hazards,* **4,** 433, 1991.

Okal, E. A., Use of the mantle magnitude M_m for the reassessment of the seismic moment of historical earthquakes. I: Shallow events, *Pure Appl. Geophys.,* **139,** 17–57, 1992.

Okal, E. A., Normal modes energetics for far-field tsunamis generated by dislocations and landslides, *Pure Appl. Geophys.,* **160,** 2189–2221, 2003.

Okal, E. A., Seismic records of the 2004 Sumatra and other tsunamis: A quantitative study, *Pure Appl. Geophys.,* **164,** 325–353, 2007a.

Okal, E. A., The generation of T waves by earthquakes, *Adv. Geophys.,* **49,** 1–65, 2007b.

Okal, E. A., and H. Hébert, Far-field modeling of the 1946 Aleutian tsunami, *Geophys. J. Intl.,* **169,** 1229–1238, 2007.

Okal, E. A., and S. H. Kirby, Energy-to-moment ratios for damaging intraslab earthquakes: Preliminary results on a few case studies, *USGS Open File Rept.,* **02–328,** 127–131, 2002.

Okal, E. A., and D. R. MacAyeal, Seismic recording on drifting icebergs: Catching seismic waves, tsunamis and storms from Sumatra and elsewhere, *Seismol. Res. Letts.,* **77,** 659–671, 2006.

Okal, E. A., and A. V. Newman, Tsunami earthquakes: The quest for a regional signal, *Phys. Earth Planet. Inter.,* **124,** 45–70, 2001.

Okal, E. A., and B. A. Romanowicz, On the variation of b-value with earthquake size, *Phys. Earth Planet. Inter.,* **87,** 55–76, 1994.

Okal, E. A., and C. E. Synolakis, Theoretical comparison of tsunamis from dislocations and landslides, *Pure Appl. Geophys.,* **160,** 2177–2188, 2003.

Okal, E. A., and C. E. Synolakis, Source discriminants for near-field tsunamis, *Geophys. J. Intl.,* **158,** 899–912, 2004.

Okal, E. A., and J. Talandier, M_m: A variable period mantle magnitude, *J. Geophys. Res.,* **94,** 4169–4193, 1989.

Okal, E. A., and J. Talandier, Single-station estimates of the seismic moment of the 1960 Chilean and 1964 Alaskan earthquakes, using the mantle magnitude M_m, *Pure Appl. Geophys.,* **136,** 103–126, 1991.

Okal, E. A., and V. V. Titov: M_{TSU}: Recovering seismic moments from tsunameter records, *Pure Appl. Geophys.,* **164,** 355–378, 2007.

Okal, E. A., A. Piatanesi, and P. Heinrich, Tsunami detection by satellite altimetry, *J. Geophys. Res.,* **104,** 599–615, 1999.

Okal, E. A., C. E. Synolakis, G. J. Fryer, P. Heinrich, J. C. Borrero, C. Ruscher, D. Arcas, G. Guille, and D. Rousseau, A field survey of the 1946 Aleutian tsunami in the far field, *Seismol. Res. Letts.,* **73,** 490–503, 2002.

Okal, E. A., P.-J. Alasset, O. Hyvernaud, and F. Schindelé, The deficient T waves of tsunami earthquakes, *Geophys. J. Intl.,* **152,** 416–432, 2003.

Okal, E. A., A. Sladen, and E. A.-S. Okal, Rodrigues, Mauritius and Réunion Islands, field survey after the December 2004 Indian Ocean tsunami, *Earthquake Spectra,* **22,** S241–S261, 2006a.

Okal, E. A., H. M. Fritz, R. Raveloson, G. Joelson, P. Pančošková, and G. Rambolamanana, Madagascar field survey after the December 2004 Indian Ocean tsunami, *Earthquake Spectra,* **22,** S263–S283, 2006b.

Okal, E. ., J. Talandier, and D. Reymond, Quantification of hydrophone records of the 2004 Sumatra tsunami *Pure Appl. Geophys.,* **164,** 309–323, 2007.

Pančošková P., E. A. Okal, D. R. MacAyeal, and R. Raveloson, Delayed response of far-field harbors to the 2004 Sumatra tsunami: the role of high-frequency components, *Eos, Trans. Amer. Geophys. Un.,* **87,** *(52),* U53A-0021, 2006 [abstract].

Pekeris, C. L., and H. Jarosch, The free oscillations of the Earth, **in:** *Contributions in Geophysics in honor of Beno Gutenberg,* Edited by H. Benioff, M. Ewing, B. F. Howell, Jr., and F. Press, pp. 171–192, Pergamon, New York, 1958.

Peltier, W. R., and C. O. Hines, On the possible detection of tsunamis by a monitoring of the ionosphere, *J. Geophys. Res.,* **81,** 1995–2000, 1976.

Plafker, G. L., Catastrophic tsunami generated by submarine slides and backarc thrusting during the 1992 earthquake on Eastern Flores, Indonesia, *Geol. Soc. Amer. Abstr. with Prog.,* **29,** *(5),* 57, 1997 [abstract].

Plafker, G., and J. C. Savage, Mechanism of the Chilean earthquakes of May 21 and 22, 1960, *Geol. Soc. Amer. Bull.,* **81,** 1001–1030, 1970.

Polet, J., and H. Kanamori, Shallow subduction zone earthquakes and their tsunamigenic potential, *Geophys. J. Intl.,* **142,** 684–702, 2000.

Reymond, D., and E. A. Okal, Rapid, yet robust source estimates for challenging events: Tsunami earthquakes and mega-thrusts, *Eos, Trans. Amer. Geophys. Un.,* **87,** *(52),* S14A-02, 2006 [abstract].

Romanowicz, B. A., Strike-slip earthquakes on quasi-vertical transcurrent faults; inferences for general scaling relations, *Geophys. Res. Letts.,* **19,** 481–484, 1992.

Ruff, L. J., and H. Kanamori, Seismicity and the subduction process, *Phys. Earth Planet. Inter.,* **23,** 240–252, 1980.

Rundle, J. B., Derivation of the complete Gutenberg-Richter magnitude-frequency relation using the principle of scale invariance, *J. Geophys. Res.,* **94,** 12337–12342, 1989.

Saito, M., Excitation of free oscillations and surface waves by a point source in a vertically heterogeneous earth, *J. Geophys. Res.,* **72,** 3895–3904, 1967.

Satake, K., Effects of bathymetry on tsunami propagation: Application of ray tracing to tsunamis, *Pure Appl. Geophys.,* **126,** 28–35, 1988.

Scharroo, R., W. H. F. Smith, V. V. Titov, and D. Arcas, Observing the Indian Ocean tsunami with satellite altimetry, *Geophys. Res. Abs.,* **7,** 230, 2005 [abstract].

Schwarz, H.-U., *Subaqueous slope failures — Experiments and modern occurrences,* E. Schweizerbart'sche Verlagsbuchhandlung, 116 pp., Stuttgart, 1982.

Seno, T., and K. Hirata, Did the 2004 Sumatra-Andaman earthquake involve a component of tsunami earthquakes?, *Bull. Seismol. Soc. Amer.,* **97,** S296–S306, 2007.

Solov'ev, S. L., *Zemletryaseniya i tsunami 13 i 20 oktyabrya 1963 goda na Kuril'skikh ostrovakh,* Akad. Nauk SSSR, Sibirs. Otdel., 105 pp., Yuzhno- Sakhalinsk, 1965 [in Russian].

Solov'ev, S. L., and M. D. Ferchev, Summary of data on tsunamis in the USSR, *Bull. Council Seismol. Acad. Sci. USSR,* **9,** 23–55, transl. by W. G. Van Campen, Hawaii Inst. Geophys. Transl. Ser., 37 pp., Honolulu, 1961.

Solov'ev, S. L., and Ch. N. Go, Catalogue of tsunamis on the Western Shore of the Pacific Ocean, *Can. Transl. Fish. Aquat. Sci.,* **6077,** 437 pp., 1984.

Stein, S., and E. A. Okal, Size and speed of the Sumatra earthquake, *Nature,* **434,** 581–582, 2005.

Stein, S., and E. A. Okal, Ultra-long period seismic study of the December 2004 Indian Ocean earthquake and implications for regional tectonics and the subduction process, *Bull. Seismol. Soc. Amer.,* **97,** S279–S295, 2007.

Steketee, J. A., On Volterra's dislocations in a semi-infinite elastic medium, *Can. J. Phys.,* **36,** 192–205, 1958.

Synolakis, C. E., The runup of long waves, *Ph.D. Dissertation,* Calif. Inst. Technol., 228 pp., Pasadena, 1986.

Synolakis, C. E., and E. A. Okal, 1992–2002: Perspective on a decade of post-tsunami surveys, **in:** *Tsunamis: Case studies and recent developments,* ed. by K. Satake, *Adv. Natur. Technol. Hazards,* **23,** pp. 1–30, 2005.

Synolakis, C. E., and E. A. Okal, Far-field tsunami risk from mega-thrust earthquakes in the Indian Ocean, *Eos, Trans. Amer. Geophys. Un.,* **87,** *(52),* U53A-0040, 2006 [abstract].

Synolakis, C. E., J.-P. Bardet, J. C. Borrero, H. L. Davies, E. A. Okal, E. A. Silver, S. Sweet, and D. R. Tappin, The slump origin of the 1998 Papua New Guinea tsunami, *Proc. Roy. Soc. (London), Ser. A,* **458,** 763–789, 2002.

Synolakis, C. E., J. C. Borrero, H. M. Fritz, V. V. Titov, and E. A. Okal, Inundation during the 26 December 2004 tsunami, *Coastal Engineering 2006,* ed. by J. McKee Smith, pp. 1625–1637, Word Scientific, Singapore, 2007.

Tadepalli, S., and C. E. Synolakis, The runup of *N* – waves, *Proc. Roy. Soc. London, Ser. A,* **445,** 99–112, 1994.

Tadepalli, S., and C. E. Synolakis, Model for the leading waves of tsunamis, *Phys. Rev. Letts.,* **77,** 2141–2145, 1996.

Talandier, J., and E. A. Okal, Human perception of *T* waves: the June 22, 1977 Tonga earthquake felt on Tahiti, *Bull. Seismol. Soc. Amer.,* **69,** 1475–1486, 1979.

Talandier, J., and E. A. Okal, An algorithm for automated tsunami warning in French Polynesia, based on mantle magnitudes, *Bull. Seismol. Soc. Amer.,* **79,** 1177–1193, 1989.

Tang, L., M. C. Spillane, M. Eble, R. Weiss, V. V. Titov, and E. N. Bernard, The Tonga tsunami of May 3, 2006 : A comprehensive test for developing the NOAA tsunami forecast system, *Eos, Trans. Amer. Geophys. Un.,* **87,** *(52),* T-21F06, 2006 [abstract].

Tanioka, Y., and K. Satake, Tsunami generation by horizontal displacement of ocean bottom, *Geophys. Res. Letts.,* **23,** 861–864, 1996*a.*

Tanioka, Y., and K. Satake, Fault parameters of the 1896 Sanriku tsunami earthquake estimated from tsunami numerical modeling, *Geophys. Res. Letts.,* **23,** 1549–1552, 1996*b.*

Tanioka, Y., L. J. Ruff, and K. Satake, What controls the lateral variation of large earthquake occurrence along the Japan trench?, *Island Arc,* **6,** 261–266, 1997.

Tinti, S., and A. Armigliato, Single-force point-source static fields: an exact solution for two elastic half-spaces, *Geophys. J. Intl.,* **135,** 607–626, 1998.

Titov, V. V., and C. E. Synolakis, Numerical modeling of tidal wave run-up, *J. Waterw. Port Ocean Coastal Eng.,* **124,** 157–171, 1998.

Tolstoy, M., and D. R. Bohnenstiehl, Hydroacoustic constraints on the rupture duration, length and speed of the great Sumatra-Andaman earthquake, *Seismol. Res. Letts.,* **76,** 419–425, 2005.

Tsai, V. C., M. Nettles, G. Ekström, and A. M. Dziewonski, Multiple CMT analysis of the 2004 Sumatra earthquake, *Geophys. Res. Letts.,* **32,** *(17),* L17304, 4 pp., 2005.

Tsuji, Y., F. Imamura, H. Matsutomi, C. E. Synolakis, P. T. Nanang, Jumadi, S. Harada, S. S. Han, K. Arai, and B. Cook, Field survey of the East Java earthquake and tsunami of June 3, 1994, *Pure Appl. Geophys.,* **144,** 839–854, 1995*a.*

Tsuji, Y., H. Matsutomi, F. Imamura, M. Takeo, Y. Kawata, M. Matsuyama, T. Takahashi, Sunarjo, and P. Harjadi, Damage to coastal villages due to the 1992 Flores Island earthquake tsunami, *Pure Appl. Geophys.,* **144,** 481–524, 1995*b.*

Vvedenskaya, A. V., Opredelenie polej smeshchenii pri zemletryaseniyakh s pomoshchyu teorii dislokatsii, *Izv. Akad. Nauk SSSR, Ser. Geofiz.,* **6,** 277–284, 1956 [in Russian].

Ward, S. N., Relationship of tsunami generation and an earthquake source, *J. Phys. Earth,* **28,** 441–474, 1980.

Ward, S. N., On tsunami nucleation: I. A point source, *J. Geophys. Res.,* **86,** 7895–7900, 1981.

Ward, S. N., On tsunami nucleation: II. An instantaneous modulated line source, *Phys. Earth Planet. Inter.,* **27,** 273–285, 1982.

Weinstein, S. A., and E. A. Okal, The mantle wave magnitude M_m and the slowness parameter Θ: Five years of real-time use in the context of tsunami warning, *Bull. Seismol. Soc. Amer.,* **95,** 779–799, 2005.

Wells, D. L., and K. J. Coppersmith, New empirical relationships among magnitude, rupture length, rupture width, rupture area, and surface displacement, *Bull. Seismol. Soc. Amer.*, **84**, 974–1002, 1994.

Wiggins, R. A., A fast, new computational algorithm for free oscillations and surface waves, *Geophys. J. Roy. astr. Soc.*, **47**, 135–150, 1976.

Woodhouse, J. H., and F. A. Dahlen, The effect of a general aspherical perturbation on the free oscillations of the Earth, *Geophys. J. Roy. astr. Soc.*, **53**, 335–354, 1978.

Woods, M. T., and E. A. Okal, Effect of variable bathymetry on the amplitude of teleseismic tsunamis: a ray-tracing experiment, *Geophys. Res. Letts.*, **14**, 765–768, 1987.

Wu, F. T., and H. Kanamori, Source mechanism of the February 4, 1965 Rat Island earthquake, *J. Geophys. Res.*, **73**, 6082–6092, 1973.

Yuen, P. C., P. F. Weaver, R. K. Suzuki, and A. S. Furumoto, Continuous, traveling coupling between seismic waves and the ionosphere evident in May 1968 Japan earthquake data, *J. Geophys. Res.*, **74**, 2256–2264, 1969.

Zachariasen, J., K. Sieh, F. W. Taylor, R. L. Edwards, and W. S. Hantoro, Submergence and uplift associated with the giant 1833 Sumatran subduction earthquake; evidence from coral microatolls, *J. Geophys. Res.*, **104**, 895–919, 1999.

Chapter 6. TSUNAMI GENERATION: OTHER SOURCES

GALEN R. GISLER

Physics of Geological Processes, University of Oslo

Contents

1. Introduction

Earthquakes are overwhelmingly dominant as a source mechanism for tsunamis, whether directly—by producing a rapid change in the seafloor topography that is communicated directly to the water—or indirectly—by triggering another event such as a landslide that in turn generates a water wave. In this chapter I shall be concerned with the coupling to water waves of rock motions that are associated with events other than earthquakes. To be specific, I refer to sub-aerial and submarine landslides, volcanic eruptions (pyroclastic flows and caldera collapses), and asteroid impacts. This chapter deals with the generation process only, not with propagation or run-up. I focus precisely on the process by which the kinetic energy of large-scale rock motion is transformed into the kinetic energy of water waves. Further, what are the conditions for the production of a long-wavelength tsunami as opposed to other sorts of water waves?

My perspective is that of a numerical modeler, and I shall therefore include several examples from modeling and a discussion of what techniques should be used. The biases I have come from my background in fully-compressible gas hydrodynamics, but I hope that my perspective helps to illuminate some issues in tsunami science.

The Sea, Volume 15, edited by Eddie N. Bernard and Allan R. Robinson
ISBN 978–0–674–03173–9 ©2009 by the President and Fellows of Harvard College

2. Examples of non-Earthquake Tsunamis

The National Geophysics Data Center (NGDC) of the US National Oceanic and Atmospheric Administration (NOAA) maintains a Tsunami Event Database on the web at http://www.ngdc.noaa.gov/seg/hazard/tsevsrch_idb.shtml. At present this database consists of over 2400 entries, of which over 1500 have earthquake sources. We point here to a few which do not, by way of illustration only.

The La Fossa volcano on the island of Vulcano suffered a landslide of about 2×10^5 cubic meters on the 20th April 1988. A local tsunami with amplitude about a meter, persisting for a few minutes, was observed by eyewitnesses, but no casualties or damage was reported (Tinti *et al.*, 1999).

A landslide at Heggura near Tafjord, Norway on the 7th April 1934 released 3×10^6 cubic meters of rock into a narrow fjord. The resulting tsunami had a run-up of 62 meters, washed away several communities, and killed about 40 people (Braathen *et al.*, 2004; Panthi and Nilsen, 2006).

The largest amplitude tsunami on record was produced in Lituya Bay, Alaska, on 8th July 1958 when 3×10^7 cubic meters of schists slid down the steep side of Gilbert Inlet and caused a wave that ran up on the opposite side to a height of 524 m. This landslide was indeed triggered by a 7.7 magnitude earthquake, but it is included in this listing of non-earthquake tsunamis because of its significance in the subject area of landslide tsunamis. Besides its immense size it is noteworthy because it has proved amenable both to physical modeling (Fritz et al., 2001) and numerical simulation (Mader and Gittings, 2002; Quecedo *et al.*, 2004). The successes of these efforts have established this event as a major benchmark that numerical models of tsunami generation must meet.

The construction of a reservoir in the Vaiont Valley in northern Italy led to the deadliest landslide-induced tsunami on record. While the reservoir was being filled, on 9th October 1963, some 3×10^8 cubic meters of rock slid into the lake at high speed, producing a massive wave that flooded one village upstream, then overtopped the dam and destroyed five villages downstream, killing some 2000 people in all. This disaster was probably triggered by rising pore pressures in a thin clay layer (due to the newly introduced water) that led to a catastrophic loss of strength (Semenza and Ghirotti, 2000).

In Skagway Harbor, Alaska, on the 3rd November 1994, a destructive tsunami wave with a maximum height of 7 to 9 meters killed one person and caused over $20 million in damages. An early suggestion as to the cause was that a cruise-ship wharf under construction collapsed, leading to an underwater landslide that produced the tsunami (Kulikov *et al.*, 1996). A more careful reconstruction of the chronology of events proved conclusively that the tsunami caused the dock to collapse and not the other way around (Campbell and Nottingham, 1999). Subsequent bathymetry of the Taiya Inlet revealed three slide regions separated by ridges, involving a total volume of more than 1.5×10^7 cubic meters of sediments. Modeling of the tsunami resulting from these slides using both shallow-water-wave and Navier-Stokes models has proven very successful in matching the observations (Mader, 2004). The slides were in turn caused by a heavy sediment load carried by the Skagway River during a major flood at an extremely low tide. Particularly notable about this event was a cross-inlet seiche with a 3-minute period and an amplitude of 1 meter that persisted for nearly an hour.

The Aniakchak Volcano in Alaska may have triggered a tsunami 3500 years ago via a pyroclastic flow (Waythomas and Watts, 2003). Both the explosive vaporization of water when contacted by hot lava and the physical displacement of the water contribute to the formation of the wave, but the physics is complex. An interesting modern, but rather confusing, case is the Soufriere Hills Volcano on Montserrat Island (Pararas-Carayannis, 2004), which has been in almost continual eruption since 1995, devastating the southern portion of the island including the destruction of its capital. Among its many modes of activity, it generates pyroclastic flows and tsunamis, which are likely associated with each other, although flank collapses may also contribute (Pelinovsky et al., 2004).

Before the devastating 2004 Sumatra tsunami, the deadliest known tsunami was that produced by the volcano Krakatau in 1883. The mechanism for the generation of the tsunami is still debated, but possibilities include a submarine explosion, a caldera collapse, or the sudden emplacement of pyroclastic flow deposits. In any case, the total displacement of water probably exceeded 10^{10} cubic meters. Near-field tsunami heights exceeded 15 meters, and some 40,000 people lost their lives (Nomanbhoy and Satake, 1995; Mader and Gittings, 2006).

The great Storegga submarine slide off the western coast of Norway released 2.4×10^{12} cubic meters with a run-out distance of 450 kilometers (Løvholt et al., 2005; Haugen et al., 2005; DeBlasio et al., 2003). This area has been thoroughly studied on account of natural gas resources in the nearby Ormen-Lange field. Evidence of tsunami deposits from this event 7300 years ago have been found on nearby shores, including Scotland and Norway, with tsunami runups of 10–15 meters.

The Eltanin asteroid impact occurred in deep water in the Bellingshausen Sea, 1300 km west of the southern tip of South America, some 2.5 million years ago (Gersonde et al., 1997, 2002, 2003). Estimates of the projectile size vary, but the debris that has been collected locally point to a minimum diameter of about 1000 kilometer (Kyte, 2002a, b). Wave heights of several hundred meters have been calculated for this event within a few tens of kilometers of the impact point (Shuvalov, 2003), but these die off quickly with distance from the source. No unambiguous tsunami deposits have been recovered on land from this event. However, some extinct Cenozoic microfossils in Antarctica may be ejecta from the impact (Gersonde, 1997).

The Mjølnir impact structure was formed in the late Jurassic in relatively shallow water off the northwest coast of Norway (Tsikalas et al., 2002). The asteroid is thought to be of 1.6 km diameter, and the impact would also have produced waves hundreds of meters in amplitude near the impact point, decaying quickly away with distance (Glimsdal et al., 2005).

These examples illustrate several of the possible ways in which non-earthquake tsunamis are generated: subaerial landslides (La Fossa, Tafjord, Lituya Bay, Vaiont), submarine landslides (Skagway, Storegga), volcanic eruptions (Krakatau, Aniakchak, Soufriere Hills), and asteroid impacts (Eltanin, Mjølnir).

3. Experimental Work and Comparative Analyses

Experimental modeling of landslide-generated waves has been undertaken in a number of laboratories internationally. I briefly mention here some of the more recent developments.

Hermann Fritz and collaborators (Fritz *et al.*, 2003a, b) studied sub-aerial landslides using slurries of granular matter impacting into water and diagnosed with particle-image velocimetry. A notable conclusion of this work is that the water displacement volume significantly exceeds the landslide volume, owing to the energy dissipated by turbulence in the multi-phase mixing of slurry, air, and water. Solid-block modeling of submarine landslides have been done by a number of groups, and recently reviewed by Sue *et al.* (2006) who find from their own work that greater maximum wave heights are achieved with rapid deceleration of the slide material, thus making simple volume *vs.* wave height comparisons problematic or misleading.

In fact Murty (2003) has attempted to make a correlation between landslide volume and the maximum wave heights of the resultant tsunamis, with little success.

4. Numerical Modeling

Numerical work has also been done by a very large number of groups worldwide, and new developments in algorithms and computational facilities have enabled significant progress in this area. Conceptually the tsunami modeling problem is usually divided into three stages: the generation of a wave by an initial disturbance, the propagation of the wave over distances that may be very long, and the final stage of run-up and inundation. Traditionally, the non-linear shallow-water wave equations (*e.g.* Mader, 2004) are used for all but the generation phase. Often (perhaps usually) the generation phase is done simply by translating an ocean bottom displacement instantaneously into a similar displacement of the ocean surface and using that as input for a propagation model.

One of the best known of the simulation ensembles is the MOST suite (Titov and Gonzalez, 1997) developed and maintained by the Pacific Marine Environmental Laboratory. This consists of separate generation, propagation, and run-up models, and has been used very successfully in modeling a wide variety of events, most famously in the case of the 2004 December 26 Sumatra earthquake-driven event (Titov *et al.*, 2005). This suite is currently being tuned for operational use in many parts of the world, incorporating sophisticated models of harbors and ports and potential regions of slumps and slides.

George and Leveque (2006) have proposed a new adaptive-mesh model for computing global propagation and run-up using a finite-volume approach for solving the nonlinear shallow-water wave equations. Within a single code, this method allows for zooming in on coastal regions where variations in the local bathymetry strongly influence the flow. This method was applied to the 2004 December 26 Sumatra event at three levels of refinement for the coasts of Sri Lanka and India.

Global modeling of the propagation of waves from the Sumatra event has also been done by a new global code (Kowalik, 2005) using vertically averaged equations of motion including Coriolis forces and bottom friction. The difference equations are solved in a spherical system for the large-scale motion relative to the geopotential on a grid extending from 80° S to 69° N latitude and resolution of 1' (1 nautical mile, or 1.85 km). Parallel processing is required as the number of grid points exceeds 10^8, and with a time step of 2 seconds (to resolve the shallow-water wave speed for the deepest part of the ocean), 50 hours of tsunami propagation time was simulated using 9 hours of computer time on 40 processors. Arrival times showed good agreement with observations.

Boussinesq treatments (see, for example Watts *et al.,* 2003), which preserve density changes due to gravity, are superior to shallow-water treatments for propagation in basins with uneven bottoms. These have also been used with some success for tsunami generation in the case where the initial disturbance is relatively slow. An example of such an application that begins to look in some detail at the generation of tsunami waves from a non-earthquake source is provided by Waythomas and Watts (2003) who examine ancient pyroclastic flows from the Aniakchak Volcano. Unfortunately, they do not address the complex dynamics of the turbulent lava-water interaction, and therefore cannot reliably treat the near-field effects.

But while shallow-water equations and other vertically-averaged approximations work reasonably well for long-distance propagation, the generation region and the near-field effects, especially in violent cases, are poorly served by these approximations.

5.   Full Hydrodynamics

It has been pointed out (Trifunac and Todorovsk, 2002) that an important difference between earthquake and slide sources for tsunamis is the speed of the disturbance in the dislocation of the solid surface. For slides and slumps the dislocation velocities are comparable to the long-wavelength tsunami velocity, while for earthquakes the dislocation velocities are often considerably larger and the direct coupling to waves is correspondingly less efficient. The coseismic displacement in an earthquake is effectively instantaneous and, like a dam break, represents a new configuration to which the water surface must adjust. As I argue below, this consideration calls for a more complete treatment than is usually given.

Okal *et al.* (2003) point out that "tsunami earthquakes" are often deficient in seismic T waves, indicating slow rupture velocities in general for these kinds of events. Further, Okal and Synolakis (2003) specifically compare seismic dislocations and underwater slumps that are comparable in total energy, finding that slumps are more efficient, *per volume of material moved,* at generating tsunamis because slumps move the solid material farther. Additionally, because of the dipolar nature of the slumping source, near field waves are higher, and far field lower, than for seismic sources.

The issue here is the coupling of rock motion to water motion. These two dissimilar media are in the one case weakly deformable but mainly elastic, and in the other weakly elastic and highly mobile. While their dominant modes are often so distant from one another in phase space that they are usually decoupled, in the violent interactions that produce tsunamis the coupling of these modes is of considerable interest. We therefore argue here for a full hydrodynamic treatment in the tsunami generation region.

A tsunami results when a large volume of solid (block or granular) material moves and couples its kinetic energy very quickly to a comparable or larger volume of water. Water in normal experience is incompressible, and so tsunami calculations often begin with the water volume already displaced in some way, assuming in effect that the disturbance is instantaneously propagated from the point at which it occurs (often at great depth) to the surface. The water motion under the influence of gravity then follows using conventional incompressible hydrodynamics, a Boussinesq-type approximation, the simpler shallow-water wave treatments, or another depth-averaged approach. There are several reasons, however, why water should not be regarded as incompressible in calculations of the tsunami source regions, and why, in general, a fully compressible multi-fluid hydrodynamical approach is preferable.

From Landau and Lifshitz (1959) a fluid may be regarded as incompressible provided two conditions are met. The first is that the fluid velocity v should be much smaller than the speed of sound c:

$$v \ll c. \tag{1}$$

This comes from Bernoulli's observation that pressure changes are comparable to ρv^2 while the pressure itself is of order ρc^2. Thus in incompressible flow we ignore contributions of order v^2/c^2. The speed of a tsunami (shallow-water speed $\sqrt{gd}$, where g is the acceleration due to gravity and d is the ocean depth) in the deep ocean is ~200 m/s, while the sound speed in water is ~1500 m/s, so the incompressible approximation ignores contributions of order 2%, which would seem to be permissible, at least in the regime of tsunami propagation.

However, the second condition is that the time τ for a significant change in the flow should be much greater than the time l/c taken for a sound wave to cross the region over which that change occurs:

$$\tau \gg l/c. \tag{2}$$

This condition turns out to be much more stringent indeed. If we identify l with the local water depth (and not, even more stringently, with the basin size), and consider a perturbation induced in a tenth of a second, then this condition is not satisfied in water only 150 meters deep. Since earthquakes, landslide impacts, and explosive eruptions can all produce significant changes in the rock-water boundary in less than a tenth of a second, the problem of tsunami generation cannot be treated with incompressible methods, in general. The usual assumption of instantaneous displacement of the water surface, of course, manifestly violates condition (2). Nosov (1999) also makes this point. I'll elaborate this argument further with some phenomenological specifics.

Firstly, the coupling of materials having different sound speeds requires considerable care. An earthquake, for example, begins with the failure of the fault-locking mechanism at some point, and the rupture propagates along the fault at a speed that is most likely above the speed of sound in water. If the earthquake produces a sizable vertical displacement of rock, it pushes the water out of the way locally, but the water as a body can respond only as rapidly as the signal is propagated by acoustic waves. Regions of water not in acoustic causal contact respond independently to local rock motion, and depending on the geometry and the rupture speed, shock waves arise when the signals conflict. The resulting surface water wave is a superposition of the waves generated continuously along the rupture, but this superposition is strongly nonlinear because of the shocks that propagate through the water. This results in a surface wave that is augmented in some places and diminished in others from what it would be in a direct translation of the bottom disturbance to the surface.

Secondly, many tsunami generation mechanisms involve a thermodynamic phase change in one or more of the components of the problem. For example, in a pyroclastic flow or an asteroid impact, the explosive vaporization of water helps to drive the initial disturbance that evolves into the wave. This disturbance is plainly supersonic with respect to the liquid water, and the onset of turbulence cannot be neglected. It is important to conserve total energy, to keep track of how energy is partitioned among the internal and kinetic energies of the various components of the problem, to take the latent heats into account, and to track the production of entropy. Good equations of state are required for all the important components of the problem.

Even apart from the depth-over-sound-speed limit on the disturbance time, the great depth of typical ocean water, and gradients in salinity and temperature, mean that the water density varies considerably within the domain of the tsunami's generation region, and incompressible treatments are therefore inadequate. When the coupling between rock motion and water motion occurs at great depth while the resultant water wave is obtained at the surface, a fully compressible scheme is required. A Boussinesq treatment does not address this problem.

The notion of a thermodynamic equation of state needs generalizing from its usual equilibrium connotation. By "equation of state" I wish to include all of the relations that obtain between certain intensive and extensive properties of a particular material, such as temperature, pressure, internal energy, stress state, and density. The phase of a material (solid, liquid, or gas) is a function of these properties, and the ability of a material to sustain asymmetric stresses is a function in turn of the phase, and must therefore be included in a generalized equation of state. Nonequilibrium thermodynamics and mechanics are closely bound together, particularly when dissimilar materials are involved. The mechanical coupling between materials with different sound speeds and different latent heats of vaporization, as is the case for a tsunami that results from violent motion of rock, is typically not well treated even in multi-fluid hydrocodes, and we need better ways of doing this. Because the equations of state of geological materials are not very well constrained at present, more laboratory experimental work is also needed.

A slope failure that leads to a landslide is a generalized kind of phase change. A pile of grains on a slope is stable if it can sustain the asymmetric stress due to its own weight under gravity. A perturbation due to an earthquake, a change in pore pressure, a gas release, an injection of thermal energy, or some other kind of additional loading, is like a latent heat that causes a phase change: the granular assembly begins to act as a fluid when it can no longer sustain shear stress. As the slide progresses, the motion of granules relative to one another produces friction, fragmentation, melting, lubrication, and mixing with an ambient fluid (air or water or both). These processes often lead to a catastrophic reduction in viscosity, producing very long distance runouts. The large volumes involved can produce dramatically large tsunamis. All the energies must be accounted for: the latent heat involved in triggering the instability, the fragmentation and melting energies, and the decrease in dissipation due to lubrication. Importantly, the mixing of water with the granules and the granular melt material cannot be treated incompressibly, even though the flow speed is subsonic and the disturbance of long duration.

Finally, all tsunamis involve interaction with the atmosphere as well as motion of the water itself. In the source region, where the waves are not well formed and can be highly turbulent, this interaction is important, even though the density of the air is a thousand times less than that of water. Kelvin-Helmholtz instabilities associated with the relative motion of the two will induce turbulence and significantly affect the form of the final wave.

We start with the simplest form of the equations of motion for a single inviscid fluid, the Euler equations, consisting of the equation of continuity (conservation of mass),

$$\frac{\partial \rho}{\partial t} + \nabla \cdot (\rho \mathbf{v}) = 0; \tag{3}$$

the equation for the conservation of momentum,

$$\frac{\partial (\rho v)}{\partial t} + \nabla \cdot \rho \mathbf{v} \mathbf{v} + \nabla p = 0; \tag{4}$$

and the equation for the conservation of energy,

$$\frac{\partial (\rho \varepsilon)}{\partial t} + \nabla \cdot ((\rho \varepsilon + p)\mathbf{v}) = 0. \tag{5}$$

The symbols have their ordinary meaning: the density of the fluid is ρ, its velocity is $\mathbf{v}$, its specific energy and pressure are ε and p. I have omitted from the momentum equation (4) body forces such as the acceleration due to gravity $\mathbf{g}$ because their inclusion would require the addition to equation (5) of potential energy terms which are scale and geometry dependent. I have also omitted terms for heat flux and energy dissipation through friction or viscosity. The above general equations suffice for the purposes of discussion, though not for calculation.

The three scalar variables ε, p and ρ are related to one another, and to the auxilliary thermodynamic variable T (temperature) through equations of state. While there are useful analytic expressions for equations of state under some specific approximations (for example, the ideal gas law and the van der Waals equation of state), for geophysical materials these relations are usually found in the form of look-up tables.

Solid materials can sustain shear stresses, and therefore the scalar pressure in equations (4) and (5) is not an adequate description of the stress state. These two equations must be generalized as:

$$\frac{\partial (\rho v)}{\partial t} + \nabla \cdot \rho \mathbf{v} \mathbf{v} + \nabla \cdot \boldsymbol{\sigma} = 0; \tag{6}$$

and

$$\frac{\partial(\rho\varepsilon)}{\partial t} + \nabla \cdot (\rho\varepsilon\mathbf{v}) + \nabla \cdot (\boldsymbol{\sigma}\mathbf{v}) = 0, \tag{7}$$

with $\boldsymbol{\sigma}$ the stress tensor. For a fluid, the pressure p is the trace of $\boldsymbol{\sigma}$ and the off-diagonal components are zero, so equations (4) and (5) are exactly recovered. In this case, the equation of state must also be generalized, as mentioned above, to provide good approximations to the bulk modulus, shear modulus, yield strength, and melt energy of the solid so that its elastic behavior (Hooke's law), its deformability (plastic flow, for example), and changes of phase are recovered under the appropriate conditions of stress and temperature.

In geophysical situations, and particularly in regions in which tsunami generation occurs, material exists in all three phases (solid, liquid, and gas), and mixing occurs. While the general conservation laws (equations 3, 6, and 7) can be expected to hold for mixtures, the equation of state becomes even more complex, yet still amenable to numerical treatment, under appropriate approximations.

Beyond the tsunami generation region, once the surface water wave is well developed, it is reasonable to use traditional incompressible hydrodynamic techniques. In this case one might ignore the contribution of the atmosphere or the seafloor to the dynamics by using appropriate boundary conditions, and then replace the equation of continuity (3) by the divergence-free condition on the velocity:

$$\nabla \cdot \mathbf{v} = 0. \tag{8}$$

The addition of the viscosity stress tensor to the momentum flux density tensor in equation (4) then yields the Navier-Stokes equation for incompressible viscous flow. For following the propagation of tsunamis over very long distances, full three-dimensional hydrodynamics becomes prohibitively expensive and then it is appropriate to use a depth-averaged approach such as the shallow-water equations. However, it is the detailed physics of the source generation and interaction region that determines the initial form of the wave that is eventually propagated.

Following this philosophical preamble I move to discuss some tsunami simulations that my colleagues and I have done with a particular code that solves equations (3, 6 and 7) for multiple materials using tabular equations of state.

6. The Sage hydrocode

The Sage hydrocode is a multi-material, multi-phase, adaptive-grid Eulerian code with a high-resolution Godunov scheme originally developed by Michael Gittings for Science Applications International (SAIC) and Los Alamos National Laboratory (LANL). The grid refinement is continuous, cell-by-cell and cycle-by-cycle throughout the problem run. Refinement occurs when gradients in physical properties (density, pressure, temperature, material constitution) exceed user-defined limits, down to minimum cell sizes specified by the user. With the computing power concentrated on the regions of the problem which require higher resolution,

veŕy large computational volumes and substantial differences in scale can be simulated at low cost.

Sage runs in several geometries: 1-D Cartesian and spherical, 2-D Cartesian and
cylindrical, and 3-D Cartesian. Because modern supercomputing is often done on
clusters of many identical processors, the parallel implementation of the code is
supremely important. For portability and scalability, Sage uses the widely available
Message Passing Interface (MPI). Load leveling is accomplished through the use
of an adaptive cell pointer list, in which newly created daughter cells are placed
immediately after the mother cells. Cells are redistributed among processors at
every time step, while keeping mothers and daughters together. With M cells and
N processors, this gives very nearly M/N cells per processor. As neighbor-cell
variables are needed, the MPI gather/scatter routines copy those neighbor variables into local scratch.

In a multi-material code like Sage, every cell in the computational volume can
contain all the materials defined in the problem, each with its own equation of
state and strength model, as appropriate. There are a number of equations of state
available, analytical and tabular. In general we use the LANL Sesame tabular
equations of state for most materials in the problem, and for water we use a special
tabular equation of state produced by SAIC. For the rock material in these problems, we use a simple elastic-plastic strength model, specified by constant parameters for bulk modulus, shear modulus, yield strength, tensile failure, and melt
energy. A variety of boundary conditions are available, the most important being
reflective boundary walls, reflective internal boundaries, and "freeze regions"
which allow specified inflows and unrestricted outflows of material.

Although Sage is a multi-material and multi-phase code, it is a single-fluid code,
meaning that in every cell of the computational problem a single velocity, density,
pressure, and internal energy are defined. It does not therefore treat porous media
unless the pores are resolved by the grid, and matrix and fluid explicitly defined.
The Eulerian method in the Sage code is a hybrid that advances the conservation
equations (3, 6, and 7) by an intermediate Lagrangian half-step to construct cell-
face Eulerian fluxes before the full-step advance is taken by solving the Riemann
problem for the characteristics. The time step is set automatically by criteria to
ensure stability and accuracy, most importantly the Courant-Friedrich-Levy limit.
Conservation of mass, momentum, and energy are monitored and routinely
achieved to machine accuracy. The determination of unique values of the cell
variables is done through the assumption that local mixing within a cell results in
local thermodynamic equilibrium. Density, momentum, and internal energy are
advanced according to the conservation equations with known mass fractions for
each material, then partial pressures are calculated from the equations of state for
all the materials within the cell in a fast iterative procedure under a constraint
requiring a unique temperature in the cell as a result.

Interfaces between materials are resolved only by the computational grid, and
there is therefore no uniquely defined surface (for example, of water) to be
tracked. The advantage of this technique is that wave breaking is handled without
difficulty, and physical mixing of materials is explicitly allowed. The disadvantage
is that the Eulerian grid is inherently diffusive: if a particular material enters a cell,
it has access to all that cell's neighbors. The diffusion is controlled by the adaptive
refinement method; gradients in material properties at interfaces (and elsewhere)

cause the code to refine those cells to limits set by the user. Interface steepening methods are also available, and can be invoked to inhibit mixing, if desired.

The Sage code and its descendent Rage (which incorporates radiation transport) have been, and continue to be, subjected to extensive verification and validation studies at Los Alamos. These studies and more details on the mathematics and mechanics of the codes are found in Gittings (2006).

6.1. *Submarine volcano simulations*

Water and magma make a highly explosive combination, particularly at water depths less than about 130 meters. The explosive vaporization of water, heated by contact with magma at 1200 C or hotter, produces an instantaneous pressure of ~50 kBar, that can have extremely dangerous consequences. It is of interest to discuss whether there is a significant danger of tsunami from the submarine volcano Kick-em Jenny in the eastern Caribbean. We conclude here that there is not. The simulations that we have performed, in an axisymmetric geometry resembling Kick-em Jenny, suggest that only for very much more energetic events are significant waves generated, and that even these waves do not propagate as classical tsunami. These results are consistent with conclusions we have drawn from simulations of other explosively-generated waves.

Kick-em Jenny, located 8 km north of the island of Grenada in the volcanic arc of the Lesser Antilles, is one of the most active volcanos in the region. It has erupted a dozen times since 1939, and is a known hazard to shipping because of volcanic missiles that are sometimes projected to altitudes of a few hundred meters, and from gases emitted into the seawater from the volcano. No significant tsunami have been observed to arise from these eruptions.

Smith and Shepherd (1993, 1995, 1996) investigated the tsunami hazard from Kick-em Jenny using linear theory. Their worst case scenario included run-ups as high as 46 meters on the northern shore of Grenada for a VEI=6, or Krakatau-like event, considered as potentially likely on a 1000-year scale. Their more recent unpublished analyses (Shepherd 2003, private communication) reflect regional opinion that the danger from tsunamis caused by this volcano is insignificant. It is nevertheless of interest to study the role of underwater volcanic explosive eruptions in producing tsunamis with particular application to this very interesting case.

For the sake of simplicity, we performed our simulations (Gisler, Weaver, and Gittings 2006b) in two-dimensions (cylindrical geometry) only, ignoring the very real three-dimensional character of the Kick-em Jenny volcano. We model the volcano as a simple geometrical frustum, with a base of 5 km diameter, a top of 100 m diameter, and a height of 1.4 km. The cone has a hot magma core of 20 m diameter. We take the water depth to be 1.5 km, so that the submerged top of the frustum is only 100 m below the water surface, significantly shallower than the true cone summit. We use three materials in the problem, air for the atmosphere, water for the ocean, and basalt for the seafloor, the volcanic cinder cone, and the hot magma core.

To model an explosive eruption, we took the end-member case of an instantaneous explosion near the top of the cone. Because the strongest coupling to the water motion will be from the motion of rock, we place the explosion at a depth of 150 m below the summit.

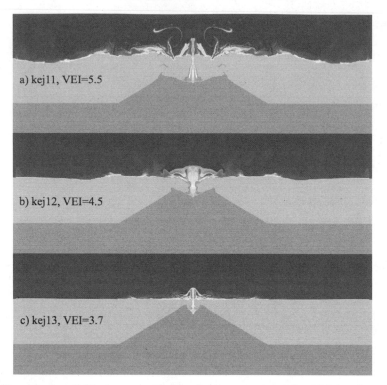

Fig. 6.1 – (See color insert) Final wave profiles for the three representative Kick-'em Jenny runs. The domain illustrated is 12 km in length and 4 km in height. The volcanic explosive index, VEI, is indicated for the three runs. The extrapolated wave heights at 10 km distance are 37, 21, and 2.7 meters respectively, declining steeply with distance. These wave heights are significantly lower than reported by Smith and Shepherd (1993). At most, 2% of the source energy is coupled to the water wave.

In Fig. 6.1 we illustrate three representative runs from our Kick-em Jenny by showing their final density configurations. The explosive energy is sourced in instantaneously at the beginning of the calculation. A hot crater quickly opens in the basalt, and the explosive vaporization of the water in contact with this crater produces a large transient water cavity. A "debris curtain" or rim wave makes a precursor tsunami that dies off very quickly. The main wave is produced by the collapse of the transient water cavity, and the strong water currents modify the shape of the basalt crater produced in the explosion. The wave produced by the cavity collapse is very turbulent and dissipative, propagates slowly, and dissipates strongly. When the main wave leaves the computational domain (6 km from the center), we terminate the simulation, though the center is still hot and turbulent.

6.2.   Sub-aerial landslide simulations

6.2.a.   Lituya Bay landslide tsunami

The Lituya Bay event of 8th July 1958, mentioned in the Introduction, was modeled experimentally at a scale of 1/675 by Hermann Fritz and collaborators in a laboratory in the Swiss Federal Institute of Technology at Zurich (Fritz *et al.*, 2001). A pneumatic landslide generator was used to generate a high-speed granular slide. This slide created a crater upon impact with the water and an enormous nonlinear wave that ran up to a scaled elevation of 530 meters, which compares extremely well to the observed run-up elevation of 524 meters. The flow is complex with all three-phases (air, water, and granular material) showing distinct boundary layers before massive mixing occurs.

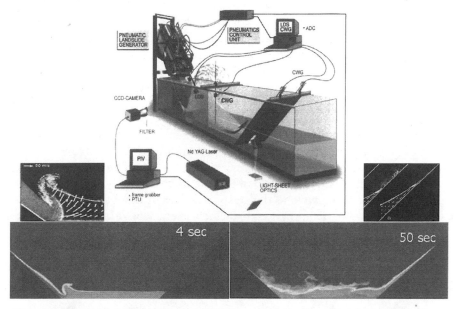

Fig. 6.2 – Experimental setup of Fritz *et al.* (2001) to model the tsunami at Lituya Bay. The set up includes laser distance sensors (LDS), capacitance wave gauges (CWG) and particle image velocimetry (PIV). Black-and-white insets at left and right show the impact crater formed by the slide hitting the water surface and the maximum run-up, respectively, from the experimental results. Below that are color density plots corresponding to those times from the Sage code simulations of Mader and Gittings (2002).

Mader and Gittings (2002) reproduced the Fritz results with the Sage code in a geometry scaled to the true Lituya Bay size, and achieved a run-up of 580 meters, some 10% greater than observed. Viscous dissipation and bottom friction, not included in the Sage calculations, would account for this modest overestimate. A sketch of the Fritz set-up and sample results from that experiment and from the Sage calculations are shown in Fig. 6.2.

6.2.b. Hypothetical lateral flank collapse on La Palma

A lateral flank collapse of the Cumbre Vieja volcano on the island of La Palma in the Canary Islands was considered by Ward and Day (2001) as likely to produce a dangerous Atlantic-ocean-wide tsunami. That work was criticized by Mader (2001) because a long-period wave is unlikely to result even from a total flank collapse, by Pararas-Carayannis (2002) because such a collapse is unlikely, and by Wynn and Masson (2003) because the record of turbidite deposits in the vicinity of the Canary Islands does not support such catastrophic events in the past.

It is nevertheless of general interest to calculate the production of tsunamis from such an event, and we therefore performed these in a simplified geometry resembling that of Cumbre Vieja (Gisler, Weaver, and Gittings, 2006a). The slide region, consisting of fluidized basalt, is wedge shaped, deepest at the top, tapering to a point or toe at the bottom, and having an offset towards the back side of the peak. Below this, to the bottom of the computational grid, is a rigid reflecting internal boundary representing the unchanging basement rock. The motion is initiated solely by the gravitational acceleration of the fluidized rock in the slide region. The two-dimensional calculations are done in Cartesian coordinates, thus effectively assuming infinite extent in the direction perpendicular to the simulation plane. We ran several variants of this geometry, with slide volumes (assuming a transverse extent of 20 km) ranging from 300 to 500 km^3. The near-field wave height in the two-dimensional runs is very high, up to 1.5 km, and the wavelength is up to 60 km. The initial near-field wave height in the three-dimensional run is smaller, about 500 m, and the back-propagating wave is devastating locally.

These waves are far more impressive than the waves from our simulations of Kick-em Jenny, reflecting the more efficient coupling of the relatively slower rock motion to the water wave. Nevertheless, the waves are of relatively short wavelength and highly dissipative because of turbulent interaction with the atmosphere. We find no significant damage to the western shores of the Atlantic Ocean from a maximal collapse at La Palma, though such an event would be extremely hazardous locally.

Fig. 6.3 – (See color insert) Density raster plot in the La Palma run with largest slide volume at a time of 180 seconds after the start of the slide. The reflective region representing the unchanging basement of La Palma is at left in black, the basalt fluid slide material is red, water is orange, and air is blue. Intermediate shades represent the mixing of fluids. The mixing of water and rock is readily apparent in the form of vortices induced by the fluid-fluid interaction instability These lead eventually to the deposition of turbidite layers in remnant landslide deposits. The water wave leads the bullnose of the slide material by a small amount; the forward-rushing slide material (with a peak velocity of 190 m/s almost matching the wave velocity) continues to pump energy into the wave. The wave height at this time is 1500 meters, and the wavelength is roughly 60 km. This figure has a width of 50 km, representing less than half of the computational domain, which extends 120 km to the right.

An impression of the character of the wave is obtained from a graphic rendering of density at a time early in the calculation (Fig. 6.3). This figure shows a small portion of the computational domain; the box extends to the right out to a distance of 120 km from the island. The fluidized rock has accelerated into the water, initially producing a cavity, then a substantial wave. The slide speed in this inviscid calculation reaches as high as 190 m/s before decelerating as the slide gives up energy to the water. The peak slide speed almost matches the shallow-water wave speed ($\sqrt{gd}$ = 198 m/s) so the bullnose of the slide material falls behind the water wave only gradually, and continues to pump energy into the water wave. A single broad wave is thus produced, with trailing higher-frequency components that are excited by the eddy currents in the turbulent mixing and entrainment of slide material and water. As the slide material decelerates, the water wave detaches and outruns it, but no further low-frequency modes are produced.

The wave period and wavelength depend only weakly on the slide volumes for the runs in our study (roughly as the 1/3 and 1/2 power, respectively), but both the period and wavelength of these waves are short compared to those of classical teletsunamis. The long-distance propagation of these waves is therefore not as effective as for classical long-wavelength tsunamis. There are no low-frequency components arising in the water wave, though a cascade to higher frequency is apparent in data from Lagrangian tracer particles in these runs.

Output simulation data from a three dimensional run that included bathymetry and topography of the Canary Islands, reported in Gisler, et al. (2006a) has subsequently been imported into a depth average treatment for study of transatlantic propagation (Løvholt *et al.,* 2008). The western Atlantic effects of the extreme case considered are substantially less than those predicted by Ward and Day (2001). Of greater interest and concern are the local effects in the Canary Islands and the nearby shores of Africa and Iberia, which are treated at some depth in the latter paper.

6.3   Submarine landslides

As part of the work reported in the previous section, we examined deviations from our usual inviscid case (justified by the very long run out distances seen in many tsunami deposits) by treating the fluidized rock as a plastic flow, with a shear modulus giving an effective viscosity. We are investigating this now by focusing strictly on submarine slides, and examining the effects of various rheologies. We note that Watts (2003) used a Bingham rheology as a superposition upon a center-of-mass treatment for submarine landslides.

Fig. 6.4 – (See color insert) Density raster plots for 4 two-dimensional runs of a submarine landslide with successively increasing spatial resolution at the same physical time of 158 seconds after the slide start. The speed of the fluidized rock slide increases substantially as resolution improves, and the turbidity waves become more distinct and developed. The dimensions of the box are 10 km vertical by 60 km horizontal. From the top, the finest spatial resolution in the AMR grid is: 62.5 m, 31.3 m, 15.6 m, and 7.8 m. The average velocities of the toe of the slide up to this point are 32 m/s, 51 m/s, 63 m/s, and 68 m/s. Extrapolating to infinitely fine spatial resolution would give an average slide velocity of 73 m/s.

Importantly, we have done a study of the effect of numerical resolution on the speed of the slide and the wavelength of the turbidity waves that result, and show results in Fig. 6.4. The studied case, in two dimensions, is a massive slope failure on a 12° continental slope with the top at 900 m depth and base at 4000 m depth. A wedge-shaped region was fluidized at the start of the calculation and accelerated under gravity alone. This is done in a computational volume of size 10 km vertical by 60 km horizontal, and water depth 5 km. Slide speed is found to increase as the spatial resolution is improved, and extrapolates to an average velocity of 73 m/s at infinite resolution.

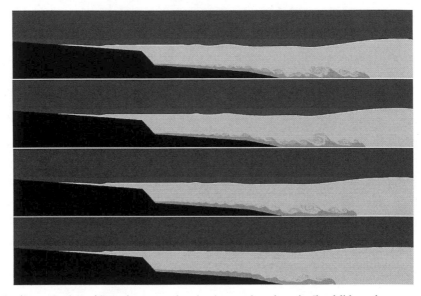

Fig. 6.5 – (See color insert) Density raster plots for 4 runs of a submarine landslide at the same physical time (225 seconds after slide initiation) with successively increasing shear modulus for the slide, modeled as a plastic flow. The dimensions are the same as in Fig. 6.4. From the top we show the inviscid rock slide with the same spatial resolution as the third plot in Fig. 6.4 (but at a later time, so the turbidity waves are more developed), then slides at the same resolution with shear moduli of 1 bar, 3 bars, and 10 bars. Note the increasing deceleration of the rock slide with increasing shear modulus, and the corresponding lower spatial frequency of the turbidity waves. These runs were all performed with a finest spatial resolution of 16.3 m.

We took our second-highest resolution inviscid run, in which the slide velocity differs from the extrapolated infinite-resolution value by 14%, as a baseline. Then in subsequent runs we treated the rock slide material as a plastic flow, using a nominal yield criterion and chosen shear modulus, to study how the speed of the slide and the turbidity waves depend upon the shear modulus or effective viscosity. These results are shown in Fig. 6.5 at a time of 225 seconds. The average velocity up to this time in the inviscid case was 80 m/s, and 56 m/s in the case with a 10 bar shear modulus. The spatial wavelength of the turbidity waves varied from 1.25 km in the inviscid case to 1.55 km in the stiffest case. From runs such as this, one might possibly diagnose the rheology of past submarine slides from the present disposition of turbidite deposits on the seafloor.

6.4. Asteroid impact simulations

While asteroid impacts as tsunami sources are much less frequent than endogenous sources of tsunamis, they are expected to occur frequently enough in earth's history to produce traces in the geological record. The Mjølnir event, mentioned before, is an example of a shallow-water impact that left tsunami traces, as is the impact associated with the Cretaceous-Tertiary boundary at Chicxulub, Mexico.

We have performed a large number of two-dimensional simulations of ice, stony, and iron asteroidal projectiles of various diameters and speeds plunging vertically into deep (5 km) water. We have additionally performed a few three-dimensional simulations to examine the effects of impact angle (Gisler, Weaver, Mader, and Gittings, 2003).

A three-dimensional simulation of a 1-km diameter iron asteroid impacting the ocean at a 45-degree angle at 20 km/s is illustrated in a montage in Fig. 6.6. The asteroid, started 30 km above the ocean surface, is not substantially decelerated until it reaches the water. Explosive vaporization of water produces a crater that begins asymmetrically but rapidly symmetrizes. Water vapor expands vertically out of the crater, and liquid water pushed aside forms a nearly vertical splash or crown. The crown collapse produces a precursor rim wave. The maximum crater diameter of 25 km is reached at 37 seconds, after which the cavity collapses and a central jet is produced, rising to a height of about 40 km. It is the collapse of this jet that gives rise to the principal tsunami wave, which expands out from the impact point in circular symmetry, independent of the angle of impact.

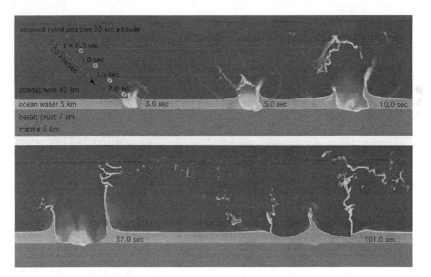

Fig. 6.6 – (See color insert) Montage of 9 separate images from a 3-d run of the impact of a 1-km iron bolide at an angle of 45 degrees into an ocean of 5-km depth. These are density raster graphics in a two-dimensional slice in the vertical plane containing the asteroid trajectory. Note the initial asymmetry of the water crater and its disappearance in time as the debris curtain collapses.

The asteroid enters the water at a speed much greater than that of sound, and the rapid dissipation of its kinetic energy is very much like an explosion. Shocks propagate outward from the cavity in the water, in the basalt crust and in the mantle beneath. Multiple reflections of shocks and acoustic waves between the material interfaces complicate the dynamics severely, and induce turbulence that steals energy from the tsunami.

To examine the dynamics at better resolution, we have done calculations in two dimensions, cylindrical symmetry, for asteroids of various compositions and diameters impacting the ocean vertically at various speeds (Gisler, Weaver, Mader, and Gittings, 2003; Gisler, 2007).

For all compositions, diameters, and speeds, we find that the mass of water displaced during the formation of the cavity scales as the ¾ power of the asteroid kinetic energy. A fraction of this displaced mass is actually vaporized during the explosive phase of the encounter, while the rest is pushed aside by the pressure of the vapor to form the crown and rim of the transient cavity.

The initial tsunami amplitude is also found to scale as the ⅔ power of the asteroid kinetic energy, but it evolves in a complex manner, eventually decaying rather faster than $1/r$, where r is the distance of propagation from the impact point. The wave trains are initially highly complex (Fig. 6.7) because of the multiple shock reflections within the water and interactions with the seismic waves propagating through the crust and with atmospheric motions.

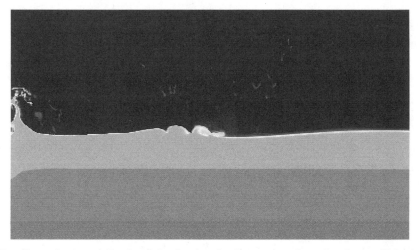

Fig. 6.7 – (See color insert) Portions of a density plot from a two-dimensional asteroid impact run, shortly after the collapse of the transient crater at left, illustrating the complexity of the wave train. The phenomena are influenced by reflections and interactions of multiple shocks propagating through the water and the basalt crust.

Because of the steeply decaying amplitude of asteroid-induced tsunamis with distance from the impact point, there is a threshold at which concern over an ocean-wide tsunami is justified. For stony asteroids impacting at 20 km/s, the diameter threshold is about 1 km. Stony asteroids of ~200m diameter would cause dangerous tsunamis only if they impact within ~100 km of an inhabited coastline.

The asteroid that produced the Eltanin deep-water impact, which was probably greater than a kilometer in diameter, should thus have left tsunami deposits. None have as yet been found, but they may be difficult to recognize.

7. Conclusions

Much progress has been made recently in understanding tsunamis from non-earthquake sources, but more needs doing: experimental work with both slurries of appropriately scaled rheology and with solid blocks, field surveys of previous slide events with particular attention to the spacing of turbidite beds and estimates of sediment rheologies, and numerical work with appropriately detailed treatment of all the aspects of physical importance. In addition, careful monitoring of sites of potential slide or slump danger is of the highest importance. The monitoring of the Tafjord/Åknes slide areas in Norway is worth imitating in areas of similar sensitivity. Detailed bathymetry, repeated at regular intervals of steeply sloping continental shelf-edge regions should be implemented. This could possibly be supplemented by the emplacement of strain gauges in areas found to be near critical conditions, where they are also close to shore populations or facilities of importance.

Acknowledgments

I wish to thank my Los Alamos collaborators Robert Weaver and Michael Gittings for stimulating my interest in this research and for providing the computational model that made my own work on this subject possible, and Charles Mader for starting me down this road and providing inspiration and critical insights. Los Alamos National Laboratory, where I worked for twenty-four years, provided computational horsepower and an environment for healthy exploration in science in the public interest. Over the few years of my involvement in this field, I have learned a great deal from conversations with many of the experts in this field, and I acknowledge their help, and pay tribute to the efforts they have made in saving lives around the world. Referee reports from Mader and from Eric Geist improved the quality of this paper. I gratefully acknowledge the present and continuing support of a Center of Excellence grant from the Norwegian Research Council to Physics of Geological Processes at the University of Oslo, my present employer.

References

A. Braathen, L. H. Blikra, S. S. Berg, and F. Karlsen, "Rock-slope failures in Norway; type, geometry and hazard", *Norwegian Journal of Geology* **84**, 67–88 (2004).

B. A. Campbell and D. Nottingham, "Anatomy of a landslide-created tsunami at Skagway, Alaska", *Science of Tsunami Hazards* 17, 19–45 (1999).

F. V. DeBlasio, D. Issler, A. Elverhøi, C. B. Harbitz, T. Ilstad, P. Bryn, R. Lien, and F. Løvholt, in *Submarine Mass Movements and Their Consequences,* edited by J. Locat, J. Mienert, and L. Boisvert (Kluwer Academic Publishers, Dordrecht/Boston/London, 2003).

H. M. Fritz, W. H. Hager, and H.-E. Minor, "Lituya Bay case: rockslide impact and wave run-up", *Science of Tsunami Hazards* **19,** 3–23 (2001).

H. M. Fritz, W. H. Hager, and H.-E. Minor, "Landslide generated impulse waves. 1. Instantaneous flow fields", *Experiments in Fluids* **35,** 505–519 (2003a).

H. M. Fritz, W. H. Hager, and H.-E. Minor, "Landslide generated impulse waves. 2. Hydrodynamic impact craters", *Experiments in Fluids* **35,** 520–532 (2003b).

D. L. George and R. J. LeVeque, "Finite volume methods and adaptive refinement for global tsunami propagation and local inundation", *Science of Tsunami Hazards* **24,** 319–328 (2006).

R. Gersonde, F. T. Kyte, U. Bleil, B. Diekmann, J. A. Flores, K. Gohl, G. Grahl, R. Hagen, G. Kuhn, F. J. Sierro, D. Völker, A. Abelmann, and J. A. Bostwick, "Geological record and reconstruction of the late Pliocene impact of the Eltanin asteroid in the Southern Ocean", *Nature* **390,** 357–363 (1997).

R. Gersonde, A. Deutsch, B. A. Ivanov, and F. T. Kyte, "Oceanic impacts – a growing field of fundamental geology", *Deep-Sea Research II* **49,** 951–957 (2002).

R. Gersonde, F. T. Kyte, T. Frederichs, U. Bleil, and G. Kuhn, "New data on the late Pliocene Eltanin impact into the deep southern ocean", in *Third International Conference on Large Meteorite Impacts,* Nördlingen Germany, 2003.

G Gisler, R Weaver, C Mader, M Gittings, "Two and three dimensional simulations of asteroid ocean impacts", *Science of Tsunami Hazards,* **21,** 119 (2003).

G Gisler, R Weaver, M Gittings, "Sage Calculations of the tsunami threat from La Palma", *Science of Tsunami Hazards,* **24,** 288–301 (2006a).

G Gisler, R Weaver, M Gittings, "Two-dimensional simulations of explosive eruptions of Kick-em Jenny and other submarine volcanos", Science of Tsunami Hazards, **25,** 34 (2006b).

G Gisler, "Tsunamis from asteroid impacts in deep water", *Planetary Defense Conference,* invited paper, Washington DC (2007).

Gittings ML, Weaver RP, Clover M, Betlach T, Byrne N, et al. "The RAGE radiation-hydrodynamic code", *Los Alamos Unclassif. Rep. LA-UR-06-0027,* Los Alamos Natl. Lab., Los Alamos, New Mexico (2006).

S. Glimsdal, G. Pedersen, V. V. Shuvalov, H. Dypvik, H. P. Langtangen, and Ø. Kristiansen, "Tsunami Generated by the Mjølnir impact", *Lunar and Planetary Science* **36** (2005).

K. B. Haugen, F. Løvholt, and C. B. Harbitz, "Fundamental mechanisms for tsunami generation by submarine mass flows in idealized geometries", *Marine and Petroleum Geology* **22,** 209–217 (2005).

Z. Kowalik, W. Knight, T. Logan, and P. Whitmore, "Numerical modeling of the global tsunami", *Science of Tsunami Hazards* **23)**, 40–56 (2005).

E. A. Kulikov, A. B. Rabinovich, R. E. Thomson, and B. D. Bornhold, "The landslide tsunami of November 3, 1994, Skagway Harbor, Alaska", *Journal of Geophysical Research* **101,** 6609–6615 (1996).

F. T. Kyte, "Iridium concentrations and abundances of meteoritic ejecta from the Eltanin impact in sediment cores from *Polarstern* expedition ANT XII/4 ", *Deep-Sea Research II* **49,** 1049–1061 (2002a).

F. T. Kyte, "Unmelted meteoritic debris collected from Eltanin ejecta in *Polarstern* cores from expedition ANT XII/4", *Deep-Sea Research II* **49,** 1063–1071 (2002b).

L. D. Landau and E. M. Lifshitz, *Fluid Mechanics* (Pergamon Press, Oxford, 1959).

Jan M. Lindsay, John B. Shepherd and Doug Wilson, "Volcanic and scientific activity at Kick 'em Jenny submarine volcano 2001–2002: implications for volcanic hazard in the Southern Grenadines, Lesser Antilles", *Natural Hazards,* **34:**1–24 (2005).

F. Løvholt, C. B. Harbitz, and K. B. Haugen, "A parametric study of tsunamis generated by submarine slides in the Ormen Lange/Storegga area off western Norway", *Marine and Petroleum Geology* **22,** 219–231 (2005).

F. Løvholt, G. Pedersen, and G. Gisler, "Modeling of a potential landslide generated tsunami at La Palma Island", in press *Journal of Geophysical Research—Oceans* (2008).

Charles L. Mader, "Modeling the La Palma landslide tsunami", *Science of Tsunami Hazards,* **19,** 160 (2001).

C. L. Mader and M. L. Gittings, "Modeling the 1958 Lituya Bay mega tsunami , II", *Science of Tsunami Hazards* **20,** 241–250 (2002).

C. L. Mader, *Numerical Modeling of Water Waves* (CRC Press, 2004).

C. L. Mader and M. L. Gittings, "Numerical model for the Krakatoa hydrovolcanic explosion and tsunami", *Science of Tsunami Hazards* **24**), 174–182 (2006).

T. S. Murty, "Tsunami wave height dependence on landslide volume", *Pure and Applied Geophysics* **160,** 2147–2153 (2003).

N. Nomanbhoy and K. Satake, "Generation mechanism of tsunamis from the 1883 Krakatau eruption", *Geophysical Research Letters* **22,** 509–512 (1995).

Nosov MA. 1999. Tsunami generation in a compressible ocean by vertical bottom motions. *Izvestiya, Atmospheric and Oceanic Physics,* 36:661—669

Nosov MA, Skachko SN. 2001. Nonlinear tsunami generation mechanism. *Natural Hazards and Earth System Sciences,* 1:251—253

E. A. Okal, P.-J. Alasset, O. Hyvernaud, and F. Schindelé, "The deficient T waves of tsunami earthquakes", *Geophysical Journal International* **152,** 416–432 (2003).

E. A. Okal and C. E. Synolakis, "A theoretical comparison of tsunamis from dislocations and landslides", *Pure and Applied Geophysics* **160,** 2177–2188 (2003).

K. K. Panthi and B. Nilsen, "Numerical analysis of stresses and displacements for the Tafjord slide, Norway", *Bulletin of Engineering Geology and the Environment* **65,** 57–63 (2006).

George Pararas-Carayannis, "Evaluation of the threat of mega tsunamis generation from postulated massive slope failures of island stratovolcanoes on La Palma, Canary Islands, and on the island of Hawaii", *Science of Tsunami Hazards,* **20,** 251, 2002.

George Pararas-Carayannis, "Volcanic tsunami generating source mechanisms in the eastern Caribbean region", *Science of Tsunami Hazards* **22,** (2), 74–114 (2004).

E. Pelinovsky, N. Zahibo, P. Dunkley, M. Edmonds, R. Herd, T. Talipova, A. Kozelkov, and I. Nikolkina, "Tsunami generated by the volcano eruption on July 12–13, 2003 at Montserrat, Lesser Antilles", *Science of Tsunami Hazards* **22,** (1), 44–57 (2004).

M. Quecedo, M. Pastor, and M. I. Herreros, "Numerical modeling of impulse wave generated by fast landslides", *International Journal for Numerical Methods in Engineering* **59,** 1633–1656 (2004).

V. V. Shuvalov, "Numerical modeling of the Eltanin impact", *Lunar and Planetary Science* **34** (2003).

E. Semenza and M. Ghirotti, "History of the 1963 Vaiont slide: the importance of geological factors", *Bulletin of Engineering Geology and the Environment* **59,** 87–97 (2000).

T. Simkin, L. Siebert, L. McClelland, D. Bridge, C. Newhall, and J. H. Latter, *Volcanoes of the World,* Smithsonian Institution, Hutchinson Ross, Stroudsburg , 1981.

Martin S. Smith and John B. Shepherd, "Preliminary investigations of the tsunami hazard of Kick 'em Jenny submarine volcano," *Natural Hazards,* **7:**257–277 , 1993.

Martin S. Smith and John B. Shepherd, "Potential Cauchy-Poisson waves generated by submarine eruptions of Kick 'em Jenny volcano," *Natural Hazards,* **11:**75–94 , 1995.

Martin S. Smith and John B. Shepherd, "Tsunami waves generated by volcanic landslides: an assessment of the hazard associated with Kick 'em Jenny", In: W. J. McGuire, A. P. Jones, and J. Neuberg, (eds), Volcano Instability on the Earth and Other Planets, *Geol. Soc. Spec. Pub. No. 110,* pp. 115–123, 1996.

L. P. Sue, R. I. Nokes, and R. A. Walters, "Experimental modeling of tsunami generated by underwater landslides", *Science of Tsunami Hazards* **24**), 267–287 (2006).

S. Tinti, E. Bortolucci, and A. Armigliato, "Numerical simulation of the landslide-induced tsunami of 1988 on Vulcano Island, Italy", *Bulletin of Volcanology* **61,** 121–137 (1999).

V. V. Titov and F. J. Gonzalez, "Implementation and testing of the method of splitting tsunami (MOST) model," NOAA Technical Memorandum ERL-PMEL-112 (1997).

V. Titov, A. B. Rabinovich, H. O. Mofjeld, R. E. Thomson, and F. I. Gonzalez, "The global reach of the 26 December 2004 Sumatra tsunami", *Science* **309,** 2045–2048 (2005).

M. D. Trifunac and M. I. Todorovsk, "A note on differences in tsunami source parameters for submarine slides and earthquakes", *Soil Dynamics and Earthquake Engineering* **22,** 143–155 (2002).

F. Tsikalas, S. T. Gudlaugsson, J. I. Faleide, and O. Eldholm, "The Mjølnir marine impact crater porosity anomaly", *Deep-Sea Research II* **49,** 1103–1120 (2002).

Steven N. Ward and Simon Day, "Cumbre Vieja Volcano—potential collapse and tsunami at La Palma, Canary Islands", *Geophysical Research Letters* **28,** 397–400, 2001.

P. Watts and S. T. Grilli, "Underwater landslide shape, motion, deformation, and tsunami generation", *Proceedings of the Thirteenth (2003) International Offshore and Polar Engineering Conference,* 364–371 (2003).

P. Watts, S. T. Grilli, J. T. Kirby, G. J. Fryer, and D. R. Tappin, "Landslide tsunami case studies using a Boussinesq model and a fully nonlinear tsunami generation model", *Natural Hazards and Earth System Sciences* **3,** 391–402 (2003).

C. F. Waythomas and P. Watts, "Numerical simulation of tsunami generation by pyroclastic flow at Aniakchak Volcano, Alaska", *Geophysical Research Letters* **30,** 1751–1754 (2003).

R. B. Wynn and D. G. Masson, "Canary Island landslides and tsunami generation: can we use turbidite deposits to interpret lansdlide processes?" pp325–332 in: *Submarine Mass Movements and their Consequences* (eds. J. Locat and J. Mienert). Kluwer Academic Publishers, Dordrecht, Boston, London, 2003.

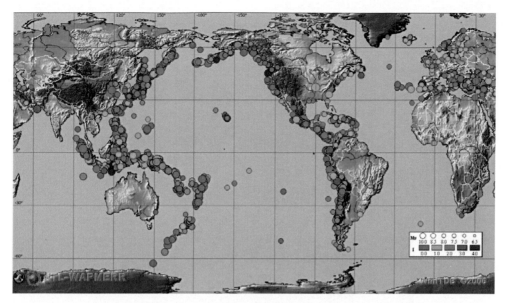

Fig. 2.1 – Visualization of the NTL/ICMMG global tsunami catalog. 1,990 tsunamigenic events with identified sources are shown for the period from 2000 BC to present time. Size of circles is proportional to event magnitude (for seismically induced tsunamis), color represents tsunami intensity on the So-loviev-Imamura scale. Gray tone shows the events with unknown intensity.

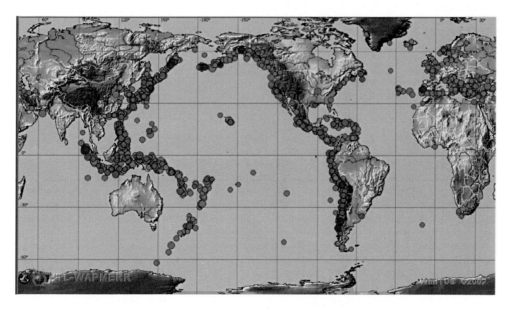

Fig. 2.2 – The same data as in Fig.2.1. but presented with 1,990 historical tsunamigenic events having identified sources divided into three groups: 1 – transoceanic tsunami (red) (see text for definition), 2 – regional tsunami resulting in human fatalities (magenta), 3 – all other tsunamis (blue).

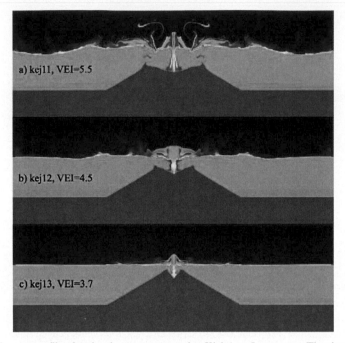

Fig. 6.1 – Final wave profiles for the three representative Kick-'em Jenny runs. The domain illustrated is 12 km in length and 4 km in height. The volcanic explosive index, VEI, is indicated for the three runs. The extrapolated wave heights at 10 km distance are 37, 21, and 2.7 meters respectively, declining steeply with distance. These wave heights are significantly lower than reported by Smith and Shepherd (1993). At most, 2% of the source energy is coupled to the water wave.

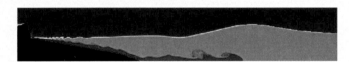

Fig. 6.3 – Density raster plot in the La Palma run with largest slide volume at a time of 180 seconds after the start of the slide. The reflective region representing the unchanging basement of La Palma is at left in black, the basalt fluid slide material is red, water is orange, and air is blue. Intermediate shades represent the mixing of fluids. The mixing of water and rock is readily apparent in the form of vortices induced by the fluid-fluid interaction instability These lead eventually to the deposition of turbidite layers in remnant landslide deposits. The water wave leads the bullnose of the slide material by a small amount; the forward-rushing slide material (with a peak velocity of 190 m/s almost matching the wave velocity) continues to pump energy into the wave. The wave height at this time is 1500 meters, and the wavelength is roughly 60 km. This figure has a width of 50 km, representing less than half of the computational domain, which extends 120 km to the right.

Fig. 6.4 – Density raster plots for 4 two-dimensional runs of a submarine landslide with successively increasing spatial resolution at the same physical time of 158 seconds after the slide start. The speed of the fluidized rock slide increases substantially as resolution improves, and the turbidity waves become more distinct and developed. The dimensions of the box are 10 km vertical by 60 km horizontal. From the top, the finest spatial resolution in the AMR grid is: 62.5 m, 31.3 m, 15.6 m, and 7.8 m. The average velocities of the toe of the slide up to this point are 32 m/s, 51 m/s, 63 m/s, and 68 m/s. Extrapolating to infinitely fine spatial resolution would give an average slide velocity of 73 m/s.

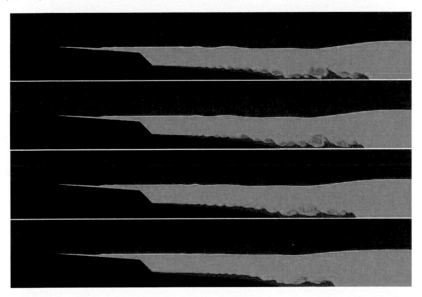

Fig. 6.5 – Density raster plots for 4 runs of a submarine landslide at the same physical time (225 seconds after slide initiation) with successively increasing shear modulus for the slide, modeled as a plastic flow. The dimensions are the same as in Fig. 6.4. From the top we show the inviscid rock slide with the same spatial resolution as the third plot in Fig. 6.4 (but at a later time, so the turbidity waves are more developed), then slides at the same resolution with shear moduli of 1 bar, 3 bars, and 10 bars. Note the increasing deceleration of the rock slide with increasing shear modulus, and the corresponding lower spatial frequency of the turbidity waves. These runs were all performed with a finest spatial resolution of 16.3 m.

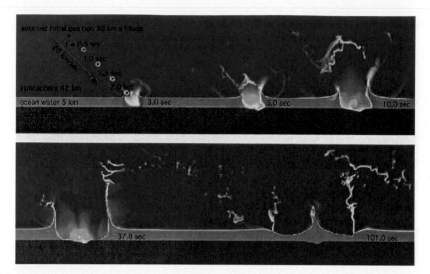

Fig. 6.6 – Montage of 9 separate images from a 3-d run of the impact of a 1-km iron bolide at an angle of 45 degrees into an ocean of 5-km depth. These are density raster graphics in a two-dimensional slice in the vertical plane containing the asteroid trajectory. Note the initial asymmetry of the water crater and its disappearance in time as the debris curtain collapses.

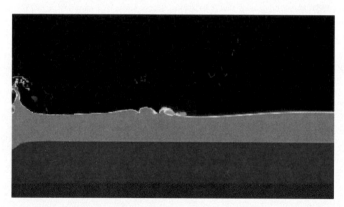

Fig. 6.7 – Portions of a density plot from a two-dimensional asteroid impact run, shortly after the collapse of the transient crater at left, illustrating the complexity of the wave train. The phenomena are influenced by reflections and interactions of multiple shocks propagating through the water and the basalt crust.

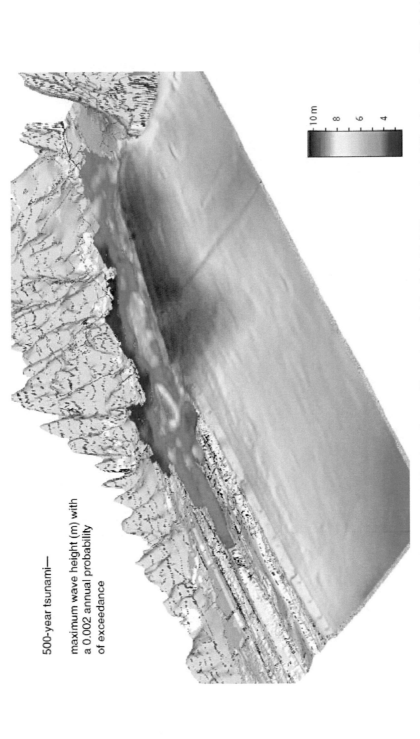

500-year tsunami—

maximum wave height (m) with a 0.002 annual probability of exceedance

10 m

8

6

4

Fig. 8.17 – 500-year tsunami wave heights (m) in the Seaside/Gearhart, Oregon. Tsunami wave heights (m) with 0.2% annual probability of exceedance. Wave heights include the effects of tides. View looks southeastward with vertical exaggeration of 10 times. After Tsunami Pilot Study Working Group (2006).

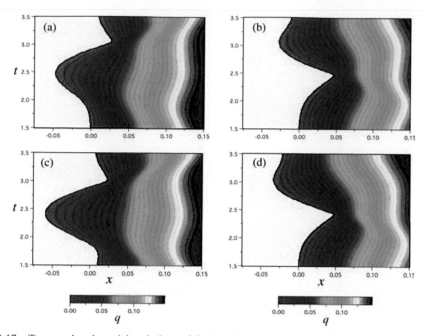

Fig. 11.17 – Temporal and spatial variations of the inundation depth for a) the initial waveforms of the Gaussian shape (Case a), b) the negative Gaussian shape (Case b), c) the leading depression N-wave (Case c), and d) the waveform caused by the submarine landslide (Case d). (after Carrier, Wu and Yeh, 2003).

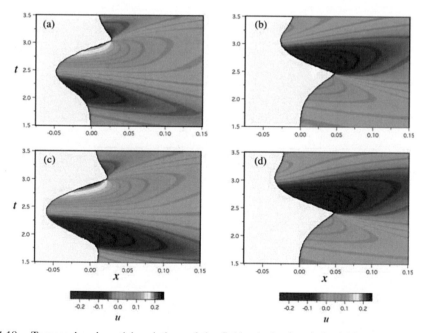

Fig. 11.18 – Temporal and spatial variations of the fluid velocity for a) the initial waveforms of the Gaussian shape (Case a), b) the negative Gaussian shape (Case b), c) the leading depression N-wave (Case c), and d) the waveform caused by the submarine landslide (Case d). (after Carrier, Wu and Yeh, 2003).

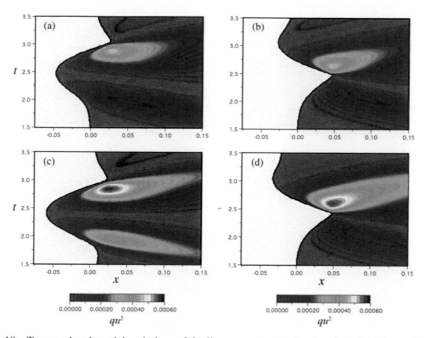

Fig. 11.19 – Temporal and spatial variations of the linear-momentum flux for a) the initial waveforms of the Gaussian shape (Case a), b) the negative Gaussian shape (Case b), c) the leading depression N-wave (Case c), and d) the waveform caused by the submarine landslide (Case d). (after Carrier, Wu and Yeh, 2003).

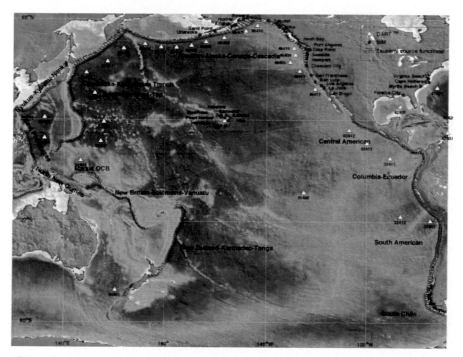

Fig. 12.1 – Components of the Short-term Inundation Forecast for Tsunamis in the Pacific. △ are DART locations; • are U.S. forecast sites; ⊞ are tsunami source functions.

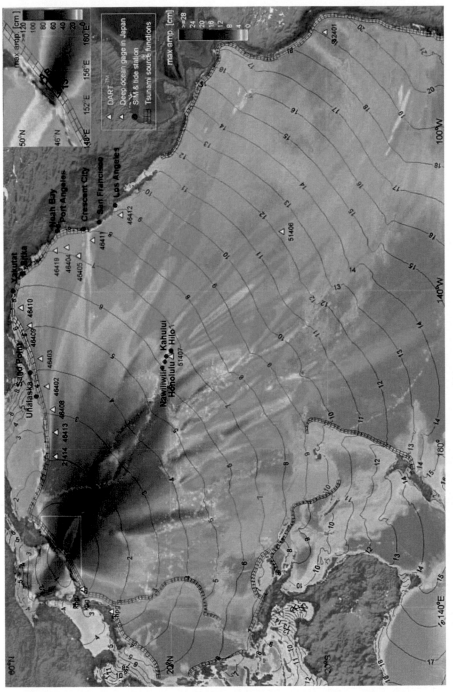

Fig. 12.7 – Test forecast for the November 15, 2006 Kuril Island tsunami. The symbols are the same as in Figure 12.4.

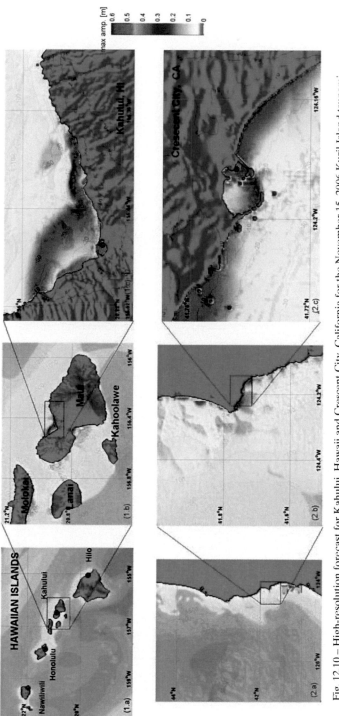

Fig. 12.10 – High-resolution forecast for Kahului, Hawaii and Crescent City, California for the November 15, 2006 Kuril Island tsunami.

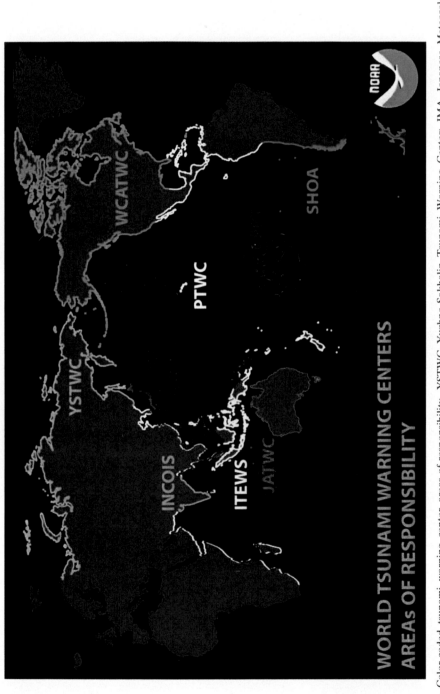

Fig. 13.3 – Color-coded tsunami warning center areas-of-responsibility. YSTWC=Yuzhno-Sakhalin Tsunami Warning Center; JMA=Japanese Meteorological Agency; NWPTAC Northwest Pacific Tsunami Advisory Center; INCOIS=Indian National Centre for Ocean Information Services; ITEWS=Indonesia Tsunami Early Warning System; JATWC=Joint Australian Tsunami Warning Centre; CPPT= Centre Polynésien de Prévention des Tsunamis (Polynesian Tsunami Warning Center); SHOA= Servicio Hidrográfico y Oceanográfico de la Armada (Chilean Navy Hydrographic and Oceanographic Service).

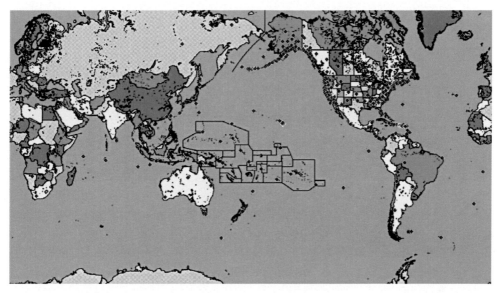

Fig. 13.6 – Seismometers which transmit data to U.S tsunami warning centers as of November, 2007 are shown as diamonds.

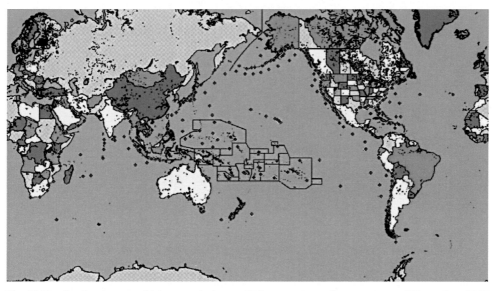

Fig. 13.10 – Sea level gages which transmit data to U.S. tsunami warning centers as of November, 2007 are shown as diamonds.

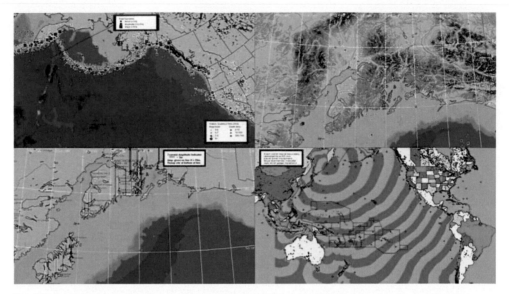

Fig. 13.14 – Example outputs from the EarthVu GIS. The upper-left map shows an overall location map with historic earthquake and tsunami overlain. The upper-right section shows an expanded view of the epicentral region with contours and other physical overlays. The lower-left section displays historic tsunami and the runups from the 1964 Alaska tsunami. The lower-right section is a tsunami travel time map for an event in the Gulf of Alaska. Contour intervals are one hour.

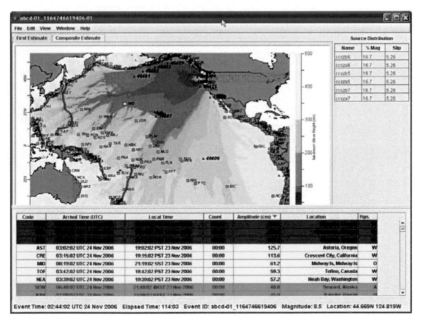

Fig. 13.19 – Example of the PMEL SIFT tsunami forecasting user interface. This display shows the expected coastal amplitudes and propagation pattern for a hypothetical 8.5 earthquake off the coast of Washington.

Chapter 7. TSUNAMI MEASUREMENTS

HAROLD O. MOFJELD

National Oceanic and Atmospheric Administration (Retired)

Contents

1. Introduction

In-situ instruments have provided tsunami measurements to the research community for over 150 years, and these observations are now essential to the operation of tsunami warning systems. Within the past decades, bottom pressure sensors deployed offshore have greatly increased this capability. As with every other aspect of tsunami research, operations and mitigation, the 2004 Indian Ocean tsunami continues to have a major impact on the field of tsunami measurements. This includes an international effort to greatly expand the tsunami observing networks, in support of tsunami warning systems in the major oceans and adjacent seas. Following the 2004 tsunami, many nations have also created or enhanced national tsunami programs. This rapid expansion in capability is leading to a new global observing network to serve the needs of both the tsunami warning systems and tsunami research.

Besides observations by water level and pressure gages, tsunami measurements include post-tsunami field surveys that are carried out soon after tsunamis have struck. These measurements include the extent of inundation, maximum wave heights, the number of damaging waves, measures of current strength, and the transport of debris, sediment, rock and coral. Together with the estimated number of casualties and the cost of damage, the results of such surveys are the primary data available in many tsunami databases to document the impacts of tsunamis (Chapter 11).

The Sea, Volume 15, edited by Eddie N. Bernard and Allan R. Robinson
ISBN 978–0–674–03173–9 ©2009 by the President and Fellows of Harvard College

Tsunami measurements are closely linked to the other aspects of tsunami research described in this volume and are essential for the validation of tsunami models (Chapters 8–10). Conversely, the interpretation of in-situ measurements is largely done via numerical models that are tuned to the measurements. For example, this frequently provides insight into the tsunami sources with finer detail than is available from analyses of seismic data and other source information alone (Chapters 5–6). The on-site measurements taken during post-tsunami field surveys are combined with those made by remote sensing to give a more complete description of tsunami inundation (Chapters 2 and 11) and are used as a basis for interpreting paleotsunami evidence (Chapter 3).

This chapter provides a brief history of tsunami measurements, a description of the present state-of-the-art, and issues to be resolved as the international tsunami observing networks grow. The focus is on a general description of the instrumentation, the acquisition of scientifically credible measurements, and the availability of the measurements to the research community. The peer-reviewed literature and Web-based information on in-situ measurements continues to increase rapidly. The specific examples in this chapter are therefore meant to be illustrative. It is assumed that the reader will seek more detailed and updated information from the rapidly growing literature and from Web resources.

2. History

Beginning in ancient times, eye-witness accounts of the loss of life and destruction caused by tsunamis have come down to us as written accounts and dramatic images (Chapter 2). These often focused on events of great historical significance, due to huge loss of life and their impact on civilizations. More recently, systematic reporting of tsunamis began that included the arrival times of waves, maximum wave heights and the extent of inundation along impacted coastlines (e.g., Lander and Lockridge, 1989; Lander et al., 1993; Lockridge et al., 2002; see National Geophysical Data Center Website). These records also extended the documentation to modest-sized and local events, providing more insight into the frequency of tsunamis and their relationship to local and distant earthquakes. For example, the written records of the January 26, 1700 tsunami in Japan were key to confirming that the Cascadia Subduction Zone off the West Coast of North America is an active tectonic region capable of generating great earthquakes and devastating tsunamis (Atwater et al., 2005).

In presenting a brief history of in-situ tsunami measurements, it is convenient to divide the topic into subsections according to the type of instrumentation: coastal water level (tide) gages, autonomous pressure gages deployed in the open ocean, and submarine cable-based pressure systems deployed on or near the continental shelf and slope. A large number of scientists, engineers and technicians working over many years have contributed to the effort to develop accurate and reliable tsunami instrumentation. Even when the technology underlying some of these systems has been superseded, they provided essential stepping stones along the path to the modern systems, which continue to evolve themselves.

2.1 Early Coastal Water Level Gage Observations

The first instrumental records of tsunamis were taken by water level (tide) gages that had been installed to support safe navigation and accurate mapping. In the United States, the first detailed tsunami reports based on in-situ measurements were on the December 1854 Ansei-Tokai and Ansei-Nankai tsunamis generated off Japan, as recorded by recently installed gages at San Diego and San Francisco, California. The resulting communication to the American Association for the Advancement of Science by Superintendent A. D. Bache (1856a,b) of the Coast Survey (now the National Oceanic and Atmospheric Administration, or NOAA) anticipates in many ways the modern approach to measuring and analyzing tsunami time series. Because of this, it is worth summarizing the results of this early research in some detail.

Lt. W. P. Trowbridge, in charge of the West Coast gages, had noticed irregularities in the 1854 analog record (maximum: 0.50 ft, 15.2 cm) from the San Diego gage that he ascribed to the effects of submarine earthquakes, since they could not be accounted for by any local meteorological events. Further investigation revealed similar fluctuations (maximum: 0.65 ft, 19.8 cm) in the San Francisco record also occurred on December 23 and 25, 1854. The ship log from the Russian frigate Diana, then in Shimoda harbor, Japan showed that major earthquakes had generated large tsunamis on these two days, each devastating part of Japan's Pacific Coast. Having this information, Bache interpreted the observed fluctuations on the records as groups of gravity waves that had propagated across the Pacific Ocean from Japan to the U.S. West Coast.

Bache then used estimates of the times of the earthquake and the arrival times of the tsunami waves at the gages—together with the average wave periods observed in the tide records (25 min for the first wave group and 31 min for the second group), the great-circle distance from the earthquakes to the gages, and Airy's formula for wave speed—to estimate the mean depth of the Pacific Ocean along the propagation paths from the tsunami sources to the tide stations. The mean depths were estimated to be 2100–2500 fathoms (3840–4572 m). A range of values was given to account for uncertainties in the location and times of the tsunami sources and the range of the observed wave periods. Since their publication, Bache's results have been widely quoted in the oceanographic literature as examples of applying wave theory to measurements in order to infer important properties of the ocean.

There are a number of important issues concerning in-situ tsunami observations brought forth in this remarkable contribution (Bache, 1856a,b) that are still of major interest today. These are:

- The need to understand instrumental effects (noise introduced by the water level float of the San Diego gage),
- Proper attention to data processing (the requirement for adequate temporal sampling when digitizing analog records to resolve the details of the tsunami waves and the need to remove the tides before further analysis),
- The value of comparing observations of the same tsunami from several locations (comparing the arrival times, amplitudes, and periods of tsunami waves

at San Diego and San Francisco to gain insight as to the location of the tsunami sources),

- The careful documentation of auxiliary information (the occurrence of local
  or distant earthquakes that may have generated the tsunamis, possible meteorological forcing immediately before and during the tsunamis, and the pro-
  gression of tidal stages during the tsunamis),
- Addressing the effects on tsunami propagation of regional topography and
  bathymetry near observation sites (attenuation caused by the Channel Islands off Southern California seaward of the San Diego tide gage and the
  major attenuation by the Columbia River Bar seaward of the Astoria gage in
  Oregon Territory),
- The need for the international research community to share tsunami measurements (the hope expressed by Bache that the publication of his findings
  would lead to such collaboration for the 1854 tsunamis).

Japan was the first country to begin comprehensive studies of (what they called)
"secondary undulations of tides", i.e., short-frequency oscillations in the tidal
records due to tsunamis. Omori (1902) was the first to formulate the fundamental
principle that "Tsunami waves from different events recorded at the same site are
similar, while the same tsunami recorded at different sites are strongly different"
and "The properties (observed periods) of tsunami waves recorded on the coast
are mainly determined by resonant (eigen) properties of the local (bay, inlet or
harbor) rather than by the source characteristics". These principles have been
confirmed by numerous studies (e.g., Miller, 1972; Rabinovich, 1997). Honda et al.
(1908) examined long wave properties in 68 bays and harbors along the coast of
Japan (this classic study was published in English almost 100 years ago). Their
work became the standard for others working on seiches, harbor oscillations and
coastal tsunami measurements and was responsible for stimulating studies in Japan
and other countries. Nakano and Unoki (1962) summarized these studies for Japan
and examined long waves at 45 tide gages.

The role of the gages in measuring tsunamis changed fundamentally during the
mid-Twentieth Century with the advent of real-time communications. With this
new capability, the measurements could be used to support tsunami warning systems by reporting the amplitude of a tsunami in time for staff to decide whether to
issue a tsunami watch or warning (Chapter 13). Initially, tidal observers read the
analog record and reported the heights to the centers; more recently, digital gages
allow the centers to query the instruments directly.

Much more detailed documentation of tsunamis began in the mid-and late
Twentieth Century. The tsunami inundation maps at Hilo, Hawaii for the 1946
Aleutian, 1960 Chile and 1964 Alaska events remain important test cases that are
used for the verification of numerical tsunami models and as a basis for tsunami
mitigation. They had a major impact internationally, and the research community
responded by quickly publishing comprehensible reports of these events based on
the best available data. These reports included information on the earthquakes,
water level data, and observations from post-event field surveys.

The first earthquake and tsunami that initiated detailed reports (e.g., Takahashi
and Hatori, 1961; Berkman and Symons, 1964) was the 1960 Chilean event. The
tsunami from this event killed people in such far away areas as Hawaii and Japan

and initiated creation of the International Tsunami Warning System (see Chapter 13 for a detailed history of tsunami warning systems). Because it was generated in Alaska and caused widespread damage both there, Hawaii and the West Coast, the 1964 Alaska earthquake and tsunami led to a set of particularly detailed reports (e.g., Spaeth and Berkman, 1967; Wilson and Törum, 1968; Berg et al., 1970). Together with reports for other major tsunamis, these provided much of the in-situ tsunami measurements for scientific and engineering books on earthquakes and tsunamis in the following decade (e.g., Wiegel, 1970, 1976; Murty, 1977).

The analysis of tsunami measurements was greatly enhanced by the application of digital technology that in turn necessitated the digitization of analog tsunami records. Probably the best set of hand-digitized tsunami series were by Van Dorn (1984, 1987) and his students. These include time series at Hawaiian, U.S. West Coast and Pacific island stations for the 1946, 1960, 1964 tsunamis as well as the destructive 1952 Kamchatka and 1957 Aleutian Tsunamis. Detailed procedures for digitizing analog tsunami records are given by Wigen (1983).

In the 1980s, analog tide gages in major harbors began to be replaced by digital gages. During this time, Van Dorn (1984) and others expressed deep concern that these new digital gages would not sample water levels frequently enough to be adequate for tsunami measurement. Fortunately, the requirements of the tsunami warning centers were strong enough that the replacement tide gages were capable of sampling and reporting data with 1-min sampling or shorter. Removable memory cartridges were also installed in many U.S. Pacific gages that could be retrieved by tidal observers following tsunami events and sent to the NOAA/Pacific Marine Environmental Laboratory for analysis. The cartridges contained five days of the water level measurements with 15-s sampling, providing adequate resolution for even small amplitude tsunamis.

Observations of major tsunami events in the Pacific Ocean demonstrated some limitations of standard water level gages for measuring tsunamis. The first was that a gage could be destroyed by the impact of moving debris or by the tsunami waves themselves. Essential measurements of the largest tsunami waves would then be lost for the communities that had been most impacted by the tsunami. The second limitation is that water level could be measured only as high as the tops of the gages, possibly missing the opportunity to measure the height of the largest and most destructive waves, as took place during the 1960 Chile tsunami (Berkman and Symons, 1964).

2.2   Open Ocean Observations—Autonomous Instruments

The rationale for placing bottom pressure stations in the open ocean for observing tsunamis and reporting the data in real-time to tsunami warning centers was put forward by Saxena and Zielinski (1981), Zielinski and Saxena (1983a, 1983b) and Poplavsky et al. (1988). These stations can be placed near potential tsunami sources and thereby report direct tsunami measurements, often hours before the tsunamis can propagate to major coastal communities. Bottom pressure sensors are able to measure tsunamis because tsunamis are barotropic gravity waves with wavelengths much longer than the ocean depth. They therefore have bottom pressure fluctuations directly proportional to the time-varying sea surface elevation induced by the tsunamis as they propagate past the station. Observing tsunamis in the open ocean also has the advantage that they are made before the tsunamis encounter the complicating, often fine-scale, effects of the continental shelf and the coastal region as well as resonance effects in bays and harbors. This permits a more straightforward interpretation of the tsunami waves and their relationship to their sources.

The technology to make precise observations of bottom pressure in the deep ocean was developed in the 1960–1970s. Much of this early development focuses on overcoming a host of instrumental problem associated with making precise pressure observations at the bottom of the deep ocean (high pressure, temperature influence, instrumental drift, frequency instability, power supply, storage, data transmission, aliasing, instrument retrieval, etc.). This effort was aided by the parallel development of shallower-water pressure gages for the continental shelf and slope (e.g., Beardsley et al., 1977). A host of scientists contributed to the development of the deep-sea capability using a variety of different pressure sensors and instrument platforms (e.g., Eyries et al.,1964; Hicks et al.,1965; Vitousek, 1965; Nowroozi et al., 1966; Caldwell et al., 1969; Snodgrass, 1968, 1969; Filloux, 1970; Irish and Snodgrass, 1972; Nowroozi, 1972; UNESCO, 1975). Some of the first applications of these instruments were in the study of tides and lower frequency pressure fluctuations in the deep sea (e.g., these previous references, and Brown et al., 1975; Zetler et al., 1975; Mofjeld and Wimbush, 1977).

After retrieving an autonomous pressure recorder deployed in deep water near the entrance to the Gulf of California, Filloux (1982, 1983) discovered that the instrument had recorded a small tsunami clearly generated by the March 14, 1979 Petatlan earthquake off the nearby Pacific Coast of Mexico. By analyzing the tsunami time series in the bottom pressure record, he obtained more detailed information about the earthquake than was available from the seismic data alone and found that a simple shelf-resonance of the continental shelf was not consistent with the observed characteristics of the tsunami. He also correctly identified background noise as due to apparent pressure fluctuations caused by seismic surface (Rayleigh) waves from earthquake sources, as well as meteorological induced fluctuations (meteotsunamis; see e.g. Monserrat et al., 2006).

Through the 1980s, the development of autonomous tsunami instruments for the open ocean accelerated. This effort was carried out by the NOAA/Pacific Marine Environmental Laboratory with the goal of creating a network of real-time reporting bottom pressure systems to improve tsunami forecasting (Eble and González, 1991; González et al., 1991; González and Kulikov, 1993; Ritsema et al., 1995; Bernard et al., 2001). Pilot projects were carried out in the Gulf of Alaska using internally recording gages. This was the first time that high-resolution tsunami models were used together with bottom pressure measurements to study the details of deep-water earthquakes, including the effects of finite rupture time on tsunami generation (González et al., 1987, 1991).

The 1990s saw the transition to modern tsunami observing systems, encouraged by the United Nations as part of the International Decade of Natural Hazard Reduction. A number of regional tsunamis occurred during the decade that provided the opportunity of international teams to develop and refine on-shore and offshore survey techniques. These teams were also instrumental in helping the impacted nations create or expand their own capability to address the tsunami threat. During this decade, numerical models evolved to be the principal tools by which tsunami measurements are interpreted.

This decade also saw the first deployment of the prototype array of real-time reporting DART (Deep-ocean Assessment and Reporting of Tsunamis) systems to provide open-ocean tsunami measurements to tsunami warning centers (Bernard et al., 2001; González et al., 2005; Meinig et al., 2005a,b; Titov et al., 2005a). The proof-of-concept came with the 2003 Rat Islands tsunami when DART measurements taken near the earthquake in the North Pacific were used to tune a tsunami forecast model and done so soon enough to contribute to the important operational decision on whether to issue a tsunami warning for Hawaii (Titov et al., 2005a).

2.3 Submarine Cable Observations

The first long-term open-ocean cable station OBS-II was deployed by the Lamont Observatory 100 miles offshore from northern California (in 4 km depth) in April 1965 (Nowroozi et al., 1966). This station worked for ~ four months. In May 1966, the OBS-III was deployed in the same region and operated for more than seven years (Nowroozi, 1972). S. Soloviev initiated similar work in Russia. For many years the Sakhalin Complex Research Institute, Yuzhno-Sakhalinsk, Russia (at present, the Institute of Marine Geology and Geophysics) provided cable measurements of long waves in the tsunami frequency band on the shelves off the Kuril Islands, Sakhalin and Kamchatka. On 23 February 1980, the first deep-sea records of tsunami waves were obtained on the shelf of Shikotan Island (Dykhan et al., 1983). Successful two-year long cable measurements on the southwestern shelf of Kamchatka were provided by Kovalev et al. (1991). All of these deployments were relatively inexpensive and conducted for the research purposes. An important aspect of these studies was the analysis of simultaneous tsunami observations in the deep-ocean and coastal instruments to estimate a spectral transfer function between the deep ocean and the coast (e.g., Rabinovich et al., 2006b).

Regional earthquakes and tsunamis are of special significance to Japan since there is a long history of them causing enormous loss of life and property damage to its coastal communities. In response to these threats, Japan has developed regional cable-based systems providing continuous tsunami measurements in real-time for operational purposes as well as for research. The first Japanese cable and TK1 station were initiated by the Japan Meteorological Agency in August in 1978 (Taira et al., 1985). Several tsunamis have been recorded by this station (cf. Okada, 1991, 1993, 1995). The tsunami measurements from this system have been used to study the spatial patterns of vertical ground deformation due to regional earthquakes, in which model tsunami waveforms from candidate source distributions are fitted to the observed time series (e.g., Hino et al., 2001; Hirata et al., 2003; Baba et al., 2004; Tanioka et al., 2004; Matsumoto and Mikada, 2005; Satake et al., 2005). The success of its first cable-based system encouraged Japan to deploy six more cable lines.

3. Present In-Situ Instrumentation

This section is devoted to a general overview of in-situ instrumentation presently being used to measure tsunami time series. The instruments consist of coastal gages measuring water levels and/or pressure and offshore systems measuring bottom pressure.

3.1   Coastal Water Level Gages

Coastal water level gages provide direct measurements of tsunami wave heights. Most often, these systems are maintained on a permanent basis to report water levels in support of safe navigation, to allow highly accurate determination of tidal datums for nautical charting and for coastal-zone mapping. They also monitor long-term changes in sea level and relative land elevation in support of climate and engineering studies (NOAA, 2006). For the United States, the NOAA/National Ocean Service maintains the gages as part of the National Water Level Observing Network. They are also integral components of warning systems that forecast and monitor tsunamis and storm surges. Many gages have been in place for decades and have thereby provided detailed records of numerous tsunamis. To improve their long-term chances of survival, the gages are most often placed inside protected ports and harbors. These locations also allow for efficient maintenance and communications. In addition, the U.S. tsunami warning centers maintain regional networks of tide gages in Alaska and Hawaii.

Unfortunately, many gages in protected regions don't measure tsunamis. A good example of this is for the east coast of the US and Canada. Thomson et al. (2007) examined 107 tide gage records and found that those from stations in channels and inlets did not see the 2004 Indian Ocean tsunami while those on the outer coast did. The increased awareness of tsunamis following this tsunami has caused many nations in the Indian and Atlantic Regions to consider improvements in their coastal tide gages so that these are able to resolve tsunamis in time through more rapid sampling and to report the measurements in real-time. Many gages are near major population centers and critical infrastructure along the coast that might be impacted by tsunamis. However, such locations are not necessarily optimal for measuring tsunamis in support of scientific research.

The instruments used by NOAA are typical of modern water level gages (Fig. 7.1). Each gage consists of: a set of water level and other sensors, a microprocessor to control the sampling sequence, data processing and internal recording of the measurements, a redundant communication system to receive instructions and to transmit both routine and event data, a reliable power supply, and a robust housing to protect delicate components.

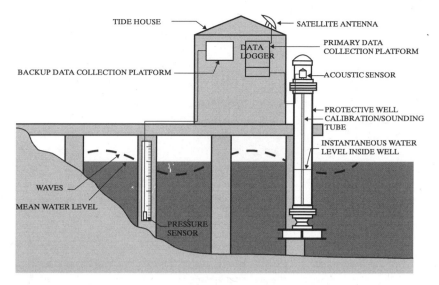

Fig. 7.1 – Schematic diagram for a coastal water level gage. Shown is a pier installation of a modern water level system using an acoustic method to measure the water level inside a stilling well, with a pressure sensor that makes back-up measurements. The observations are transmitted in real-time as well as stored internally for later access. Not shown is a telephone land-line that is often used to connect the gage to a tsunami warning center. From the NOAA/National Ocean Service Tides and Currents Website.

The primary water level measurements are made inside a vertical pipe, called a stilling well, that connects near its bottom to the water outside via a small orifice (NOAA, 2006). The purpose of the orifice is to mechanically filter out high-frequency fluctuations in water level caused by ship wakes, wind waves and swell. A sound transducer mounted at the top of the pipe sends rapid 1-Hz pulses down-ward that reflect off the water surface and return back to the transducer. The net travel time is then a measure of the water level at that moment. After correcting the speed of sound inside stilling well using directly measured air temperature and humidity, the travel times are converted to distances between the transducer and water level and averaged over 15 seconds. The gages are self-calibrating to ensure accuracy. The 15-s averages are converted to water levels relative to a gage reference level and stored in a 22-day buffer. The water levels are subsequently converted to heights relative to tidal and geodetic datums for use in navigation and charting, tracking long-term trends in relative sea level, and in support of warning systems. When combined with GPS observations, they also provide trends in absolute sea level.

The gages routinely report water level and other data. Sets of six 1-min water levels, each an average of four 15-s values, are sent via GOES satellites and/or telephone lines. The regularly scheduled transmissions are limited to 1-min data because of power considerations at remote sites and the slow data-rates of the GOES systems. Earlier versions of the gages reported hourly, except during events when the gages are set into rapid reporting mode. This was done either by an operator sending commands to the gages or automatically when a gage detects a sufficiently rapid change in water level. The 1-min sampling is regarded as adequate for operational purposes and for the purpose of measuring large tsunamis with long periods. However the 15-s data are preferred for detailed scientific analysis and for observing tsunamis with short wave periods; the 15-s data can now be retrieved remotely upon command. Work is underway to also use the Iridium satellite system for more capable communication.

It has also become clear that the restricted flow through the small (sometimes biologically or sediment fouled) orifices into the stilling wells of tide gages can non-linearly distort the heights and shapes of large tsunami waves when the tsunami-induced flow through the orifices is sufficiently strong (Shipley, 1963; Lennon, 1971; Noye, 1974a,b,c; Braddock, 1980; Wigen, 1983; Satake et al., 1988). This occurs during large tsunami events but is much less of an issue for small tsunamis. A partial solution is to estimate the frequency response of the stilling well and then adjust the tsunami measurements accordingly. However, the correction factors for a given gage will vary over time as fouling changes the effective orifice. Hence there will always be some uncertainty in tsunami measurements made within stilling wells as compared with measuring the tsunami water levels directly. In practice, the instrumental effects on tide gage measurements of strong tsunamis are often ignored.

An alternate method to observing coastal water levels replies on pressure measurements. At 5–7% of the U.S. tide gage locations, bubbler pressure sensors are installed instead. Pressure gages can be deployed quickly on the shore of active volcanos when there is the imminent danger of an eruption or landslide causing a tsunami. Such a system has been deployed recently on Augustine Island in Cook Inlet, Alaska (Burgy and Bolton, 2006). These sensors measure the pressure at the landward end of tubes that extend just offshore; a slight overpressure supplied by gas tanks prevents water from entering the underwater end of the tubes. Pressure sensors also serve as secondary sensors in the standard NOAA water level gages (Fig. 7.1). The pressure is converted to water levels using the local atmospheric pressure, acceleration of gravity and water density, where the latter is adjusted for water temperature and salinity measured concurrently at the site. The conversion to water level is done under the hydrostatic assumption that the measured pressure is equal to the atmospheric pressure plus the product of gravity, density, and the water column height.

For measuring tsunamis, pressure sensors outside the stilling wells have advantages over sensors inside the wells. The first advantage is that these pressure sensors can measure large tsunami crests that are higher than the top of the stilling well, which is typically at pier height. This prevents clipping of tsunami wave crests in which the actual heights of overtopping waves are not measured. They also avoid non-linear distortions and attenuation of strong tsunamis measured inside the well.

Besides shore-bound systems, autonomous self-reporting pressure gages have been available commercially for many years that have the potential to measurement tsunamis on the continental shelf and slope. In principle, energy-efficient electronics and large data storage capacity would make these gages useful for monitoring pressure for tsunamis. Since tsunamis are rare and unpredictable at present, the deployment and maintenance of such research gages are probably best justified as part of broader research programs. Conversely, the data obtained from autonomous gages deployed for other purposes may contain valuable tsunami measurements.

Another way to directly measure water levels is by a downward-looking radar mounted on a pier. The West Coast/Alaska Tsunami Warning Center presently has five of these gages in Alaska (Shemya, Amchitka, and Akutan in the Aleutian Islands, Old Harbor on Kodiak Island, and Craig on Prince of Wales Island in Southeast Alaska). The radar systems are relatively inexpensive and can be placed in locations where drifting ice in winter would destroy a stilling well.

Despite the advantages of exposed sensors for tsunami observations, the great majority of water level gages presently in operation are stilling-well systems in harbors. An exception is the NIWA Sea-level Network of coastal pressure gages that New Zealand has installed along its coast for the expressed purpose of measuring tsunamis at remote exposed sites. The measurements from such open coast locations often do contain substantial storm wave and infragravity fluctuations, producing background noise that can interfere with the measurement of tsunamis. Hence no type of site can be regarded by ideal for measuring tsunamis in the coastal region, and it is therefore necessary to trade-off a number of factors in choosing coastal station locations and instrumentation.

3.2   Deep-Ocean Bottom Pressure Gages

The tsunami instruments (Fig. 7.2) presently deployed in the deep ocean consist of a bottom-mounted unit (tsunameter) that measures the fluctuations in bottom pressure and a nearby surface buoy that relays both the data via satellite to shore and instructions from operators down to the bottom unit (Meinig et al., 2005a,b; NOAA, 2006). Analogous to coastal water level gages, each bottom unit has pressure and temperature sensors, a microprocessor, a communication system, a power supply using batteries, and pressure housings to protect components. The bottom unit also has an anchor-platform, acoustic release and floatation to allow it to be recovered along with the data it has recorded internally. The battery capability of the DART II systems is sufficient to allow servicing every 24 months.

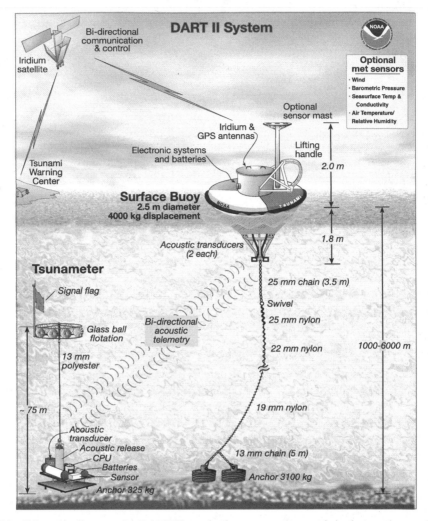

Fig. 7.2 – Schematic diagram of a DART II system that measures tsunamis in the open ocean. Shown are the bottom unit (tsunameter) that measures bottom pressure and a separate surface buoy that communicates with the bottom unit using an acoustic modem and with shore installations via the Iridium satellite system.

The bottom unit and surface buoy communicate with each other using acoustic modems that send digital information by sound. The buoy reformats the bottom pressure, temperature and engineering data from the bottom unit and transmits these via satellite to the NOAA/Data Buoy Center and tsunami warning centers. The DART II systems use the global Iridium satellite system, which allows for efficient two-way communication. To communicate reliably, the bottom and surface units must be within the footprint of the other's acoustic transducer. For redundancy, each has two acoustic transducers. There is also redundancy between sequential data messages during tsunami events for each transmission channel, to ensure that the data are received on shore even if a block of data is lost.

The duration of the data messages in time translates to a physical length that can be a significant fraction of the water column. Also determining the physical length of a message is the carrier frequency used for transmission. Since the attenuation of sound by seawater decreases with frequency, this frequency needs to be set low enough to reliable data transmission without placing excessive power requirements on the bottom unit. To avoid surface and bottom reflections of a message's first part arriving at a receiving transducer at the same time the latter is still receiving the last past of the message, it is necessary to set a minimum deployment depth in order to get sufficient spatial separation between the surface and bottom transducers as compared with the length of the data messages. For the DART II systems, these constraints set the minimum deployment depth to 1000 m. The maximum deployment depth is 6000 m, set by the capability of the pressure housings and fittings that protect the components of the bottom unit.

The Paroscientific Digiquartz™ pressure sensor in each DART bottom unit contains a quartz oscillator that is mechanically connected to the outside water and thereby varies with frequency as the pressure changes. Counting pulses from the transducer over 15 seconds, the microprocessor converts the average oscillator period to pressure using a conformance equation, with corrections made to remove temperature effects. The values of the parameters in the equation are determined by a pre-deployment calibration. Given the transducer sensitivity, the 15-s average provides a sub-millimeter resolution of equivalent changes in sea surface elevation.

The bottom units are able to resolve tsunamis as small as 1 cm in amplitude because the background noise in pressure is so low at the bottom of the oceans. This high-frequency noise is due to seismic Rayleigh waves propagating from regional and distant earthquakes, microseisms and other meteorologically-generated fluctuations, and micro-tsunamis propagating throughout the ocean basins (Filloux, 1982, 1983; Dykhan et al., 1983; Kulikov et al., 1983; Webb, 1998). The bottom units are deployed much deeper than the depth of influence of wind waves and swell, although groups of these waves induce small pressure fluctuations that reach the bottom. The seismic pressure is caused by accelerations of the water column that are induced by vertical movements of the ocean bottom as the waves pass by. While it is substantially attenuated by the 15-s averaging of bottom pressure, some seismic noise still remains even after this filtering. Studies of DART measurements in the Gulf of Alaska indicated that the noise level there was generally less than 1 cm. The seismic noise does exceed this amplitude immediately following Magnitude ≥ 6.0 earthquakes in the region (Eble and González, 1991; Eble et al., 2001; Mofjeld et al., 2001).

In monitoring mode each DART II system reports every six hours a set of pressure values spaced every 15 minutes (NOAA, 2006). When attached to previous values and plotted in sequence over the past few days, the result is a time series of pressure (Fig. 7.3) that is dominated by the tides. These data, and those obtained during tsunami events, are processed and made available by the NOAA/National Data Buoy Center on its Website as preliminary data; this Center also has the responsibility for the deployment and maintenance of the U.S. DART Network. Such plots (Fig. 7.3) provide an operator and staff at tsunami warning centers with an immediate check on whether a system is operating and in calibration, based on a comparison with predicted tides for that station.

Station 46403 – 230 NM Southeast of Shumagin Island, AK

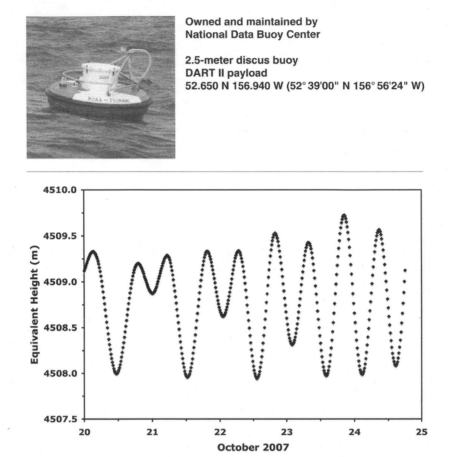

**Owned and maintained by
National Data Buoy Center**

**2.5-meter discus buoy
DART II payload
52.650 N 156.940 W (52° 39'00" N 156° 56'24" W)**

Fig. 7.3 – Typical bottom pressure (converted to equivalent water depth) time series of a DART II system in monitoring mode. The measurements were taken during October 20–24, 2007 and transmitted every hour via the Iridium satellite system to shore for processing and display by the NOAA/National Data Buoy Center, which deploys and maintains the U.S. DART Network. The station is located near the Alaska-Aleutian Subduction Zone (AASZ), which generated the 1964 Alaska Tsunami. From the NOAA/National Geophysical Data Center DART Website.

A DART II system goes into tsunami event mode either when an operator at a tsunami warning center sends the appropriate command or when the automatic detection algorithm inside the bottom unit is triggered by a pressure fluctuation. Experience has shown that the automatic detection algorithm is tripped by earth-quake-generated seismic surface waves reaching the systems ahead of any tsunami waves generated by the same event. This threshold is usually set to 3.0 cm but can be changed remotely by an operator anywhere in the range 3–9 cm to allow for greater background noise. Seismic fluctuations occasionally cause the automatic tsunami detector within the bottom units to set DART systems into event reporting mode, thereby providing useful if unplanned tests of the system's performance.

The great M9.3 Sumatra earthquake in December, 2004 set nearly all the DART systems in the Pacific Ocean into event mode, including those deployed in the northern Gulf of Alaska (Titov et al., 2005b).

In event mode, one-minute averaged pressure measurements are transmitted in overlapping data blocks to provide redundancy. These follow an initial block of 15-s values to better define the triggering fluctuation. An operator can initiate the event mode at a time computed from pre-computed tsunami travel time between the location of the source and the observing station. This provides real-time data not only for tsunami warning centers but also for scientists analyzing the new event. However, the length of the time series is limited to several 1-hr blocks of data, rather than days, in order to conserve battery resources. Each bottom unit has recorded the full 15-s time series internally for the entire deployment, and this series is recovered when the station is serviced.

3.3   Cable-Mounted Bottom Pressure Systems

Cable-mounted pressure systems have advantages over autonomous system that must rely on self-contained sources of power. A continuous supply of power to sensors and electronics eliminates many complications in sampling and communications. A continuous stream of measurements sampled at high frequency can be transmitted to shore. Japan presently operates seven such systems also using quartz oscillator-based pressure sensors, as well as ocean bottom seismometers. Information on these systems and their operation are provided in the references (Iwasaki et al., 1997; Hirata et al., 2002, 2003; Hirata and Baba, 2006; Momma et al., 1996, 1997, 1998).

Submarine cables for operational tsunami measurement are laid in trenches and then buried to protect them from fishing activities and other hazards. Further offshore, cables have been damaged by submarine landslides moving down the continental slope and rise. For example, twelve telegraph cables were cut off Eastern Canada during the 1929 Grand Banks earthquake (M7.2), submarine landslide and tsunami (Bryant, 2001; Fine et al., 2005). More recently, submarine landslides triggered by earthquakes damaged submarine cables in the Mediterranean Sea in 2003 and off Taiwan in 2006. However, submarine cables can be made less vulnerable to hazards through careful siting and installation.

3.4   Instrument Improvements and New Technologies

The large cost of deploying and maintaining real-time observing networks creates the impetus for efficiencies and cost-sharing with other programs. This has always been true with coastal water level gages. Examples of these are the reliable cable systems deployed by Japan. Including bottom pressure sensors in cable-linked ocean observatories with broad scientific missions, such as the new NEPTUNE Program off the West Coast of North America, is a way to greatly decrease the cost of cable-mounted systems to the tsunami community (Garrett, 2004).

For the open-ocean network, a major ongoing cost is for the use of large oceanographic research vessels needed to handle the large surface buoys, e.g. of the DART systems. Making these buoys more compact would allow smaller, and therefore less costly, vessels to be used; a compact surface buoy is presently under development by NOAA. Increasing endurance of the surface and bottom units allows less frequent servicing, although this delays access to internally-recorded data.

Observations of the 2004 tsunami in the Indian Ocean by the JASON 1, Topex/Poseidon and Envisat satellites (e.g., Gower, 2005; Ambrosius et al., 2005; Fujii, and Satake, 2007) show that altimetric satellites are capable is providing detailed spatial profiles of a tsunami's instantaneous (nearly) surface elevation pattern along the satellite track. This is if the satellite is passing over the tsunami at that time and location. An increase in the number of altimetric satellites would increase the chances of this happening.

Another type of technology being developed for offshore tsunami measurement utilizes global positioning system (GPS) receivers on surface buoys (Kato et al., 2000, 2005). The advantage is that no bottom unit is needed, since all the measurements are made on the surface buoy. The disadvantage is that the ocean surface is a very noisy environment in which the buoys respond in response to a wide variety of non-tsunami movements. Filtering these out to accurately measure tsunamis, especially small ones, is a challenge.

A real-time tsunami inundation detection system has been developed and deployed along the coast of the Island of Hawaii (Walker and Cessaro, 2002). At each station any flooding over 1 ft (0.3 m) above the ground is detected and immediately communicated to warning centers via cellular phone technology. Because it observes inundation directly, the system is able to warn against local landslide-generated tsunamis that are not associated by earthquakes, as well as those that are. There is a history of both occurring around the Island. As cellular and other communication systems become more widely available, this type of system will provide a very inexpensive supplement to tsunami warning systems (Walker and Cessaro, 2002).

Except for the cable system on the Japanese continental shelf, there is presently a gap in tsunami measurements between the coast and the open ocean and in shallow seas. Offshore instrumentation is needed that is capable of long-term monitoring in the water depths that are shallower than the 1000-m limitation of the DART systems.

An emerging technology has been demonstrated to measure and report near-shore tsunami currents in real-time using the cable-based Kilo Nalu Near-Shore Reef Observatory near Honolulu, Hawaii (Bricker et al., 2007). An ADCP current sensor deployed 400 m off the shore in 12 m of water provided time series of currents during the November 15, 2006 Kuril Islands tsunami. Using the combination of water level and current measurements, Bricker et al. (2007) show that the incident tsunami generated groups of edge waves propagating in both alongshore directions around Oahu. Hence, having simultaneous observations of water level and currents increases the value of each when interpreting tsunami measurements.

4. Observational Networks

The 2004 Indian Ocean tsunami demonstrated the need for a more extensive tsunami observational network to improve the speed and accuracy of tsunami warning systems. New international tsunami warning systems will be established in the Indian Ocean, the Caribbean Region, and a region covering the Northeastern

Atlantic Ocean, Mediterranean and associated seas. The new tsunami observing networks are being created under the auspices of the United Nations to support the regional tsunami warning systems (Chapter 13). Each network will be planned and operated under a UNESCO/Intergovernmental Coordination Group, with special working groups devoted to instrumentation.

4.1   Coastal Water Level Gages

A global network of coastal gages was already in place before the 2004 Indian Ocean tsunami, because these instruments provide ongoing measurements for a wide range of purposes. As a result of the tsunami, additional gages have also been added. However the main emphasis has been on upgrading the existing system to provide adequate tsunami measurements. The expanded capability for coastal water level gages is being achieved largely by upgrading the existing GLOSS network (Fig. 7.4). The gages will report 1-min measurements in real-time using the World Meteorological Organization's (WMO) Global Telecommunications System (GTS). Together with standardized data processing procedures and transmission format, using the existing GTS is efficient and shortens the time needed to become operational.

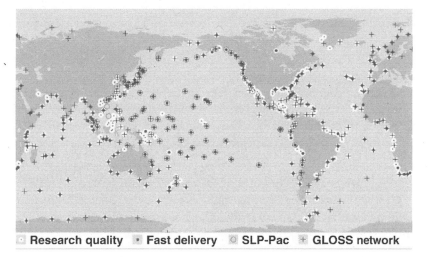

Fig.7.4 – Tide gage locations for the Global Sea Level Observing System (GLOSS) and other climate-related sea level networks. For routine purposes the data from "Fast delivery" network are available within a month of observation, while the data from stations designated as "SLP-Pac" have been adjusted for atmospheric pressure. Presently most of the indicated gages in the Pacific Ocean are tsunami-capable; and there are plans to upgrade stations in this and other oceans in support of regional tsunami warning systems. From the website of the University of Hawaii Sea Level Center.

4.2   Open Ocean DART Systems

Prior to the Indian Ocean tsunami, a limited DART network existed in eastern Pacific Ocean. As a result of the 2004 Indian Ocean tsunami, the network of offshore tsunami stations is now expanding rapidly. The United States has completed a major expansion of the DART Network to a total of 39 DART systems (Fig. 7.5) deployed in the Pacific and Atlantic Regions (NOAA, 2006). Chile and Australia have also deployed DART systems in the South Pacific. New DART networks are being planned for other oceans. The first two operational DART systems in the Indian Ocean have been deployed by Thailand, and India has announced the deployment of a new network as well.

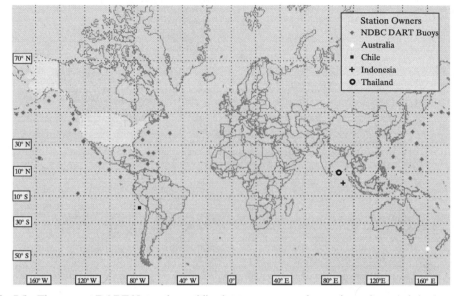

Fig. 7.5 – The present DART Network providing bottom pressure observations of tsunamis in the open ocean to tsunami warning centers and the research community. From the NOAA/National Data Buoy Center Website.

To provide a perspective on how tsunami measurement networks are designed, the following focuses on design criteria for DART systems deployed in the open ocean. This is with the understanding that similar criteria exist for the networks of coastal gages and offshore cable systems. Each type of instrumentation can in fact be regarded as complementing the others in the trade-offs between capability, cost and flexibility.

A number of factors need to be considered in designing a DART network and siting individual stations (Bernard et al., 2001; Mofjeld et al., 2001; González et al., 2005). DART systems are usually placed near active subduction zones that have been identified as sources of dangerous tsunamis. They may also be placed further offshore where passing tsunamis must still propagate for many hours before they can reach coastal communities. Overlapping coverage of DART stations is required so that the loss of a single station does not lead to a spatial gap in coverage. The candidate DART station locations are determined from model studies using the tsunami source distributions for the region. The stations cannot be too close to sources, or the seismic surface waves generated by the earthquake will still be strong enough when the first tsunami waves arrive to interfere strongly with their interpretation. Experience and modeling has shown that the minimum distance between an open-ocean DART station and the nearest source region should ideally correspond to about the 0.5-hour contour on a tsunami travel time chart for the source region.

However, there is also impetus to deploy some DART stations at locations closer to tsunami sources. This is in order to provide earlier cancellation of tsunami warnings for communities in the vicinity of the sources. An example along the U.S. West Coast is the June 2005 California earthquake and resulting tsunami warning, in which the need for more rapid warning cancellation was strongly voiced by emergency management agencies. A detailed study of the tsunami is given by Rabinovich et al. (2006a). For regional purposes, a station is located near the Island of Hawaii to detect landslide-generated tsunamis from the Island. Another is located west of the Channel Islands off Southern California to provide the last open-ocean measurement of tsunamis before they enter the Southern California Bight. Hence DART networks are required to serve regional needs, as well as measuring tsunamis near the sources.

The bottom at DART stations needs to be relatively level so that the bottom units and surface buoys can communicate reliably; this requires that site selection be based on accurate bottom topography and often entails a detailed bathymetric survey during the deployment cruise. The stations should be sufficiently far away from islands and seamount chains that scatter tsunami waves. Otherwise, this scattering creates complicated, short-scale wave patterns that are not resolved by numerical models being tuned to the measurements.

As mentioned previously, the acoustic link requires water depth ≥ 1000 m, although this limitation could be overcome by using a different type of data link. The pressure limit of the bottom units' housings requires that the depth be ≤ 6000 m, a requirement readily achievable in the world's ocean and seas. Without compromising the operational goal of the network, the regional DART arrays should be designed for logistical efficiency such as locating stations near meteorological and oceanic buoys deployed by other programs. Avoidance of areas subject to vandalism is also desirable. Finally, it is expedient to locate DART stations in international waters to respect national concerns and to avoid procedural delays in getting permission for deployments and maintenance.

4.3   Submarine Cable Systems

At present, seven submarine cables (Fig. 7.6) with pressure sensors are deployed on the continental shelf off of Japan's Pacific coast for the primary purpose of measuring tsunamis. The pressure sensors are located along the path that both regional and distant-source tsunamis must take on their way to the open coast and to embayments with major metropolitan areas. The data are sent to the Japanese Meteorological Agency.

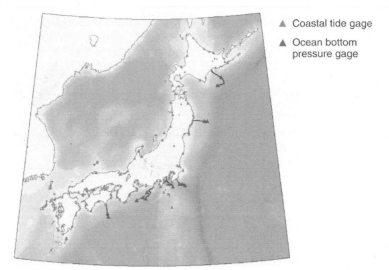

▲ Coastal tide gage

▲ Ocean bottom
  pressure gage

Fig. 7.6 – Present network of cable-mounted pressure sensors off Japan. Also shown are coastal tide gages reporting tsunami observations in real-time (see the Japan Meteorological Agency Website for details).

A cable-mounted system is being installed off the Pacific Coast of Canada and the United States as part of the multi-purpose NEPTUNE Program (Garrett, 2004). The initial phase (NEPTUNE Canada) will have two optic cables running offshore from Vancouver Island and connecting on the far side of the deep Cascadia Basin, forming a continuous loop with side-branches on the Juan de Fuca Ridge and Explorer Plate. Instrumented nodes along the cables will contain bottom pressure and a variety of other sensors. There are plans to expand the network of cables southward to the full NEPTUNE array in order to also instrument the continental margin off Washington and Oregon.

5. Tsunami Time Series

The purpose of in-situ tsunami measurement systems is to provide accurate time series. Ideal time series of water levels or pressure are sequences of measurements in which the time of each measurement is known precisely, the instruments are accurately calibrated, and the reference water level or mean pressure is also known with precision. Accurate timing is needed to estimate, via numerical models, the location of the tsunami sources. Since analysis software typically requires data that are spaced evenly in time, irregularly spaced data are interpolated to even spacing after they have been corrected for instrumental data spikes and data gaps. The temporal resolution must be fine enough to resolve the details of the tsunami waveform. It is usually assumed that the tsunami fluctuations are linearly superimposed on the background tides and other fluctuations with frequencies below the tsunami band (wave periods of roughly 2–120 min), as well as high-frequency noise. The total water level η at a time t is then assumed to be the sum

$$\eta(t) = \eta_{tsu}(t) + \eta_{Low}(t) + \eta_{Noise}(t) + \eta_{ref}, \qquad (1)$$

where η_{tsu} is the tsunami water level at time t relative to mean sea level, η_{Low} is the water level elevation due to the tides and lower frequency fluctuations, η_{Noise} is the noise in the high-frequency and tsunami bands, and η_{ref} is the height of mean sea level relative to the standard datum for the tide gage. Implicit in the linear sum (1) is the assumption that the tsunami can be separated out from the other fluctuations in water level and analyzed in isolation from them.

Care must be taken when relating the total water levels $\eta(t)$ to land elevations since they are often referred to a different datum. For instance along the U.S. coastline, the water levels are referred to mean lower low water (MLLW), whereas the land elevations are given relative to mean high water (MHW). Other nations may use different reference datums for their water level data, land elevations and water depths. The differences in height between these datums range from a few 10s of centimeters to several meters, depending on the local tidal range. Having the reference level is also an essential part of the metadata needed to interpret runup heights from post-tsunami field surveys and compare these with the results of tsunami inundation models.

Pressure time series are also assumed to be sums of terms, with the addition of atmosphere pressure P_{atmos} when the absolute pressure is recorded

$$P(t) = P_{tsu}(t) + P_{Low}(t) + P_{Noise}(t) + P_{atmos}(t) + P_{ref}, \tag{2}$$

Converting from pressure to water level requires assumptions about the tsunami wave dynamics. Except for very high-frequency tsunamis, the hydrostatic assumption is valid in which the vertical acceleration of the water can be neglected. Then, the pressure deviation (force per unit area) due to the tsunami is equal to the weight per unit area caused by the tsunami's sea surface elevation immediately above the sensor. Then at each time t,

$$P_{tsu}(t) = \rho \, g \, \eta_{tsu}(t) \tag{3}$$

where ρ is the in-situ water density, averaged over the water column, and g is the local acceleration of gravity. For high-precision tidal research, g should also be averaged over the water column (Mofjeld et al., 1995).

When converting deep-ocean measurements of bottom pressure to equivalent water level (e.g., Fig. 7.3), the in-situ density takes into account the increase in density with depth. It can be computed from the equation of state, using climatological profiles of temperature and salinity for the station's location. Operationally it is convenient to convert pressure measured by DART systems to water level using a nominal conversion factor (presently 670.0 mm/psi, or 0.09718 mm/Pa). However the measurements should also be available in the original pressure units at data centers, since pressure is what was actually measured by the instrumentation (NOAA, 2006).

5.1 Empirical Analysis of Tsunami Time Series

The study of the 2004 Indian Ocean tsunami by Rabinovich et al. (2006b) is representative of the present state of the art in the empirical analysis of tsunami time

series, as observed at coastal water level gages. Merrifield et al. (2005) and Nagarajan et al. (2006) have also published analogous studies for tide gage series within
the Indian Ocean. Rabinovich et al. (2006b) obtained time series from 200 tide
stations in the Indian, Pacific and Atlantic Oceans. After data editing and converting to 1-min sampled series (a 2-min low pass applied to 15-s data), predicted tides
were subtracted and the resulting series (Fig. 7.7) filtered to focus on the tsunami
frequency band. Specifically, a 4-hr high pass filter was applied to the 1-min sampled series in order to remove high-frequency noise due to infragravity waves and
fluctuations below the tsunami band, respectively. Digital Kaiser-Bessel filters
were used, which are optimal in simultaneously limiting the width of the transition
band and filter overshoot.

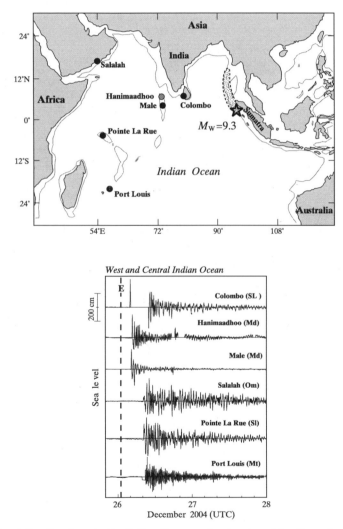

Fig. 7.7 – Tide station locations in the Indian Ocean and epicenter of the December, 2004 Sumatra
earthquake (top) and observed water level time series (tides removed) for the resulting tsunami. The
symbol E indicates the time of the earthquake. Modified from Rabinovich et al. (2006b).

A wavelet method was used to track the changes of tsunami frequency with time and to identify persistent frequency concentrations that are characteristic of the tide station. The study by Rabinovich et al. (2006b) shows that amplitude and frequency distribution of the 2004 Indian Ocean tsunami varied greatly between tide stations along the same coast, as also observed in other studies of Pacific tsunamis (e.g., Rabinovich, 1997; Mofjeld et al., 1997, 1999; also see Subsection 2.1). The behavior of the tsunami as observed at the stations were strongly influenced by the presence or absence of local resonances, which cause the temporal and frequency patterns for different tsunamis at a given tide station to be similar. Rabinovich et al. (2006b) give an empirical formula to retain a tsunami's amplitude at a coastal tide gage to that immediately offshore in the deep ocean, as a function of frequency.

Going beyond empirical analyses requires a comparison of the observed tsunami time series with numerical models of the event (e.g., Kowalik et al., 2005; Titov et al., 2005b; Arcas and Titov, 2006; Geist et al., 2006). For instance, comparisons show that the observed distribution of the 2004 Indian Ocean tsunami along the world's coastlines are consistent with model simulations (Kowalik et al., 2005; Titov et al., 2005b) in which the mid-ocean ridges act as waveguides to channel tsunami energy preferentially at certain locations in other oceans. Fine et al. (2005) used tide gage data from the Indian Ocean to identify a dual source for this tsunami.

The first wave in an observed tsunami time series is particularly important in any analysis related to the tsunami source. This is because it is the least affected by reflections, scattering and wave interference that complicate later waves. The timing, amplitude, period and sign (leading trough or crest) of this first wave provide the best information about the source's pattern of vertical ground displacement. Operational forecast systems use the first tsunami waves observed at DART stations (Fig. 7.8) to refine the preliminary tsunami sources by inverting the data back to the source using pre-computed model simulations (Titov et al., 2005a).

Tsunamis in the major oceans form irregular wave patterns that decay slowly over many days (Miller et al., 1962; Van Dorn, 1984, 1987; Mofjeld et al., 2000). Near the source, partial wave trapping on the continental shelf and slope, multiple reflections at the coast and shelf break, and scattering by local islands and shoals transform the initial spatial pattern of the tsunami into a series of waves that propagate out of the source region. When these waves encounter islands, seamount chains, and shelf regions, they generate an even more complex pattern of waves in time (Mofjeld et al., 2001, 2004; Rabinovich et al., 2006b). Tsunami amplitudes also decrease as tsunamis spread out over the broad surface of the ocean, propagate into other oceans, and dissipate due to friction. Taking a statistical approach to the amplitude decrease in later waves was justified as a result of random scattering and reflection by earlier waves (Miller et al., 1962). The e-folding time for amplitude was found to be 44 hours (22 hours for energy) for the Pacific Ocean (Van Dorn, 1984), where this time is how long it takes to decay to 37.8% of the initial value. Based on similar analyses of observed series, the e-folding time is found to be shorter for the Indian Ocean and Mediterranean Sea (Van Dorn, 1987).

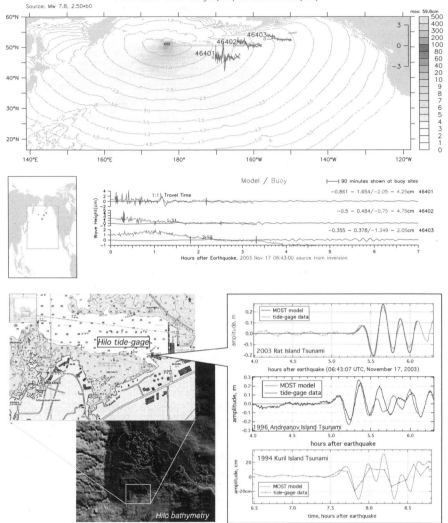

Fig. 7.8 – Observed time series at Gulf of Alaska DART stations (top) and at the Hilo, Hawaii water level gage (bottom) for the 2003 Rat Islands tsunami. Also, shown are observed series for the 1996 Andreanov Is. and 1994 Kuril Is. tsunamis together with model series that were tuned to the observations. From Titov et al. (2005a).

Proximity to the tsunami source (Fig. 7.7) also contributes to a more rapid temporal decrease in amplitude (Rabinovich et al., 2006b). The slow decay of tsunamis is observed at open-ocean bottom pressure stations, although the energy decay in the open ocean has not been investigated in detail. As tsunami models become more accurate at simulating tsunamis over many hours, the deterministic results will help explain unusually large wave groups that sometimes punctuate the generally monotonic pattern of decreasing tsunami energy.

5.2 Access to Tsunami Time Series Measurements

Nations have an obvious need to share real-time tsunami data with tsunami warning systems serving them; Pacific nations have done this successfully for many years as members of the International Tsunami Warning System. The 2004 Indian Ocean tsunami has also led to an international demand to provide rapid access to tsunami time series from coastal tide gages and oceanic systems for research purposes. Tide gage data for the 2004 Indian Ocean tsunami are available from the University of Hawaii Sea Level Center (Merrifield et al., 2005) and the Institute of Ocean Sciences (Rabinovich et al., 2006b). At present, the NOAA/National Data Buoy Center provides Web-access to monitoring (15 min) and selected DART event data (1 min). In the near future the NOAA/National Geophysical Data Center (NGDC) will provide Web-access to the full series (15 sec) for DART stations as part of its function as a World Data Center.

It has been common practice for tsunami scientists to collegially share in-situ tsunami measurements, typically soon after a tsunami, when it is the focus of the research community. This has been done without any formal institutional structures in place to permanently provide quality-controlled time series. Greatly facilitating the communication between tsunami scientists, as well as sharing information and plans immediately following tsunami events, is the electronic Tsunami Bulletin Board developed by the NOAA/Pacific Marine Environmental Laboratory and now administered by the International Tsunami Information Center.

Presumably, unexamined tsunami data from digital water level gages may still exist for at least some past events, although much may have been lost unless specific requests were made at the time that the data be saved. Access to early analog records is now often limited to figures in articles and reports. An important exception is the digital dataset created from analog records by Van Dorn (1984, 1987). Inspection of other analog records from this time period reveals that they have faded to the point where the tsunami series will be unreadable in the near future. The tsunami research community would benefit from programs to rescue tsunami data before they disappear.

6. Post-Tsunami Surveys

Post-event field surveys provide on-site measurements of tsunamis within the impact areas. They document the extent of inundation, the distribution of maximum wave heights, sand deposition, and the physical effects of the tsunamis on the landforms, vegetation, and built structures (Chapter 11). Together with photographs and videos, these measurements serve to document the impact of the tsunamis. When the tsunami source is in the region, the surveys frequently include direct measurements of co-seismic ground displacement.

The first phase of a post-tsunami survey must be done quickly because much of the evidence will disappear rapidly as debris is removed and recovery operations proceed. At the same time, the scientific surveys must not interfere with emergency response and recovery operations. Later surveys help fill in data gaps; they also make use of latent plant and coral damage that does not become apparent until some time has passed. The on-site measurements provide essential ground truth for remotely sensed data, test cases for numerical tsunami models (Chapter

8), and insight into the relationship between newly formed tsunami sand deposits and those left by paleo-tsunamis (Chapter 3).

In the 1940s, land survey methods began in the United States to quantitatively map the extent of tsunami inundation and runup, which is the maximum elevation of tsunami waves at the farthermost points of inundation. Also included was evidence of the tsunami currents. Of particular importance were the maps generated at Hilo, Hawaii, after the 1946 Aleutian, the 1960 Chile and 1964 Alaska tsunamis (NGDC Website). The data from these and later tsunamis have become the standards against which tsunami inundation models are tested and certified. The documentation of the 1958 Lituya Bay tsunami in Alaska has served the same purpose for extreme landslide-generated tsunamis. Such maps have become an integral part of reports documenting tsunami events. During the 1990s, post-tsunami survey methods evolved rapidly as international teams responded to a number of regional tsunamis. Fourteen such events occurred during this decade, ranging in location from Indonesia, Papua New Guinea, Vanuatu, The Philippines, Japan, Russia, Mexico, Nicaragua, and Peru to the Sea of Marmara in Turkey (see Satake and Imamura, 1995; NGDC and ITIC Websites).

One of the most detailed surveys was done on Okushiri Island in the Sea of Japan in response to the 1993 Hokkaido Nansei-oki earthquake (Hokkaido Tsunami Survey Group, 1993; Shimamoto et al., 1995; Shuto and Matsutomi, 1995). The high-resolution measurements of runup (Fig. 7.9) along rugged coastal areas of the island provided a definitive test of the resolution needed by numerical inundation models to replicate the observed distributions (Takahashi, 1996; Titov and Synolakis, 1997). The survey results allowed valuable comparisons of inundation extent and runup with remotely sensed measurements based on aerial and satellite photogrammetry. The most complete surveys measure sediment distribution patterns that can sometimes be used, via tsunami models, to constrain the earthquake source (Chapter 3); an example is the sediment mapping done in the inundation area of the (Mw7.5) 1996 Chimbote, Peru tsunami (Bourgeois et al., 1999).

Increasingly, high-resolution bathymetric surveys are conducted in parallel with those on land to assess whether co-seismic landslides contributed to the tsunamis; the surveys also provide accurate bathymetry for numerical tsunami models. The value of offshore surveys was clearly demonstrated by the research done of the regional 1992 Nicaragua and 1998 Papua New Guinea tsunamis. Each had significantly higher amplitudes than would be expected from the magnitudes of the earthquakes. The bathymetric survey off Nicaragua showed no strong evidence for a substantial co-seismic landslide. Instead, the enhanced source strength was found to be due to a tsunami earthquake mechanism (slow-release), which is more efficient in generating a tsunami than the apparent earthquake magnitude would suggest (Imamura et al., 1993; Piatanesi et al., 1996; see Chapter 5). It did identify a nearshore ridge and thereby helped to rectify model predictions with the observed alongshore pattern of tsunami amplitude. The offshore surveys after the 1998 Papua New Guinea tsunamis showed evidence for a recent landslide seaward of the impact area. Tsunami modeling based on a combined source strongly suggested that a submarine slump contributed significantly to the 1998 tsunami (e.g., Synolakis et al., 1997, 2002).

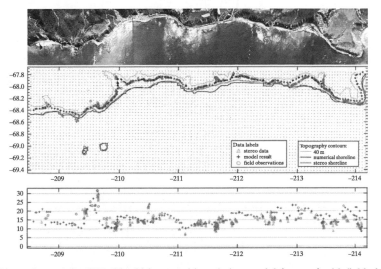

Fig. 7.9 – Comparison of the 1993 Okushiri tsunami inundation model (crosses) with field observations (circles) and stereo photo data (triangles). Top frame shows an aerial photo of the modeled area used for the stereo analysis of the inundation data. Middle frame illustrates the numerical grid used for the simulation of the same area (dots are computational nodes, contours show topography data) and compares inundation distances. Bottom frame compares maximum vertical runup for the same shoreline locations. Figure and caption from Titov et al. (2005a).

The experience from the post-tsunami field surveys showed the need for pre-event planning and the availability of survey equipment on short notice. It also led to the development of a field manual that gives standardized methods for carrying out post-tsunami field surveys. Coordination of tsunami field teams has also been greatly aided by the use of the e-mail based Tsunami Bulletin Board of the International Tsunami Information Center. This experience provided a useful foundation for the numerous surveys that were done immediately after the 2004 Indian Ocean tsunami (e.g., Borrero, 2005a,b; Choi et al., 2006; Fujima et al., 2006; Imamura, 2006; Liu et al., 2005; Narayan et al., 2006; Obura, 2006; Tomita et al., 2006 and other papers in the Vol. 48, No. 2 issue of the Coastal Engineering Journal on surveys following the 2004 Indian Ocean tsunami; Tsuji et al., 2006 and other papers on Vol. 58, No. 2, issue of Earth, Planet and Space on The 2004 Great Sumatra Earthquake and Tsunami).

In areas of co-seismic uplift or subsidence, GPS measurements made before and during post-event surveys document the tsunamigenic ground displacement and serve as point measurements to adjust and verify aerial and satellite-based measurements of the displacement fields. Continuing GPS measurements are made during the succeeding months and years to measure the tectonic rebound of the source area, ideally measured relative to pre-event elevations. Such an effort is underway to understand why the great (Mw8.7) Nias-Simeulue earthquake that occurred off Sumatra on 28 March, 2005 produced only a locally damaging tsunami (e.g., Briggs et al., 2006; Gahalaut and Catherine, 2006; Hsu et al., 2006; Kreemer et al., 2006).

7. Conclusions

As a result of the 2004 Indian Ocean tsunami, there is an international commitment to create a global tsunami observing network of coastal and open-ocean instruments. The primary use of this network is to assist tsunami warning systems in producing timely and accurate tsunami warnings. It is also recognized that in-situ tsunami measurements must serve the needs of the research community, if the community is to advance tsunami science and thereby contribute to improvements in warning systems and tsunami mitigation programs.

The capability to make tsunami measurements is taking advantage of networks that were originally deployed for other purposes. An example is the global GLOSS tide gage network to monitor sea level rise. Because tsunamis can occur infrequently, long-term monitoring is required to acquire measurements of them. There are major logistical advantages to co-locating tsunami, meteorological and oceanographic stations in order to reduce the cost of sustained networks. This requires coordination of tsunami networks with other operational and research programs.

There is considerable interest in developing new technologies to improve tsunami instrumentation and to reduce the cost of maintaining the observational networks. This is true for the coastal tide gages, autonomous pressure systems and submarine cable observatories, each of which complements the other in forming the full tsunami measurement network. International agreements under the Global Environment Observing Systems of Systems (GEOSS) program will facilitate efficient operational networks and data sharing. A data management plan has recently been completed by NOAA (2006) and could serve as a model for the international community. It is also important to provide training for making and interpreting tsunami measurements to scientists from developing countries who are entering the tsunami field.

The creation of world data centers for tsunamis will standardize tsunami measurements and provide improved access to more comprehensive datasets of tsunami time series and post-tsunami field measurements. This will complement the existing requirement that tsunami measurements be of sufficient quality and documentation to be publishable in the peer-reviewed literature.

Acknowledgments

The author wishes to thank A. B. Rabinovich, R. E. Thomson, and P. M. Whitmore for their many helpful comments and additions to the history and other sections.

References

Ambrosius, B., R. Scharroo, C. Vigny, E. Schrama, and W. Simons, 2005. The 26 December 2004 Sumatra Earthquake and Tsunami seen by satellite altimeters and GPS. In *Geo-Information for Disaster Management,* P. van Oosterom, S. Zlatanova, and E. M. Fendel, eds. Springer, 323–336.

Arcas, D., and V. Titov, 2006. Sumatra tsunami: lessons from modeling. *Surv. Geophys.,* **27,** 679–705.

Atwater, B. F., S. Musumi-Rokkaku, K. Satake, Y. Tsuji, K. Ueda, and D. K. Yamaguchi, 2005. *The Orphan Tsunami of 1700.* USGS Prof. Paper 1707, Washington, 133 pp.

Baba, T., K. Hirata, and Y. Kaneda, 2004. Tsunami magnitudes determined from ocean-bottom pressure gauge data around Japan. *Geophys. Res. Lett.,* **31**(8), L08303, doi:10.1029/2003GL019397.

Bache, A. D., 1856a. Notice of earthquake waves on the western coast of the United States, on the 23d and 25th of December, 1854. *Amer. J. Sci. Arts, 2nd Series,* **21,** 37–43.

Bache, A. D., 1856b. Notice of earthquake waves on the western coast of the United States, on the 23d and 25th of December, 1854. In *Report of the Superintendent of the Coast Survey showing the progress of the Survey during the year 1855, 34th Congress, 1st Session,* 342–346 and Appendix J, No. 9.

Beardsley, R. C., H. O. Mofjeld, M. Wimbush, C. N. Flagg, and J. A. Vermersch, Jr., 1977. Ocean tides and weather induced bottom pressure fluctuations in the Middle-Atlantic Bight. *J. Geophys. Res.,* **82,** 3175–3182.

Berg, E., C. Cox, A. S. Furumoto, K. Kinjira, H. Kawasumi, and E. Shima, 1970. *Field survey of the tsunamis of 28 March 1964 in Alaska, and conclusions as to the origin of the major tsunami,* Hawaii Inst. Geophys. Tech. Rprt. HIG-70-2, 49 pp. plus bibliography and Appendix.

Berkman, S. C., and J. M. Symons, 1964. *The Tsunami of May 22, 1960 as Recorded at Tide Stations,* United States Department of Commerce, Coast and Geodetic Survey, Washington, D.C., 79 pp.

Bernard, E. N., F. I. González, C. Meinig, and H. B. Milburn, 2001. Early detection and real-time reporting of deep-ocean tsunamis. In *Proceedings of the International Tsunami Symposium 2001,* Seattle, WA, 7–10 August 2001, 97–108.

Borrero, J. C., 2005a. Field survey of Northern Sumatra and Banda Aceh, Indonesia after the Tsunami and earthquake of 26 December 2004. *Seismo. Res. Lett.,* **76**(3), 312–320.

Borrero, J. C., 2005b. Field data and satellite Imagery of tsunami effects in Banda Aceh. *Science,* **308**(5728), 1596. DOI: 10.1126/science.1110957.

Bourgeois, J., C. Petroff, H. Yeh, V. Titov, C. E. Synolakis, B. Benson, J. Kuroiwa, J. Lander, and E. Norabuena, 1999. Geologic setting, field survey and modeling of the Chimbote, northern Peru, tsunami of 21 February 1996. *PAGEOPH,* **154**(3–4), 513–540.

Braddock, D., 1980. Response of a conventional tide gauge to a tsunami. *Mar. Geod.,* **4**(3), 223–236.

Bricker, J. D., S. Munger, C. Pequignet, J. R. Wells, G. Pawlak, and K. F. Cheung, 2007. ADCP observations of edge waves off Oahu in the wake of the November 2006 Kuril Islands tsunami. *Geophys. Res. Lett.,* **34,** L23617, doi:10.1029/2007GL032015.

Briggs, R. W., K. Sieh, A. J. Meltzner, D. Natawidjaja, J. Galetzka, B. Suwargadi, Y. J. Hsu, M. Simons, N. Hananto, I. Suprihanto, D. Prayudi, J. P. Avouac, L. Prawirodirdjo, and Y. Bock, 2006. Deformation and slip along the Sunda Megathrust in the great 2005 Nias-Simeulue earthquake. *Science,* **311**(5769), 1897–1901.

Brown, W., W. Munk, F. Snodgrass, H. Mofjeld, and B. Zetler, 1975. MODE Bottom Experiment. *J. Phys. Oceanogr.,* **5,** 75–85.

Bryant, E., 2001. *Tsunami—The Underrated Hazard.* Cambridge, New York, 319 pp.

Burgy, M., and D. K. Bolton, 2006. New coastal tsunami gauges: Application at Augustine Volcano, Cook Inlet, Alaska. *EOS Trans. AGU,* abs V51C-1693–2006 Fall Meeting.

Caldwell, D. R., F. E. Snodgrass, and M. H. Wimbush, 1969. Sensors in the deep sea. *Physics Today,* **22**(7), 34–42.

Choi B. H., S. J. Hong, and E. Pelinovsky, 2006. Distribution of runup heights of the December 26, 2004 tsunami in the Indian Ocean. *Geophys. Res. Lett.,* **33**(13): Art. No. L13601 204.

Dykhan, B. D., V. M. Jaque, E. A. Kulikov, S. S. Lappo, V. N. Mitrofanov, A. A.Poplavsky, A. V. Rodionov, A. A. Shishkin, and S. L. Soloviev, 1983. Registration of tsunamis in the open ocean. *Mar. Geod.,* **6,** 303–310.

Eble, M. C., and F. I. Gonzalez, 1991. Deep-ocean bottom pressure measurements in the Northeast Pacific. *J. Atmos. Ocean. Tech.,* **8**(2), 221–233.

Eble, M. C., S. E. Stalin, and E. F. Burger, 2001. Acquisition and quality assurance of DART data. In *Proceedings of the International Tsunami Symposium 2001,* Session 5–9, Seattle, WA, 7–10 August 2001, 625–632.

Eyries, M., M. Dars, and L. Erdely, 1964. Mareographie par grand fond. *Cahiers Oceanogr.,* **16**(9), 781–798.

Filloux, J. H., 1970. Bourdon tube deep see tide gauges. In: *Tsunamis in the Pacific Ocean,* University Press, Honolulu, 223–238.

Filloux, J. H., 1982. Tsunami recorded on the open ocean floor. *Geophys. Res. Lett.,* **9,** 25–28.

Filloux, J. H., 1983. Pressure fluctuations on the open-ocean floor off the Gulf of California: tides, earthquakes and tsunamis. *J. Phys. Oceanogr.,* **13,** 783–796.

Fine, I. V., A. B. Rabinovich, B. D. Bornhold, R. E. Thomson, and E. A. Kulikov, 2005. The Grand Banks landslide-generated tsunami of November 18, 1929: preliminary analysis and numerical modeling. *Mar. Geol.,* **215,** 45–57.

Fine, I. V., A. B. Rabinovich, and R. E. Thomson, 2005. The Dual Source Region for the 2004 Sumatra Tsunami. *Geophy. Res. Lett.,* **32,** L16602–06. doi:10,1029/2005GL023821.

Fujii, Y., and K. Satake, 2007. Tsunami Source of the 2004 Sumatra–Andaman Earthquake Inferred from Tide Gauge and Satellite Data. *Bull. Seis. Soc. Amer.,* **97,** 1A, S192-S207; doi: 10.1785/0120050613.

Fujima, K., Y. Shigihara, T. Tomita, et al., 2006. Survey results of the Indian Ocean Tsunami in the Maldives. *Coastal Eng. J.,* **48**(2), 81–97.

Gahalaut, V. K., and J. K. Catherine, 2006. The rupture characteristics of 28 March 2005 Sumatra earthquake from GPS measurements and its implication for tsunami generation. *Earth Planet. Sci. Lett.,* **249**(1–2), 39–46.

Garrett, J., 2004. *NEPTUNE Canada Ocean Observing Systems Workshop 1 Report.* May 3–5, 2004, University of Victoria, Canada, 107 pp.

Geist, E. L., S. L. Bilek, R. D. Arcas, and V. V. Titov, 2006. Differences in tsunami generation between the December 26, 2004 and March 28, 2005 Sumatra earthquakes. *Earth Planet Space,* **58,** 1–9.

González, F. I., E. N. Bernard, C. Meinig, M. Eble, H. O. Mofjeld, and S. Stalin, 2005. The NTHMP tsunameter network. *Nat. Hazards,* **35**(1), Special Issue, U.S. National Tsunami Hazard Mitigation Program, 25–39.

González, F. I., and Y. A. Kulikov, 1993. Tsunami dispersion observed in the deep ocean. In: *Tsunamis in the World,* S. Tinti, ed. Kluwer, Dordrecht, pp. 7–16.

González, F. I., E. N. Bernard, and H. B. Milburn, 1987. A program to acquire deep ocean tsunami measurements in the North Pacific. In *Proceedings of Coastal Zone* **87,** WW Div., ASCE, Seattle, WA, 26–29 May 1987, 3373–3381.

González, F. I., C. L. Mader, M. C. Eble, and E. N. Bernard, 1991. The 1987–88 Alaskan Bight Tsunamis: Deep ocean data and model comparisons. Special Issue on Tsunami Hazard, E. N. Bernard, ed. *Nat. Hazards,* **4**(2,3), 119–139.

Gower, J. (2005), Jason 1 detects the 26 December 2004 Tsunami. *Eos Trans. AGU,* **86**(4), 37.

Hicks, S. D., A. J. Goodheart, and C. W. Iseley, 1965, Observations of the tide on the Atlantic continental shelf. *J. Geophys. Res.,* **70**(8), 1827–1830.

Hino, R., Y. Tanioka, T. Kanazawa, S. Sakai, M. Nishino, and K. Suyehiro, 2001. Micro-tsunami from a local interplate earthquake detected by cabled offshore tsunami observations in northeastern Japan. *Geophys. Res. Lett.,* **28,** 3533–3536.

Hirata, K., M. Aoyagi, H. Mikada, K. Kawaguchi, Y. Kaiho, R. Iwase, S. Morita, I. Fujisawa, H. Sugioka, K. Mitsuzawa, K. Suyehiro, and H. Kinoshita, 2002. Real-time geophysical measurements on the deep seafloor using submarine cable in the Southern Kurile Subduction Zone. *IEEE J. Oceanic Eng.,* **27,** 170–181.

Hirata, K., and T. Baba, 2006. Transient thermal response in ocean-bottom pressure measurements. *Geophys. Res. Lett.,* **33,** L10606, doi:10.1029/2006GL026084.

Hirata, K., H. Takahashi, E. Geist, E. Satake, Y. Tanioka, H. Sugioka, and H. Mikada, 2003. Source depth dependence of micro-tsunamis recorded with ocean-bottom pressure gauges: the January 28, 2000 Mw 6.8 earthquake off Nemuro Peninsula, Japan. *Earth. Planet. Sci. Lett.,* **208,** 305–318.

Hokkaido Tsunami Survey Group, 1993. Tsunami devastates Japanese Coastal Region. *Eos Trans. AGU,* **74,** 417, 432.

Honda, K., T. Terada, Y. Yoshida, and D. Isitani, 1908. An investigation on the secondary undulations of oceanic tides. *J. College Sci., Imper. Univ. Tokyo,* 108 pp.

Hsu, Y. J., M. Simons, J. P. Avouac, J. Galetzka, K. Sieh, M. Chlieh, D. Natawidjaja, L. Prawirodirdjo, and Y. Bock, 2006. Frictional afterslip following the 2005 Nia-Simeulue earthquake, Sumatra. *Science,* **312**(5782), 1921–1926.

Imamura, F., 2006. Special issue—2004 Indian Ocean tsunami—Preface. *Coastal Eng. J.,* **48**(2), III–IV.

Imamura, F., N. Shuto, S. Ide, Y. Yoshida, and K. Abe, 1993. Estimate of the tsunami source of the 1992 Nicaraguan earthquake from tsunami data. *Geophys. Res. Lett.,* **20**(14), 1515–1518.

International Tsunami Information Center Website, http://ioc3.unesco.org/itic/

Irish, J. D., and F. E. Snodgrass, 1972. Quartz crystals as multipurpose oceanographic sensors. 1. Pressure. *Deep-Sea Res.,* **19** (2), 165–169.

Iwasaki, S. I., T. Eguchi, Y. Fujinawa, E. Fujita, I. Watabe, E. Fukuyama, and H. Fujiwara, 1997. Precise tsunami observation system in deep ocean by an ocean bottom cable network for the prediction of earthquakes and tsunamis. In *Perspectives on Tsunami Hazard Reduction,* G. Hebenstreit, ed. Kluwer, Dordrecht, 47–66.

Japan Meteorological Agency Website, http://www.jma.go.jp/jma/indexe.html.

Kato, T., Y. Terada, M. Kinoshita, H. Kakimoto, H. Isshiki, M. Matsutishi, A. Yokoyama, and T. Tanno, 2000. Real-time observation of tsunami by RTK-GPS. *Earth Planets Space,* **52,** 841–845.

Kato, T., Y. Terada, K. Ito, et al., 2005. Tsunami due to the 2004 September 5th off the Kii peninsula earthquake, Japan, recorded by a new GPS buoy. *Earth Planets Space,* **57**(4), 297–301.

Kovalev, P. D., A. B. Rabinovich, and G. V. Shevchenko, 1991. Investigation of long waves in the tsunami frequency band on the southwestern shelf of Kamchatka. *Nat. Hazards,* **4,** 141–159.

Kowalik Z., W. Knight, and P. Whitmore, 2005. Numerical modeling of the global tsunami: Indonesian tsunami of 26 December 2004. *Sci. Tsu. Haz.,* **23**(1), 40–56.

Kreemer, C., G. Blewitt, and F. Maerten, 2006. Co- and postseismic deformation of the 28 March 2005 Nias M-w 8.7 earthquake from continuous GPG data. *Geophys. Res. Lett.,* **33**(7), L07307.

Kulikov, E. A., A. B. Rabinovich, A. I. Spirin, S. L. Poole, and S. L. Soloviev, 1983. Measurement of tsunamis in the open ocean. *Mar. Geod.,* **6,** 311–329.

Lander, J. F., and P. A. Lockridge, 1989. *United States tsunamis: (including United States possessions) : 1690–1988.* U.S. Dept. of Commerce, Boulder, 41–2, 265 pp.

Lander, J. F., P. A. Lockridge, and M. J. Kozuch, 1993. *Tsunamis affecting the West Coast of the United States, 1806–1992,* U.S. Dept. of Commerce, Boulder, 242 pp.

Lennon, G. W., 1971. Sea level instrumentation, its limitations and the optimization of the performance of conventional gauges in Great British. *Intern. Hydrogr. Rev.,* **48**(2), 129–147.

Liu, P. L., P. Lynett, H. Fernando, B. E. Jaffe, H. Fritz, B. Higman, R. Morton, J. Goff, and C. Synolakis, 2005. Observations by the international tsunami survey team in Sri Lanka. *Science,* **308**(5728),1595–1595.

Lockridge, P. A., L. S. Whiteside, J. F. Lander, 2002. Tsunamis and tsunami-like waves of the eastern United States. *Sci. Tsu. Haz.,* **20**(3), 120–157.

Matsumoto, H., and H. Mikada, 2005. Fault geometry of the 2004 off the Kii peninsula earthquake inferred from offshore pressure waveforms. *Earth Planets Space,* **57**(3), 161–166.

Meinig, C., S. E. Stalin, A. I. Nakamura, F. González, and H. G. Milburn, 2005a. Technology Developments in Real-Time Tsunami Measuring, Monitoring and Forecasting. In *Oceans 2005 MTS/IEEE*, 19–23 September 2005, Washington, D.C.

Meinig, C., S. E. Stalin, A. I. Nakamura, and H. B. Milburn, 2005b. Real-Time Deep-Ocean Tsunami Measuring, Monitoring, and Reporting System: The NOAA DART II Description and Disclosure, NOAA/PMEL/NOAA Center for Tsunami Research Website, http://nctr.pmel.noaa.gov/Dart /dart_ref.html.

Merrifield, M. A., Y. L. Firing, T. Aarup, et al., 2005. Tide gauge observations of the Indian Ocean tsunami, December 26, 2004. *Geophys. Res. Lett.*, **32**(9), L09603, doi:10.1029/2005GL022610.

Miller, G. R., 1972. Relative spectra of tsunamis. *Hawaii Inst. Geophys. HIG-72-8*, Honolulu, 7 pp.

Miller, G. R., W. H. Munk, and F. E. Snodgrass, 1962. Long-period waves over California's borderland. Part II: Tsunamis. *J. Mar. Res.* **20**(1), 31–41.

Mofjeld, H. O., F. I. González, E. N. Bernard, and J. C. Newman, 2000. Forecasting the heights of later waves in Pacific-wide tsunamis. *Nat. Hazards,* **22,** 71–89.

Mofjeld, H. O., F. I. González, and J. C. Newman, 1997. Short-term forecasts of inundation during teletsunamis in the eastern North Pacific Ocean. In *Perspecives on Tsunami Hazard Reduction,* G. Hebenstreit, ed., Kluwer, Dordrecht, 145–155.

Mofjeld, H. O., F. I. González, and J. C. Newman, 1999. Tsunami prediction in U.S. coastal regions. In *Coastal Ocean Prediction, Coastal and Estuarine Studies* **56,** C. Mooers, ed., Chapter 14, *Amer. Geophys. Union,* 353–375.

Mofjeld, H. O., V. V. Titov, F. I. González, and J. C. Newman, 2001. Tsunami scattering provinces in the Pacific Ocean. *Geophys. Res. Lett.,* **28**(2), 335–337.

Mofjeld, H. O., P. M. Whitmore, M. C. Eble, F. I. González, and J. C. Newman, 2001. Seismic-wave contributions to bottom pressure fluctuations in the North Pacific—Implications for the DART Tsunami Array. In *Proceedings of the International Tsunami Symposium 2001,* Session 5–10, Seattle, WA, 7–10 August 2001, 633–641.

Mofjeld, H. O., C. M. Symons, P. Lonsdale, F. I. González, and V. V. Titov, 2004. Tsunami scattering and earthquake faults in the deep Pacific Ocean. *Oceanography,* **17**(1), 38–46.

Mofjeld, H. O., F. I. González, M. C. Eble, and J. C. Newman, 1995. Ocean tides in the continental margin off the Pacific Northwest Shelf. *J. Geophys. Res.,* **100**(C6), 10,789–10,800.

Mofjeld, H. O., and M. Wimbush, 1977. Bottom pressure observations in the Gulf of Mexico and Caribbean Sea. *Deep-Sea Res.,* **24,** 987–1004.

Momma, H., H. Kinoshita, N. Fujiwara, Y. Kaiho, and R. Iwase, 1996. Recent and future developments of deep sea research in JAMSTEC. *Int. J. Offshore Polar Engineering,* **6** (4), 11–16.

Momma, H., N. Fujiwara, K. Kawaguchi, R. Iwase, S. Suzuki, and H. Kinoshita, 1997. Monitoring system for submarine earthquakes and deep-sea environment. *Proc. MTS/IEEE OCEANS '97,* Vol. 2, 1453–1459.

Momma, H., R. Iwase, K. Mitsuzawa, Y. Kaiho, and Y. Fujiwara, 1998. Preliminary results of a three-year continuous observation by a deep seafloor observatory in Sagami Bay, Central Japan, *Special Issue: Seafloor Observatories and Geophysical Networks., Phys. Earth Planet. Inter.,* **108,** 263–274.

Monserrat, S., I. Vilibić, and A. B. Rabinovich, 2006. Meteotsunamis: atmospherically induced destructive ocean waves in the tsunami frequency band. *Nat. Hazards Earth Syst. Sci.,* **6,** 1035–1051.

Murty, T. S., 1977. *Seismic Sea Waves: Tsunamis. Bull. Fish. Res. Board Canada,* No. 198, 337 pp.

Nagarajan, B., I. Suresh, D. Sundar, R. Sharma, A. K. Lal, S. Neetu, S. S. C. Shenoi, S. R. Shetye, and D. Shankar, 2006. The Great Tsunami of 26 December 2004: A description based on tide-gauge data from the Indian subcontinent and surrounding areas. *Earth Planets Space,* 58(2), 211–215.

Nakano, M., and S. Unoki, 1962. On the seiches (secondary undulations of tides) along the coast of Japan. *Records Oceanogr. Works Japan,* Spec. No. 6, 169–214.

Narayan, J. P., M. L. Sharma, and B. K. Maheshwari, 2006. Tsunami intensity mapping along the coast of Tamilnadu (India) during the deadliest Indian Ocean tsunami of December 26, 2004. *PAGEOPH,* **163**(7), 1279–1304.

NOAA, 2006. *Tsunami Data Management—An initial report on the management of data required to minimize the impact of tsunamis in the United State*s, Version 1.0, 91 pp.

NOAA/National Data Buoy Center Website, http://www.ndbc.noaa.gov/.

NOAA/National Geophysical Data Center Website, http://www.ngdc.noaa.gov/seg/hazard/tsu.shtml.

NOAA/National Ocean Service Tides and Currrents Website, http://www.tidesandcurrents.noaa.gov/.

Nowroozi, A. A., G. Sutton, and B. Auld, 1966. Oceanic tides recorded on the sea floor. *Ann. Geophys.,* **22**(3), 512–517.

Nowroozi, A. A., 1972. Long-term measurements of pelagic tidal height off the coast of northern California. *J. Geophys. Res.,* **77**(3), 512–517.

Noye, B. J., 1974a. Tide-well systems I: Some non-linear effects of the conventional tide well. *J. Mar. Re*s., **32**(2), 129–153.

Noye, B. J., 1974b. Tide-well systems II: The frequency response of a linear tide-well system. *J. Mar. Res.,* **32**(2), 155–181.

Noye, B. J., 1974c. Tide-well systems III: Improved interpretation of tide-well records. *J. Mar. Res.,* **32**(2), 183–194.

Obura, D., 2006. Impacts of the 26 December 2004 tsunami in Eastern Africa. *Ocean & Coastal Manag.,* **49**(11), 873–888.

Okada, M., 1991. Ocean bottom pressure gauge for the Tsunami Warning System in Japan, *Second UJNR Tsunami Workshop,* Honolulu, Hawaii, 1991.

Okada, M., 1993. Tsunami observation by ocean bottom pressure gauge, In: *Proc. IUGG/IOC Intern. Tsunami Symp.,* Wakayama, Japan, 1993, 385–396.

Okada, M., 1995. Tsunami observation by ocean bottom pressure gauge, In: *Tsunami Hazard Reduction,* G. Hebenstreit, ed., Kluwer, Dordrecht, 1995, 287–303.

Omori F., 1902. On tsunamis around Japan (in Japanese), *Rep. Imp. Earthq. Comm.,* **34,** 5–79.

Piatanesi, A., S. Tinti, and I. Gavagni, 1996. The slip distribution of the 1992 Nicaragua earthquake from tsunami run-up data. *Geophys. Res. Lett.,* **23**(1), 37–40.

Poplavsky, A. A., E. A. Kulikov, and L. N. Poplavskaya, 1988. *Methods and Algorithms of Automatic Tsunami Warning* (in Russian), Moscow, Nauka, 128 pp.

Rabinovich, A. B., 1997. Spectral analysis of tsunami waves: Separation of source and topography effects. *J. Geophys. Res.,* **102**(C6), 12,663–12,676.

Rabinovich, A. B., F. E. Stephenson, and R. E. Thomson, 2006a. The California tsunami of 14 June 2005 along the coast of North America. *Atmos. Ocean,* **44**(4), 415–427.

Rabinovich, A. B., R. E. Thomson, and F. E. Stephenson, 2006b. The Sumatra tsunami of 26 December 2004 as observed in the North Pacific and North Atlantic oceans. *Surv. Geophys.,* **27,** 647–677.

Ritsema, J., S. N. Ward, and F. I. González, 1995. Inversion of deep-ocean tsunami records for 1987–88 Gulf of Alaska earthquake parameters. *Bull. Seismol. Soc. Am.,* **85,** 747–754.

Satake, K., T. Baba, K. Hirata, et al., 2005. Tsunami source of the 2004 off the Kii peninsula earthquakes inferred from offshore tsunami and coastal tide gauges. *Earth Planets Space,* **57**(3), 173–17.

Satake, K., and F. Imamura, 1995. Introduction to Tsunamis 1992–1994: Their Generation, Dynamics and Hazard. *PAGEOPH,* **145,** 373–379.

Satake, K., M. Okada, and K. Abe, 1988. Tide gauge response to tsunamis: Measurement at 40 tide gauges in Japan. *J. Mar. Res.,* **46,** 557–571.

Saxena, N., and A. Ziclinski, 1981. Deep-ocean system to measure tsunami wave height. *Mar. Geodesy,* **5**(1), 55–62.

Shimamoto, T., A. Tsutsumi, E. Kawamoto, M. Miyawaki, and H. Sato, 1995. Field survey report on tsunami disasters caused by the 1993 Southwest Hokkaido earthquake. *PAGEOPH,* **145,** 665–691.

Shipley, A. M., 1963. On measuring long waves with tide gauge. *Deut. Hydrogr. Zeitschr.,* **16,** 136–140.

Shuto, N., and H. Matsutomi, 1995. Field survey of the 1993 Hokkaido Nansei-oki earthquake tsunami. *PAGEOPH,* **144,** 649–663.

Snodgrass, F. E., 1968. Deep sea instrument capsule. *Science,* **162**(3849), 78–87.

Snodgrass, F. E., 1969. Study of ocean waves, 10^5 to 1 Hz, Inst. Geophys. Planet. Phys., University of California, San Diego, Surv. Paper No. 8, 34 pp.

Spaeth, M. G., and S. C. Berkman, 1967. The Tsunami of March 28, 1964, as Recorded at Tide Stations. *U.S. C&GS Tech. Bull.,* **33,** Rockville, 86 pp.

Synolakis, C. E., J. P. Bardet, J. Borrero, H. Davis, E. Okal, E. Sylver, J. Street, and D. Tappin, 2002. Slump origin of the 1998 Papua New Guinea tsunami. *Proc. R. Soc. Lond., A,* **458**(2020), 763–789.

Synolakis, C. E., P. Liu, H. Yeh, and G. Carrier, 1997. Tsunamigenic seafloor deformations. *Science,* **278**(5338), 598–600.

Taira, K., T. Teramoto, and S. Kitagawa, 1985. Measurements of ocean bottom pressure with quartz sensor. *J. Oceanogr. Soc. Japan,* **41**(3), 181–192.

Takahashi, R., and T. Hatori, 1961. A summary report on the Chilean tsunami of 1960, in The Committee of the Chilean Tsunami of 1960, Report on the Chilean tsunami of May 24, 1960, as observed along the coast of Japan, Maruzen Co., Ltd, 23–24.

Takahashi, T., 1996. Benchmark problem 4. The 1993 Okushiri tsunami—Data, conditions and phenomena. In *Long Wave Runup Models,* H. Yeh, P. Liu, and C. Synolakis, eds. World Scientific, Singapore, 384–403.

Tanioka, Y., K. Hirata, R. Hino, et al., 2004. Slip distribution of the 2003 Tokachi-oki earthquake estimated from tsunami waveform inversion. *Earth Planets Space,* **56**(3), 373–376.

Thomson, R. E., A. B. Rabinovich, and M. V. Krassovski, 2007. Double jeopardy: Concurrent arrival of the 2004 Sumatra tsunami and storm-generated waves on the Atlantic coast of the United States and Canada. *Geophys. Res. Lett.,* **34,** L15607, doi:10.1029/2007GL030685.

Titov, V. V., F. I. González, E. N. Bernard, M. C. Eble, H. O. Mofjeld, J. C. Newman, and A. J. Venturato, 2005a. Real-time tsunami forecasting: Challenges and solutions. *Nat. Hazards,* **35**(1), Special Issue, U.S. National Tsunami Hazard Mitigation Program, 41–58.

Titov, V. V., A. B. Rabinovich, H. O. Mofjeld, R. E. Thomson, and F. I. González, 2005b. The global reach of the 26 December 2004 Sumatra Tsunami. *Science,* **309**(5743), 2045–2048.

Titov, V. V., and C. E. Synolakis, 1997. Extreme inundation flows during the Hokkaido-Nansei-Oki tsunami. *Geophys. Res. Lett.,* **24**(11), 1315–1318.

Tomita, T., F. Imamura, T. Arikawa, T. Yasuda, and Y. Kawata, 2006. Damage caused by the 2004 Indian Ocean Tsunami on the southwestern coast of Sri Lanka. *Coastal Eng. J.,* **48**(2), 99–116.

Tsuji, Y., Y. Namegaya, H. Matsumoto, S.-I. Iwasaki, W. Kanbua, M. Sriwichai, and V. Meesuk, 2006. The 2004 Indian tsunami in Thailand: Surveyed runup heights and tide gauge records. *Earth Planets Space,* **58**(2), 223–232.

UNESCO, 1975. An intercomparison of open sea tidal pressure sensors, Techn. Papers in Marine Sciences, No.21, 67 pp.

University of Hawaii Sea Level Center Website, http://uhslc.soest.hawaii.edu/.

Van Dorn, W. G., 1984. Some tsunami characteristics deductible from tide records. *J. Phys. Oceanogr.,* **14,** 353–363.

Van Dorn, W. G., 1987. Tide gage response to tsunamis. Part II: Other oceans and smaller seas. *J. Phys. Oceanogr.,* **17,** 1507–1516.

Vitousek, M. J., 1965. An evolution of the vibrotron pressure transducer as a mid-ocean tsunami gage. *Hawaii Inst. Geophys., HIG-65-13,* Univ. Hawaii, 12 pp.

Walker, D. A., and R. K. Cessaro, 2002. Locally generated tsunamis in Hawaii: A low cost, real time warning system with world wide applications. *Sci. Tsu. Haz.*, **20**(4), 177–186.

Webb, S. C., 1998. Broadband seismology and noise under the ocean. *Rev. Geophys.*, **36**, 105–142.

Wiegel, R. L., 1970. Tsunamis. In *Earthquake Engineering*, R. L. Wiegel, ed. Prentice-Hall, Englewood Cliffs, 253–306.

Wiegel, R. L., 1976. Tsunamis. In *Seismic Risk and Engineering Decisions*, C. Lomnitz and E. Rosenblueth, eds. Elsevier, Amsterdam, 225–286.

Wigen, S. O., 1983. Digitization of tsunamigrams. In *Tsunamis—Their Science and Engineering*, K. Iida, and T. Iwasaki, eds. Terra Scientific, Tokyo, 213–224.

Wilson, B. W., and A. Törum, 1968. *Engineering damage from the tsunami of the Alaskan earthquake of March 27, 1964.* Coastal Engr. Res. Center Tech Memo. 33, U.S. Army Corps of Engineers, 401 pp.

Zetler, B. D., W. Munk, H. Mofjeld, W. Brown, and F. Dormer, 1975. MODE tides. *J. Phys. Oceanogr.*, **5**, 430–441.

Zielinski, A., and N. Saxena, 1983a. Rationale for measurement of midocean tsunami signature. *Mar. Geodesy,* **6**(3–4), 331–337.

Zielinski, A., and N. Saxena, 1983b. Tsunami detectability using open-ocean bottom pressure fluctuations. *IEEE J. Oceanic Eng.,* **OE-8**(4), 272–280.

Chapter 8. TSUNAMI MODELING: DEVELOPMENT OF BENCHMARKED MODELS

COSTAS E. SYNOLAKIS[1,2] AND UTKU KÂNOĞLU[3]

[1]*Technical University of Crete*

[2]*University of Southern California*

[3]*Middle East Technical University*

Contents

1. Introduction

The term tsunami made its grand entrance in most world languages on 26 December 2004. Its origin (harbor wave, in Japanese) belies its early observations. Even a relatively small tsunami entering a port or bay can trigger substantial water level oscillations. Not only it is not uncommon for these harbor waves to reach substantial heights, but also water motions can persist for many hours, as most recently observed in the seiching at Crescent City, California harbor from the 15 November 2006 Kuril Islands tsunami (Uslu et al., 2007). Ports have always been centers of commercial activity and the points of contact with the sea, and probably eyewitnesses first observed these unusual waves in harbors, hence the name. In Japan, fairly systematic historical documentation of tsunamis dates back to the 9th Century AD. However, the first historical report of coastal inundation by tsunamis refers to the eruption of the Thera volcano in the eastern Mediterranean, now believed around 1620BC (Bruins et al., 2007). While Marinatos (1939) posited this

The Sea, Volume 15, edited by Eddie N. Bernard and Allan R. Robinson
ISBN 978–0–674–03173–9 ©2009 by the President and Fellows of Harvard College

volcanic tsunami as the primary agent for the demise of the Minoans in Crete, his hypothesis was not favored in later decades because of differences in the relative dating of the eruption with the destruction of the palaces. New discoveries of Bronze-age flooded debris layers in excavations in Crete have revived the Marinatos hypothesis, and it is now believed that tsunami destroyed the Minoan ports and ships and flooded their fields and warehouses. The tsunami precipitated the demise of the Minoans, very much as the 1755 tsunami did in Portugal and ended its imperial power. The 1755 earthquake and tsunami killed almost one in ten thousand of the estimated world population of its time and had such profound impact in the philosophical thinking of the time, that later quite often the reported tsunami impacts had been believed exaggerated. In the aftermath of the 2004 Boxing Day tsunami, it is no longer difficult to comprehend the impact of such natural disasters in human history.

Tsunamis are long waves generated by impulsive geophysical events forcing the seafloor or coastal topography such as earthquakes, submarine/subaerial mass failures (Liu et al., 1991; Synolakis and Bernard, 2006). More spectacular, but far less common generation mechanisms are volcanic eruptions and bolide impacts (Morrison, 2006). Tsunamis evolve substantially through three dimensional spreading and as they propagate over shallow bathymetry. The forecasting the inundation/runup and of the forces on coastal structures is the basic objective of tsunami hazard mitigation.

Tsunamis propagate across the oceans as a series of long, low-crested waves, usually less than one meter high –even the Boxing Day tsunami was less than 0.75 m as recorded by satellite altimetry across the Indian Ocean (Smith et al., 2005). Within a tsunami train, individual crests are separated from each other by tens or hundreds of kilometers and periods of tens of minutes. In contrast to tsunamis, wind waves vary in height from tiny ripples on the sea surface to the rogue waves featured in the movie *The Perfect Storm.*

The unavailability of instrumental recordings of tsunamis in the deep ocean – until recently (Titov et al., 2005a; Bernard et al., 2006)– resulted in tsunami science and engineering evolving slower than studies in other extreme natural hazards. While the basic equations for analysis have been known for decades, the existing synthesis leading to a real time forecast had to await the development of sophisti-cated modeling tools, the large-scale laboratory experiments in the 1980s-1990s and the tsunameter recordings of 2003 and since. The field survey results in the 1990s served as crude proxies to free-field tsunami recordings –field surveys typi-cally provide measurements of the maximum onland penetration of the tsunami or of the maximum flow depth at select locations– and allowed for the validation and verification of numerical procedures. This synthesis is still in progress. State-of-the-art inundation and forecasting codes have evolved through a painstaking proc-ess of careful validation and verification (Synolakis et al., 2007). Operational tsu-nami forecasting was only made possible through the availability of deep ocean measurements, which also allowed for closure, e.g. Titov et al. (2005a) and Titov (this volume). Since 2003, every event in the Pacific Ocean has posed a new but diminishing forecast challenge at least for farfield tectonic tsunamis. We will de-scribe here this journey from development of the basic field equations to forecasts, through the scientific milestones that served as benchmarks and reality checks.

In the last twenty years, consensus emerged that tsunami propagation and their terminal effects can be adequately described by depth-averaged equations, which are approximations of the Navier-Stokes equations. The latter govern all incompressible fluid motions. They are three dimensional and notoriously difficult to solve for free-surface geophysical flows and are too computationally intensive for inundation mapping and forecasting. In applied tsunami modeling practice, the Navier-Stokes equations are averaged from the seafloor to the free surface, the pressure is assumed hydrostatic, and viscous stresses are either eliminated, or presumed to follow simple bottom friction laws. Liu (this volume) provides an excellent discussion of the current formulation of the different approximations. For here, it suffices to be reminded that the depth-averaged Navier-Stokes equations result into the nonlinear shallow water-wave equations (NSW); eliminating the nonlinear terms results into the linear shallow water-wave equations (LSW). Peregrine (1967) provides a comprehensive exposition of the orders of approximation through a perturbation expansion.

In every scientific endeavor, a critical mass of scientists and accumulated incremental research is precursory to major advances. Before describing the development of tsunami hydrodynamics and the evolution of benchmarks for testing computational models, it is important to revisit the state-of-the-art twenty years ago. Synolakis and Bernard (2006) have discussed the evolution of tsunami hydrodynamics beginning with the devastation of Hilo, Hawaii following the April 1, 1946 Aleutian tsunami. Significant advances between 1950 and 1990 included the pioneering experiments of Hall and Watts (1953), Wiegel's (1955) landslide experiments, Carrier and Greenspan's (1958) hodograph transformation for the NSW, the Raichlen (1966) harbor resonance formulation, the Peregrine (1966) derivation of the long wave equations, the Madsen and Mei (1969), Hammack (1972) and Goring (1978) laboratory experiments, the Hibberd and Peregrine (1979) runup computational algorithm, the Houston and Garcia (1974a, b) first attempt to develop tsunami flooding maps, the Ward (1980) and Okal (1982) normal mode theory, the Pedersen and Gjevik (1983) computations of one dimensional runup through a Boussinesq equation and the Synolakis (1986, 1987) analytical results for the runup of solitary waves that established the runup invariance between linear and nonlinear theories. Also, Synolakis' (1987) comparison of the LSW and NSW predictions with laboratory measurements provided the necessary confidence to adopt the NSW equations as a suitable field model for studying runup.

As research in live networks –where problems and solution ideas arise spontaneously– tsunami hydrodynamic modeling had several facets, by necessity, and was driven by milestone scientific meetings, and post tsunami surveys that kept identifying novel problem geometries and previously unrecognized phenomena. We will describe below these developments and we note that Synolakis and Bernard (2006) discuss tsunami hydrodynamics from a historical time perspective; Shuto (2003) also provides a very useful historic review of tectonic tsunamis. Here, first, we describe the key workshops that led developments, then summarize landmark field observations and how they helped not only posing new questions but establishing reality checks for rapidly evolving computational codes. Then, we present analytical results, followed by laboratory studies and discuss how they helped benchmark numerical model development. Finally, the necessary validation and verification steps for numerical codes to be used for inundation mapping, design and operations are presented.

2. Key scientific workshop milestones

Despite the substantial individual advances briefly summarized above, by the end of the 1980s, there was not even a clear realization of the scales of tsunamis of geophysical interest. With the exception of Synolakis (1987) who identified the differences in the runup variation for breaking and non-breaking waves, laboratory studies remained largely disjoint from analytical results. Field-scale numerical computations remained an exotic undertaking which, even when attempted, it was admired for its audacity and effort and not for its realism. Even the term *runup* (or *run-up*) promulgated since Hibberd and Peregrine's (1979) work was used interchangeably with tsunami vertical rise or tsunami height –the latter sometimes referred to the offshore height of the incoming wave at an often unidentified location, or to the height at the initial shoreline.

Advances in tsunami hydrodynamics in the 1990s were jumpstarted through two landmark scientific meetings, one in Novosibirsk, Russia in 1989, the other in Twin Harbors, Catalina Island, in California in 1990. The 1989 Novosibirsk workshop was organized by V. K. Gusiakov of the Computing Center of the Siberian branch of Union of Soviet Socialist Republics Academy of Sciences. While it was not focused solely on hydrodynamics, it was clear then that computational efforts were developing rapidly towards modeling the two-plus-one (2+1, refers to evolution in two horizontal space dimensions and time) propagation of long waves from the source to the target, but that coastal evolution and inundation remained unexplored. For the US National Science Foundation (NSF), this development was a wake up call. The US scientific efforts had concentrated on analytical results and laboratory experiments, and in academia, there was no modeling capability at geophysical scales to match the competition. Most of the hydrodynamics improvements presented in Novosibirsk were also described in the Catalina workshop (Liu et al., 1991).

In Catalina, the emphasis was on examining how to best validate one-plus-one (1+1, refers to evolution in one space dimension and time) numerical and analytical tools under development at the time. As Synolakis and Bernard (2006) wrote, E. Pelinofsky demonstrated, with a different transformation than Synolakis (1987), the runup invariance between linear and nonlinear theories. C. E. Synolakis discussed the generalization of Green's law, a classic result for the amplitude evolution of periodic waves on sloping beaches, to solitary waves (Synolakis, 1991). D. H. Peregrine presented novel computations of colliding breaking waves and inferred that the associated fluid accelerations when a breaking wave collapses on a vertical wall can reach $1000g$. C. C. Mei presented new results with a multiple-scale perturbation theory to study harbor resonance. H. Yeh presented laboratory measurements of a bore climbing a sloping beach. J. Ramsden and F. Raichlen presented laboratory experiments on interaction of bores with vertical walls and a methodology for calculating the force (Ramsden and Raichlen, 1990).

In terms of numerical modeling, V. V. Titov presented the Novosibirsk's Computing Center's finite-difference 2+1 algorithm that allowed for efficient calculations of the farfield evolution of tsunami-looking waves, a computation that became the basis of the code Method of Splitting Tsunami (MOST) now used by the National Oceanic and Atmospheric Administration (NOAA) for tsunami inundation forecasting. N. Shuto presented an amazing animation of a long wave flooding rectangular shaped structures on dry land, which however did not include actual inundation calculations. Yet, it showed what the future target was. Z. Kowalik presented a 1+1 computation for bore runup. P. L.-F. Liu presented an animation for a three-dimensional potential flow in a rectangular basin.

F. Gonzalez presented the first ever measurement of waves with period greater than 5 min in the open ocean at a resolution of >2 mm, with a novel bottom pressure recorder (BPR) deployable at depths up to 7 km. This was another glimpse into the future that was not fully appreciated then for what it was, being too fanciful for the zeitgeist. This instrument was the basis of what is now known as tsunameters or Deep-ocean Assessment and Reporting of Tsunamis (DART) buoys (Bernard et al., 2006). E. N. Bernard suggested the importance of both site- and source-specific inundation maps that would depict the flooding hazard from scenario events emphasizing the link between modeling and civil defense strategies. R. T. Guza presented compelling evidence that even when wind waves climb up a sloping beach, the maximum runup is controlled by long period waves, thus underscoring the wide application of the emerging ideas in long wave runup beyond just tsunami geophysics.

While within the tsunami field the 1990 Catalina meeting –in terms of excitement and international talent in attendance– was analogous to the International Congress of Mathematicians in Paris in 1900 where David Hilbert posed his infamous 23 problems in mathematics, the understanding of real tsunamis that emerged was not all together entirely different than twenty years earlier. Substantial improvements had been achieved but only in the least realistic problems, which would not rapidly lead to tsunami forecasting or inundation mapping. Liu et al. (1991) summarized the conclusions of the Catalina workshop: one, the runup of a single non-breaking wave could be computed analytically and numerically; two, the NSW model was determined adequate for applications of geophysical interest; three, there was a pervasive need for laboratory data for long waves propagating in two directions to allow further progress in computational models particularly for 2+1 runup computations with hindsight they should had concluded that there was no realization of the importance of the seafloor/water-surface coupling and the solitary wave as the initial wave of tsunamis remained the standard for runup studies.

By the middle 1990s, the tsunami landscape had changed substantially beginning with the Nicaraguan tsunami that struck on September 2, 1992. Another two events followed, the 12 December 1992 Flores Island, Indonesia and the 12 July 1993 Hokkaido-Nansei-Oki (HNO), Japan; another three events in 1994, in East Java, Indonesia (June 3, 1994), the Kuril Islands, Shikotan (October 4, 1994), and Mindoro Island, Philippines (November 11, 1994) and finally the 9 October 1995 Manzanillo, Mexico tsunami. They all provided a wealth of field observations and sense of urgency to improve modeling capabilities as tsunamis appeared to be increasing in frequency, and were being reported at a rate of one a year or more.

The 1995 Friday Harbor, Washington workshop also organized by the NSF of the US provided a forum for exploring and most importantly validating the newly emerging computational codes. The findings are described by Yeh et al. (1996). The three organizers had prepared four benchmark problems: one, the propagation of edge waves; two, solitary wave runup on piecewise linear Revere Beach, Massachusetts bathymetry; three, conical island runup; and four, the field runup measurements from the 1993 HNO tsunami. The initial data of the corresponding benchmark experiments were provided to the participants without the runup measurements, except as published already. As perhaps expected, there were several 1+1 computations that predicted the runup of solitary waves, but only three codes were able to reproduce more than one of the 2+1 runup benchmarks. Kânoğlu and Synolakis (1996) and Fujima (1996) were able to compute the runup around a conical island using analytical tools, and without numerical calculations beyond simple integral evaluations.

In 1997, Tsunamigenic Seafloor Deformations, a meeting of similar impact, was held in Santa Monica, California, again funded by the NSF. While by then, at least one code (MOST) had demonstrated a satisfactory capability in modeling even the extreme runup observed in HNO tsunami in 1993, the initial condition for the computation had been fairly well prescribed, through a combination of inverse modeling and seafloor mapping of the deformed ocean crust off Okushiri Island, Japan (Takahashi, 1996). Otherwise, comparisons of predictions with earlier field data appeared to have been in need of improvement, and runup computations remained temperamental. When measurements and predictions disagreed, geophysicists would blame the hydrodynamics calculations while the modelers would point to inadequate definition of the initial condition by the geophysicists. The meeting's focus was on understanding better the seafloor-free ocean surface interaction, and is described in Synolakis et al. (1997a). One very prophetic conclusion was the lack of any consistent methodology to estimate the waves from landslides or understanding of the failure criteria of submarine slopes. Slow earthquakes (as described by E. A. Okal) were identified as another show stopper in terms of accurately calculating the seafloor deformation, and to a large extend they still are. Another advance foretold in this meeting was the soon to take place deployment of the first real time BPRs in the Northern Pacific by NOAA's Pacific Marine Environmental Laboratory (PMEL). Before then, BPRs were deployed and recovered months later for data collection.

In 1998, an earthquake of the size that happens on the average more than once monthly somewhere around the world generated a tsunami in Papua New Guinea (PNG) with extreme runup and a death toll of over 2,100. No single tsunami had more than 1,200 casualties since the 1933 Showa-Sanriku, Japan tidal wave that killed more than 20,000 people. While the landslide trigger hypothesis was originally proposed by E. A. Okal on the field during the first PNG survey it was eventually proven by Synolakis et al. (2002a). It was clear then that substantial research was needed to understand landslide induced free-surface waves. Landslide tsunamis had earlier been referred to as backfill waves in the geology community.

An NSF workshop on the Prediction of Underwater Landslide and Slump Occurrence and Tsunami Hazards off of Southern California to discuss submarine mass failures and their consequences took place in 2000 at the University of Southern California in Los Angeles. In terms of landslide tsunamis, it jumpstarted the field as the 1990 Catalina workshop did for tectonic tsunamis. The participants examined the state-of-the-art in modeling of earthquake induced ground motions and their characteristics and modeling of onland and offshore landslides. A central observation was that landslide tsunamis were definite hazards to the world coastlines (Bardet et al., 2003a) and that multidisciplinary studies are needed to develop even rudimentary predictive capabilities. Echoing E. N. Bernard's comments in 1990, again the need for coastal hazards maps was recognized. Many of the workshop presentations appear in a volume edited by Bardet et al. (2003b). The 1998 PNG tsunami was the benchmark model, but with a basic difference with respect to earlier workshops: it had been impossible to specify an initial condition a priori. In other words, the handful who attempted to model it, used their own individual ad hoc initial waves. To date there is still no widely accepted and robust computational method for the specifying initial waves from landslides from geometric characteristics of mass failures. A companion NATO Advanced Research Workshop on *Underwater Ground Failures on Tsunami Generation, Modeling, Risk and Mitigation* with similar objectives but wider international representation took place in Istanbul, Turkey in 2001 (Yalçiner et al., 2003).

In 2004, the last-to-date intermodel and code validation workshop took place, again, on Catalina Island, Los Angeles. The organizers pre-assigned four benchmark problems. One, the prediction of the shoreline motion from the sliding wedge on a slope experimental data (Liu et al., 2005a); two, the prediction of the runup distribution measured in a laboratory model of the locale where extreme runup of >32 m occurred during the 1993 HNO tsunami (Takahashi, 1996); three, the prediction of the shoreline evolution as in thin landslide exact solution (Liu et al., 2003); and four, the prediction of the shoreline position and velocity for an initial wave introduced over a sloping beach –an initial value problem (IVP)– (Carrier et al., 2003). Some workshop results are presented in Liu et al. (2008), and only briefly summarized here. R. A. Dalrymple presented an entirely novel model of solving the benchmark problems using smooth particle hydrodynamics (SPH) which describes a fluid by replacing its continuum properties with locally smoothed quantities at discrete Lagrangian locations and hence it is meshless. S. Ward presented an evolution computation based on small amplitude dispersive theory for both tectonic and landslide tsunamis (Ward, 2001; Ward and Day, 2001 and 2003). However, most computational models were only capable of solving accurately the Carrier et al. (2003) benchmark –1+1 IVP of a long wave initially over a sloping beach– whether solving the shallow water-wave (SW), Boussinesq or Euler's equations. (Here it needs to be noted that the difference between the Carrier et al. (2003) and the Synolakis (1987) analytical solutions is that only the latter solves the canonical problem of a long wave propagating first over a flat ocean floor and then climbing on a sloping beach.) There was little difference in this regard from the 1995 Friday Harbor findings. A few computations were able to predict reasonable results compared with the predictions of the Liu et al. (2003) exact linear analytical solution for 1+1 thin landslides. Practically, all 2+1 directional models presented had attempted the Okushiri-based laboratory simulation with varying degrees of success. Only the DNS code developed by Liu et al. (2005a) appeared capable of modeling the wedge 2+1 landslide data, but at extreme computational cost. At the conclusion of this workshop, it was quite clear that operational and research applications would require different models, but that both would need to be comprehensively validated.

The 2004 Boxing Day tsunami followed six months later. There have been numerous tsunami meetings since, but none focused on benchmarking –the recent emphasis has been on using existing modeling tools to simulate the 2004 rupture and settle conclusively questions regarding the fault length and resolution needed for satisfactory predictions in the Indian Ocean. Given the past history, it is anticipated that future benchmark oriented workshops will identify new validation exercises. Code validation is a continuing process.

3. Field observations

We proceed discussing the field observations since 1992 that helped changed perspectives. The September 2, 1992 Nicaraguan tsunami is the dawn of modern tsunami observations. Beginning in 1992, post-tsunami surveys have been carried out by the highly interdisciplinary so called International Tsunami Survey Teams (ITSTs), immediately following each event. Fortuitously, following this event, tsunamis started being reported at a rate of almost one per year. Dr. Cliff Astill, then with the Natural Hazards Mitigation Program of the NSF of the US, aggressively encouraged field measurements in the immediate aftermath of tsunami catastrophes. The surveys generated valuable datasets that have already been used for code development and will continue to be useful for their validation. The discussion here follows Synolakis and Okal (2005) with minor changes.

The $M_w = 7.7$ parent tremor which triggered the 1992 Nicaragua tsunami had the clear nature as a tsunami earthquake (Kanamori, 1972). With more than 160 people killed, it motivated a number of detailed surveys and studies. Recall that a tsunami earthquake is associated with seafloor displacements which trigger tsunamis with far greater amplitude than expected from an earthquake with a similar size conventional magnitude, while its characteristic slow motion is often not reported to be felt (Okal, 1993).

There were two major observations regarding the 1992 Nicaraguan tsunami. The first observation, was that for coastal residents the only harbinger of the incoming tsunami attack was the shoreline recession, which locals regrettably (as tens of thousands during the 2004 Boxing Day tsunami) did not identify it as such. Tsunamis sometimes arrive to the target coastlines as a leading depression N-wave (LDN), i.e., they cause the shoreline to retreat first and for sometimes hundreds of meters resembling a fast ebbing tide (Fig. 8.1). Even though it was suspected that the leading tsunami wave could be an LDN (Mei, 1983), before 1992, LDNs were believed to be hydrodynamically unstable, the crest was supposed to quickly overtake the trough. This 1992 report of the manifestation of the initial shoreline withdrawal led Tadepalli and Synolakis (1994) to propose a model for the leading depression N-wave (Fig. 8.1). Yet, even the additional reports of the LDNs striking the south coast of Java, Shikotan Island and Mindoro Island, three more tsunamigenic earthquakes in 1994, the reports did little to widen the acceptance of the LDNs as a physical phenomenon by the mainstream hydrodynamics community. Even in 1995, LDNs were still thought by some to be figments of scientific imagination developed only to facilitate analytical solutions and, sometimes, LDN aficionados were heckled at meetings.

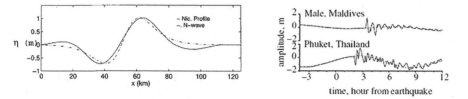

Fig. 8.1 – (left) Leading depression N-wave at the toe of the beach for the 1992 Nicaragua event (Tadepalli and Synolakis, 1996). (right) Two tige gage recordings during the 2004 Boxing Day tsunami (Satake, 2007). In Male, Maldives (upper trace), the tsunami manifested itself as a leading elevation N-wave (LEN), as elsewhere to the west of the Sumatran subduction zone. In Phuket, Thailand (lower trace), it manifested itself as an LDN, as elsewhere to the east of the subduction zone. This was consistent to the inferences of Tadepalli and Synolakis (1994, 1996).

On 9 October 1995, an $M_w = 8.0$ earthquake struck Manzanillo, Mexico and generated a moderate tsunami with runup ranging mostly up to 4 m and 11 m extreme values on steep coastal cliffs. Borrero et al. (1997) was able to acquire two series of photographs from eyewitnesses. The first was taken by a family sitting in a veranda of their hotel in the steep coastal cliffs on the south end of La Manzanilla. As soon as they felt the ground shake, they noted the Manzanillo Bay *emptying* and took a photograph. This photograph is believed to be the first documented observation of an LDN causing shoreline recession. In the other series of photographs, taken by the local tortilla maker in a cart at the square in La Manzanilla, one can see three men running away from the tsunami, which reached a maximum runup of 2 m on the mild beach fronting the town. What is remarkable is that despite the thin advancing tsunami front, and its small overall size, the eyewitnesses could not outrun it. The observation was not quantitatively analyzed at the time. Fritz et al. (2006a) estimated tsunami overland flow velocities for the 2004 Boxing Day tsunami at Banda Aceh. They applied a cross-correlation based particle image velocimetry and rectified video images to determine instantaneous tsunami flow velocities on dry land in the range of 2 to 5 m/s.

The second key observation from the 1992 Nicaraguan tsunami was the initially puzzling variability of inundation over short distances along the shoreline. While the runup ranged up to 11 m in the coastal resort town of El Transito, the most devastated locale in this catastrophe, in the adjoining Playa Hermosa even the beach umbrellas had been left standing. This was widely noted then in local newspapers. This observation underscored the difficulty with using tsunami intensity scales which had been proposed by analogy to seismic intensity scales. Here it is interesting to note that the European Seismological Commission (ESC) in its meeting in Jena in 1962 established a subcommittee on tsunamis which N. Ambraseys chaired for a number of years. In 1964, the subcommittee proposed a tsunami intensity scale, the *ESC-Tsunami scale*. Eventually it became known as the Sieberg-Ambraseys scale which has six points, from very light to disastrous. If the 1992 event had taken place in before this century, most probably only the El Transito catastrophe would had been survived in historic reports and the tsunami would had been a level 6. Yet it was a level 1 on the adjacent beach. It was for this reason that the Sieberg-Ambraseys scale has not been in wide use. However, the latter is far more useful for assigning a single numeric value to historic damage reports than the 12 point scale, i.e., the scale introduced resolution that is not resolvable even for recent events. Most historic reports simply do not have the information needed to resolve damage and inundation to 6 points, much less to twelve. See Ambraseys and Synolakis (in review) for a review of intensity scales.

A subsequent offshore bathymetric and onshore sediment survey by J. Bourgeois, C. E. Synolakis and J. C. Borrero in 1995 identified the origin of the unusual runup distribution as the geometry of the coral reef in central Nicaragua. The reef fronting the devastated El Transito had an opening to allow for easier navigation, hence its rapid development as a fishing village. The adjacent Playa Hermosa that was largely spared did not. This fact was ignored in the rush to explain the unusual distribution with nonuniform slip models (Piatanesi et al., 1996) through least squares fits of the runup measurements with threshold models stopping the computations at the 10 m depth offshore –the reef was at a depth of about 5 m. Nonetheless, the runup variation remains puzzling, as the Nicaraguan event has not been computationally studied at sufficient resolution to allow realizations of the effects of the narrow opening <5 m. The runup of extreme long waves had been earlier believed not dependent on such small scale coastal features, and several pre-1980 investigations assumed that the shoreline resembled a vertical wall, no matter what its slope or geomorphology. While these earlier studies did not have the modeling capability to account for these shoreline details, the rationalization nonetheless had been that they were not believed important. The 1992 Nicaraguan event led these earlier speculations to rest.

Further, during the 1994 East Java and 1996 Peruvian tsunamis, it had been observed that coastal dunes limited the amount of tsunami penetration. In the 2004 Indian Ocean tsunami, it was observed by Dalrymple and Kriebel (2005) that in Karon Beach, Thailand, a low sand dune protected the area behind it. A resort hotel proprietor had removed some of the dune seaward of his property for the purpose of better scenic views in Yala, Sri Lanka. Substantially larger flow depths and greater damage were observed in the hotel grounds as compared to neighboring areas behind unaltered dunes and the hotel was entirely razed to the ground (Liu et al., 2005b). The analytical work of Synolakis et al. (1997b), Kânoğlu (1998) and Kânoğlu and Synolakis (1998) had been suggestive that in many cases the last topographic slope long waves encountered as they attacked composite beaches affected the runup to first order. In addition, it had been inferred that other small-scale features might do so as well. Unfortunately, this speculation had not been quantitatively confirmed through field observations until the 2004 catastrophe; see also Fernando et al. (2005) and Danielsen et al. (2005).

Fig. 8.2 – (top left) Babi Island and (bottom left) catastrophe on the back side of it (Yeh et al., 1994). (top right) Top view and (bottom right) side view of the laboratory manifestation of a solitary wave attacking a conical island of slope 1:2 to model the catastrophe in Babi Island in 1992 (Kânoğlu and Synolakis, 1998).

The Flores 1992 earthquake hit and triggered a large tsunami; almost three months post the Nicaraguan event. At the time, the confluence of the two events was deemed as a coincidence, after all the last major tsunami known widely before Nicaragua 1992 was the 1983 Sea of Japan tsunami. Flores 1992 is mostly remembered by the catastrophe in Babi Island and overland flow in Wuhring. Babi is a volcanic island of conical shape with a diameter at the shoreline of about 2 km, situated between the epicentral region and Flores Island. While the tsunami attacked Babi traveling from the epicentral region from the north, most of the inundation was in the south, where two fishing villages were located. Their location had been strategic as most of swell waves also originated from the north in the Flores Sea, so the south coastline was the lee side and provided protection to fishermen. The two villages were completely inundated and runup values ranged up to 7 m (Fig. 8.2). A few survivors who had to swim to Flores about 5 km away to get help as their boats were destroyed described particularly gruesome scenes with human remains left dangling on trees (Yeh et al., 1993). Wuhring, a 400 m-long, 200 m-wide densely populated peninsula next to Maumere, was completely inundated with overland flow of 3 m depth and more deaths per 100 persons than anywhere else other than Babi Island. Photographs from Wuhring are quite reminiscent of images from Banda Aceh, including the surviving mosque (Yeh et al., 1994). Note that an alternative explanation for the extreme runup in Riangkroko, Flores has been proposed by G. Plafker and is briefly discussed in Okal et al. (2003).

The tsunami generated by the July 12, 1993 HNO, Japan $M_w = 7.8$ earthquake produced the worst local tsunami-relate death toll in fifty years in Japan, with estimated 10–18 m/s overland flow velocities and 30 m runup. These extreme values are the largest recorded in Japan this century and remain post 2004 among the highest ever documented for non-landslide generated tsunamis per Synolakis and Okal (2005). This event was surveyed extensively and produced densely distrib-uted field measurements. With high resolution bathymetric data and reasonable ground deformation data available (Takahashi, 1996), this event was used as one of the benchmark problems during the 1995 Friday Harbor, Washington work-shop, see Yeh et al. (1996), Titov and Synolakis (1997) and Titov (1997). Later, a laboratory experiment with geometry closely representing the bathymetry off Okushiri was constructed at the Central Research Institute for Electric Power Industry (CRIEPI) in Abiko, Japan, at a scale of 1:400. The initial wave was gen-erated by a piston wavemaker and was chosen to best represent the model results to the field runup measurements. These laboratory measurements were used as a benchmark problem during the 2004 Catalina Island workshop (Liu et al., 2008).

The event, however, that shook tsunami science on a scale analogous to the 2004 megatsunami, was the 17 July 1998 Papua New Guinea (PNG) tsunami. De-spite the relatively small size of the parent earthquake with $M_w = 7.0$, this tsunami resulted in over 2,100 fatalities, officially surpassed in the twentieth century only by the 1933 Showa-Sanriku, Japan tsunami. Before the megatsunami, it was thought of as *the* extreme tsunami. The field survey (Kawata et al., 1999) con-firmed exceptional runup heights, reaching 15 m, concentrated on a 25 km stretch of coastline outside which the effects of the tsunami were minimal. As documented in detail in Synolakis et al. (2002a), the combination of excessive amplitude and concentration of the runup was quickly recognized as incompatible with the excita-tion of the tsunami by a seismic dislocation; in addition, the earthquake did not feature a slow source comparable to those of documented tsunami earthquakes so excessive runup was not expected. Finally, eyewitness reports generally indicated that the tsunami had arrived at least 10 min later than predicted by all acceptable models of propagation. However, the key piece of evidence was the identification of a hydroacoustic record in Wake Island by Okal (2003), which demonstrated beyond doubt the characteristic signature of the landslide through comparisons with the acoustic signatures at the same locale from the earthquake and its main aftershock. Without identifying the timing correctly, marine geology findings of a recent landslide scarp had remained just speculation. With E. A. Okal's fundamen-tal contribution, the source of 1998 PNG tsunami was identified as an underwater landslide (Synolakis et al., 2002a).

The PNG 1998 tsunami resulted in renewed sensitivity for the hazards created by underwater landslides, as identified in the 2000 NSF Los Angeles workshop. Before then, the understanding of the mechanics of landslide tsunamis was so lacking, that a simple arithmetic error of two orders of magnitude in an empirical formula had remained unnoticed, leading to substantial underpredictions of the height of the leading wave from a landslide off Palos Verdes, California. This was first noticed by Borrero (2002). With the wrong 0.15 m estimate, many had dis-missed the local landslide hazard in Southern California altogether. The PNG analysis allowed for the resolution of an earlier bitter dispute concerning the land-slide trigger of the 1994 Skagway, Alaska tsunami (Synolakis et al., 2002b). De-spite being largely ad-hoc, the initial condition from the PNG simulation is still in use in 2007 to model initial waves when assessing the impact of landslide tsunamis. With the exception of Ward (2001) there is no other comprehensive landslide wave generation theory. Ward's methodology is not in wide use as it is considered unconventional by some despite having been peer reviewed and applied in many instances. Apparently, the rediscovery of the wheel is by its definition appealing, particularly when done with more seemingly complex methods than Ward's. Other investigators have made what we consider outlandish claims about the efficacy of their own methods in predicting initial waves from landslides. Regrettably, the calculation of the initial wave from a submarine landslide remains an art and is still shrouded with controversy.

The 17 August 1999 Izmit, Turkey and 26 November 1999 Vanuatu earthquakes are believed to have triggered mass movements that generated the larger than otherwise expected observed coastal inundation. These events, along with the 13 September 1999 Fatu Hiva, Marquesas aseismic tsunami generated by an aerial slump due to the collapse of weathered cliffs (Okal et al., 2002), further focused interest on landslide tsunamis, as summarized by Bardet et al. (2003a). In the aftermath of this series of nearshore mass movements, marine geology surveys that were then published suggested that some continental margins in California, Oregon and even the East Coast of the US and elsewhere were especially susceptible to submarine landslides due to the combination of offshore faults and nearshore submarine canyons with stored sediment, as well as bottom material on relatively steep slopes. Even the East Coast of the US, long believed to be tsunami-free, was evaluated for its landslide tsunami hazard. As a result, the level of hazard posed by relatively moderate earthquakes (typically at the moment magnitude 6 level) started to be re-examined upwards, on a case-by-case basis (Borrero et al., 2001 and 2004).

According to Synolakis and Bernard (2006), before 2004, some claimed that as many as one-third of the tsunamis ever historically reported were due to landslides, and earlier events were re-interpreted to account for what appeared to the zeitgeist as anomalous. Gutenberg's (1939) aphorism that *submarine landslides were the chief causes, if not indeed the major cause of tsunamis* was rediscovered with a vengeance. Research studies started focusing at landslide tsunamis. Fortuitously, once again, the $M_w = 7.6$ Wewak, Papua New Guinea earthquake of 9 September 2002 occurred 135 km northwest of the 1998 event and allowed for less speculative revisionism. The moment of the 2002 event was seven times larger than in 1998, yet the tsunami generated was much smaller, with no associated tsunamigenic landslide. The field survey of Borrero et al. (2003) revealed a strikingly different runup distribution from the 1998 event, see Fig. 8.3.

Okal and Synolakis (2004) fitted Gaussian distributions to the respective longshore runup measurements for both real events and the computed values from runup over idealized geometries. For tectonic (dislocation-triggered) tsunamis, they proposed two dimensionless parameters $I_1 = b/\Delta u$ and $I_2 = b/a$, where b, Δu and a are measures of the maximum of the runup distribution, the seafloor vertical displacement and the width of the runup distribution in the longshore direction respectively. The invariant I_1 is replaced with $I_3 = b/\eta_{min}$, where η_{min} is the height of the leading dipole wave in the LDN, for landslides. I_1, I_2 and I_3 behave as invariants. These invariants characterize the tsunami source landslide versus dislocation, yet they are largely independent of the exact parameters describing the respective sources. Hence, they can be used as discriminants to identify the nature of the tsunami source. Fig. 8.3 also shows Okal and Synolakis (2004) numerical simulations using MOST to augment the historic data set.

Note that one fundamental observation linking the tsunami source with inundation, although never published as such, is G. Plafker's. While surveying the aftermath of the 1964 Great Alaskan tsunami (Pararas-Carayannis, 1972), he proposed that the maximum runup locally does not exceed twice the height of deformed seafloor offshore (Plafker, 1965; Plafker et al., 1969). This is now referred to as *the Plafker rule*, and has largely withstood the test of time, remaining unrecognized for three decades (Synolakis and Okal, 2005). The Okal and Synolakis (2004) parameter $I_1 < 2$ which is precisely what the empirical Plafker rule suggests. The 2004 tsunami did not contradict this rule, even at Longa in Aceh. However, local effects such as bathymetric focusing might result in significant amplification and consequent higher runup values.

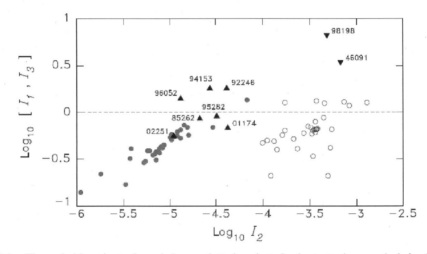

Fig. 8.3 – The scaled invariants I_1 and I_3 are plotted against I_2, for tectonic numerical simulations (solid dots) and landslide simulations (circles) and as derived from field measurements (triangles) from near-field events from 1992 to 2001 identified by Julian date, e.g. 98198 and 02251 identify the 1998 and 2002 PNG events, respectively. Note that all landslide events feature aspect ratios with $\log I_2 > -4.2$. After Okal and Synolakis (2004).

One vexing question raised by the PNG landslide wave was whether the coastal evolution of landslide tsunamis could be effectively modeled with the NSW equations, or whether the Boussinesq equations would have to be invoked. Landslide waves are steeper and shorter than tectonic tsunamis and hence disperse more rapidly. Lynett et al. (2003) compared results from Boussinesq and NSW models with the field measurements and concluded that while the nearshore evolution was predicted to be different, the runup predictions were not. In the process, they also differentiated between runup and flow depth measurements in the field observations. Before 2000, the term runup, which refers to the elevation of the most inward penetration of the wave, the same term was used to refer to the flow depth as well, introducing further challenges in the interpretations of older field datasets. In fact, it was the tsunami surveys of the 1990s that allowed for perspective in evaluating historical reports of onshore measurements.

Despite the once-a-year incidence of tsunamis in the decade prior to December 26, 2004, it was the Boxing Day event that rendered tsunami in the vernacular of most languages. A summary of field survey observations is given in Synolakis and Kong (2006) and detailed reports appear in a special issue of Earthquake Spectra (2006). In terms of basic hydrodynamic science, to date, four noteworthy phenomena unobserved or undocumented before 2004 tsunami have been identified (Synolakis and Bernard, 2006). One, as shown in Fig. 8.1, the Boxing Day tsunami manifested itself as a leading elevation N-wave (LEN), i.e., elevation wave in front and an LDN in coastlines west and east of the subduction zone respectively. This behavior was implicit in the work of Tadepalli and Synolakis (1994, 1996) but had not been observed in nature earlier; while practically all tsunamis in the earlier decade had manifested themselves in the nearby shorelines as LDNs, they didn't have farfield impact to evaluate the other aspect of the theory, i.e., the propagation of the LEN offshore. Two, the waveguide type effect from mid-ocean ridges that appears to have funneled the megatsunami away from the tip of Africa (Titov et al., 2005b). Three, the sparing of the Maldives, an archipelago with coral atolls with elevations <2 m from mean water level. The islands rise from the seafloor as pillared structures with no continental shelf, hence there was no significant wave amplification. While the reef fronting the islands determined the extent of inundation, there is little question that the Maldives experienced a tsunami with heights closer to the free-field tsunami height. This was implicit in earlier analytic work by Lautenbacher (1979) and Kânoğlu and Synolakis (1998). Four, the comparison between the 28 March 2005 and the 26 December 2004 Sumatra tsunamis suggested that the presence of the islands of Nias and Simeulue reduced the effective water mass set into motion during the later event, thereby drastically reducing the size and the impact of the generated tsunami (see figure of Arcas and Synolakis in Kerr (2005) and Geist et al. (2006a, b)). This effect is also implicit in the analytic work of Kânoğlu and Synolakis (2006).

In most affected coastlines of the Indian Ocean, observations of tsunami damage to buildings were unsurprisingly similar. In Sri Lanka, churches and Buddhist temples were left standing in numerous locations. Closer to the beach in Aceh, the only conspicuously standing structures in midst of the devastated wasteland of coastal areas were mosques. In all locales, residents credited divine intervention. A more scientific explanation may well be that places of worship are more carefully constructed to a higher standard than surrounding buildings (Synolakis and Bernard, 2006; Synolakis et al., 2005). Also, the open architecture of mosques in Southeast Asia, with multiple circular columns on the ground floor and no surrounding walls, allowed the tsunami to flow freely through it, with little impact on the superstructure above them. Yet, few, if any, knew to evacuate to upper floors of surviving structures. In Vilufushi, the most severely impacted atoll in the Maldivian archipelago, while the local school and city hall survived with no damage and were two storey structures (Fritz et al., 2006b), people in surrounding non-engineered houses perished. While the further survival of buildings left standing from the shaking might not be known a priori, when there is no other choice, vertical evacuation is the only choice.

Another result whose usefulness was not entirely recognized until the wide availability of amateur videos of the 26 December 2004 megatsunami was the acceleration of the wavefront past the initial shoreline. In a video taken near the Grand Mosque in Aceh, one can infer that the wavefront first moved at speeds less than 2 m/s, then accelerates to about 5 m/s. The same phenomenon is probably responsible for the mesmerization of victims during tsunami attacks, apparent with hindsight in a series of photographs of the 1946 Aleutian· tsunami approaching Hilo, Hawaii, and noted again in countless photographs and videos from the 2004 megatsunami. The wavefront appears slow as it approaches the shoreline, leading to a sense of false security. It easy to imagine one can outrun it, but tens of seconds or just seconds later (depending on location), the wavefront accelerates rapidly as the main disturbance arrives. This effect of wavefront acceleration was first shown by Synolakis (1986) analytically and is implicit in the work of Carrier et al. (2003) and Kânoğlu (2004). Zelt and Raichlen (1991) also studied front velocities experimentally and numerically over an impermeable bed by incident solitary waves.

The fortuitous presence of satellite based altimeters from almost orbits over the Bay of Bengal contemporaneous with the tsunami evolution provided a new type of field observation that served as a benchmark problem for models. Indeed Smith et al. (2005) removed most *permanent* water level features observed in earlier passes of the altimeters in the previous month and compared their normalized altimeter data predictions with MOST. They comment that *the signal of the leading edge two hours after the earthquake is particularly prominent, with an amplitude of 60 cm and two narrow peaks* very close to where the NOAA tsunami model forecast shows two overlapping peaks coalescing into one broad (250 km) crest. Their results model the 60 cm deep water height of the tsunami satisfactorily. These satellite altimeter data provided the first direct measurement of the height of a large tsunami in the deep ocean and have confirmed the long speculated maxim that tsunamis in the open seas have heights less than 1 m.

In terms of field observations, it is worthwhile to discuss two major advances: one, deep-ocean measurements, and two real time forecasts based on these measurements. No tsunameters (Bernard et al., 2006) existed in the Indian Ocean at the time of the megatsunami, having only been deployed in the Pacific. Only a very limited number of deep-ocean tsunami measurements by several research tsunameters –without real-time data transmission– had been made until the 17 November 2003 Rat Island tsunami event, i.e., the 10 June 1996 Andreanov Islands (Tanioka and Gonzalez, 1998) and the 4 October 1994 Kuril Islands (Yeh et al., 1995) events. NOAA PMEL's experimental forecast methodology was tested against DART records for these events (Titov et al., 2005a).

On 17 November 2003, the $M_w = 7.8$ earthquake occurred on the shelf near Rat Islands, Alaska and generated a tsunami that was detected by three tsunameters located along the Aleutian Trench. This real-time tsunameter data combined with pre-computed propagation database were used to produce a real-time tsunami forecast for the first time, i.e., tsunami height predictions were obtained during the tsunami propagation before the waves reached target coastlines. The West Coast/Alaska Tsunami Warning Center (TWC) made available preliminary earthquake parameters (location and magnitude $M_w = 7.5$) about 15–20 min after the earthquake. The initial offshore experimental forecast was made immediately after preliminary earthquake parameters became available. This initial forecast provided expected tsunami time series at tsunameter locations and once the closest tsunameter (Sta. 46401-D171) recorded the first tsunami wave, the model prediction was compared with the measurement which resulted into the adjustment of the earthquake magnitude ($M_w = 7.7 - 7.8$). The new forecast was obtained approximately 1 hour 20 minutes after the earthquake. The adjusted earthquake magnitude satisfactorily predicted the tsunami records at other tsunameter locations, as later confirmed by the US Geological Survey ($M_w = 7.8$) and the Harvard Seismology Group ($M_w = 7.7$). This forecast was performed in an experimental mode and was not part of the TWCs operation, but it provided a real test of PMEL's forecast methodology. As a hindcast, the offshore model forecast was used as an input for a high resolution inundation model for Hilo Bay, Hawaii, with three nested grids and highest spatial resolution of 30 m, inside Hilo Bay. The comparison of the model predictions with the Hilo tide gage data was excellent for the forecasting for the amplitude, arrival time and periods of several of the first waves of the tsunami wave train (Titov et al., 2005a).

Since the 2003 Rat Island tsunami, NOAA PMEL's experimental forecasting system demonstrated great potential estimating the farfield impact for five additional events in real time, i.e., the 3 May 2006 Tonga, 15 November 2006 and 13 January 2007 Kuril Islands, 1 April 2007 Solomon Islands and 15 August 2007 Peru events. Again, PMEL's experimental forecast system is based on the inversion of the high-quality DART data to constrain the tsunami source. Once the source is constrained site specific tsunami models are used to produce efficient and reliable forecasts, e.g., a specific example is given in Wei et al. (2008) for the 15 August 2007 Peru event. This experimental forecast was produced within 2 hours of tsunami generation and would had provided enough lead time for potential evacuation or warning cancellation for farfield, i.e., Hawaii and the US West Coast. When implemented operationally, such forecasts could be obtained even faster. A similar end to end system is currently under development for the Indian Ocean, with progress limited only from funding and intergovernmental agreements to allow for tsunameter deployments.

As a final note of field data as benchmarks for validating tsunami inundation models used in inundation mapping, Satake et al. (1996) and Atwater et al. (2005) reconstructed the effects of the 1700 Cascadia tsunami across the Pacific in Japan using available historical archives. These records helped constrain the magnitude of the 1700 event, and were instrumental in debunking rush estimates for the same event when it was rediscovered by some in the aftermath of the 2004 megatsunami. The similarities between possible earthquakes from the Cascadia and Sumatra Subduction Zones were thus exploited by the same in a public relations fury with runup estimates for Cascadia ranging up to four times what has been measured in the Cascadia field studies (Synolakis and Bernard, 2006).

Fig. 8.4 – Scotch Cap lighthouse remains after the April 1, 1946 Aleutian Island tsunami (National Geophysical Data Center's tsunami picture).

Along similar lines of using pre-present tsunami observations to improve models, interest in surveys of historic events –where it was likely to still find eyewitnesses alive– arose. The 1946 tsunami was triggered by a deceptively small $M_w = 8.1$ earthquake, yet eradicated Scotch Cap lighthouse in the nearfield (Fig. 8.4). In the farfield, it killed 159 people in Hawaii and wrought significant damage in the Marquesas, with some reports that it was noted as far as Antarctica. Okal et al. (2003) compiled farfield and nearfield runup measurements through the interview of 69 witnesses aged 59–89 years, with two conclusions. In the nearfield, the maximum runup was revised upward to 42 m at Scotch Cap, rapidly decayed along the coast of Unimak Island. Their measurements require the involvement of a major underwater landslide. The longshore runup distribution along the south west shore of Unimak suggests an $I_2 = 6.7 \times 10^{-4}$ and a ratio of maximum runup to seismic slip $I_1 = 3.4$, both in excess of those theoretically acceptable for any dislocation. In the farfield, E. A. Okal and his coworkers documented a very pronounced directivity based on the field measurements at Juan Fernandez Island, which cannot be reconciled with a landslide source, and requires a substantial dislocation source. The 1946 controversy appears to start being reconciled with the hypothesis that nearfield at Unimak most of the inundation is due to a landslide source, while farfield it is due to a tectonic source.

In summary, as Synolakis and Bernard (2006) discuss, the 1992–2004 field surveys were instrumental in providing field measurements that helped not only validate numerical codes in contemporaneous development in this period, but also establish realistic geophysical constraints for the initial condition of tsunamis of geophysical scales, i.e., they served as proxies of the tsunameter and altimeter data that only became available since 2003 and 2004, respectively.

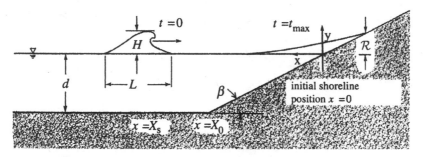

Fig. 8.5 – Definition sketch for long wave propagation over a canonical consideration on analytical studies.

4. Analytical advances

We discuss here advances in analytical methods that provided the first benchmarks for the development of numerical codes for studying coastal effects of tsunamis.

Hall and Watts (1953) presented the first study of a single long wave approaching a sloping beach, in the laboratory. They introduced what is now known as the canonical problem of one-dimensional long wave theory, i.e., propagation of long wave over a constant depth first then sloping beach as in Fig. 8.5. They used empirically generated water waves which resembled the classical solitary wave shape, as described by Russel (1845). Using dimensional analysis, they concluded that $R/d = \alpha(\beta)(H/d)^{\lambda(\beta)}$, with $\alpha(\beta)$ and $\lambda(\beta)$ were empirical coefficients dependent on the beach slope β. They normalized the offshore height H and runup R using the offshore constant depth d. However, they did not distinguish between breaking and nonbreaking waves, and attempted to fit all their measurements into a single relationship.

In 1958, Carrier and Greenspan (1958) introduced a powerful nonlinear transformation, since known as the Carrier–Greenspan transformation. They were able to transform the coupled NSW equations into a single linear Bessel-type ordinary differential equation. Nonetheless, while there is conservation of difficulty, the moving shoreline is fixed in the transform space and the wave evolution can be predicted with the usual tools of linear theory. However, their solution was limited to specific initial wave profiles which they were unable to generalize to more realistic waveforms due to the complexity of the transform. Further translating their solution back to physical coordinates was a daunting task they never attempted. Thus their work remained largely unrecognized and unused outside applied mathematics for decades, until rediscovered by Synolakis (1986).

In the same year, while working on shock wave propagation through nonuniform flow regions, Whitham (1958) suggested a method to circumvent the difficulty of evaluating differential relationships at a discontinuity –the shock wave is the classic example of a discontinuous propagating front. Even though not recognized just then, the Whitham bore rule proved useful in the development of inundation algorithms. Extending earlier results, Shen and Meyer (1963) calculated the bore front path approximately after the bore collapsed. Using the LSW equations, Keller and Keller (1964) first derived a Bessel-type equation for monochromatic wave 1+1 evolution over a sloping beach, which peculiarly was the same as the one derived through the Carrier–Greenspan transformation for the NSW equations. They then matched this inner solution to an outer solution of periodic waves over constant depth at the transition point, i.e., the toe of the beach, and derived an amplification factor for monochromatic waves.

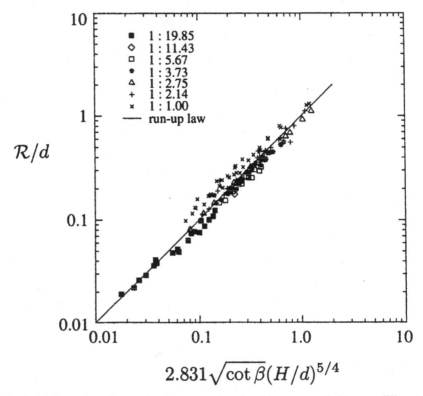

Fig. 8.6 – Laboratory data for maximum runup of nonbreaking waves climbing up different beach slopes: ■ 1:19.85 (Synolakis, 1986); ◇ 1:11.43, □ 1:5.67, * 1:3.73, + 1:2.14, × 1:1.00 (Hall and Watts, 1953); △ 1:2.75 (Pedersen and Gjevik, 1983). Solid line represents *the runup law* (R/d = 2.831 $\sqrt{\cot\beta}\,(H/d)^{5/4}$).

The work of Synolakis (1986, 1987) illustrated further the characteristics of breaking and nonbreaking solitary waves and their runup. Synolakis (1986, 1987) was able to mathematically justify the empirical relationship of Hall and Watts (1953) with exact coefficients, using the exact solution of the LSW equations. He solved the IVP solution of the LSW equations and runup of Boussinesq solitary wave first propagating over constant depth and then evolving over a sloping beach (Fig. 8.5). To achieve this, Synolakis (1986, 1987) matched the linear theory solution at the constant depth with the solution over the sloping beach as in the Keller and Keller (1964) formalism to derive the solution over the sloping beach for the linear problem.

In continuation, a key advance was the development of a mathematical formalism that allowed for direct contour integration of the resulting Fourier transforms and the derivation of McLaurin series that could be summed up asymptotically, in parameter ranges of geophysical interest. For example, for the runup R of solitary waves of offshore height H, propagating over depth d and then climbing up a beach of angle β, with profile $\eta(x,0) = H\text{sech}^2[\gamma(x - x_1)]$; $\gamma = \sqrt{(3/4)(H/d^3)}$, Synolakis (1986) derived asymptotically the maximum runup where the coefficient and the exponent are exact, i.e., $R/d = 2.831\sqrt{\cot\beta}\,(H/d)^{5/4}$ which is known as *the runup law* (Fig. 8.6).

Then, Synolakis (1986, 1987) linearized the Carrier–Greenspan transformation at the transition point, arguing that far off the beach nonlinear effects were expected to be small and proceeded to derive a new boundary value problem (BVP) solution of the Carrier–Greenspan equation for general offshore initial conditions and calculated the evolution of a solitary wave with nonlinear theory, through the entire runup and rundown process. Unlike Carrier and Greenspan (1958), Synolakis (1986) was able to back calculate the evolution of the wave into the physical (x,t) space, and successfully compared the analytical predictions with laboratory measurements. Synolakis (1986, 1987) was then able to demonstrate that for the same initial wave offshore, and despite the fact that linear and nonlinear theory predicted different wave evolutions nearshore, their predictions for the maximum runup were mathematically identical. Carrier (1966) suggested this invariance first without accounting for reflection off the beach even though reflections can drastically change the wave shape.

By calculating the Jacobian of the Carrier–Greenspan transformation, Synolakis (1986) derived a breaking criterion to identify among his own runup data and Hall and Watts' (1953) and compared the predictions from the asymptotic result to the data. The problem of scaling solitary wave runup was solved and the differences in functional variations for the runup of breaking and non-breaking waves were identified. Most of the later works with numerical solutions for solitary wave runup have used *the runup law* for validation.

By noting that the first arrival in Nicaragua and Flores were waves that caused the shoreline to recede before advancing, Tadepalli and Synolakis (1994) proposed a model for the leading wave of tsunamis with an N-wave shape, in analogy to dipole waves in gas dynamics. To facilitate asymptotic analysis, they chose a certain class of dipolar waveforms with amplitude given by $\eta(x,0) = \varepsilon\, d\, H(x - x_2)\text{sech}^2[\gamma(x - x_1)]$, where γ is a wavenumber parameter that defines the spread of the wave by analogy to the solitary wave profile; $\gamma = \sqrt{(3/4)(H/d^3)}$; and ε is a scaling parameter to allow for direct comparison with solitary waves. Tadepalli and Synolakis (1994) used the methodology of Synolakis (1987) to evolve these N-waves in the canonical geometry and found that LDNs run up higher than LENs. They derived asymptotic results for different families of N-waves and showed that LENs run up higher than the equivalent solitary waves.

As described earlier, the work of Tadepalli and Synolakis (1994) was controversial at the time, for it was not reconcilable with the solitary wave paradigm. LDN waves were believed hydrodynamically unstable, the crest was supposed to quickly overtake the trough. Later, Tadepalli and Synolakis (1996) forced the LSW equation with a step function motion to imitate fault rupture, and showed that N-waves were one particular exact solution for the forced LSW. Using the Inverse Scattering theory, known as INS after Gardner et al. (1967), they determined that LDN waves of geophysical scales would have to evolve more than around the perimeter of the Earth to fission into solitons. LDN waves of steepness less than 0.0001 were found stable over 1000 km propagation distances. Tadepalli and Synolakis (1994, 1996) also implied that because of the nature of the dipolar seafloor deformation unzipping in subduction zones, LDN waves would strike the adjacent shoreline of the subsiding plate, while LEN waves would move towards the open ocean. Using their asymptotic formulae on a representation of the initial 1992 Nicaraguan tsunami (Fig. 8.1) derived from seismic models, and a simplified form of the long Nicaraguan shoreline, they showed that they could predict the observed runup analytically within a factor of two. Geist (1998) confirmed these inferences by demonstrating how subduction zone events generated N-waves and developed approximate formulae based on the asymptotic results of Tadepalli and Synolakis (1994).

Kânoğlu and Synolakis (1998) extended the Synolakis (1986) formalism to wave propagation over piecewise linear topographies. Requiring continuity of wave amplitude and its slope at the transition point between adjacent linear topographic segments, Kânoğlu and Synolakis (1998) were able to present a matrix formulation which could be applied on different topographies. In addition, they were able to evaluate maximum runup for continental shelf and slope bathymetries typical of real beaches. They benchmarked their formulation using laboratory data from an experiment that modeled wave evolution in front of a seawall at Revere Beach, Massachusetts, as further described in section 5. They derived a simple formulation which shows that the maximum runup is exquisitely dependent on the water depth at the shoreline. Their conclusion that the runup is governed by the topographic features closest to the shoreline, this has recently been recognized for its usefulness in assessing to first order the effect of rising sea level due to climate change, see Ewing (2008).

Kânoğlu and Synolakis (1998) then extended their analytical solution to evaluate the evolution of solitary waves around a conical island. They used the same method as for 1+1 propagation over piecewise linear bathymetries. Using the existing solution for LSW equation for a sill, i.e., a cylinder sitting over a constant depth of water, they were able to construct a conical island from sills. They thus constructed an analytic solution by combining solutions. The methodology produced good comparison with laboratory data in terms of time histories of surface elevation and maximum runup distribution around the island and was presented in the 1995 Friday Harbor NSF workshop on Long-Wave Runup Models, see Kânoğlu and Synolakis (1996).

The 1+1 initial value problem (IVP) of the NSW was revisited by Carrier et al. (2003) with a hodograph transformation for a uniformly sloping beach, in contrast to the canonical problem, whose existing BVP solution is for an initial wave at constant depth and then evolving over a sloping beach and is more realistic. The results showed how N-waves not only have different runup, but different onland velocity distributions. The Carrier et al. (2003) solution includes integrals with singularities and ad hoc linearizations of the transformation, issues addressed shortly thereafter by Kânoğlu (2004). Later, Kânoğlu and Synolakis (2006) were able to present a more general solution to the IVP solution of the NSW equations. Their solution includes initial conditions where the initial wave has nonzero velocity. Unlike Carrier et al. (2003), Kânoğlu and Synolakis (2006) were able to solve the case with initial velocity without any difficulty for larger initial wave heights.

In terms of solving for 2+1 wave evolution, Carrier and Noiseux (1983) presented a solution to a weakly two-dimensional problem. Realistic initial conditions have yet to be applied to their practice, at least as to the time of this writing. Brocchini and Peregrine (1996) presented weakly nonlinear solutions for wave propagation over a sloping beach to discuss flow properties in the swash zone. Using a similarity transformation to relate the longshore coordinate y with t and considering small angle of incidence which is a realistic case, they were able to transform the 2+1 problem into a solvable 1+1 Synolakis (1987) formulation. The Brocchini and Peregrine's solution remains largely unexplored with the exception of Brocchini (1998) who extended it for multiple-solitary-pulses. Carrier and Yeh (2005) presented a solution of the evolution of a finite crest Gaussian shape wave over a flat bottom. The resultant integrals are singular, as in Carrier et al. (2003), especially for large values of spatial and temporal variables. However, they were able to obtain a self-similar solution and discussed the directivity in energy radiation.

Another kind of coupling of the seafloor with the free surface was investigated in 1+1 propagation by Tuck and Hwang (1972). They derived a solution to a forced wave equation for a deforming sloping beach, and suggested that it represented a fault rupture on a continental shelf. No inundation results were shown as their attention was focused on the waves propagating offshore, as their initial condition was an exponential decaying from the shoreline seaward. The forced linear long wave equation was revisited by Liu et al. (2003) with an exponential-shaped block sliding from the initial shoreline. Their exact solution developed is useful for code-validation when calculating the impact of landslide tsunamis. Numerically, near the initial shoreline, landslides pose a vexing computational challenge, for, not only the shoreline retreats and then advances, but also the seafloor is deforming at a rate that cannot be ignored in simulations, as when calculating the evolution of tectonic tsunamis. The analytic solution allowed for a direct comparison of predictions and guided shoreline algorithm improvement. The variation of the 1+1 runup with the triggering fault generation mechanism based on one-dimensional approximations was accomplished by Tinti and Tonini (2005); it remains largely a scientific curiosity as most real faults generate waves that spread in two dimensions, hence it is not useful per se in evaluating the effects of fault parameters, see Okal and Synolakis (2004).

Ward (1980) followed an entirely different modeling methodology. He introduced the concept of tsunami waves as a particular branch of the spheroidal family of normal modes, or free oscillations, of the Earth. His formalism is traceable to Love's (1911) monograph for the computation of the eigenfunctions of the Earth's modes (Synolakis and Bernard, 2006). Ward asserted that it is directly applicable to the case of tsunamis, as long as the Earth model includes an oceanic layer of constant depth, and the computation is carried in the full six-dimensional space. As discussed later by Okal (1982, 1988), this methodology allows the direct and seamless handling of the coupling between the ocean water layer and any vertically heterogeneous solid Earth structure. While the normal modes methodology is not limited to long water waves, it is by nature linear. It also requires a spherically symmetric structure involving an ocean covering the whole planet with a uniform depth. Until recently, it had remained largely unused in inundation modeling.

Another new methodology for modeling landslide tsunamis was pursued by Ward (2001). If the equations of motion are not depth averaged, but only linearized, the resulting formulation involves solving Laplace's equation. Ward re-derived Green's functions and proceeded to describe complex landslide sources through their linear superposition, as discussed earlier by Mei (1983) in a semi analytic fashion. One advantage is that the formalism is by definition dispersive. A disadvantage is that for some of the extreme waves considered in later work, e.g. volcanic collapse waves of 900 m height in 2000 m depth, these waves are highly nonlinear. Nonlinear effects start to become important in tsunami evolution when the height-to-depth ratio exceeds 0.1, and they rapidly reduce the wave height of larger and steeper waves. Occasionally, dispersion and nonlinearity balances and the resulting wave does not evolve rapidly as is the case with soliton-like waves. This remains to be established for waves of the size and steepness often considered by Ward in subsequent work. Recently, Ward (2004) has calculated the evolution of tectonic tsunamis quantitatively correct over fairly complex bathymetry, where the limitations above are less onerous.

5. Advances in laboratory

Physical models at small scales have been used in coastal engineering when designing harbors or coastal structures, long before the availability of numerical codes. The objective is to visualize flows and measure wave heights or currents in the laboratory and then scale the predictions to the prototype. Even today and with numerical tools widely available, laboratory models are often used to confirm different flow details observed in computations and/or validate the particular numerical model used in the design. The model is validated or its limitations explored for a particular set of parameters, it can then be used to study different physical geometries numerically. Therefore validation of a numerical model through comparisons with results from laboratory experiments is essential to both better understand tsunami hydrodynamics and to enhance confidence in any particular model.

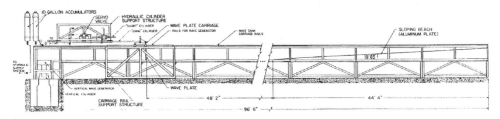

Fig. 8.7 – The California Institute of Technology wave tank most often used in tsunami laboratory studies. Figure from Synolakis (1986).

Up until the late 1960s, water-waves were generated in laboratory channels using a horizontally vertical paddle displacing the entire fluid column, i.e. Hall and Watts (1953). In a landmark study, Hammack (1972) used a novel generation method in the 36.6 m-long, 0.381 m-wide and 0.61 m-deep California Institute of

Technology (Caltech), Pasadena, California wave tank, described also in Goring (1978) and Synolakis (1986, 1987), see Fig. 8.7. The flume and its appurtenances had been constructed by F. Raichlen, who first introduced quantitative laboratory tsunami studies. The Caltech tank is constructed with a bottom of a structural steel channel with glass sidewalls throughout. Carriage rails run along the whole length of the tank, permitting the horizontal movement of instrument carriages where wave gages measured the time history of free surface elevations. One end of the channel had a short bottom section that could be impulsively raised or down-dropped, to model the thrust-type seafloor motion that mostly triggers tsunamis. This can be seen in Fig. 8.7 at the extreme left, along with the horizontal wave generator attached to the two hydraulic cylinders.

By displacing the movable bottom of his tank, J. Hammack measured very precisely the evolution of the resulting waveforms over constant depth. He then used the laboratory measurements to validate Peregrine's (1966) solution method for the nonlinear and dispersive Korteweg–de Vries equation, one form of the depth-averaged equation of motion. Hammack then related the initial wave height to the wave motion at large distances, by using the INS algorithm. The INS methodology developed by Gardner et al. (1967) for solving certain nonlinear equations with soliton solutions, predicted that fairly arbitrary initial conditions would generate a sequence of solitons at infinity (In water, solitons are also known as solitary waves, and in theory they do not change their shape as they propagate over constant depth.)

The advantage of the soliton-INS hypothesis was overwhelming. If indeed arbitrary *long* free-surface disturbances would eventually fission into series of solitary waves, then it was no longer necessary to fully understand the source details, farfield they were irrelevant. This was quite convenient, also because the seafloor motion was then usually imagined only as vertical uplift. For example, in their detailed calculations of wave heights off California from farfield earthquakes, Houston and Garcia (1974a, b) used an elliptic-shaped uplift area of constant displacement, a two-dimensional analog to Hammack's tsunami generator.

The generation issue thought solved, or at least not necessary for further progress, the research focus in the 1970s thus shifted from generation to evaluating the transformation of solitary waves as they propagate over the continental shelf. This was accomplished by Goring (1978) who not only developed an algorithm for driving a laboratory wavemaker to generate precise solitary waves (later extended by Synolakis (1988) for arbitrary long wave generation), but also Goring derived a finite-element solution to the 1+1 Boussinesq equations and validated it with the laboratory data, thus confirming earlier theoretical predictions by Madsen and Mei (1969), suggestive of soliton fission at transitions in depth.

We need to emphasize that the basic premise of insensitivity to source details as suggested by the INS algorithm is not entirely unphysical. Okal and Synolakis (2008) found that characteristics of the tsunami farfield appear remarkably robust with respect to perturbations in the properties of the parent earthquake, as long as the seismic moment M_0 (or in the case of a change of dip δ, as the product $M_0 \times \sin \delta$ remains constant), all quite reminiscent of Saint-Venant's principle in linear elasticity. The latter states that the detailed distribution of stress farfield does not depend on the details of the source, but only on its total load (Knowles, 1966).

The evolution over the continental shelf thought understood, following Hammack (1972) and Goring (1978), Synolakis (1986) used the same Caltech wave tank with a 1:19:85 sloped ramp installed one end to model the bathymetry of the canonical problem, that of a constant-depth region adjoining a sloping beach, as in Fig. 8.5. The ramp was sealed to the tank side walls and its toe distant 14.95 m from the rest position of the piston wave generator. A total of more than 40 series of experiments with solitary waves running up the sloping beach were performed, with wave depths ranging from 6.25 cm to 38.32 cm and normalized heights H/d ranging from 0.02 to 0.62 –recall that solitary waves are uniquely defined by their maximum height to depth ratio H/d and the depth d. Breaking occurred when $H/d > 0.045$, for this particular beach. This is the same set of experiments used to validate the maximum runup analytical predictions presented in section 4.

In Synolakis's laboratory practice the initial location where the measurement for the definition of the solitary wave height H/d changed with different wave heights, because solitary waves of different H/d have different effective wavelengths. If H/d for a given wave was inferred from a wave gage at the toe of the beach, the runup of this H/d wave would be different from that of another solitary wave with H/d inferred from a gage far offshore, as the solitary wave evolves as soon as its front *feels* the toe of the beach. Also, in the laboratory, even waves which initially have solitary wave shape dissipate as they propagate over constant depth. If H/d is measured far offshore and used as an initial condition for non-dissipative computations to model runup, the comparisons will be less meaningful, unless H/d is defined at the same location in both laboratory and numerics and the depth is the same. By keeping the same relative offshore distance for defining the initial condition, meaningful comparisons are assured. This practice has become the standard for solitary wave laboratory tests.

A measure of the effective wavelength of a solitary wave is the distance between the point x_f on the front and x_t on the tail where the local wave height is 1% of the maximum, i.e., $\eta(x_f, t=0) = \eta(x_t, t=0) = (H/d)/100$. The distance that Synolakis used for defining the offshore height is at an offshore location that only 5% of the solitary wave is already over the beach, so that scaling can work. Therefore, in the laboratory experiments, the initial wave heights H/d were identified at a distance of L, $(L/d)/2 = \mathrm{arccosh}\sqrt{20}/\sqrt{(3H)/(4d)}$, from the toe of the beach.

While Synolakis used only ten wave gages to measure free surface elevations $\eta(t)$ in each experimental run, the generation process was extremely repeatable. As experiments were re-run for specific H/d, the gage array was moved to different locations, and the same wave generated again, until a sufficient number of data points existed to resolve the entire wave profile in x. In Synolakis (1987), two different comparisons are presented, one is the amplitude variation at specific x-locations, and second, the amplitude variation at specific t-times. This set of laboratory data became available at time that 1+1 runup computations started being attempted. The original runup algorithms were ad hoc, and without any laboratory reality checks, it was not easy to constrain the inherently temperamental shoreline computation. The Synolakis data set has thus been used extensively for code validation, for example in Synolakis (1987), Zelt (1991), Titov and Synolakis (1995), Grilli et al. (1997), Li and Raichlen (2000, 2001 and 2002).

By the mid 1990s, there was growing confidence in the shoreline computations over a sloping beach, but little was known about their robustness in more realistic situations, such as a seawall fronting a composite beach. Revere Beach is located approximately six miles northeast of Boston, Massachusetts. To address beach erosion and severe flooding problems, a seawall was contemplated, and a physical model of the beach was constructed at the Coastal Engineering Laboratory of the US Army Corps of Engineers, Vicksburg, Mississippi facility, earlier known as Coastal Engineering Research Center (CERC). The model consisted of three piecewise linear slopes of 1:53, 1:150 and 1:13 from seaward to shoreward (Briggs et al., 1996). At the shoreline there was a vertical wall. In the laboratory, the wavemaker was located 23.22 m away from the seawall and tests were done at two still water depths, at 18.8 cm and at 21.8 cm.

In the experiments, solitary waves of different heights were generated, water surface elevations at ten locations and the maximum runup on the vertical wall were measured. Following Synolakis (1986), the initial location where the solitary wave was measured for defining its initial height was at different distances from the toe of the slopes. The results were used as benchmark data to validate 1+1 numerical codes in the 1995 Friday Harbor workshop, as discussed in Yeh et al. (1996), and by Kânoğlu (1998) and Kânoğlu and Synolakis (1996, 1998) when validating their analytical formulation. These are complementary validation data to the single beach slope wave runup data set. Some shoreline algorithms may predict the vertical rise on a seawall adequately, but not on a sloping beach, and vice versa, hence comparisons of predictions with both data sets are needed for confidence. Further, when the solitary wave breaks near the seawall, it produces substantial splash up, which is another excellent test for advanced shoreline algorithms.

In response to recommendations from the 1990 Catalina workshop, to establish a high quality runup data set for waves on a two-dimensional beach, large-scale 2+1 laboratory experiments had been planned at the 30 m-long, 27 m-wide and 60 cm-deep wave basin of CERC. While Briggs et al. (1995) considered propagation of solitary waves with different wave-crest length over a constant depth then a sloping beach, Briggs et al. (1994) considered propagation of solitary wave around a conical island.

Solitary waves were created by a horizontal wave generator with 60 different 45 cm-wide paddles each moving fairly independently –a.k.a. a *snake* generator. By the time the experiments were starting the 1992 Flores tsunami had hit Babi Island, see section 3, and controversy had surrounded the explanation of the extreme runup on the lee side of the island, some arguing that it was reflection off Flores, others that it was unusual amplification due to local bathymetric trapping. To shed light, given that numerical models were still not considered reliable enough to provide satisfactory answers, a laboratory model of a 7.2 m base-diameter conical island was constructed with a slope angle of 14°. The experiments demonstrated that once the wave hits the side of the island across from the generator, the crest splits into two waves (see Fig. 8.2). These two waves move with the crest perpendicular to the shoreline, propagate around the island and collide behind it, in a spectacular demonstration of constructive interference (Yeh et al., 1994). Neither long waves of smaller crest length nor periodic storm waves do so. This appears to be the first laboratory visualization of an observed tsunami catastrophe.

The circular island data measurements were then used to refine shoreline 2+1 algorithms under development. Preliminary modeling results were published by

Yeh et al. (1994), and more comprehensive analyses by Liu et al. (1995), Kânoğlu (1998) and Kânoğlu and Synolakis (1998). Titov (1997) and Titov and Synolakis (1998) comprehensively validated the numerical model MOST using circular island data set and MOST is now used by NOAA for real-time forecasting. The experimental data formed the basis of one of the four benchmark problems used in the 1995 NSF Friday Harbor workshop for inter-model comparison and code validation.

The 12 July 1993 HNO tsunami hit northern Japan and devastated the island of Okushiri west of Hokkaido. While the field measurements of runup and inundation were one of the benchmark models in the 1995 Friday Harbor workshop (Yeh et al., 1996), and, by 1998, the extreme runup observed in Monai had been modeled, it was felt that a large scale laboratory experiment would help better visualize what took place and help better define resolution thresholds for numerical codes. Thus, a 1/400 laboratory model of Monai was constructed in a 205 m-long, 6 m-deep, and 3.5 m-wide tank at Central Research Institute for Electric Power Industry (CRIEPI) in Abiko, Japan. Because of excellent bathymetry data, the experimental basin closely resembles the actual bathymetry. Hence, this data set is another benchmark test for model validation and it was used in the 2004 Catalina Island, California NSF Long-Wave Runup Models workshop, see Liu et al. (2008).

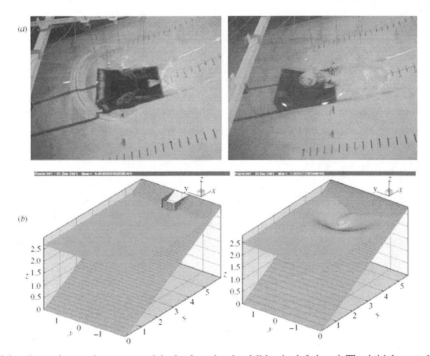

Fig. 8.8 – A moving wedge as a model of submarine landslide. (a, left inset) The initial wave from a submerged slide, immediately after motion started; note the LDN. (a, right inset) The wave about 1 s later, as the runup forms per Raichlen and Synolakis (2003). (b) Animation stills from Liu et al.'s (2005a) simulations, i.e., DNS modeling of the landslide experiments.

While the extreme inundation from landslide waves was brought into frontline focus by the 1998 PNG tsunami (Synolakis et al., 2002a), it is instructive to briefly discuss the history of landslide experiments, at least from the point of view of classical hydrodynamics. Wiegel (1955) first used a wedge-shaped box sliding down a plane beach in the laboratory and established the standard model of a moving block still used in many landslide wave studies. Wiegel only measured the waves propagating offshore. Based on approximate estimates of the characteristics of these waves, he concluded that only about 1% of the energy in the landslide motion is converted to wave energy. Watts (1997) used a similar set up and measured the evolution of waves offshore.

In an effort to better understand the generation and runup of waves from submarine and subaerial slides, Raichlen and Synolakis (2004) conducted large scale experiments at the Oregon State University, Corvallis, Oregon 104 m-long, 3.66 m-wide and 4.47 m-deep wave channel with a plane slope (1:2) located at one end of the tank. Following Wiegel's (1955) experiments, they first used a 91.44 cm-long, 62.25 cm-wide with a 45.72 cm-high vertical face wedge block as in Fig. 8.8 (made of 1.27 cm welded aluminum plate), and then hemi-spherical and rectangular boxes of equal volume. For all experiments, the water depth in the wave tank was 2.44 m. Two configurations of the wedge on the slope were used, one with the front face of the wedge vertical and another with the wedge turned *end-for-end* so that for this orientation the top and front faces were neither horizontal nor vertical. The hemisphere was 91.44 cm in diameter constructed also of 1.27 cm thick aluminum plate.

By varying the weight of the blocks, Raichlen and Synolakis (2004) were able to vary the initial acceleration, while the initial position varied from totally aerial to totally submerged. A photograph of the wedge slide experiment is displayed in Fig. 8.8, showing that a leading positive seaward propagating wave is generated by the sliding wedge after its release and the water surface above the wedge is depressed causing the shoreline to retreat first. The experiments revealed a three-dimensional depression forming over the wedge as motion initiates. For the aspect ratio of their boxes, they observed that wave generation became rapidly inefficient as the submergence of the blocks exceeded one block height. These experiments have since been used as another benchmark test for model validation.

The runup on the slope was measured in two ways. Video images of the leading edge of the runup tongue were recorded and the maximum runup was estimated visually. In addition to the video imaging method, three resistance wave gages were installed along the slope to record the time history of the runup. For the wedge slides, the gage rods were about one millimeter above the slope, while for the hemisphere slides the arrangement were improved so that the top of the gage rods flush with the slope. A sufficient number of wave gages were used to determine the seaward propagating waves, the waves propagating to either side of the sliding bodies, and for the submerged case, the water surface-time history over the body.

Hydrodynamic computations of the free surface motions from landslide waves are very challenging. Not only is the seafloor continuously deforming, but the impact of the solid slide with the water surface, during subarieal generation may trigger local breaking. Breaking can also happen if the slide moves rapidly and it is located very close underneath the surface. Further the wedge geometry has sharp corners which trigger further instabilities even in full DNSs of the parent Navier-

Stokes equations. Hence extreme-nearshore or on-shore landslide-tsunami generation in the laboratory remains a vexing calculation. The results from these experiments defined one of the four benchmark problems in the 2004 NSF Catalina
workshop, and were the reality check and validation exercise of the full DNS simulation of Liu et al. (2005a).

In terms of wave breaking relevant to tsunami hydrodynamics, computational
advances are reviewed in the chapter by P. L.-F. Liu in this volume. It is worthwhile however to reproduce a figure from Petroff (1993) which shows a series of
photographs of a breaking solitary wave on a sloping bed in a companion wave
tank to the one described earlier (Fig. 8.9). C. Petroff and F. Raichlen were at-

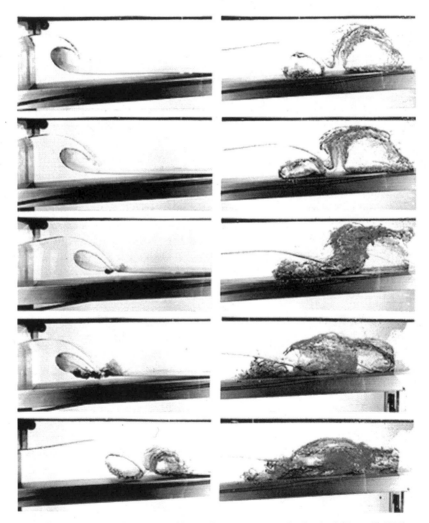

Fig. 8.9 – A series of photographs of a breaking solitary wave on a sloping bed (Petroff, 1993).

tempting to document the details of breaking long waves to provide a laboratory data set for model validation. The dynamics are complex and still quite beyond computational methods, at least in terms of the collapse of the impinging jet ahead of the breaking wave. Mercifully, in most applications of geophysical interest in tsunami hydrodynamics such flows are rare; most 2004 videos in the nearfield suggest spilling breakers or undular bores of the variety that are modeled to first order with depth averaged equations, using shock fitting methods. See also the discussion in Madsen et al. (in review). Nonetheless the Petroff (1993) analysis is a reminder of how much more needs to be understood before claims that computations model the real phenomena correctly can be defensible.

In terms of the current work in deformable slides, the state-of-the-art remains at the experiments of Fritz (2002a, 2002b). Granular subaerial landslide impact generated tsunami waves were investigated in a two-dimensional physical laboratory model at the Swiss Federal Institute of Technology in Switzerland. The granular rockslide impact experiments were conducted in a rectangular prismatic 11 m-long, 0.5 m-wide and 1 m-deep water wave channel with varying depths $h = 0.30$, 0.45 and 0.675 meters. At the front end of the channel a 3 m long hill slope ramp was built into the channel. The landslides were modeled with a granular material (PP-BaSO4) to match rockslide characteristics and provide water quality for laser access. H. Fritz and his collaborators found out that the relevant parameters for free surface wave generation were the granular slide mass m_s, the slide impact velocity v_s, the stillwater depth h and slide thickness s. Three different measurement techniques were built into the physical model: laser distance sensors (LDS), particle image velocimetry (PIV) and capacitance wave gauges (CWG), see Fig. 8.10 (Fritz and Moser, 2003; Fritz et al., 2003a and 2003b). Four wave types were determined: weakly non-linear oscillatory wave, non-linear transition wave, solitary-like wave and dissipative transient bore. Most of the generated impulse waves were located in the intermediate water depth regime. Nevertheless the propagation velocity of the leading wave crest closely followed the theoretical approximations for solitary waves. The slide Froude number ($F = v_s / \sqrt{gh}$) was identified as a dominant parameter. The physical model results were then compared to the giant rockslide generated impulse wave which struck the shores of the Lituya Bay, Alaska in 1958 (Miller, 1960). That slide produced runup in excess of 400m along the steep fjord walls. (This type of geometry was not the kind for which Synolakis (1991) argued that solitary waves were the limiting waveforms in long wave runup.) A cross-section of Gilbert Inlet was rebuilt at small scale and the measured wave runup matched the trimline of forest destruction (Fritz et al., 2001). The Lituya Bay experiments highlighted the formation of an impact crater, as the slide plunged into the fjord, thus increasing the water displacement. The PIV runup measurements revealed a non-breaking short wave propagation and runup of a solitary wave exceeding the breaking limit by a factor of 1.5 due to the narrow Gilbert inlet and the steep headland. The measured wave runup closely matched the empirical Hall and Watts (1953) formulation and the analytical solution of Synolakis (1987). Examples from the results of H. Fritz are shown in Fig. 8.11 through Fig. 8.16.

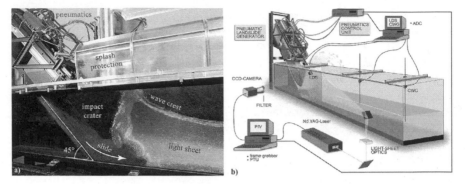

Fig. 8.10 – (a) Landslide impact experiment at $F = 3.3$, $m_s = 108$ kg, $h = 0.45$ m over a slope of 45°; (b) measurement setup with pneumatic installation and the three measurement systems: laser distance sensors, capacitance wave gages and laser based digital particle image velocimetry (PIV). Experimental setup of H. Fritz.

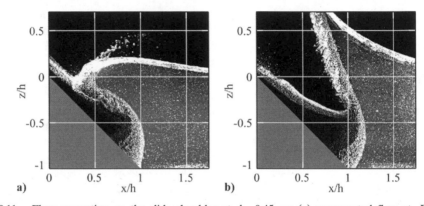

Fig. 8.11 – Flow separation on the slide shoulder at $h = 0.45$ m: (a) unseparated flow at $F = 1.4$, $V = 0.35$, $S = 0.23$ and $t(g/h)^{1/2} = 1.13$; (b) separated flow at $F = 2.6$, $V = 0.35$, $S = 0.21$ and $t(g/h)^{1/2} = 0.79$. Here, for dimensional slide volume V_s, slide width b, slide thickness s and stillwater depth h the dimensionless slide volume and slide thickness are defined as $V = V_s/(bh^2)$ and $S = s/h$ respectively. Results of H. Fritz.

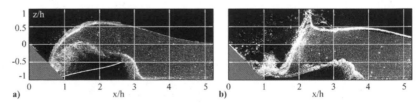

Fig. 8.12 – (a) Backward collapsing impact crater at $F = 2.8$, $V = 0.79$, $S = 0.34$ and $h = 0.3$ m; (b) outward collapsing impact crater at $F = 3.2$, $V = 0.79$, $S = 0.31$ and $h = 0.3$ m. See caption of Fig. 8.10 for the definitions of V and S. Results of H. Fritz.

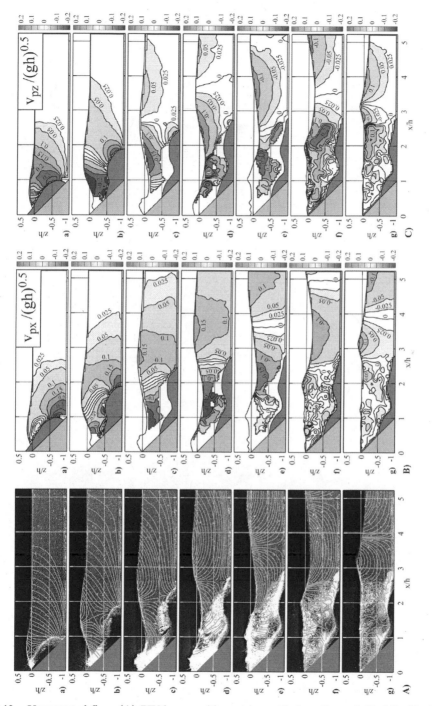

Fig. 8.13 – Unseparated flow: (A) PIV-images with superimposed streamlines at $F = 1.7$, $V = 0.39$, $S = 0.19$ and $h = 0.3$ m and $t(g/h)^{1/2}$ with (a) 0.93, (b) 2.07, (c) 3.22, (d) 4.36, (e) 5.88, (f) 7.41 and (g) 9.7; (B) horizontal particle velocity fields $v_{px}/(gh)^{1/2}$; (C) vertical particle velocity fields $v_{pz}/(gh)^{1/2}$. See caption of Fig. 8.11 for the definitions of V and S. Results of H. Fritz.

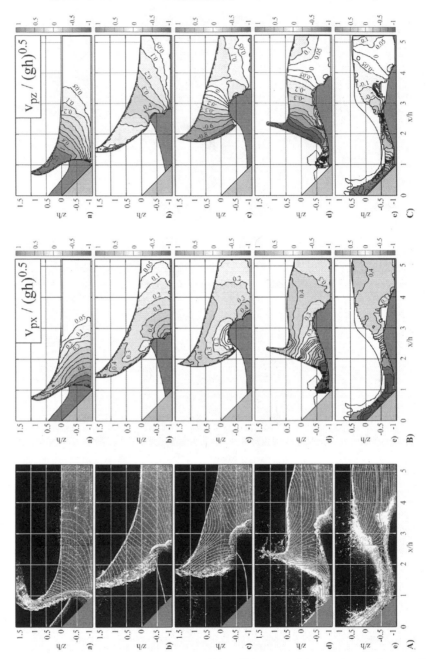

Fig. 8.14 – Outward collapsing impact crater: (A) PIV-images with superimposed streamlines at $F = 3.2$, $V = 0.79$, $S = 0.31$, $h = 0.3$ m and $t(g/h)^{1/2}$ with (a) 0.58, (b) 1.73, (c) 2.49, (d) 3.25 and (e) 4.01; (B) horizontal particle velocity fields $v_{px}/(gh)^{1/2}$; (C) vertical particle velocity fields $v_{pz}/(gh)^{1/2}$. See caption of Fig. 8.11 for the definitions of V and S. Results of H. Fritz.

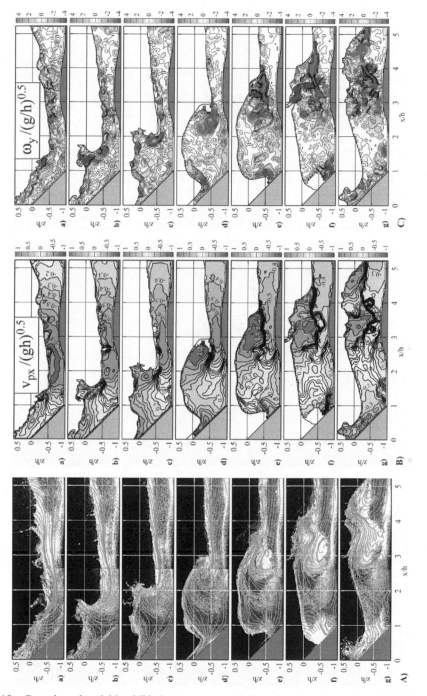

Fig. 8.15 – Granular subaerial landslide impact generated. Secondary bore formation: (A) PIV-images with superimposed streamlines at $F = 4.7$, $V = 0.39$, $S = 0.17$, $h = 0.3$ m and $t(g/h)^{1/2}$ with (a) 6.97, (b) 7.73, (c) 8.49, (d) 9.25, (e) 10.01, (f) 10.78 and (g) 11.51; (B) horizontal particle velocity fields $v_{px}/(gh)^{1/2}$; (C) out-of-plane vorticity fields $\omega_y/(g/h)^{1/2}$. See caption of Fig. 8.11 for the definitions of V and S. Results of H. Fritz.

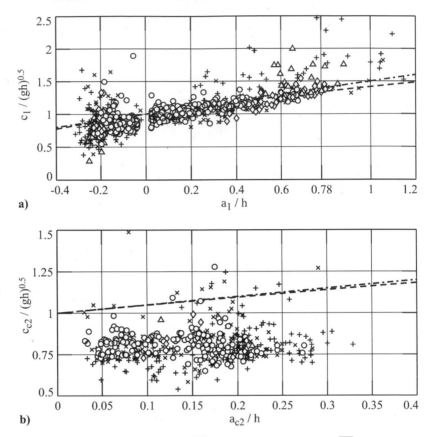

Fig. 8.16 – Wave celerity: (a) first crest $c_{c1}/\sqrt{gh}$ and trough velocities $c_{t1}/\sqrt{gh}$ versus amplitude a/h with (+) first and (×) second gauge, (○) non-breaking wave, (◇) spilling breaker, (□) transient bore, (- -) $c_{c1}/\sqrt{gh} = 1+(a_{c1}/2h)$, (· - ·) $c_{c1}/\sqrt{gh} = \sqrt{1+(a_{c1}/h)}$; (b) second crest velocity $c_{c2}/\sqrt{gh}$, symbols as in (a). After Fritz et al. (2004).

6. Numerical advances leading to inundation mapping and forecasting

Early advances in one dimensional computations are described in Synolakis and Bernard (2006). By the mid 1970s, 1+1 calculations were concentrating on solving the nonlinear dispersive Boussinesq equations, while 2+1 calculations were attempting to solve the SW equations. The latter set the basis for the eventual production of the first tsunami flooding maps, while the former assisted in the development of the shoreline evolution algorithms that are now in use in the inundation models.

Houston and Garcia (1974a, b) in an attempt to develop flooding maps for the Federal Emergency Management Agency (FEMA) calculated the 2+1 deep-ocean propagation of tsunamis from Chile and Alaska evolving towards the West Coast of the US. They used a hybrid method and first solved a linear form of the spherical long wave equations to propagate the tsunami from the source to the edge of the continental shelf, using a finite-difference model; at the continental shelf, they matched the outer amplitudes to an equivalent sinusoid and determined a simple amplification factor to determine wave heights closer to shore. Bernard and Vastano (1977) and Houston (1978) further refined the method to hindcast the tsunami response of the Hawaiian Islands, using steady-state models. By the late 1970s, numerical calculations were capable of solving the nonlinear dispersive equations, i.e., the Boussinesq equations, and calculate the 1+1 wave evolution over depth transitions or the continental shelf (Madsen and Mei, 1969; Goring, 1978).

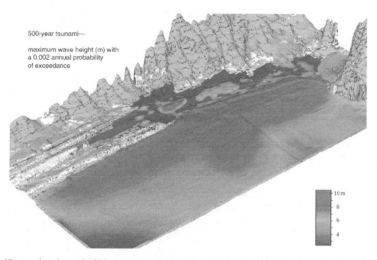

Fig. 8.17 – (See color insert) 500-year tsunami wave heights (m) in the Seaside/Gearhart, Oregon. Tsunami wave heights (m) with 0.2% annual probability of exceedance. Wave heights include the effects of tides. View looks southeastward with vertical exaggeration of 10 times. After Tsunami Pilot Study Working Group (2006).

The Houston and Garcia (1974a, b) works were groundbreaking, not only because they tried to do a transpacific propagation calculation, but also because they attempted to use more geophysically realistic initial conditions, than 1+1 evolution models did. They argued that, at that time of their study, the only reliable data for defining source characteristics were from the 1964 Alaskan and the 1960 Chilean earthquakes. They thus inferred an initial ground deformation by a 1000 km long hypothetical uplift mass of ellipsoidal shape, with a 10 m maximum vertical uplift. They then divided the Aleutian trench into 12 segments and calculated the wave evolution from each segment, and repeated the procedure for tsunamis from the Peru–Chile trench, and convolved their results with tidal cycles. They had thus attempted the first ever probabilistic tsunami hazard analysis, a feat not since repeated until 2006 (Tsunami Pilot Study Working Group, 2006), see Fig. 8.17 (color insert). While Houston and Garcia's (1974a, b) predictions for the offshore wave height have been repeatedly interpreted as runup –recall that, until recently, there was little differentiation in the emergency management community between runup and tsunami height– and their predictions were later inferred to differ with detailed runup computations, in some cases, by a factor of two (e.g. Synolakis et al., 1997a, b), the scope of their work was pioneering.

We note here the work of Hibberd and Peregrine (1979) who proposed an entirely new shoreline algorithm to calculate the evolution of strong bores over a sloping beach and then computed runup using the NSW equations. The determination of the shoreline evolution was the missing link towards connecting tsunami heights estimates offshore with inundation. Typically, even in 1+1 models, the computations would stop at some threshold offshore depth, not only because there was no methodology to calculate the moving shoreline motion, but also to avoid having to consider wave breaking, for which anyway criteria didn't exist, see for example Piatanesi et al. (1996). While entirely ad hoc, speculative and not computationally robust, the real utility of Hibberd and Peregrine (1979) recipe was the demonstration that the runup calculation was possible, and became the basis or inspiration of many subsequent models. To wit, Kim et al. (1983) solved a 1+1 potential flow equation and Pedersen and Gjevik (1983) a 1+1 Boussinesq equation in Lagrangian coordinates. Both studies compared their predictions with laboratory data as a reality check, with the latter study correctly differentiating between breaking and nonbreaking waves when validating their results. Zelt (1986) solved a 2+1 Boussinesq equation and calculated the runup in a parabolic basin, again checking his results with laboratory measurements.

The concurrent analytical results and laboratory data of Synolakis (1986, 1987) allowed for the cross-check of the shoreline algorithm predictions with runup data derived from laboratory, analytical calculations and other numerical solutions, hence there was enhanced confidence in the newly discovered runup calculations. The latter are still missing even in current 2+1 propagation models owing to the perceived complexity when inundation models are applied over physical bathymetries. While regrettable, since some studies have suggested that runup inferences from threshold models may differ from models with shoreline algorithms by factors up to two, and sometimes even more, it is understandable. Real shorelines can present significant challenges to structured grid models (for which the shoreline algorithm works), and until the advent of the finite volume methods, computations with unstructured grids were not a viable option. Most laboratory-benchmarked algorithms, for now, use structured grids.

Pre-1990s, 2+1 models of tsunamis of geophysical scale did not include runup modeling, hence hazard zone identification was based on qualitative inferences from estimates of tsunami heights offshore. The importance of nearfield and landslide sources had not been recognized; the paradigm threat remained a teletsunami attacking Hawaii or California. The calculations performed at very coarse computational grid coverage and with no realization of the importance of the onland bathymetry in inundation estimates. By comparison, post-1990 models did include runup modeling and, when required, hazard zone identification based on calculated inundation distances and onland currents. Also, beginning in 1992 with the Nicaraguan tsunami, the hazard from nearshore sources to the proximal coastlines was rediscovered –before then only Japan and Chile had been focusing on the nearfield. Post the 1998 PNG tsunami, submarine landslides were also rediscovered –see section 3 on field surveys. In the same period, the large scale laboratory experiments of Briggs et al. (1994, 1995) led Kânoğlu and Synolakis (1998) to show the importance of the nearshore bathymetry on extreme inundation. Since then, significant effort has been spent on matching topographic and bathymetric data to the same reference system and producing high resolution grids for computations. This effort has led to more meaningful inundation studies, see for example, Borrero et al. (2004) and Legg et al. (2004).

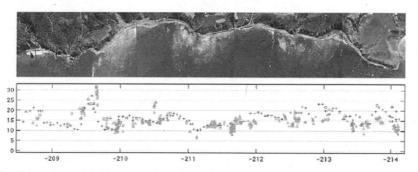

Fig. 8.18 – Comparison of field measurements from the Okushiri 1993 tsunami with measurements from (top) aerial photos and (bottom) model inundation predictions using MOST. Different symbols represent numerical modeling with different resolutions –for details see Titov and Synolakis (1997). To predict the extreme runup of 33 m successfully, a resolution of 5 m was necessary.

The tsunami generated by the 12 July 1993 HNO $M_w = 7.8$ earthquake caused the worst local tsunami-relate death toll in fifty years in Japan, with estimated 10–18 m/s overland flow velocities and 30 m runup. These extreme values are the largest recorded in Japan at this century and are among the highest ever documented for non-landslide generated tsunamis. The modeling of the event by Titov and Synolakis (1997) confirmed the estimated overland flow velocities, inundation including extreme runup (Fig. 8.18), current velocities and overland flow. There were two main factors for the extraordinary comparison with field data using current state-of-the-art shallow water-wave models: one, reasonable ground deformation data was inferred; two, quality of bathymetric/topographic data was available to resolve small features. The results of Titov and Synolakis (1997) qualitatively suggest that –for this event– coastal inundation is more correlated with inundation velocities than with inundation heights, explaining also why threshold-type modeling has substantially underpredicted coastal inundation for this and other recent events.

While more details for advances in conventional hydrodynamic codes are discussed by P. L.-F. Liu (this volume, chapter 9), it is interesting to examine advances in modeling extreme waves, as for example generated by meteorite impacts. One of the key advances in understanding earth's geologic history was the hypothesis of L. Alvarez in 1980 that the mass extinction benchmarking the Cretaceous-Tertiary boundary (K/T) –geologic history boundary was the result of massive meteoroid impact (Alvarez et al., 1980). The hypothesis was later challenged by Hansen (1988), but nonetheless it remains one of the great science controversies of all times, particularly because of the provocative evidence of

Bourgeois et al. (1988) of sediment deposits in Texas associated with the generated tsunami. While the characterization *tsunamis* is as appropriate to bolide-triggered waves as it is for landslide tsunamis, we will describe computational advances that are applicable to both.

Early work by Hills et al. (1994) and Ward and Asphaug (2000) indicated that these meteorite impact waves can be a substantial threat for surrounding coastlines. On the other hand Melosh (2003) proposed that impact-generated tsunami waves are an overrated hazard. Sparks (2008) has re-iterated that volcanic eruptions with or without tsunamigenesis remain a far more common natural hazard. Neither point of view had been supported by a comprehensive modeling of the entire problem, i.e., of impact cratering, propagation and inundation.

Weiss et al. (2006) were the first to simulate the influence of a meteorite impact on the ocean and the oceanic crust, and evaluate the evolution of waves using a Boussinesq-type propagation model. The inundation was computed using the MOST model. The impact model was iSALE, a code validated and verified with shock wave experiments and analytical solutions (Weiss, 2005). iSALE is by some accounts the state-of-the-art methodology for providing initial conditions for tsunami modeling because it can handle three different material in a single grid cell with *sophisticated* algorithms to track phase boundaries (Weiss et al., 2006). Fig. 8.19 (left) illustrates the computational domains of the different models Weiss et al. (2006) used and Fig. 8.19 (right) gives the runup distribution along shoreline for three different distances from the impact center to the still-water shoreline. This example is based on an impact of a 800 m-diameter meteorite in 5000 m-deep water and shows runup values over a beach slope of 1:25 in the 1000, 2000 and 3000 km distances from the impact area. Results are of the same order as measured in the aftermath of the 2004 Boxing Day tsunami in Indonesia, Thailand, India and Sri Lanka (Synolakis and Kong, 2006).

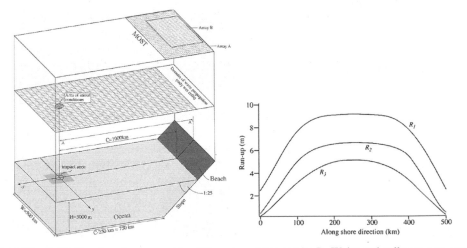

Fig. 8.19 – (left) The domains covered by different models used by R. Weiss and colleagues to compute tsunamigenesis from bollide impacts. (right) Runup values in the shoreward direction over a beach slope of 1:25 for three different distances of meteoroid impacts off the coast, R_1 = 1000 km, R_2 = 2000 km and R_3 = 3000 km.

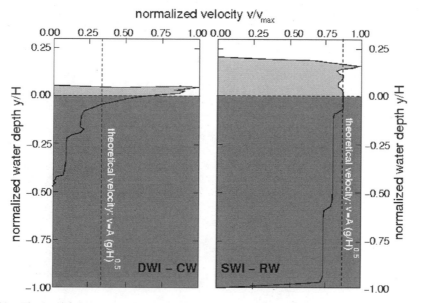

Fig. 8.20 – The particle/water velocity for (left) deep water and (right) shallow water impacts. H is the total water depth. After Wünnemann et al. (2007).

A surprising observation was made by Wünnemann et al. (2007) by studying the generation processes of tsunami waves caused by oceanic meteorite impact in greater detail. While large waves are always generated for impacts in deep water, the water particle velocity substantially varies with the depth (Fig. 8.20). This means the impact water waves cannot be considered as tsunami waves in the classical sense of earthquake-generated tsunamis. It remains to be seen whether the farfield decay of such waves is proportional to $1/r$, as the classical theory of fields suggests, or $1/r^n$, with $2 < n < 4$ as some recent unpublished results imply. Future challenges for impact-generated tsunami waves remain the quantification of wave dissipation. Fig. 8.21 highlights the generated velocity field in deep and shallow water and figures on the left are suggestive of turbulence.

As pointed out in section 2 of this chapter, one challenging task is understanding the seafloor-free ocean surface interaction better. This issue along with the effect of overlying sediments on tsunamigenicity was discussed during the 1997 National Science Foundation supported workshop in Santa Monica, California (Synolakis, 1997a). At that time the standard practice in hydrodynamic modeling was to transfer seafloor displacements into the ocean surface and ignore overlying sediments. To critically examine it, Dutykh et al. (2006) and Kervella et al. (2007) considered two cases: one, passive generation where the seafloor displacement is transferred to the ocean surface, as per the standard practice; two, active generation where the dynamics of the seafloor displacement is taken into account and the seafloor is coupled to the ocean surface through the model –see Dutykh and Dias (2008a). These studies showed differences between passive and active generation, which suggest differences in the nearfield predictions of tsunami evolution. Dias (personal communication) is now examining further the changes the active generation may introduce in the farfield, if any. Dias and Dutykh (2008b) have also ex-

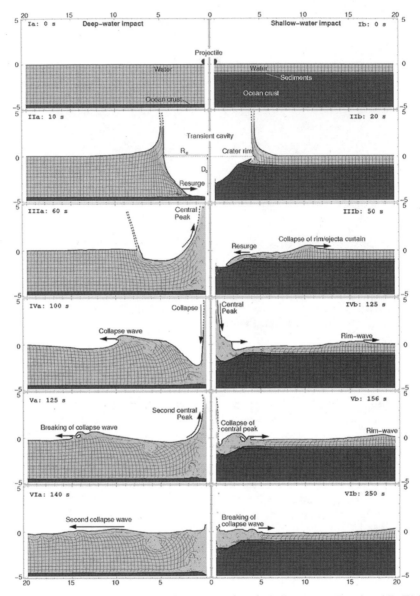

Fig. 8.21 – Impact of a meteoroid into (left) deep and (right) shallow water. Results of R. Weiss and K. Wünnemann.

amined the effects of sediment layers in amplifying locally the seabed motion. Their preliminary results suggest that sediment layers may amplify –under certain resonant conditions– the ground motion and the resulting free surface waves may be larger than otherwise expected.

While more conventional advances in numerical modeling are discussed in this volume by P. L.-F. Liu (chapter 9), it is useful to note that the code MOST used in inundation mapping and forecast in the US has been validated through extensive series of benchmark tests by Titov and Synolakis (1995, 1998). Validation remains an ongoing process. A first generation of inundation maps based mostly on validated 2+1 models now exists for the US and Japan and in some locations in South America, while even fewer exist in the Indian Ocean (Borrero et al., 2006). A high-resolution probabilistic tsunami hazard map is now available for Seaside, Oregon, integrating nearfield and farfield hazards and showing the geographical distribution of different levels of flooding with associated probabilities, including high impact zones (Tsunami Pilot Study Working Group, 2006). The latter are based on estimates of the normalized momentum flux, and account for the fact that the flow depth is not the only predictor of damage. The impact zones reflect the variation of simple Froude number type parameter $V^2/(gD)$, where V is the magnitude of the overland flow velocity and D is the overland flow depth.

7. Standards for tsunami numerical modeling

It is hopefully by now clear that tsunami inundation modeling has evolved in the last two decades through careful and explicit validation and verification, through comparisons of computed predictions with analytical solutions, laboratory experiments and field measurements. We are now at a point in time, where at least one validated and verified code (MOST) is becoming web-based. The model MOST with its interface ComMIT (Community Model Interface for Tsunami) is accessible to professionals who get trained in UNESCO sponsored classes; approximately 70 professionals from 17 Indian Ocean countries have been trained in inundation modeling with ComMIT between June 2006 and January 2008. ComMIT allows for the construction of the deep water tsunami wave field through combination of unit sources and the calculation of inundation in user supplied grids.

It is a matter of time that tsunami computations with variants of ComMIT or other tools will be performed by less trained individuals either locally or through paid companies, or through government agencies. A few *free* codes are distributed already without licensing and with no training requirements. The level of virtual reality possible with computer graphics often creates the false impression that quantitative predictions are also physically realistic. High-end or lesser animations are helpful for training and visualization, but do not improve the realism or credibility of underlying computations. There is need for vigilance in ensuring that codes used for inundation mapping or operational forecasts meet certain minimum criteria developed over the past two decades by the pre-2004 tsunami scientific community. Clearly additional criteria will evolve as the field progresses.

Numerical codes used for inundation mapping have to first be validated with analytical solutions and laboratory measurements, then verified with field data. The models need to be published in the scientific literature in peer-review journals indexed by ISI with impact factors greater than 0.5. Numerical procedures planned for tsunami forecasting by warning centers have to be verified –in addition to the verification process for inundation codes– with the existing and future tsunameter recordings, as they become available. Codes that are used for *operational* forecasting and inundation modeling have to be formally evaluated, and then formally approved before being transferred to operations. (Here, *formally* refers to the review by some national or international body.) It is emphasized again that model testing must remain a continuous process. If not done so as an event unfolds, operational models need to be tested after every tsunami event is recorded, and the results reported no matter what their veracity is, to assist the community in further improvement of models, and identification of show-stoppers, if any. While this process may appear onerous, it reflects our state of knowledge as of December 2007, and is the only defensible methodology when human lives are at stake.

While there is in principle no absolute certainty that a numerical code that has performed well in available benchmark tests will also produce realistic inundation predictions with any assigned source motions, validated codes reduce the level of uncertainty in their results to the uncertainty in the geophysical initial conditions. Further, when coupled with real-time free-field tsunami measurements from tsunameters, validated and verified codes are the only choice for realistic forecasting of inundation; the consequences of failure are too ghastly to take chances with less-validated numerical procedures. Experimenting with unvalidated/unverified codes is reprehensible when it comes to warning center operations.

Specific steps are recommended here for the approval of modeling tools, their further development and their transfer to operations. These steps can be classified into five categories: basic considerations, analytical benchmarking, laboratory benchmarking, field data benchmarking and scientific evaluations. Additional steps need to be considered for models used for real time forecasts (Synolakis et al., 2007); the latter are not just research exercises.

1. Basic considerations: A most basic step to ensure that a numerical model works satisfactorily for predicting evolution, before even checking its inundation results, is to check if the model conserves mass and if it converges to a known analytical solution as the grid resolution increases and time-steps reduced. While the conservation of mass equation is part of the equations of motion that are solved in any numerical procedure, cumulative numerical approximations can sometimes produce predictions that violate mass conservation. This is particularly the case when friction factors are used, or smoothing to stabilize inundation computations when breaking manifests itself. As the computational step size is reduced, the numerical predictions should be seen to converge to a certain value, and further reductions in step sizes should not change the results. The optimal locations to check convergence are the extreme runup and rundown locations.

2. Analytical benchmarking: Validation with analytical results ensures that the numerical code performs well over a wide range of parameters and solves the parent equations accurately. Analytical solutions are exact. Even when they involve approximations, these approximations are quantifiable, after all the equations of motion used in tsunami modeling themselves involve depth-averaging which is itself an approximation. Over two decades, analytic solutions of the SW equations of motion have been validated themselves with laboratory data and have demonstrated a remarkable and surprising capability to quantitatively model complex evolution phenomena adequately, in particular the maximum runup and maximum inundation. The maximum runup is arguably the single most important parameter in the design of coastal structures and for evaluating the inundation potential of tsunamis. Certain numerical models have been validated with both the analytical solutions and the laboratory measurements and verified with the field measurements, thus setting a golden standard for numerical code validation and verification exists.

Exact solutions of the SW equations are useful for validating the complex numerical models which are used for final design and which often involve ad-hoc assumptions, particularly during inundation computations when grid points are introduced and withdrawn during the runup process on what was dry land. Comparisons of numerical predictions with analytical solutions can identify systematic errors, as when using friction factors or dissipative terms to augment the idealized equations of motion or stabilize what is inherently an unstable computation. It is important to note that validation should take place with non-periodic waves. During runup, individual monochromatic waves reflect with phase-shifts that are seafloor slope dependent. Whereas a code may model a periodic wave well, it may not model wave-superposition equally well, as when a spectrum of waves attacks a beach. Real tsunamis can be thought of in the frequency domain as spectra. Early SW formulations did not account for reflection and while some model predictions for the Carrier–Greenspan sinusoids were satisfactory, they exhibited significant errors when modeling solitary waves or N-waves.

Furthermore, the analytical solutions are useful in evaluating the performance of numerical codes at scales of geophysical interest. This is in contrast to laboratory experiments that require modeling at small-scale, at best 1 in 100. Well-engineered numerical codes should allow for modeling of waves ranging from the laboratory to prototype scale, hence exact solutions are valuable. Exact solutions are expressed in dimensionless terms and are valid over all scales, as long as the underlying assumptions of the parent equations are not violated.

No modeling of any real tsunami should ever be undertaken with any code, before comparing its predictions with the available benchmark solutions. Therefore, 1+1 versions of the 2+1 models should first be tested with 1+1 directional analytical models; specifically Synolakis (1986, 1987) solitary wave over a canonical bathymetry; Tadepalli and Synolakis (1994) N-wave over a canonical bathymetry; IVP solutions of the NSW equations presented by Carrier et al. (2003), Kânoğlu (2004) and Kânoğlu and Synolakis (2006) for shoreline runup/rundown motion and velocity. 1+1 models that perform well with the single wave on simple beach must still be tested with the composite beach geometry, for which analytical solution exist (Kânoğlu and Synolakis, 1998), with solitary waves as inputs. Inundation computations are exceedingly difficult when the beach is deforming, as for example when a landslide is occurring. Numerical predictions of the runup from an idealized landslide of translating Gaussian shaped mass should be tested against existing analytical model (Liu et al., 2003) and should not differ more than 20%.

3. Laboratory benchmarking: It is quite clear from the earlier discussion that numerical methods have evolved through careful validation accomplished through comparisons with analytic solutions and laboratory measurements. Long before the availability of numerical codes, physical models at small scale have long been used to visualize wave phenomena in the laboratory and then predictions were scaled to the prototype. Examples include the Army Corps of Engineers (ACE) model for the Sacramento Delta and San Francisco Bay located in Sausalito and the Ports of Los Angeles (POLA), Ports of Long Beach (POLB) physical model (Fig. 8.22) in ACE's Vicksburg, Mississippi laboratories (Briggs, personal communication). The first model was designed to check numerical results of mixing of effluent plumes under tidal forcing. The POLA/POLB model was built to facilitate the design of breakwaters and pier expansion projects and for checking harbor resonance. Both Bay Delta and San Pedro basin models are deformed, in the sense that the horizontal scale differs from the vertical scale so that the model fits in the physical space where the model is located. In the POLA/POLB physical model, the vertical scale is 1:100 and the horizontal is 1:400 so as to reproduce at small scale the entire harbor area and the shoreline from Point Fermin to Huntington Beach in Southern California.

Fig. 8.22 – The physical model of Ports of Los Angeles (POLA) and Ports of Long Beach (POLB) in ACE's Vicksburg, Mississippi laboratories (Briggs, personal communication).

Scale models, in general, do not have similar bottom friction characteristics as real ocean floors or sandy beaches, but this has not turned out a severe limitation. Tsunamis are such long waves, that bottom friction tends to be less important than the inertia of the motion. Friction may be important in cases of extreme inundation, as observed in Banda Aceh with 3 km inundation distances. However, it has been observed that even with numerical codes that use friction factors, the predictions are not sensitive to first order to the choice of friction factors, within of course reasonable limits. 1+1 versions of the 2+1 models should first be tested with 1+1 directional laboratory models (Synolakis, 1986) given the smaller number of laboratory measurements on 2+1 wave basins. The solitary wave experiments on the canonical model of waves propagating over a constant-depth region and running up a 1:19.85 sloping beach should be used first (Synolakis, 1986, 1987). Numerical models should calculate the maximum runup of nonbreaking solitary waves within 5% of the measured values in the laboratory. For breaking waves, the models should produce predictions within 10% of the measured values, and they should consistently predict the runup variation shown in Fig. 8.6 and time histories of surface elevation given in Synolakis (1986).

1+1 models that perform well with the solitary wave experiments must still be tested with the composite beach geometry, for which both analytical and laboratory data exist, with solitary waves as inputs (Kânoğlu and Synolakis, 1998). This additional test will ensure that the code is stable enough for large waves which are near the breaking limit offshore. Numerical predictions should not differ by more than 5% from the experimental values for non-breaking waves, and the numerical procedure should be capable of predicting the entire runup variation.

2+1 dimensional calculations should be tested with the conical island geometry of 1:4 slope (Liu et al., 1995; Kânoğlu, 1998; Kânoğlu and Synolakis, 1998). It is important to ensure that the numerical procedure models adequately the two wave fronts that split in front of the island, and collide behind it, while remaining stable. Predictions of the runup on the back of the island where the two fronts collide should not differ by more than 20% from the laboratory measurements.

2+1 numerical computations should then be tested with the laboratory model of Monai Valley, Okushiri Island, Japan (Takahashi, 1996; Liu et al., 2008). The initial condition is a leading depression N-wave, and the entire simulation allows the identification of how well the code performs in a rapid sequence of withdrawal and runup. Comparison of results from different codes has shown that the maximum runup in these experiments can be calculated within 10%, which is thus the standard.

As discussed, landslide wave generation remains the frontier in terms of numerical modeling, particularly for aerial slides. These involve not only the rapid change of the seafloor, but also the impact of the slide on the shoreline. Therefore models that will be used to model landslide generated tsunamis need to be tested against three-dimensional landslide experiments given in Liu et al. (2005a).

4. Field data benchmarking: Verification of a model in a real-world setting is important, especially for operational models. No analytical or laboratory data comparisons (or any limited number of tests, for that matter) can assure robust model performance in operational environments. Test comparisons with real-world data provide an additional important step in the verification of a model to perform well during operational implementation. The 17 November 2003 Rat Islands tsunami has provided the most comprehensive test for forecast methodology and forecasting was performed successfully. Yet, testing with every new event is required to ensure that the inundation forecast will work for every tsunami likely to occur. Nevertheless, this first test indicated that the tsunami forecast methodology of Titov et al. (2005a) works and useful tools could be developed and implemented, as they already have (Wei et al., 2008). The data also provide one benchmark for testing operational models. Such real-world simulations are extremely important since they provide test of a model performance in operational setting.

For operational codes, testing should use the tsunameter signal of the 17 November 2003 Rat Islands tsunami to scale an estimate of the earthquake source, then provide an estimate of the inundation in Hilo, Hawaii. This is the most difficult but most realistic test for any operational model, for it involves a forecast (when performed later it becomes a hindcast) and needs to be done much faster than real time. Here, at least first four waves have to be simulated, with accuracy of at least 75% of amplitude and period, with arrival time error of not more than 3 minutes. Details are given in Synolakis et al. (2007).

An excellent benchmark is prediction of the field measurements from the HNO tsunami around Okushiri Island (Takahashi, 1996). The initial condition known as DCRC-17a is a composite fault with three segments and at least when used with MOST has been shown to model inundation measurements within 10%. Predictions of any operational model for the maximum runup at Aonae should not differ by more than 20% from the actual measurements.

5. Scientific Evaluation: Model validation and verification is a continuing process. Any model used for inundation mapping or operational forecasts needs to be presented in peer-reviewed scientific journals with impact factors greater than 0.5. The publications need to include comparisons of the model predictions with all the above benchmarks. A formal evaluation process of individual models needs to be established to avoid ad-hoc decisions as to the suitability of any given model. This process may include solicitation of additional reviews of the model's veracity by experts, or the requirement that additional testing be performed. This process will set the standard for best available practice at any given time, and it will hopefully eliminate the liability to code developing institution, states, engineers and geophysicists who collaborate on the development of inundation maps.

6. Formal operational evaluation: While the evaluation process may identify models that are realistic and computationally correct, some of them may not be sufficiently versatile for inundation mapping or operational applications, as is often the case with university based research codes. An additional approval process is helpful to assess operational factors of the model, such as special implementation hardware/software issues, ease of use, computation time, etc. If a model is not approved at this phase, it is recommended that the evaluation advises what additional steps or improvement will be helpful for its eventual approval.

8. Conclusions

To understand tsunami currents, forces, runup on coastal structures and inundation of coastlines, the evolution of tsunamis from their source region to their target must be calculated numerically. Here, the developments in tsunami hydrodynamic have been discussed in the context of field observations, analytical, laboratory and numerical advances. Steps for validating and verifying computational tools for predicting the coastal effect of tsunami are presented.

Overall, substantial progress had been achieved in 50 years of tsunami science and engineering and, while tens of models have been developed and published with varying success, less than three are currently in worldwide use. Real time forecasts are now possible using the state-of-the-art numerical tools. The tsunami inundation and forecasting model MOST has evolved in the last two decades through careful and explicit validation and verification through comparisons of its predictions with 1+1 and 2+1 analytical solutions, laboratory experiments and field measurements.

Tsunami inundation forecasts integrating seismic and tsunameter recordings in real time and leading to cancellation of warnings have been attempted successfully (Bernard et al., 2006). Based on a single reading off a NOAA tsunameter in the North Pacific for the 17 November 2003 tsunami, Titov et al. (2005a) scaled their precomputed scenario event that most closely matched the parent earthquake to evaluate the leading wave height off Hilo, Hawaii. They then proceeded with a real-time NSW computation that resulted in the first tsunami forecast and a warning was enabled.

Since then the same real time forecast took place during the 3 May 2006 Tonga, 15 November 2006 and 13 January 2007 Kuril Islands, 1 April 2007 Solomon Islands and 15 August 2007 Peru tsunamis for farfield impact. In all cases, the tsunami inundation was within 10% of the forecast. Their comparison of the forecast to the measured wave is the new golden standard for operational models. This success was only possible because MOST had been extensively through all these validation and verification steps; it developed from a propagation code in the later 1980s, to an inundation code in the 1990s, to an operational forecast code in the 2000s.

In reviewing advances for this volume and elsewhere, it became again clear that if the tsunami community appeared at first perplexed in the aftermath of the Boxing Day tsunami, it was not due to the failure of hydrodynamic paradigms, much as certain goephysical paradigms failed, but because of the unprecedented loss of life, the worst possible surprise. The wide availability of tsunameter recordings and satellite altimeter data in the next decade will further assist in improving existing models and, quite possibly new ones, more computationally efficient and convergent to higher-order approximations to the Navier-Stokes equations. Tsunami science is maturing to where operational forecasts will be relied upon for making timely evacuation decisions even for nearfield events.

As Synolakis and Bernard (2006) and Synolakis and Kong (2006) emphasized, more comprehensive educational efforts on tsunami hazard mitigation are urgently necessary. During the 26 November 1999 Vanuatu tsunami the town Baie Martelli was completely destroyed. Despite this destruction, only 5 out of the 300 people living in Baie Martelli lost their lives. The small number of casualties was due to an educational video regarding the 1998 PNG event villagers had seen a few months before the event. They sent a lookout. Once the lookout reported that the water was receding, the villagers concluded that a tsunami attack was imminent and ran to a nearby hill (Caminade et al., 2000). All International Tsunami Survey Teams should spend substantial time during their field work in public outreach and education as discussed in Synolakis and Okal (2005), Fritz et al. (2007), Fritz and Kalligeris (2008), Fritz et al. (2008), Fig. 8.23. If anything, the animal kingdom provides another strong incentive; see Foley et al. (2008) who concluded that knowledge is the dominant natural selection force in vertebrate populations.

It is the best of times to be studying tsunamis.

Fig. 8.23 – C. Synolakis educating the children in a small village during the field survey of 26 November 1999 Vanuatu tsunami.

Acknowledgments

We thank Drs. H. Fritz and R. Weiss in providing the high resolution figures. This writing was partially supported by the 'Tsunami Risk ANd Strategies For the European Region (TRANSFER)' grant of the European Union to the Institute of Applied and Computational Mathematics of the Foundation of Research and Technology of Hellas (FORTH) and Middle East Technical University (METU). We also thank Nick Kalligeris for his assistance with editing and improving the manuscript, not to mention his enthusiasm and patience.

References

Alvarez L. W., W. Alvarez, F. Assaro and H. V. Michel, 1980. Extraterrestrial cause for the cretaceous - tertiary extinction- experimental results and theoretical interpretation. *Science,* **208**(4448), 1095–1108.

Ambraseys, N. and C. E. Synolakis, in review. Tsunami catalogues for the Eastern Mediterranean, revisited. *Geophys. J. Int.*

Atwater, B. F., S. Musimi, K. Satake, Y. Tsuji, K. Ueda and D. K. Yamaguchi, 2005. *The orphan tsunami of 1700.* US Geological Survey Professional Paper 1707. Washington, DC.

Bardet, J.-P., C. E. Synolakis, H. L. Davies, F. Imamura and E. A. Okal, 2003a. Landslide tsunamis: recent findings and research directions. *Pure Appl. Geophys.,* **160**(10–11), 1793–1809. (doi:10.1007/s00024-003-2406-0)

Bardet, J.-P., C. E. Synolakis, H. L. Davies, F. Imamura and E. A. Okal (Eds), 2003b. Landslide tsunamis: recent findings and research directions. *Pure Appl. Geophys.,* **160**(10–11), Birkhäuser Basel, 435 pp. (ISBN: 978-3-7643-6033-7)

Bernard, E. N. and A. C. Vastano, 1977. Numerical computation of tsunami response for island systems. *J. Phys. Oceanogr.,* 7, 389–395. (doi:10.1175/1520-0485(1977)007!0389:NCOTRFO2.0.CO;2)

Bernard, E. N., H. O. Mofjeld, V. Titov, C. E. Synolakis and F. I. González, 2006. Tsunami: scientific frontiers, mitigation, forecasting, and policy implications. *Philos. T. R. Soc. A,* **364,** 1989–2007. (doi:10.1098/rsta.2006.1809)

Borrero, J., M. Ortiz, V. V. Titov and C. E. Synolakis, 1997. Field survey of Mexican tsunami, *EOS, Trans. Amer. Geophys. Un.,* **78**(8), 85 and 87–88. (EOS cover article).

Borrero, J. C., J. F. Dolan and C. E. Synolakis, 2001. Tsunamis within the eastern Santa Barbara Channel. *Geophys. Res. Lett.,* **28,** 643–646. (doi:10.1029/2000GL011980)

Borrero, J. C., 2002. Analysis of tsunami hazards in Southern California. Ph.D. thesis, University of Southern California, Los Angeles, California, 90089-2531, 262 pp.

Borrero, J. C., J. Bu, C. Saiang, B. Uslu, J. Freckman, B. Gomer, E. A. Okal and C. E. Synolakis, 2003. Field survey and preliminary modeling of the Wewak, Papua New Guinea earthquake and tsunami of September 9, 2002. *Seismol. Res. Lett.,* **74,** 393–405.

Borrero, J. C., M. R. Legg and C. E. Synolakis, 2004. Tsunami sources in the southern California Bight. *Geophys. Res. Lett.,* **31**(13), 643–646. (doi:10.1029/2000GL011980)

Borrero, J. C., K. Sieh, M. Chlieh and C. E. Synolakis, 2006. Tsunami inundation modeling for western Sumatra. *P. Natl. Acad. Sci. USA,* **103**(52), 19673-19677. (doi:10.1073/pnas.0604069103)

Bourgeois, J., T. A. Hansen, P. L. Wiberg and E. G. Kauffmann, 1988. A tsunami deposit on the Cretaceous–Tertiary boundary in Texas. *Science,* **241,** p. 567–570.

Briggs, M. J., C. E. Synolakis, G. S. Harkins and D. Green, 1994. Laboratory experiments of tsunami runup on a circular island. *Pure Appl. Geophys.,* **144**(3–4), 569–593. (doi:10.1007/BF00874384)

Briggs, M. J., C. E. Synolakis, G. S. Harkins and S. A. Hughes, 1995. Large scale three dimensional experiments of tsunami inundation. In *Tsunami: Progress in Prediction, Disaster Prevention and Warning* (Eds. Y. Tsuchiya and N. Shuto), pp. 129–149. Kluwer, Boston, MA.

Briggs, M. J., C. E. Synolakis, U. Kânoğlu and D. R. Green, 1996. Benchmark problem 3 Runup of solitary waves on a vertical wall. *Long-wave Runup Models* (Eds. H. Yeh, P. L.-F. Liu and C. E. Synolakis).World Scientific Publishing, Singapore, 375–383.

Brocchini, M. and D. H. Peregrine, 1996. Integral flow properties in the swash zone and averaging. *J. Fluid Mech.,* **317,** 241–273. (doi:10.1017/S0022112096000742)

Brocchini, M., 1998. The run-up of weakly-two-dimensional solitary pulses. *Nonlinear Proc. Geoph.,* **5,** 27–38.

Bruins, H. J., J. A. MacGillivray, C. E. Synolakis, C. Benjamini, J. Keller, H. J. Kisch, A. Klügel and J. van der Plicht, 2007. Geoarchaeological tsunami deposits at Palaikastro (Crete) and the Late Minoan IA eruption of Santorini. *J. Archeological Science,* in press. (doi:10.1016/j.jas.2007.08.017)

Caminade, P., D. Charlie, D., U. Kânoğlu, S.-I. Koshimura, H. Matsutomi, A. Moore, C. Ruscher, C. Synolakis and T. Takahashi, 2000. Vanuatu earthquake and tsunami cause much damage, few casualties. *EOS, Trans. Amer. Geophys. Un.,* **81**(52), 641 and 646–647. (EOS cover article).

Carrier, G. F. and H. P. Greenspan, 1958. Water waves of finite amplitude on a sloping beach. *J. Fluid Mech.,* **17,** 97–110. (doi:10.1017/S0022112058000331)

Carrier, G. F., 1966. Gravity waves of water of variable depth. *J. Fluid Mech.,* **24,** 641–659. (doi:10.1017/S0022112066000892)

Carrier, G. F. and C. F. Noiseux, 1983. The reflection of obliquely incident tsunamis. *J. Fluid Mech.,* **133,** 147–160. (doi:10.1017/S0022112083001834)

Carrier, G. F., T. T. Wu and H. Yeh, 2003. Tsunami runup and drawdown on a plane beach. *J. Fluid Mech.,* **475,** 79–99.

Carrier, G. F. and H. Yeh, 2005. Tsunami propagation from a finite source. *Comp. Mod. Eng. Sci.,* **10**(2), 113–121.

Dalrymple, R. A. and D. L. Kriebel, 2005. Lessons in engineering from the tsunami in Thailand. *The Bridge, Proc. Natl. Acad. Eng.,* **35,** 4–13.

Danielsen F., M. K. Sorensen, M. F. Olwig, V. Selvam, F. Parish, N. D. Burgess, T. Hiraishi, V. M. Karunagaran, M. S. Rasmussen, L. B. Hansen, A. Quarto and N. Suryadiputra, 2005. The Asian tsunami: a protective role for coastal vegetation. *Science,* **310**(5748), 643–643. (doi:10.1126/science.1118387)

Dutykh, D., F. Dias and Y. Kervella, 2006. Linear theory of wave generation by a moving bottom. *C. R. Acad. Sci. Paris, Ser. I,* **343,** 499–504.

Dutykh, D. and F. Dias, 2008a. Tsunami generation by dynamic displacement of sea bed due to dip-slip faulting. *Math. Comput. Simulat.,* accepted.

Dutykh, D. and F. Dias, 2008b. Influence of sedimentary layering on tsunami generation. *Comput. Method. Appl. M.,* in press.

Earthquake Spectra, 2006. *The special issue on the Great Sumatra earthquakes and Indian Ocean tsunamis of 26 December 2004 and 28 March 2005.* (Eds. D. Ballantyne, C. B. Crouse, L. Dengler, M. Greene, K. Hudnut, W. Iwan, H. Kanamori, C. Synolakis, K. Tierney and L. Wyllie), **22**(S3), 569–845.

Ewing, L., 2008. *Sea level rise: Major implications to coastal engineering and coastal management.* Handbook of Coastal and Ocean Engineering. In press.

Fernando, H. J. S., J. L. McCulley, S. G. Mendis and K. Perera, 2005. Coral poaching worsens tsunami destruction in Sri Lanka. *EOS, Trans. Amer. Geophys. Un.,* **86,** 301–304.

Folley, C., N. Petrorreli and L. Folley, 2008. Severe drought and calf survival in elephants. *Biol. Lett.-UK.* (doi:10.1098/rsbl.2008.0370)

Fritz, H. M., W. H. Hager and H.-E. Minor, 2001. Lituya Bay case: rockslide impact and wave run-up. *Science of Tsunami Hazards,* **19**(1), 3–22.

Fritz, H. M., 2002a. PIV applied to landslide generated impulse waves. In: *Laser techniques for fluid mechanics* (Eds. Adrian, R. J. et al.), 305–320. Springer, New York, Berlin, Heidelberg.

Fritz, H. M., 2002b. Initial phase of landslide generated impulse waves. In *VAW Mitteilung 178.* Versuchsanstalt für Wasserbau (Ed. H.-E. Minor), Hydrologie und Glaziologie, ETH Zürich.

Fritz, H. M. and P. Moser, 2003. Pneumatic landslide generator. *Int. J. Fluid Power,* **4**(1), 49–57.

Fritz, H. M., W. H. Hager and H.-E. Minor, 2003a. Landslide generated impulse waves, part 1: instantaneous flow fields. *Exp. Fluids,* **35,** 505–519.

Fritz, H. M., W. H. Hager, H.-E. Minor, 2003b. Landslide generated impulse waves, part 2: hydrodynamic impact craters. *Exp. Fluids,* **35,** 520–532.

Fritz, H. M., W. H. Hager and H.-E. Minor, 2004. Near field characteristics of landslide generated impulse waves. *J. Waterw. Port C-ASCE,* **130**(6), 288–302. (doi: 10.1061/(ASCE)0733–950X(2004)130:6(287))

Fritz, H. M., J. C. Borrero, C. E. Synolakis and J. Yoo, 2006a. 2004 Indian Ocean tsunami flow velocity measurements from survivor videos. *Geophys. Res. Lett.,* **33**(24), Art. No. L24605. (doi:10.1029/2006GL026784)

Fritz, H. M., C. E. Synolakis and B. G. McAdoo, 2006b. Maldives survey of the 2004 Indian Ocean tsunami. *Earthq. Spectra,* **22**(S3), 137–154.

Fritz, H. M., W. Kongko, A. Moore, B. McAdoo, J. Goff, C. Harbitz, B. Uslu, N. Kalligeris, D. Suteja, K. Kalsum, V. Titov, A. Gusman, H. Latifet, E. Santoso, S. Sujoko, D. Djulkarnaen, H. Sunendar and C. E. Synolakis, 2007. Extreme runup from the 17 July 2006 Java tsunami. *Geophys. Res. Lett.,* **34,** L12602. (doi:10.1029/2007GL029404)

Fritz H. M. and N. Kalligeris, 2008. Ancestral heritage saves tribes during 1 April 2007 Solomon Islands tsunami. *Geophys. Res. Lett.,* **35,** L01607. (doi:10.1029/2007GL031654)

Fritz, H. M., N. Kalligeris, J. C. Borrero, P. Broncano and E. Ortega, 2008. The 15 August 2007 Peru tsunami runup observations and modeling. *Geophys. Res. Lett.,* **35,** L10604. (doi:10.1029/2008GL033494)

Fujima, K., 1996. Application of linear theory to the computation of runup of solitary wave on a conical island. In *Long-wave Runup Models* (Eds. H. Yeh, P. L.-F. Liu and C. E. Synolakis). Singapore, World Scientific Publishing.

Gardner, C. S., J. M. Green, M. D. Kruskal and R. M. Miura, 1967. Method for solving the KdV equation. *Phys. Rev. Lett.,* **19,** 1095–1097. (doi:10.1103/PhysRevLett.19.1095)

Geist, E., 1998. Local tsunamis and earthquake source parameters. *Adv. Geophys.*, **39**, 117–209.

Geist, E. L., S. L. Bilek, D. Arcas and V. V. Titov, 2006a. Differences in tsunami generation between the December 26, 2004 and March 28, 2005 Sumatra earthquakes. *Earth Planets Space*, **58**, 185–193.

Geist, E. L., V. V. Titov and C. E. Synolakis, 2006b. Tsunami: wave of change. *Sci. Am.*, **294**(1), 56–63.

Goring, D. G., 1978. Tsunamis—the propagation of long waves onto a shelf. Report no. Kh-R-38, W. M. Keck Laboratory of Hydraulics and Water Resources, California Institute of Technology, Pasadena, California.

Grilli, S. T., I. A. Svenden and R. Subrayama, 1997. Breaking criterion and characteristics of solitary waves on a slope. *J. Waterw. Port C-ASCE*, **123**(2), 102–112.

Gutenberg, B., 1939. Tsunamis and earthquakes. *Bull. Seismol. Soc. Am.*, **29**, 517–526.

Hall J. V. and J. W. Watts, 1953. Laboratory investigation of the vertical rise of solitary waves on impermeable slopes, Tech. Memo. 33, Beach Erosion Board, USACE, 14pp.

Hammack, J. L., 1972. Tsunamis—a model for their generation and propagation. Report no. Kh-R-28, W. M. Keck Laboratory of Hydraulics and Water Resources, California Institute of Technology, Pasadena, California.

Hansen, T. A., 1988. Early Tertiary radiation of marine molluscs and the long-term effects of the Cretaceous-Tertiary extinction. *Paleobiology*, **14**, 37–51.

Hibberd, S. and D. H. Peregrine, 1979. Surf and run-up on a beach: a uniform bore. *J. Fluid Mech.*, **95**, 323–345. (doi:10.1017/S002211207900149X)

Hills J. G., I. V. Nemchinov, S. P. Popov and A. V. Teterev, 1994. Tsunami generated by small asteroid impact. In *Hazards from Comets and Asteroids* (Eds. T. Gehrels and A. Z. Tucson) University of Arizona Press, pp. 779–789.

Houston, J. R., 1978. Interaction of tsunamis with the Hawaiian Islands. *J. Phys. Oceanogr.*, **9**, 93–102. (doi:10.1175/1520-0485(1978)008!0093:IOTWTHO2.0.CO;2)

Houston, J. R. and A. W. Garcia, 1974a. Type 16 flood insurance study. USACE WES Report H-74-3.

Houston, J. R. and A. W. Garcia, 1974b. Type 19 flood insurance study. USACE WES Report HL-80-18.

Kanamori, H., 1972. Mechanisms of tsunami earthquakes. *Phys. Earth Planet.*, **6**, 346–359. (doi:10.1016/0031-9201(72)90058-1)

Kawata, Y., B. Benson, J. C. Borrero, J. L. Borrero, H. L. Davies, W. P. de Lange, F. Imamura and H. Letz, 1999. Tsunami in Papua New Guinea was as intense as first thought. *EOS, Trans. Amer. Geophys. Un.*, **80**(9), 101 and 104–105.

Kânoğlu, U. and C. E. Synolakis, 1996. Analytic solutions of solitary wave runup on the conical island and on the Revere Beach. In *Long-wave Runup Models* (Eds. H. Yeh, P. L.-F. Liu and C. E. Synolakis). World Scientific Publishing, Singapore.

Kânoğlu, U., 1998. The runup of long waves around piecewise linear bathymetries. Ph.D. Thesis, University of Southern California, Los Angeles, California, 90089-2531, 273 pp.

Kânoğlu, U. and C. E. Synolakis, 1998. Long wave runup on piecewise linear topographies. *J. Fluid Mech.*, **374**, 1–28. (doi:10.1017/S0022112098002468)

Kânoğlu, U., 2004. Nonlinear evolution and runup–rundown of long waves over a sloping beach. *J. Fluid Mech.*, **513**, 363–372. (doi:10.1017/S002211200400970X)

Kânoğlu, U. and C. E. Synolakis, 2006. Initial value problem solution of nonlinear shallow water-wave equations. *Phys. Rev. Lett.*, **97**(14), 148501. (doi: 10.1103/PhysRevLett.97.148501)

Keller, J. B. and H. B. Keller, 1964. Water wave run-up on a beach. ONR Research Report NONR-3828(00), Department of the Navy, Washington, DC, p. 40.

Kerr, R., 2005. Model shows islands muted tsunami. *Science*, **308**, 341. (doi:10.1126/science.308.5720.341a)

Kervella Y., D. Dutykh and F. Dias, 2007. Comparison between three-dimensional linear and nonlinear tsunami generation models. *Theor. Comput. Fluid Dyn.*, **21**, 245–269. (doi: 10.1007/s00162-007 -0047-0)

Kim, S. K., P. L.-F Liu and J. A. Liggett, 1983. Boundary integral equation solutions for solitary wave generation, propagation and run-up. *Coastal Engrg.*, **7**(4), 299–317.

Knowles, J. K., 1966. Note on elastic surface waves. *J. Geophys. Res.*, **71**(22), 5480–5481.

Lautenbacher, C. C., 1979. Gravity wave refraction by island. *J. Fluid Mech.*, **41**, 655–672. (doi:10.1017/S0022112070000824)

Legg, M. R., J. C. Borrero and C. E. Synolakis, 2004. Tsunami Hazards Associated with the Catalina Fault in Southern California. *Earthq. Spectra*, **20**, 1–34.

Li, Y. and F. Raichlen, 2000. Energy balance model for breaking solitary wave runup. *J. Waterw Port C-ASCE*, **129**(2), 47–49. (doi:10.1061/(ASCE)0733-950X(2003)129:2(47))

Li, Y. and F. Raichlen, 2001. Solitary wave runup on plane slopes. *J. Waterw. Port C-ASCE*, **127**(1), 33–44. (doi:10.1061/(ASCE)0733-950X(2001)127:1(33))

Li, Y. and F. Raichlen, 2002. Non-breaking and breaking solitary run-up. *J. Fluid Mech.*, **456**, 295–318. (doi:10.1017/S0022112001007625)

Liu, P. L.-F., C. E. Synolakis and H. H. Yeh, 1991. Report on the international workshop on long wave runup. *J. Fluid Mech.*, **229**, 675–688. (doi:10.1017/S0022112091003221)

Liu, P. L.-F., Y.-S. Cho, M. J. Briggs, U. Kanoglu and C. E. Synolakis, 1995. Runup of solitary waves on a circular island. *J. Fluid Mech.*, **320**, 259–285. (doi:10.1017/S0022112095004095)

Liu, P. L.-F., P. Lynett and C. E. Synolakis, 2003. Analytical solutions for forced long waves on a sloping beach. *J. Fluid Mech.*, **478**, 101–109. (doi:10.1017/S0022112002003385)

Liu, P. L.-F., T.-R. Wu, F. Raichlen, C. E. Synolakis and J. Borrero, 2005a. Runup and rundown generated by three-dimensional sliding masses. *J. Fluid Mech.*, **536**, 107–144. (doi:10.1017 /S0022112005004799)

Liu, P. L.-F., P. Lynett, H. Fernando, B. E. Jaffe, H. Fritz, B. Higman, R. Morton, J. Goff and C. E. Synolakis, 2005b. Observations by the International Tsunami Survey Team in Sri Lanka. *Science*, **308**, 1595. (doi:10.1126/science.1110730)

Liu, P. L.-F., H. Yeh and C. Synolakis (Eds.), 2008. Advanced Numerical Models for Simulating Tsunami Waves and Runup. *Advances in Coastal and Ocean Engineering*, **10**, 250 pp.

Love, A. E. H., 1911. *Some Problems in Geodynamics.* Dover, New York, NY.

Lynett, P. J., J. C. Borrero, P. L.-F. Liu and C. E. Synolakis, 2003. Field survey and numerical simulations: a review of the 1998 Papua New Guinea earthquake and tsunami. *Pure Appl. Geophys.*, **160**, 2119–2146. (doi:10.1007/s00024-003-2422-0)

Madsen, O. S. and C. C. Mei, 1969. The transformation of a solitary wave over an uneven bottom. *J. Fluid Mech.*, **39**, 781–791. (doi:10.1017/S0022112069002461)

Madsen, P. A., D. R. Fuhrman and H. Shaffer, in review. On the solitary wave paradigm for tsunamis. *J. Geophys. Res.-Oceans.*

Marinatos, S., 1939. The Volcanic Destruction of Minoan Crete. *Antiquity* **13**, 425–439.

Mei, C. C., 1983. *The Applied Dynamics of Ocean Surface Waves.* Wiley, New York, NY.

Melosh, H. J., 2003. Impact-generated tsunamis: An over-rated hazard (abstract #2013). 34th Lunar and Planetary Science Conference. CD-ROM.

Miller, D. J., 1960. Giant waves in Lituya Bay, Alaska. Shorter Contributions to General Geology, Geological Survey Professional Paper 354-C. United States Government Printing Office, Washington.

Morrison, D., 2006. Asteroid and comet impacts: the ultimate environmental catastrophe. *Phil. Trans. R. Soc. A*, **364**, 2041–2054. (doi:10.1098/rsta.2006.1812)

Okal, E. A., 1982. Mode-wave equivalence and other asymptotic problems in tsunami theory. *Phys. Earth Planet.*, **30,** 1–11. (doi:10.1016/0031-9201(82)90123-6)

Okal, E. A., 1988. Seismic parameters controlling far-field tsunami amplitudes. A review, *Natural Hazards,* **1,** 67–96. (doi:10.1007/BF00168222)

Okal, E. A., 1993. Seismology-Predicting large tsunamis. *Nature,* **361**(6414), 686–687. (doi:10 .1038/361686a0)

Okal, E. A., 2003. *T* waves from the 1998 Papua New Guinea earthquake and its aftershocks: Timing the tsunamigenic slump. *Pure Appl. Geophys.,* **160,** 1843–1863.

Okal, E. A., G. J. Fryer, J. C. Borrero and C. Ruscher, 2002. The landslide and local tsunami of 13 September 1999 on Fatu Hiva (Marquesas islands; French Polynesia). *B. Soc. Géol. Fr.,* **173**(4), 359–367. (doi:10.2113/173.4.359)

Okal, E. A., G. Plafker and C. E. Synolakis, 2003. Near-Field Survey of the 1946 Aleutian Tsunami on Unimak and Sanak Islands. *Bull. Seismol. Soc. Am.,* **93**(3), 1226–1234. (doi:10.1785/0120020198)

Okal, E. A. and C. E. Synolakis, 2004. Source discriminants for nearfield tsunamis. *Geophys. J. Int.,* **158,** 899–912. (doi:10.1111/j.1365-246X.2004.02347.x)

Okal, E. A. and C.E. Synolakis, 2008. Far-field tsunami risk from mega-thrust earthquakes in the Indian Ocean. *Geophys. J. Intl.,* **172**(3), 995–1015. (doi:10.1111/j.1365-246X.2007.03674.x)

Pararas-Carayannis, G. 1972. A study of the source mechanism of the Alaska earthquake and tsunami of March 27, 1964. In *Seismology and Geodesy on the Great Alaska Earthquake of 1964*, National Academy of Sciences, Washington D.C., pp 249–258.

Pedersen, G. and B. Gjevik, 1983. Runup of solitary waves. *J. Fluid Mech.,* **135,** 283–390. (doi:10.1017/S0022112083003080)

Peregrine, D. H., 1966. Calculations of the development of an undular bore. *J. Fluid Mech.,* **25,** 321–330. (doi:10.1017/S0022112066001678)

Peregrine, D. H., 1967. Long waves on a beach. *J. Fluid Mech.,* **27,** 815–827. (doi:10.1017 /S0022112067002605)

Petroff, C., 1993. The interaction of breaking solitary waves with an armored bed. Ph.D. thesis, California Institute of Technology, Pasadena, California 91125.

Piatanesi, A., S. Tinti and I. Gavagni, 1996. The slip distribution of the 1992 Nicaragua earthquake from tsunami runup data. *Geophys. Res. Lett.,* **23,** 37–40.

Plafker, G., 1965. Tectonic deformation associated with the 1964 Alaska earthquake. *Science,* **148,** 1675–1687.

Plafker, G., R. Kachadoorian, E. B. Eckel and L. R. Mayo, 1969. Effects of the earthquake of March 27, 1964, on various communities. U.S. Geol. Surv. Prof. Paper 542-G, 50 p.

Raichlen, F., 1966. Harbor resonance. In *Coastline and estuarine hydrodynamics* (Ed. A. T. Ippen), pp. 281–340. McGraw-Hill, New York, NY.

Raichlen, F. and C. E. Synolakis, 2004. Runup from three dimensional sliding masses. Long Waves Symposium, Thessaloniki, Greece (Eds. M. Briggs and Ch. Koutitas), pp. 247–256.

Ramsden, J. D. and F. Raichlen, 1990. Forces on vertical wall caused by incident bores. *J. Waterw. Port C-ASCE* **116,** 592–613. (doi:10.1061/(ASCE)0733-950X(1990)116:5(592))

Russel, J. S., 1845. Report on Waves. Rp. Meet. Brit. Assoc. Adv. Sci. 14th, 311–390, John Murray, London.

Satake, K., K. Shimazaki, Y. Tsuji and K. Ueda, 1996. Time and size of a giant earthquake in Cascadia inferred from Japanese tsunami. *Nature,* **379,** 203–204. (doi:10.1038/379246a0)

Satake, K., 2007. Tsunamis, Treatise on Geophysics (ed. G. Schubert), vol. 4. Elsevier.

Shen, M. C. and R. E. Meyer, 1963. Climb of a bore on a beach. Part 3. Runup. *J. Fluid Mech.,* **16,** 113–125. (doi:10.1017/S0022112063000628)

Shuto, N., 2003. Tsunamis of seismic origin, in *Submarine Landslides and Tsunamis* (Eds. A. C. Yalciner, E. Pelinovsky, C. E. Synolakis and E. Okal). Kluwer Academic Publishers. Printed in Netherlands.

Smith, W. H. F., R. Scharroo, V. V. Titov, D. Arcas and B. K. Arbic, 2005. Satellite Altimeters Measure Tsunami. *Oceanography,* **18**(2), 10–12.

Sparks, S., 2008. Volcanoes and their impact on human society. 33rd International Geological Congress, Oslo, Norway.

Synolakis, C. E., 1986. The runup of long waves. Ph.D. thesis, California Institute of Technology, Pasadena, California 91125, p. 228.

Synolakis, C. E., 1987. The runup of solitary waves. *J. Fluid Mech.,* **185**, 523–545. (doi:10.1017 /S002211208700329X)

Synolakis, C. E., 1988. Long wave generation in the laboratory. *J. Waterw. Port C-ASCE,* **116**(2) 252–266.

Synolakis, C. E., 1991. Green's law and the evolution of solitary waves. *Phys. Fluids A-Fluid,* **3**(3), 490–491. (doi:10.1063/1.858107)

Synolakis, C. E., P. L.-F. Liu, G. Carrier and H. Yeh, 1997a. Tsunamigenic sea-floor deformations. *Science,* **278**, 598–600. (doi:10.1126/science.278.5338.598)

Synolakis, C. E., R. McCarthy and E. N. Bernard, 1997b. Evaluating the tsunami risk in California. In *Proc. ASCE California and the World Ocean '97,* San Diego, California, pp. 1225–1236.

Synolakis, C. E., J. P. Bardet, J. C. Borrero, H. Davies, E. A. Okal, E. A. Silver, S. Sweet and D. R. Tappin, 2002a. Slump origin of the 1998 Papua New Guinea tsunami. *Proc. R. Soc. A,* **458**, 763–789. (doi:10.1098/rspa.2001.0915)

Synolakis, C. E., A. Yalciner, J. Borrero and G. Plafker, 2002b. Modeling of the November 3, 1994, Skagway, Alaska tsunami. Proc. of Solutions to Coastal Disasters 2002, ASCE, 915–927.

Synolakis, C. E. and E. A. Okal, 2005. 1992–2002: perspective on a decade of post tsunami surveys. *Adv. Nat. Technol. Hazards,* **23**, 1–30. (doi:10.1007/1-4020-3331-1_1)

Synolakis, C. E., E. A. Okal and E. N. Bernard, 2005. The megatsunami of December 26, 2004. *The Bridge,* **35**(2), 26–35.

Synolakis, C. E. and E. N. Bernard, 2006. Tsunami science before and after Boxing Day 2004. *Phil. Trans. R. Soc. A,* **364**, 2231–2265. (doi:10.1098/rsta .2006.1824)

Synolakis, C. E. and L. Kong, 2006. Runup measurements of the December 26 2004 Indian Ocean tsunami. *Earthq. Spectra,* **22**(S3), 67–91.

Synolakis, C. E., E. N. Bernard, V. V. Titov, U. Kânoğlu and F. González, 2007. Standards, criteria, and procedures for NOAA evaluation of tsunami numerical models. NOAA OAR Special Report, Contribution No 3053, NOAA/OAR/PMEL, Seattle, WA, 55 pp.

Tadepalli, S. and C. E. Synolakis, 1994. The runup of N-waves on sloping beaches. *P. R. Soc. A,* **445**, 99–112.

Tadepalli, S. and C. E. Synolakis, 1996. Model for the leading waves of tsunamis. *Phys. Rev. Lett.,* **77**, 2141–2144. (doi:10.1103/PhysRevLett.77.2141)

Takahashi, T., 1996. Benchmark problem 4; the 1993 Okushiri tsunami—Data, conditions and phenomena. In *Long-Wave Runup Models*, (Eds. H. Yeh, P. L.-F. Liu and C. E. Synolakis). World Scientific, Singapore, 384–403.

Tanioka, Y. and F. I. Gonzalez, 1998. The Aleutian earthquake of June 10, 1996 (Mw 7.9) ruptured parts of both the Andreanof and Delarof segments. *Geophys. Res. Lett.,* **25**(12), 2245–2248. (doi:10.1029/98GL01578)

Tinti, S. and R. Tonini, 2005. Analytical evolution of tsunamis induced by nearshore earthquakes on a constant-slope ocean. *J. Fluid Mech.,* **535**, 33–64. (doi:10.1017/S0022112005004532)

Titov, V. V. and C. E. Synolakis, 1995. Modeling of breaking and non-breaking long-wave evolution and runup using VTCS-2. *J. Waterw. Port C-ASCE,* **121,** 308–316. (doi:10.1061/(ASCE)0733-950X(1995)121:6(308))

Titov, V. V., 1997. Numerical modeling of long wave runup. Ph.D. Thesis, University of Southern California, Los Angeles, California, 90089-2531, 141 pp.

Titov, V.V. and C. E. Synolakis, 1997. Extreme inundation flows during the Hokkaido–Nansei–Oki tsunami. *Geophys. Res. Lett.,* **24**(11), 1315–1318. (doi:10.1029/97GL01128)

Titov, V. V. and C. E. Synolakis, 1998. Numerical modeling of tidal wave runup. *J. Waterway Port Ocean Coast. Eng.* **124,** 157–171. (doi:10.1061/(ASCE)0733-950X(1998)124:4(157))

Titov, V. V., F. I. Gonzalez, E. M. Bernard, M. C. Eble, H. O. Mofjeld, J. C. Newman and A. J. Venturato, 2005a. Real-time tsunami forecasting: challenges and solutions. *Nat. Hazards* **35,** 45–58. (doi:10.1007/s11069-004-2403-3)

Titov, V. V., A. B. Rabinovich, H. O. Mofjeld, R. E. Thomson and F. I. Gonzalez, 2005b. The global reach of the 26 December 2004 Sumatra tsunami. *Science,* **309,** 2045–2048. (doi:10.1126/science.1114576)

Tsunami Pilot Study Working Group, 2006. Seaside, Oregon Tsunami Pilot Study-Modernization of FEMA flood hazard maps. NOAA OAR Special Report, Contribution No 2975, NOAA/OAR/PMEL, Seattle, WA, 83 pp. + 7 appendices.

Tuck, E. O. and L. S. Hwang, 1972. Long wave generation on a sloping beach. *J. Fluid Mech.,* **51,** 449–461. (doi:10.1017/S0022112072002289)

Uslu, B., J. C. Borrero, L. A. Dengler and C. E. Synolakis, 2007. Tsunami inundation at Crescent City generated by earthquakes along the Cascadia Subduction Zone. *Geophys. Res. Lett.,* **34**(20), L20601. (doi:10.1029/2007GL030188).

Ward, S. N., 1980. Relation of tsunami generation and an earthquake source. *J. Phys. Earth,* **28,** 441–474.

Ward S. N. and E. Asphaug, 2000. Asteroid impact tsunami: A probabilistic hazard assessment. *Icarus,* **145,** 64–78. (doi:10.1006/icar.1999.6336)

Ward, S. N., 2001. Landslide tsunami. *J. Geophys. Res.,* **106,** 11201–11216. (doi:10.1029/2000JB900450)

Ward, S. N. and S. Day, 2001. Cumbre Vieja Volcano—potential collapse and tsunami at La Palma, Canary Islands. *Geophys. Res. Lett.,* **28,** 397–400. (doi:10.1029/2001GL013110)

Ward, S. N. and S. Day, 2003. Ritter Island Volcano—lateral collapse and the tsunami of 1888. *Geophys. Res. Lett.,* **154,** 891–902. (doi:10.1046/j.1365-246X.2003.02016.x)

Ward, S. N., 2004. Earthquake simulation by restricted random walks. *Bull. Seismol. Soc. Am.,* **94**(6), 2079–2089.

Watts, P., 1997. Water waves generated by underwater landslides. PhD thesis, California Institute of Technology, Pasadena, California.

Wei Y., E. N. Bernard, L. Tang, R. Weiss, V. V. Titov, C. Moore, M. Spillane, M. Hopkins and U. Kânoğlu, 2008. Real-time Experimental Forecast of the Peruvian tsunami of August 2007 for U.S. Coastlines. *Geophys. Res. Lett.,* **35,** L04609. (doi:10.1029/2007GL032250)

Weiss, R., 2005. Modeling of generation, propagation and run-up of tsunami waves caused by oceanic impacts. Ph.D. thesis. Geologisch-Paläontologsiches Institut und Museum, WWU Münster, p. 96.

Weiss R., K. Wünnemann and H. Bahlburg, 2006. Numerical modelling of generation, propagation and run-up of tsunamis caused by oceanic impacts: model strategy and technical solutions. *Geophys. J. Int.,* **167,** 77–88. (doi:10.1111/j.1365-246X.2006.02889.x)

Whitham, G. B., 1958. On the propagation of shock waves through regions of nonuniform area of flow. *J. Fluid Mech.,* **4,** 337–360. (doi:10.1017/S0022112058000495)

Wiegel, R. L., 1955. Laboratory studies of gravity waves generated by the movement of a submerged body. *Trans. AGU,* **36,** 759–774.

Wünnemann, K., R. Weiss and K. Hofmann, 2007. Characteristics of oceanic impact-induced large water waves – Re-evaluation of the tsunami hazard. *Meteorit. Planet. Sci.*,**42**(11), 1–11.

Yalçiner, A. C., E. N. Pelinovsky, E. Okal and C. E. Synolakis (Eds), 2003. *Submarine Landslides and Tsunamis*. NATO Science Series, 21, Kluwer Academic Publishers, 328 pp.

Yeh, H., F. Imamura, C. Synolakis, Y. Tsuji, P. Liu and S. Shi, 1993. The Flores Island tsunamis, *EOS, Trans. Amer. Geophys. Un.*, **74,** 369 and 371–373.

Yeh, H., P. L.-F. Liu, M. Briggs and C. E. Synolakis, 1994. Tsunami catastrophe in Babi Island. *Nature,* **372,** 6503–6508.

Yeh, H., V. Titov, V. Gusiakov, E. Pelinovsky, V. Khramushin and V. Kaistrenko, 1995. The 1994 Shikotan Earthquake Tsunamis. *Pure Appl. Geophys.* **144**(3–4), 855–874. (doi:10.1007/BF00874398)

Yeh, H., P. L.-F. Liu and C. E. Synolakis (Eds), 1996. *Long-wave Runup Models*. World Scientific Publishing, Singapore, pp. 403.

Zelt, J.A., 1986. *Tsunamis. The Response of Harbors with Sloping Boundaries to Long Wave Excitation,* California Institute Technology, Pasadena, CA, W. M. Keck Laboratory of Hydraulics and Water Resources, Rept. No. KH-R-47, 318 pp.

Zelt, J. A., 1991. The run-up of breaking and nonbreaking solitary waves. *Coast. Eng.,* **15**(3), 205–246. (doi:10.1016/0378-3839(91)90003-Y)

Zelt, J. A. and F. Raichlen, 1991. Overland Flow from Solitary Waves. *J. Waterw. Port C-ASCE,* **117**(3), 247–263. (doi:10.1061/(ASCE)0733-950X(1991)117:3(247))

Chapter 9.   TSUNAMI MODELING: PROPAGATION

PHILIP L.-F. LIU

Cornell University

Contents

1.   Overview

Successful simulation of tsunami propagation and accurate prediction of the arrival time and wave height at different locations rely on a correct estimate of the fault plane mechanism. The interplate fault in subduction zones is responsible for most of the large tsunamis in history. For interplate fault ruptures, the resulting seafloor displacement can be estimated approximately using the linear elastic dislocation theory (*e.g.,* Mansinha and Smylie 1971, Okada 1985). Several parameters defining the geometry and strength of the fault rupture need to be specified. The mean fault slip (final displacement after the rupture), D, is calculated from the seismic moment, M_0, where $M_0 = \mu D A$ with A being the rupture area and μ the rigidity of the earth at the source. The seismic moment can be determined from the seismic data recorded worldwide and is usually reported as the Harvard CMT (Centroid Moment Tensor) solution. The rupture area is estimated from the aftershock data and is usually approximated as a rectangle with length L and width W. The Harvard CMT solution also provides the following fault plane characteristics: the focal depth d, measuring the depth of the upper rim of the fault plane, the dip angle δ (against the horizontal), the strike angle ϕ (against north), and the rake (or slip) angle λ of the dislocation on the fault plane measured from the horizontal axis. Once the seafloor displacement is determined, the initial free surface profile is assumed to take the same configuration, based on the assumptions that the upward

The Sea, Volume 15, edited by Eddie N. Bernard and Allan R. Robinson
ISBN 978–0–674–03173–9 ©2009 by the President and Fellows of Harvard College

seafloor movement is impulsive and seawater is incompressible. For more sophisticated fault models, non-uniform stress-strength fields (i.e., faults with various
kinds of barriers, asperities, etc.; *e.g.,* Kanamori 1977) are expected, so the actual
seafloor displacement may be very complicated compared with the smooth displacement computed from the mean dislocation field on the fault. Although several numerical models have considered geometrically complex faults, complex slip
distributions, and elastic layers of variable thickness, they have not yet been applied in tsunami research (Geist 1998). One reason for this is that our knowledge
of tsunami source parameters, heterogeneity, and non-uniform slip distribution is
too incomplete to justify using such complex models.

In recent years, significant advances have been made in developing analytical
and numerical models to describe the propagation and run-up of a tsunami event
(*e.g.,* Yeh *et al.* 1996, Geist 1998). These models are based primarily on the shallow-water wave equations, which consider the nonlinearity, but ignore the frequency dispersion. The resulting wave system is non-dispersive, in which the wave
celerity (or phase speed) is independent of wave frequency. The typical wavelength of earthquake generated tsunamis depends on the size of the source region
and can range from 20 km to 200 km. If the earthquake occurs in a depth of 4 km,
the measure of frequency dispersion (*i.e.,* the depth to wavelength ratio) is

$$\mu^2 = \left(\frac{h}{L}\right)^2 = 4.0 \times 10^{-4} \sim 4.0 \times 10^{-2}.$$

Therefore, the shallow water (or long wave) approximation is suitable here. The
nonlinearity is also usually not important in the tsunami generation region, because the water depth in most cases is relatively deep in comparison with the wave
amplitude, $A \sim 1.0$ m, *i.e.,*

$$\varepsilon = \frac{A}{h} = 2.5 \times 10^{-4}.$$

As tsunamis propagate across an ocean basin, wave amplitudes should decrease
slightly because of the spreading of the wave energy to a larger physical space.
Using the simple (linear wave theory) frequency dispersion relationship and shoaling formula (Mei 1989), one can estimate the variation of nonlinearity and frequency dispersion as tsunamis propagate from deep ocean to shallow water coastal
region. As shown in Figure 9.1, the nonlinearity becomes increasingly important as
the tsunami moves into continental shelf and coastal zone. However, the frequency
dispersion remains weak, if it is small in the source region, and continuously decreases as water depth becomes shallower. Consequently, as shown in Figure 9.1
for different scenarios, if the tsunamis in the source region can be described by the
linear and nondispersive wave theory (*i.e.,* linear shallow water wave theory), most
likely the linear shallow water model is a reasonable model describing wave
propagation until the tsunami reaches the coastal zone, in which the nonlinear
shallow water (MSW) theory is more adequate. On the other hand, if the source
region is relatively narrow and the frequency dispersion is mildly important, a
linear and weakly dispersive wave theory (linear version of the Boussinesq equa-

tion) is needed (The combination of 3 and b in Figure 9.1). There is a small region on the continental shelf where both frequency dispersion and nonlinearity are weakly important (*i.e.,* Boussinesq equation model). We should point out here that once the nonlinearity becomes important, both super- and sub-harmonics can be generated. Therefore, the frequency dispersion might become significant for the superharmonic wave components in shallow water.

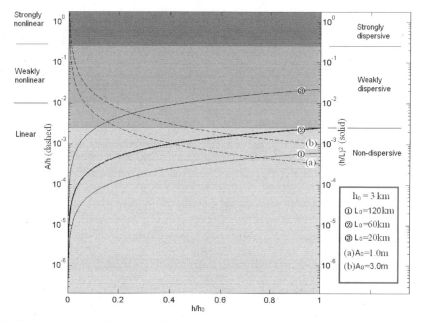

Fig. 9.1 – The magnitudes of frequency dispersion and nonlinearity from deep ocean basin to coast with different initial tsunami characteristics. The dashed lines are for nonlinearity and the solid lines for frequency dispersion.

Moreover, the frequency dispersion can also become important when a tsunami propagates for a long time, $t_d > (h/g)^{1/2}(L/h)^3$, and over a long distance, $x_d > (gh)^{1/2}t_d = L^3/h^2$ (Mei 1989). Obviously, for a tsunami with shorter wavelength (*e.g.,* $L \sim$ 20km), this distance, $x_d \sim 5 \times 10^2$km, can be reached quite easily. Therefore, in modeling the transoceanic tsunami propagation, frequency dispersion might need to be considered. Consequently, a complete model that can describe the entire process of tsunami generation, propagation, and runup needs to consider both frequency dispersion and nonlinearity.

More recently, because of the devastating Papua New Guinea tsunami in 1998, more attentions have been paid to the modeling of landslide generated tsunamis. The duration of a submarine landslide event is usually much longer than that of seismic seafloor displacement. A slide event could last for several to tens of minutes. Hence the time history of the seafloor movement needs to be included in the model. Secondly, the effective size of the landslide region is usually much smaller than the coseismic seafloor deformation zone. Consequently, the typical wavelength of the tsunamis generated by a submarine landslide is also shorter, *i.e.,* about 1 to 10 km. Depending on the water depth at the slide location, the frequency dispersion might be important in the source region.

As a tsunami propagates into the nearshore region, the wave front undergoes a nonlinear transformation while it steepens through shoaling. If the tsunami is large enough, it can break at some offshore depth and approach land as a bore—the white wall of water commonly referenced by survivors of the Indian Ocean tsunami. Wave breaking in traditional NSW tsunami models has not been handled in a satisfactory manner. Numerical dissipation is commonly used to mimic breaking, and thus results become grid dependent. In Boussinesq models, this breaking is still handled in an approximate manner due to the fact that the depth-integrated derivation does not allow for a overturning wave; however these breaking schemes have been validated for a wide range of nearshore conditions. Being depth-integrated, NSW and Boussinesq models lack the capability of simulating the vertical details of many coastal effects, such as strong wave breaking/overturning and the interaction between tsunamis and irregularly shaped coastal structures. To address this deficiency, several 2D and 3D computational models based on Navier-Stokes equations have been developed, with varying degrees of success (*e.g.*, Lin and Liu 1998a,b, Lin *et al.* 1999). The 3D model has also been used to simulate tsunamis generated by landslides (Liu *et al.* 2005). Due to their high computational costs, full 3D models would best be used in conjunction with a depth-integrated 2DH model (*i.e.*, NSW or Boussinesq). While the 2DH model provides incident far-field tsunami information, the 3D model computes local wave-structure interactions. The results from 3D models could also provide a better parameterization of small-scale features (3D), which could then be embedded in a large-scale 2DH model. Existing work in this area of coupling hydrodynamic models is limited (*e.g.*, Fujima *et al.* 2002). While the results appear promising, further work is still needed to improve the accuracy of these hybrid models.

In this chapter, we shall only focus on the general depth-integrated model describing the propagation of tsunamis by a prescribed seafloor displacement. The model is suitable for landslide generated tsunamis and earthquake generated tsunamis. In this general model the only assumption employed is that the frequency dispersion is weak, *i.e.*, the ratio of water depth to wavelength is small or $O(\mu^2) \ll 1$. However, by choosing a proper representative velocity in the governing equations the applicability of these model equations may possibly be extended to a reasonably deep water (or short waves). Moreover, the full nonlinear effect is included in the model, *i.e.*, the ratio of wave amplitude to water depth is of order one or $\varepsilon = O(1)$. In the special case where the seafloor is stationary, the general model reduces to the model for fully nonlinear and weakly dispersive waves propagating over a varying water depth (*e.g.*, Liu 1994, Madsen and Schäffer 1998).

2.   Formulation of Tsunami Propagation Models

Let's define $\xi'(x', y', t')$ as the free surface displacement of a water wave train propagating in water depth $h'(x', y', t')$. Introducing the characteristic water depth h_0 as the vertical length scale, the characteristic length of the tsunami source region ℓ_0 as the horizontal length scale, and $\ell_0 / \sqrt{gh_0}$ as the time scale, and the characteristic wave amplitude a_0 as the scale of wave motion, we can define the following dimensionless variables:

$$(x, y) = (x', y')/\ell_0, \quad z = z'/h_0, \quad t = \sqrt{gh_0} \, t'/\ell_0,$$

$$h = h'/h_0, \quad \zeta = \zeta'/a_0, \quad p = p'/\rho g a_0,$$

$$(u, v) = (u', v')/\left(\varepsilon\sqrt{gh_0}\right), \quad w = w'/\frac{\varepsilon}{\mu}\left[\sqrt{gh_0}\right], \tag{1}$$

in which (u, v) represent the horizontal velocity components, w the vertical velocity component, p the pressure. Two dimensionless parameters have been introduced in (1), which are

$$\varepsilon = a_0/h_0, \quad \mu = h_0/\ell_0. \tag{2}$$

Since the viscous effects are usually not important, the wave motion can be described by the continuity equation and the Euler's equations, *i.e.*,

$$\mu^2 \nabla \cdot \mathbf{u} + w_z = 0, \tag{3}$$

$$\mathbf{u}_t + \varepsilon \mathbf{u} \cdot \nabla \mathbf{u} + \frac{\varepsilon}{\mu^2} w \mathbf{u}_z = -\nabla p, \tag{4}$$

$$\varepsilon w_t + \varepsilon^2 \mathbf{u} \cdot \nabla w + \frac{\varepsilon^2}{\mu^2} w w_z = -\varepsilon p_z - 1, \tag{5}$$

where $\mathbf{u} = (u, v)$ denotes the horizontal velocity vector, $\nabla = (\partial/\partial x, \partial/\partial y)$ the horizontal gradient vector, and the subscript the partial derivative.

On the free surface, $z = \varepsilon\zeta(x, y, t)$ the kinematic and dynamic boundary condition applies:

$$w = \mu^2\left(\zeta + \varepsilon\mathbf{u} \cdot \nabla\zeta\right), \quad \text{on } z = \varepsilon\zeta, \tag{6}$$

$$p = 0, \quad \text{on } z = \varepsilon\zeta. \tag{7}$$

Along the seafloor, $z = -h$, the kinematic boundary condition requires

$$w + \mu^2\mathbf{u} \cdot \nabla h + \frac{\mu^2}{\varepsilon} h_t = 0, \quad \text{on } z = -h. \tag{8}$$

We remark here that the seafloor movement, which could be caused by either an earthquake or a landslide, is prescribed. For later use, we also note here that the depth-integrated continuity equation can be obtained by integrating (3) from $z = -h$ to $z = \varepsilon\zeta$. After applying the boundary conditions (6) and (8), the resulting equation reads

$$\nabla \cdot \left[\int_{-h}^{\varepsilon\zeta} \mathbf{u}\, dz \right] + \frac{1}{\varepsilon} H_t = 0 , \tag{9}$$

where

$$H = \varepsilon\zeta + h . \tag{10}$$

We remark here that (9) is exact.

3. Two-Dimensional Governing Equations

With additional lateral boundary conditions along the open sea and the coast, the three-dimensional boundary-value problem described in the previous section can be solved numerically, in principle. However, for tsunami early warning purpose the speed of obtaining direct numerical solution of the three-dimensional problem describing transoceanic tsunami propagation is still not fast enough. Simplifications and approximations must be made so as to obtain realistic and practical results. Since tsunamis are long waves, the vertical length scale of the problem is usually smaller than the horizontal length scale. Therefore, the frequency dispersion parameter can be assumed to be a small parameter, *i.e.,*

$$O(\mu^2) << 1 \tag{11}$$

Using μ^2 as the small parameter, a perturbation analysis can be performed on the primitive governing equations. The three-dimensional boundary-vale problem is approximated and projected onto the two-dimensional horizontal plane. In this perturbation analysis, the nonlinearity is assumed to be of $O(1)$. Only the key results are shown here. However, the complete derivation is given in the Appendix.

The resulting approximate continuity equation is

$$\frac{1}{\varepsilon}h_t + \zeta_t + \nabla \cdot (H\mathbf{u}_\alpha) - \mu^2 \nabla \cdot \left\{ H\left[\left(\frac{1}{6}(\varepsilon^2\zeta^2 - \varepsilon\zeta h + h^2) - \frac{1}{2}z_\alpha^2 \right) \nabla(\nabla \cdot \mathbf{u}_\alpha) \right. \right.$$
$$\left. \left. + \left(\frac{1}{2}(\varepsilon\zeta - h) - z_\alpha \right) \nabla \left(\nabla \cdot (h\mathbf{u}_\alpha) + \frac{h_t}{\varepsilon} \right) \right] \right\} = O(\mu^4), \tag{12}$$

in which $H = h + \varepsilon\zeta$ is the total water depth and $\mathbf{u}_\alpha$ is evaluated at $z = z_\alpha(x, y, t)$, which is a function of time. The choice of z_α is made based on the linear dispersion characteristics of the governing equations (*e.g.,* Nwogu 1993, Chen and Liu 1995).

Equation (12) is one of three governing equations for ζ and $\mathbf{u}_\alpha$ The other two equations come from the horizontal momentum equation, (4), and are given in vector form as

$$\mathbf{u}_{\alpha t} + \varepsilon \mathbf{u}_\alpha \cdot \nabla \mathbf{u}_\alpha + \nabla \zeta + \mu^2 \left\{ \frac{1}{2} z_\alpha^2 \nabla (\nabla \cdot \mathbf{u}_{\alpha t}) + z_\alpha \nabla \left[\nabla \cdot (h \mathbf{u}_\alpha)_t + \frac{h_{tt}}{\varepsilon} \right] \right\}$$

$$+ \mu^2 z_{\alpha t} \left\{ z_\alpha \nabla (\nabla \cdot \mathbf{u}_\alpha) + \nabla \left[\nabla \cdot (h \mathbf{u}_\alpha) + \frac{h_t}{\varepsilon} \right] \right\}$$

$$+ \varepsilon \mu^2 \left\{ \left[\nabla \cdot (h \mathbf{u}_\alpha) + \frac{h_t}{\varepsilon} \right] \nabla \left[\nabla \cdot (h \mathbf{u}_\alpha) + \frac{h_t}{\varepsilon} \right] \right.$$

$$- \nabla \left[\zeta \left(\nabla \cdot (h \mathbf{u}_\alpha)_t + \frac{h_{tt}}{\varepsilon} \right) \right] + (\mathbf{u}_\alpha \cdot \nabla z_\alpha) \nabla \left[\nabla \cdot (h \mathbf{u}_\alpha) + \frac{h_t}{\varepsilon} \right] \qquad (13)$$

$$+ z_\alpha \nabla \left[\mathbf{u}_\alpha \cdot \nabla \left(\nabla \cdot (h \mathbf{u}_\alpha) + \frac{h_t}{\varepsilon} \right) \right] + z_\alpha (\mathbf{u}_\alpha \cdot \nabla z_\alpha) \nabla (\nabla \cdot \mathbf{u}_\alpha) + \frac{z_\alpha^2}{2} \nabla [\mathbf{u}_\alpha \cdot \nabla (\nabla \cdot \mathbf{u}_\alpha)] \right\}$$

$$+ \varepsilon^2 \mu^2 \nabla \left\{ -\frac{\zeta^2}{2} \nabla \cdot \mathbf{u}_{\alpha t} - \zeta \mathbf{u}_\alpha \cdot \nabla \left[\nabla \cdot (h \mathbf{u}_\alpha) + \frac{h_t}{\varepsilon} \right] + \zeta \left[\nabla \cdot (h \mathbf{u}_\alpha) + \frac{h_t}{\varepsilon} \right] \nabla \cdot \mathbf{u}_\alpha \right\}$$

$$+ \varepsilon^2 \mu^2 \nabla \left\{ \frac{\zeta^2}{2} [(\nabla \cdot \mathbf{u}_\alpha)^2 - \mathbf{u}_\alpha \cdot \nabla (\nabla \cdot \mathbf{u}_\alpha)] \right\} = O(\mu^4).$$

Equations (12) and (13) are the coupled governing equations, written in terms of $\mathbf{u}_\alpha$ and ζ, for fully nonlinear, weakly dispersive waves generated by a seafloor movement. We reiterate here that $\mathbf{u}_\alpha$ is evaluated at $z = z_\alpha(x, y, t)$, which is a function of time. The choice of z_α is made based on the linear dispersion characteristics of the governing equations (*e.g.*, Nwogu 1993, Chen and Liu 1995). Assuming a fixed seafloor, in order to extend the applicability of the governing equations to relatively deep water (or short waves) (up to $\mu \sim 0.3$), z_α is recommended to be evaluated as $z_\alpha = -0.531h$. Higher-order equations, in terms of μ^2, have been derived (Gobbi, *et al.* 2000). However, these equations contain higher spatial derivatives and are not easy to deal with numerically.

Once the solutions for $\mathbf{u}_\alpha$ and ζ are obtained, the vertical profiles of the horizontal velocity can be expressed as:

$$\mathbf{u} = \mathbf{u}_\alpha - \mu^2 \left\{ \frac{z^2 - z_\alpha^2}{2} \nabla (\nabla \cdot \mathbf{u}_\alpha) + (z - z_\alpha) \nabla \left[\nabla \cdot (h \mathbf{u}_\alpha) + \frac{h_t}{\varepsilon} \right] \right\} + O(\mu^4),$$

$$-h < z < \varepsilon \zeta, \qquad (14)$$

which is a quadratic function in z, the vertical coordinate. The vertical profile of the vertical velocity component can be written as:

$$w = \mu^2 \left\{ -z \nabla \cdot \mathbf{u}_\alpha - \nabla \cdot (h \mathbf{u}_\alpha) - \frac{h_t}{\varepsilon} \right\} + O(\mu^4), \quad -h < z < \varepsilon \zeta. \qquad (15)$$

The vertical velocity component varies linearly in z and the order of magnitude of the leading order term depends on the time scale of the seafloor motion. The pressure field can also be calculated from the following expression:

$$p = \left(\zeta - \frac{z}{\varepsilon}\right) + \mu^2 \left\{ \frac{1}{2}\left(z^2 - \varepsilon^2\zeta^2\right)\nabla \cdot \mathbf{u}_{\alpha t} + (z - \varepsilon\zeta)\left[\nabla \cdot (h\mathbf{u}_\alpha)_t + \frac{h_{tt}}{\varepsilon}\right] \right.$$

$$+ \frac{\varepsilon}{2}\left(z^2 - \varepsilon^2\zeta^2\right)\mathbf{u}_\alpha \cdot \nabla(\nabla \cdot \mathbf{u}_\alpha) + \varepsilon(z - \varepsilon\zeta)\mathbf{u}_\alpha \cdot \nabla\left[\nabla \cdot (h\mathbf{u}_\alpha) + \frac{h_t}{\varepsilon}\right] \quad (16)$$

$$\left. + \frac{\varepsilon}{2}\left(\varepsilon^2\zeta^2 - z^2\right)(\nabla \cdot \mathbf{u}_\alpha)^2 + \varepsilon(\varepsilon\zeta - z)\left[\nabla \cdot (h\mathbf{u}_\alpha) + \frac{h_t}{\varepsilon}\right]\nabla \cdot \mathbf{u}_\alpha \right\} + O\left(\mu^4\right), \ -h < z < \varepsilon\zeta.$$

The leading order pressure field is quasi-static as the result of the long wave approximation.

We can define the depth-averaged horizontal velocity vectors as

$$\bar{\mathbf{u}} = \frac{1}{h + \varepsilon\zeta}\int_{-h}^{\varepsilon\zeta} \mathbf{u}\, dz. \quad (17)$$

The exact continuity, (9), can be written in terms of $\bar{\mathbf{u}}$ as

$$\frac{1}{\varepsilon}h_t + \zeta_t + \nabla \cdot (H\bar{\mathbf{u}}) = 0. \quad (18)$$

The relationship between $\bar{\mathbf{u}}$ and $\mathbf{u}_\alpha$ can be given as follows (see Appendix):

$$\mathbf{u}_\alpha = \bar{\mathbf{u}} + \mu^2\left\{\left[\frac{1}{6}\left(\varepsilon^2\zeta^2 - \varepsilon\zeta h + h^2\right) - \frac{1}{2}z_\alpha^2\right]\nabla(\nabla \cdot \bar{\mathbf{u}})\right.$$

$$\left. + \left[\frac{1}{2}(\varepsilon\zeta - h) - z_\alpha\right]\nabla[\nabla \cdot (h\bar{\mathbf{u}})]\right\} + O\left(\mu^4\right). \quad (19)$$

Substituting (19) into (13), we can derive the momentum equations in terms of the depth averaged velocity, $\bar{\mathbf{u}}$.

$$\bar{\mathbf{u}}_t + \varepsilon\bar{\mathbf{u}}\cdot\nabla\bar{\mathbf{u}} + \nabla\zeta + \mu^2\left\{\frac{1}{6}h^2\nabla(\nabla\cdot\bar{\mathbf{u}}_t) - \frac{1}{2}h\nabla\bar{\Gamma}_t + h_t\left[\frac{1}{3}h\nabla(\nabla\cdot\bar{\mathbf{u}}) - \frac{1}{2}\nabla\bar{\Gamma}\right]\right\}$$

$$+ \varepsilon\mu^2\left\{\bar{\Gamma}\nabla\bar{\Gamma} - \nabla(\zeta\bar{\Gamma}_t) - \frac{1}{6}[\zeta h\nabla(\nabla\cdot\bar{\mathbf{u}})]_t + \frac{1}{2}[\zeta\nabla\bar{\Gamma}]_t + \frac{1}{6}h^2\nabla\bar{\mathbf{u}}\cdot\nabla(\nabla\cdot\bar{\mathbf{u}})\right.$$

$$\left. - \frac{1}{2}h\nabla\bar{\mathbf{u}}\cdot\nabla\bar{\Gamma} + \bar{\mathbf{u}}\cdot\nabla\left[\frac{1}{6}h^2\nabla(\nabla\cdot\bar{\mathbf{u}})\right] - \bar{\mathbf{u}}\cdot\nabla\left(\frac{1}{2}\nabla\bar{\Gamma}\right)\right\}$$

$$+ \varepsilon^2\mu^2\nabla\left\{\frac{1}{6}[\zeta^2\nabla(\nabla\cdot\bar{\mathbf{u}})]_t - \frac{1}{6}\zeta h\nabla(\nabla\cdot\bar{\mathbf{u}})\cdot\nabla\bar{\mathbf{u}} + \frac{1}{2}\zeta\nabla\bar{\mathbf{u}}\cdot\nabla\bar{\Gamma} - \frac{1}{6}\bar{\mathbf{u}}\cdot\nabla[\zeta h\nabla(\nabla\cdot\bar{\mathbf{u}})]\right. \quad (20)$$

$$\left. + \frac{1}{2}\bar{\mathbf{u}}\cdot\nabla(\zeta\nabla\bar{\Gamma}) - \frac{1}{2}\nabla(\zeta^2\nabla\cdot\bar{\mathbf{u}}_t) - \nabla(\zeta\bar{\mathbf{u}}\cdot\nabla\bar{\Gamma}) + \nabla(\zeta\bar{\Gamma}\nabla\cdot\bar{\mathbf{u}})\right\}$$

$$+ \varepsilon^2\mu^2\nabla\left\{\zeta^2\nabla(\nabla\cdot\bar{\mathbf{u}})\cdot\nabla\bar{\mathbf{u}} + \frac{1}{6}\bar{\mathbf{u}}\cdot\nabla[\zeta^2\nabla(\nabla\cdot\bar{\mathbf{u}})]\right.$$

$$\left. - \frac{1}{2}\nabla[\zeta^2\bar{\mathbf{u}}\cdot\nabla(\nabla\cdot\bar{\mathbf{u}}] + \frac{1}{2}\nabla[\zeta^2(\nabla\cdot\bar{\mathbf{u}})^2]\right\} = O(\mu^4),$$

where

$$\bar{\Gamma} = \nabla\cdot h\bar{\mathbf{u}} + \frac{h_t}{\varepsilon}$$

Although equation (20) has the same order of magnitude of accuracy as (13) written in terms of $\mathbf{u}_\alpha$, the linear dispersive characteristics of (20) are slightly worse. In other words, the model equations in terms of the depth averaged velocity can not be applied to the water depth as deep as that for (13).

3.1.   Creeping seafloor movements

Up to this point the time scale of the seafloor movement is assumed to be in the same order of magnitude as the typical period of generated water wave, i.e., $t_w = \ell_0/\sqrt{gh_0}$ as given in (1). When the seafloor movement is creeping in nature, the time scale of seafloor movement, t_c, could be much longer than t_ω. The only scaling parameter that is directly affected by the time scale of the seafloor movement is the characteristic amplitude of the wave motion. After introducing the new time scale t_c into the time derivatives of h in the continuity equation, (12), along with a characteristic change in water depth Δh, the coefficient in front of h_t becomes

$$\frac{\delta}{\varepsilon}\frac{t_w}{t_c}, \qquad (21)$$

where $\delta = \Delta h/h_0$. To maintain the conservation of mass at the leading order, the above parameter must be of order one. Thus,

$$\varepsilon = \delta\frac{t_w}{t_c} = \frac{\delta\ell_0}{t_c\sqrt{gh_0}} \qquad (22)$$

The above relationship can be interpreted as follows. During the creeping ground motion, over the time period $t < t_c$ the generated wave has propagated a distance $t\sqrt{gh_0}$. The total volume of the seafloor displacement, normalized by h_0, is $\delta\ell_0\dfrac{t}{t_c}$, which should be the same as the volume of water underneath the generated wave crest, i.e., $\varepsilon t\sqrt{gh_0}$. Therefore, over the seafloor movement duration, $t < t_c$, the wave amplitude can be estimated by (22). Consequently, nonlinear effects become important only if ε defined in (22) is $O(1)$. Since, by definition of a creeping slide, the value $l_0/t_c\sqrt{gh_0}$ is always less than one, fully nonlinear effects will be important for only the largest slides. A similar analysis was carried out by Hammack (1973), but with a very different set of scaling parameters between an impulsive and creeping movement. Interestingly, however, his conclusions concerning the importance of nonlinearity are identical to those given above.

3.2. Impulsive seafloor movements

For the cases of impulsive seafloor movements, the time scale for water depth changes is much smaller than that of water wave propagation. Following the same argument given in the previous section, the percentage change of water depth is denoted as $\delta = \Delta h/h_0$ and equation (22) is still true based on the continuity requirement, i.e.,

$$\varepsilon = \delta\frac{t_w}{t_c}$$

where t_w represents the time scale for free surface elevation changes. Therefore, for impulsive seafloor movements, $t_c \sim 0$, the only possible means to have a finite free surface response is when $t_w \sim 0$. In other words, the free surface responds instantaneously and the surface profile mimics the seafloor displacement. Hence, the seafloor displacement can be used as the initial condition for ζ and the seafloor can be treated as stationary for the entire duration of simulation. The governing equations for the case of a stationary seafloor can be reduced from the general equations (12) and (13) as

$$\zeta_t + \nabla\cdot(H\mathbf{u}_\alpha) - \mu^2\nabla\cdot\left\{H\left[\left(\frac{1}{6}\left(\varepsilon^2\zeta^2 - \varepsilon\zeta h + h^2\right) - \frac{1}{2}z_\alpha^2\right)\nabla(\nabla\cdot\mathbf{u}_\alpha)\right.\right.$$
$$\left.\left. + \left[\frac{1}{2}(\varepsilon\zeta - h) - z_\alpha\right]\nabla[\nabla\cdot(h\mathbf{u}_\alpha)]\right]\right\} = O(\mu^4).$$

$$(23)$$

$$\mathbf{u}_{\alpha t} + \varepsilon \mathbf{u}_\alpha \cdot \nabla \mathbf{u}_\alpha + \nabla \zeta + \mu^2 \left\{ \frac{1}{2} z_\alpha^2 \nabla (\nabla \cdot \mathbf{u}_{\alpha t}) + z_\alpha \nabla [\nabla \cdot (h\mathbf{u}_{\alpha t})] \right\}$$

$$+ \mu^2 z_{\alpha t} \{ z_\alpha \nabla (\nabla \cdot \mathbf{u}_\alpha) + \nabla [\nabla \cdot (h\mathbf{u}_\alpha)] \}$$

$$+ \varepsilon \mu^2 \{ [\nabla \cdot (h\mathbf{u}_\alpha)] \nabla [\nabla \cdot (h\mathbf{u}_\alpha)] - \nabla [\zeta (\nabla \cdot (h\mathbf{u}_{\alpha t})] + (\mathbf{u}_\alpha \cdot \nabla z_\alpha) \nabla [\nabla \cdot (h\mathbf{u}_\alpha)]$$

$$+ z_\alpha \nabla [\mathbf{u}_\alpha \cdot \nabla (\nabla \cdot (h\mathbf{u}_\alpha)] + z_\alpha (\mathbf{u}_\alpha \cdot \nabla z_\alpha) \nabla (\nabla \cdot \mathbf{u}_\alpha) + \frac{z_\alpha^2}{2} \nabla [\mathbf{u}_\alpha \cdot \nabla (\nabla \cdot \mathbf{u}_\alpha)] \right\} \qquad (24)$$

$$+ \varepsilon^2 \mu^2 \nabla \left\{ -\frac{\zeta^2}{2} \nabla \cdot \mathbf{u}_{\alpha t} - \zeta \mathbf{u}_\alpha \cdot \nabla [\nabla \cdot (h\mathbf{u}_\alpha) +] + \zeta [\nabla \cdot (h\mathbf{u}_\alpha)] \nabla \cdot \mathbf{u}_\alpha \right\}$$

$$+ \varepsilon^3 \mu^2 \nabla \left\{ \frac{\zeta^2}{2} [(\nabla \cdot \mathbf{u}_\alpha)^2 - \mathbf{u}_\alpha \cdot \nabla (\nabla \cdot \mathbf{u}_\alpha)] \right\} = O(\mu^4).$$

We remark here that the governing equations shown above are more general than those given in the literature (*e.g.*, Liu 1994, Madsen and Schäffer 1998). In the present equations z_α is a function of time, so that $z_{\alpha t}$ appears in the equations. In other words, $\mathbf{u}_\alpha$ can be evaluated on the free surface $z = z_\alpha = \zeta$, for example. However, it may not be the appropriate choice for optimum results. The above equations are very similar to a model presented by Kennedy *et al.* (2001), who also allowed z_α to be a function of time, with the differences between the two models being a number of nonlinear dispersive terms with the coefficient ∇z_α that are included here, but not in Kennedy *et al.* (2001).

4. Other Limiting Cases

In this section the general model is simplified for different physical conditions.

4.1. *Weakly non-linear and weakly dispersive waves*

In many situations the seafloor displacement is relatively small in comparison with the local depth, and the seafloor movement can be approximated as

$$h(x, y, t) = h_0(x, y) + \delta \bar{h}(x, y, t), \qquad (25)$$

in which δ is small. In other words, the maximum seafloor displacement is much smaller than the characteristic water depth. Since the free surface displacement is directly proportional to the seafloor displacement, *i.e.*, $O(\varepsilon \zeta) = O(\delta \bar{h})$, we can further simplify the governing equations derived in the previous section by allowing

$$O(\varepsilon) = O(\delta) = O(\mu^2) \ll 1, \qquad (26)$$

which is the Boussinesq approximation. Thus, the continuity equation, (12) can be reduced to

$$\zeta_t + \nabla \cdot (H\mathbf{u}_\alpha) + \frac{\delta}{\varepsilon}\overline{h}_t - \mu^2 \nabla \cdot \left\{ h_0 \left[\left(\frac{1}{6}h_0^2 - \frac{1}{2}z_\alpha^2 \right) \nabla (\nabla \cdot \mathbf{u}_\alpha) \right. \right.$$
$$\left. \left. - \left(\frac{1}{2}h_0 + z_\alpha \right) \nabla \left(\nabla \cdot (h_0 \mathbf{u}_\alpha) + \frac{\delta}{\varepsilon}\overline{h}_t \right) \right] \right\} = O\left(\mu^4, \mu^2\varepsilon, \delta\mu^2 \right).$$

(27)

The momentum equation becomes

$$\mathbf{u}_{\alpha t} + \varepsilon \mathbf{u}_\alpha \cdot \nabla \mathbf{u}_\alpha + \nabla \zeta + \mu^2 \left\{ \frac{1}{2}z_\alpha^2 \nabla (\nabla \cdot \mathbf{u}_{\alpha t}) + z_\alpha \nabla \left[\nabla \cdot (h_0 \mathbf{u}_{\alpha t}) + \frac{\delta}{\varepsilon}\overline{h}_{tt} \right] \right\}$$
$$+ \mu^2 z_{\alpha t} \left\{ z_\alpha \nabla (\nabla \cdot \mathbf{u}_\alpha) + \nabla \left[\nabla \cdot (h_0 \mathbf{u}_\alpha) + \frac{\delta}{\varepsilon}\overline{h}_t \right] \right\} = O\left(\mu^4, \varepsilon\mu^2, \delta\mu^2 \right).$$

(28)

In the case of impulsive seafloor movements, the governing equations can be further simplified by dropping the terms involving the time derivatives of water depth. Thus, while the continuity equation becomes

$$\zeta_t + \nabla \cdot (H\mathbf{u}_\alpha) - \mu^2 \nabla \cdot \left\{ h_0 \left[\left(\frac{1}{6}h_0^2 - \frac{1}{2}z_\alpha^2 \right) \nabla (\nabla \cdot \mathbf{u}_\alpha) \right. \right.$$
$$\left. \left. - \left(\frac{1}{2}h_0 + z_\alpha \right) \nabla (\nabla \cdot (h_0 \mathbf{u}_\alpha)) \right] \right\} = O\left(\mu^4, \mu^2\varepsilon, \delta\mu^2 \right),$$

(29)

the momentum equation can be written as

$$\mathbf{u}_{\alpha t} + \varepsilon \mathbf{u}_\alpha \cdot \nabla \mathbf{u}_\alpha + \nabla \zeta + \mu^2 \left\{ \frac{1}{2}z_\alpha^2 \nabla (\nabla \cdot \mathbf{u}_{\alpha t}) + z_\alpha \nabla \left[\nabla \cdot (h_0 \mathbf{u}_{\alpha t}) \right] \right\}$$
$$+ \mu^2 z_{\alpha t} \left\{ z_\alpha \nabla (\nabla \cdot \mathbf{u}_\alpha) + \nabla [\nabla \cdot (h_0 \mathbf{u}_\alpha)] \right\} = O\left(\mu^4, \varepsilon\mu^2, \delta\mu^2 \right).$$

(30)

These model equations are the well-known extended Boussinesq equations.
    Once again, these model equations can be recast in terms of the depth-averaged velocity $\overline{\mathbf{u}}$ and the free surface displacement ζ:

$$\zeta_t + \nabla \cdot (H\overline{\mathbf{u}}) = 0,$$

(31)

$$\overline{\mathbf{u}}_t + \varepsilon \overline{\mathbf{u}} \cdot \nabla \overline{\mathbf{u}} + \nabla \zeta + \mu^2 \left\{ \frac{1}{6}h_0^2 \nabla (\nabla \cdot \overline{\mathbf{u}}_t) - \frac{1}{2}h_0 \nabla (\nabla \cdot h_0 \overline{\mathbf{u}}_t) \right\} = O\left(\mu^4, \varepsilon\mu^2 \right).$$

(32)

4.2. Non-linear and non-dispersive shallow-water waves

In the case that the water depth is very shallow or the wavelength is very long, the governing equations, (12) and (13) or (18) and (20), can be truncated at $O(\mu^2)$.

$$\frac{1}{\varepsilon} H_t + \nabla \cdot (H\mathbf{u}) = 0(\mu^2), \tag{33}$$

$$\mathbf{u}_t + \varepsilon \mathbf{u} \cdot \nabla \mathbf{u} + \nabla \zeta = 0(\mu^2). \tag{34}$$

These equations are the well-known nonlinear shallow-water (NSW) equations in which the seafloor movement is the forcing term for wave generation. The velocity vector, $\mathbf{u}$, in the above equations could be either the depth averaged velocity, $\overline{\mathbf{u}}$, or the velocity evaluated at z_α, $\mathbf{u}_\alpha$. When the depth averaged velocity is used, the continuity equation, (33), is exact, while the truncation error for the momentum equations remains $O(\mu^2)$.

4.3. Other physical considerations

When modeling tsunami propagation over a long distance, such as the Pacific Ocean, the effects of earth rotation and the curvature of the Earth's surface might become important. Denoting that f' as the Coriolis parameter, the force generated by the earth rotation can be approximated as $f' \mathbf{k} \times \mathbf{u}'$, where $\mathbf{k}$ represents the unit vector normal to the earth surface. This Coriolis force needs to be added to all the momentum equations presented in this chapter. For example, the Boussinesq equations written in terms of the depth averaged velocity can be rewritten as:

$$\overline{\mathbf{u}}_t + \varepsilon \overline{\mathbf{u}} \cdot \nabla \overline{\mathbf{u}} + \nabla \zeta + f\mathbf{k} \times \overline{\mathbf{u}} + \mu^2 \left\{ \frac{h^2}{6} \nabla (\nabla \cdot \overline{\mathbf{u}}_t) - \frac{h}{2} \nabla \nabla \cdot (h\overline{\mathbf{u}}_t) \right\} = 0 , \tag{35}$$

where f is the dimensionless Coriolis parameter, i.e.,

$$f = \frac{f' l_0}{(gh_0)^{1/2}} . \tag{36}$$

The continuity equation remains unchanged.

In some situations, a spherical coordinate system, (ψ, φ) denoting the longitude and the latitude of the Earth, should be employed. Denoting $(P, Q) = H\overline{\mathbf{u}}$ as the components of the horizontal volume flux in the ψ- and φ- directions, respectively, the Boussinesq equations can be rewritten as follows:

$$\frac{\partial \zeta}{\partial t} + \frac{1}{R \cos \varphi} \left[\frac{\partial P}{\partial \psi} + \frac{\partial}{\partial \varphi} (\cos \varphi Q) \right] = 0 , \tag{37}$$

$$\frac{\partial P}{\partial t} + \frac{\varepsilon}{R\cos\varphi}\frac{\partial}{\partial\psi}\left(\frac{P^2}{H}\right) + \frac{\varepsilon}{R}\frac{\partial}{\partial\varphi}\left(\frac{PQ}{H}\cos\varphi\right) + \frac{H}{R\cos\varphi}\frac{\partial\zeta}{\partial\psi} - fQ$$

$$= \frac{\mu^2}{R\cos\varphi}\frac{\partial}{\partial\psi}\left[\frac{H^3}{3R\cos\varphi}\frac{\partial}{\partial t}\left\{\frac{\partial}{\partial\psi}\left(\frac{P}{H}\right) + \frac{\partial}{\partial\varphi}\left(\cos\varphi\frac{Q}{H}\right)\right\}\right], \tag{38}$$

$$\frac{\partial Q}{\partial t} + \frac{\varepsilon}{R\cos\varphi}\frac{\partial}{\partial\psi}\left(\frac{PQ}{H}\cos\varphi\right) + \frac{\varepsilon}{R}\frac{\partial}{\partial\varphi}\left(\frac{Q^2}{H}\cos^2\varphi\right) + \frac{H}{R}\frac{\partial\zeta}{\partial\varphi} + fP$$

$$= \frac{\mu^2}{R}\frac{\partial}{\partial\varphi}\left[\frac{H^3}{3R\cos\varphi}\frac{\partial}{\partial t}\left\{\frac{\partial}{\partial\psi}\left(\frac{P}{H}\right) + \frac{\partial}{\partial\varphi}\left(\frac{Q}{H}\cos\varphi\right)\right\}\right], \tag{39}$$

in which R measures the radius of the Earth.

5. Leading Waves

From (33) and (34), the *dimensional* form of the linear shallow water (LSW) equations can be written in terms of the depth-averaged velocity as:

$$\zeta_t + h_0 u_x + h_0 v_y = -\overline{h}_t, \tag{40}$$

$$u_t = -g\zeta_x, \tag{41}$$

$$v_t = -g\zeta_y, \tag{42}$$

in which the water depth, h_0, has been assumed to be a constant. To understand the tsunami generation process as described by the LSW equations, we shall only consider the one-dimensional situation in which the seafloor deforms abruptly at $t = 0$, *i.e.*,

$$\overline{h}(x, 0^-) = 0, \quad \overline{h}(x, 0^+) = H_0(x). \tag{43}$$

Consequently,

$$\overline{h}_t = H_0(x)\delta(t) \tag{44}$$

and $\delta(t)$ is the Delta function. Furthermore, the free surface displacement and the horizontal velocity are zero initially, *i.e.*,

$$\zeta(x, 0) = 0, \quad u(x, 0) = 0. \tag{45}$$

The solution for the initial boundary-value problem stated above can be found by the standard Laplace and Fourier transform methods. We first apply the Laplace transform, defined by

$$\bar{f}(s) = \int_0^\infty e^{-st} f(t) dt , \quad f(t) = \frac{1}{2\pi i} \int_\Gamma e^{st} \bar{f}(s) ds,$$

where Γ is a vertical line to the right of all singularities of $\bar{f}(s)$ in the complex s plane, to the one-dimensional LSW equations. Thus,

$$-s\bar{\zeta} + h_0 \bar{u}_x = \overline{W}, \tag{46}$$

$$s\bar{u} = g\bar{\zeta}_x, \tag{47}$$

in which the initial condition (45) has been used and

$$\overline{W} = -\int_0^\infty e^{st} H_0(x) \, \delta(t) dt = -H_0(x). \tag{48}$$

Equations (46) and (47) can be combined into

$$s^2 \bar{\zeta} - gh\bar{\zeta}_{xx} = sH_0(x). \tag{49}$$

As long as $\bar{\zeta}$ diminishes when x approaches positive and negative infinity, the above equation can be solved by applying the Fourier transformation in x, defined by

$$\tilde{f}(k) = \int_{-\infty}^\infty e^{-ikx} f(x) dx , \quad \tilde{f}(x) = \frac{1}{2\pi} \int_{-\infty}^\infty e^{ikx} \tilde{f}(k) dk .$$

Thus, (49) can be transformed into

$$\tilde{\bar{\zeta}} = \frac{s\tilde{H}_0(k)}{s^2 + k^2 gh} = \frac{s\tilde{H}_0(k)}{(s + i\omega)(s - i\omega)}; \quad \omega = k\sqrt{gh} . \tag{50}$$

Taking the inverse Fourier and Laplace transforms, we have

$$\zeta(x,t) = \frac{1}{2\pi} \int_{-\infty}^\infty dk e^{ikx} \tilde{H}_0(k) \left[\frac{1}{2\pi i} \int_\Gamma ds e^{st} \frac{s}{(s + i\omega)(s - i\omega)} \right]. \tag{51}$$

The integral inside the bracket can be readily evaluated. The integrand has two poles at $s = i\omega$ and $-i\omega$. For $t < 0$, a closed semicircular contour on the right half of s-plane (i.e., $Re(s) > 0$), is introduced and Γ in the original integral is the vertical segment of the semicircular contour. Since the factor multiplying e^{st} in the integrand vanishes uniformly as s approaches infinity, the line integral along the great semi-circular arc is zero by Jordan's lemma. By Cauchy's residue theorem, the original integral along Γ is zero, since there is no singular point within the semi-circle. Thus,

$$\zeta = 0, \quad t < 0,$$

which agrees with the original assumption. For $t > 0$ the semi-circular contour must be on the left half of the s plane, $Re(s) < 0$, so that e^{st} diminishes exponentially as $|s|$ becomes large. Once again, by Jordan's lemma, the line integral along the great semi-circle vanishes, leaving only the two residues at two poles,

$$\frac{1}{2\pi i} \int_{\Gamma} ds e^{st} \frac{s}{(s+i\omega)(s-i\omega)} = \cos \omega t, \quad t > 0.$$

Substitution of the above integral into (51) yields,

$$\zeta(x,t) = \frac{1}{2\pi} \int_{-\infty}^{\infty} dk e^{ikx} \tilde{H}_0(k) \cos \omega t = \frac{1}{2\pi} \int_{-\infty}^{\infty} dk \frac{\tilde{H}_0(k)}{2} \left[e^{i(kx+\omega t)} + e^{i(kx-\omega t)} \right]. \quad (52)$$

The first and second terms in the above equation represent left- and right-going waves, respectively. For the linear shallow-water wave, $\omega = k\sqrt{gh}$, the above expression can be evaluated by the inverse Fourier transform, *i.e.*,

$$\zeta = \frac{1}{2} \left[H_0(kx + \omega t) + H_0(kx - \omega t) \right]. \quad (53)$$

Although not surprising, the results are revealing. The impulsive ground motion will generate two water wave packages, propagating in the opposite directions with the same speed, $\sqrt{gh}$. The water surface profiles mimic the seafloor deformation. However, the magnitude is exactly reduced by half. Because of the non-dispersive characteristics of the LSW equation and the one-dimensionality of the problem, the wave form remains permanent. For instance, if the initial free surface displacement has an *N*-wave form (*i.e.*, it is elevated on one side and depressed on the other side), two *N*-waves with one-half of the initial wave amplitude will propagate away from the generation region. As these waves approach respective coastlines, one will lead by an elevated (positive) wave and other depressed (negative) wave (*i.e.*, the shoreline will withdraw first).

When tsunami travels a long distance, the frequency dispersion might become important. To examine the evolution of the leading wave of a tsunami propagating over a long distance, the linear Boussinesq equations might be more appropriate than the LSW equations. From (31) and (32), the linear Boussinesq equation in the *dimensional* form can be written as

$$\zeta_t + hu_x = -\overline{h}_t, \quad (54)$$

$$u_t = -g\zeta_x + \frac{1}{3} h^2 u_{xxt}. \quad (55)$$

Assuming the same seafloor displacement and the initial conditions, (43)–(45), and applying the Laplace-Fourier transforms as shown in the previous section, we find

$$\tilde{\tilde{\zeta}} = \frac{s\tilde{H}_0(k)}{s^2 + k^2 gh / \left(1 + \frac{1}{3}k^2 h^2\right)},$$

which can be rewritten as

$$\tilde{\tilde{\zeta}} = \frac{s\tilde{H}_0(k)}{(s + i\omega)(s - i\omega)}; \quad \omega = \frac{k\sqrt{gh}}{\sqrt{1 + \frac{1}{3}k^2 h^2}}. \tag{56}$$

We remark here that the primary difference between the LSW theory and the linear Boussinesq theory is that the latter is a dispersive wave system in which the phase velocity depends on the wave frequency. If kh is sufficiently small, the dispersion relationship that appears in (56) can be further approximated as

$$\omega = k\sqrt{gh}\left(1 - \frac{1}{6}k^2 h^2 + \dots\right). \tag{57}$$

Following the same procedure shown in the LSW theory case, the free surface displacement can be written as

$$\zeta(x,t) = \frac{1}{2\pi}\int_{-\infty}^{\infty} dk \frac{\tilde{H}_0(k)}{2}\left[e^{i(kx+\omega t)} + e^{i(kx-\omega t)}\right], \tag{58}$$

which is in the same form as (52) from the LSW equations. However, the dispersion relationship is quite different, resulting in very different wave forms.

Any seafloor deformation, $H_0(x)$, can be thought of as the sum of $H_0^e(x)$ and H_0^o that are even and odd in x, respectively. Due to the linear nature of the problem, the two parts can be treated separately first and will be combined together as the final solution. The typical N-shape seafloor displacement can be viewed as an odd function in x, $H_0^o(x)$.

Introducing

$$H_0^o(x) = \frac{dB}{dx} \tag{59}$$

so that $\tilde{H}_0(k) = ik\tilde{B}(k)$. Since $\tilde{H}_0(k)$ is odd, $\tilde{B}$ must be real and even in k. Consequently, the solution for the free surface displacement, (58), can be simplified to be (Mei 1989)

$$\zeta(x,t) = \frac{1}{2\pi}\frac{d}{dx}Re\int_{-\infty}^{\infty} dk\tilde{B}(k)\left[e^{i(kx+\omega t)} + e^{i(kx-\omega t)}\right], \tag{60}$$

where "Re" denotes the "real part of".

At large t, the behavior of the leading wave can be explored by using the method of stationary phase. Focusing on the right-going wave (in the positive x-direction), (60) can be approximated as (Mei 1983):

$$\zeta(x,t) = \frac{\tilde{B}(0)}{2} \left(\frac{2}{\sqrt{gh}h^2 t} \right)^{2/3} A'_i \left[\left(\frac{2}{\sqrt{gh}h^2 t} \right)^{1/3} \left(x - \sqrt{gh}t \right) \right], \qquad (61)$$

where

$$A'_i(z) = \frac{dA_i}{dz}.$$

Based on (61), the leading wave attenuates with time as $t^{-2/3}$. The function A'_i (z) is negative for $z > 0$ and approaches zero monotonically as $z \to \infty$. On the other hand, $A'_i(z)$ behaves oscillatorily for $z < 0$ and the magnitude and frequency increase as $z \to \infty$. We also note that

$$\tilde{B}(0) = \int_{-\infty}^{\infty} B(x)dx = \int_{-\infty}^{\infty} \int_{-\infty}^{x} H_0^o(x')dx'dx = -\int_{-\infty}^{\infty} xH_0^o(x)dx.$$

Thus, if the N-shape seafloor displacement has a depression for $x > 0$ and an elevation for $x < 0$, $\tilde{B}(0) > 0$ and the leading wave propagating to the right ($x > 0$) is a depression wave (hence withdraw from a beach). The subsequent waves increase in amplitude. On the left side ($x < 0$), the leading wave is an elevated wave.

6. Numerical Models

Based on the model equations presented in the previous sections, numerical simulations of tsunami propagation have made great progress in the last thirty years. Several tsunami computational models are currently used in the National Tsunami Hazard Mitigation Program, sponsored by the National Oceanic and Atmospheric Administration, to produce tsunami inundation and evacuation maps for the states of Alaska, California, Hawaii, Oregon, and Washington. The computational models include MOST (Method Of Splitting Tsunami), developed originally by researchers at the University of Southern California (Titov and Synolakis 1998); COMCOT (Cornell Multi-grid Coupled Tsunami Model), developed at Cornell University (Liu *et al.* 1994); and TSUNAMI2, developed at Tohoko University in Japan (Imamura 1996). All three models solve the same depth-integrated and 2D horizontal (2DH) nonlinear shallow-water (NSW) equations with different finite-difference algorithms. There are a number of other tsunami models as well, including the finite element model ADCIRC (ADvanced CIRCulation Model For Oceanic, Coastal And Estuarine Waters; *e.g.,* Priest *et al.* 1997).

Several high-order depth-integrated wave hydrodynamics models (Boussinesq models) are now available for simulating nonlinear and weakly dispersive waves, such as COULWAVE (Cornell University Long and Intermediate Wave Modeling

Package; Lynett and Liu 2002) and FUNWAVE (Kennedy *et al.* 2000). The major difference between the two is their treatment of moving shoreline boundaries. Lynett, *et al.* (2003) applied COULWAVE to the 1998 PNG tsunami with the landslide source; the results agreed with field survey data well. Recently, several finite element models have also been developed based on Boussinesq-type equations (*e.g.,* Woo and Liu 2004). Boussinesq models require higher spatial and temporal resolutions, and therefore are more computationally intensive. Moreover, most of model validation work was performed for open-ocean or open-coast problems. In other words, the models have not been carefully tested for wave propagation and oscillations in semi-enclosed regions—such as a harbor or bay—especially under resonant conditions.

6.1.   A Case Study - November 15, 2006 Central Kuril Island Tsunami

After the 2004 Indian Ocean tsunamis many successful stories have been reported on the skills of the existing numerical models in describing the tsunami propagation in deep ocean basin as long as the source region information is accurately prescribed (*e.g.,* Wang and Liu 2006). In this section, we present the simulation of the tsunami generated by the Central Kuril Island earthquake on November 15, 2006. The focus of the discussion is to demonstrate the importance of accurate bathymetry in estimating the tsunami propagation.

At 11:14 UTC on November 15, 2006, a strong earthquake of magnitude Mw = 8.3 occurred at the central Kuril Islands. A small tsunami was generated, which propagated over the entire Pacific Ocean. Five hours after the tsunami advisory for the coast of California was called off, Crescent City, CA, more than 6400 km away from the epicenter, was surprised by tsunamis with wave heights exceeding 1 m. The damages inside the Crescent City harbor were estimated between $500,000 and $ 1 million. Using the linear shallow water equation version of the COMCOT, numerical simulations of the tsunami propagation in the Pacific Ocean have been performed. Numerical results reveal that the submerged ridge associated with the Mendocino Fracture Zone (MFZ), which connects to the San Andreas Fault near Eureka, CA, serves as a low-velocity wave guide. The speed of tsunami propagation is proportional to the square root of water depth. Therefore, the wave front on the northern and southern side of the ridge is, in general, moving at a faster speed than that above the ridge. Consequently, the wave front is gradually bending towards the normal to the ridge from both sides. While the wave energy is converging from both sides of the ridge, wave energy is trapped above the ridge. Together with the shoaling effects, the wave height over the ridge becomes increasingly larger than those on its both sides. Because the averaged width of the ridge is in the same order of magnitude of the wavelength of the leading tsunami wave and the length of the ridge is long (more than ten wavelengths), the wave guide is efficient. This phenomenon can be clearly seen from snapshots of the simulated surface displacement distribution at 6.5, 7.0, 7.5 and 8.0 hours after the earthquake (figure 9.2). At the 8th hour after the earthquake, the leading tsunami wave has reached the edge of the continental shelf (the leading wave crest is about 150 km offshore of Eureka and Crescent City, CA) and the maximum wave height (measured from wave crest to wave trough) is about 0.14 m. As the tsunami waves climb up the continental shelf moving toward the coast, the wave heights grow

significantly because of the shoaling effect. At the 8.5th hour after the earthquake
the leading tsunami wave has reached Crescent City and the wave height has been
amplified to 0.67 m.

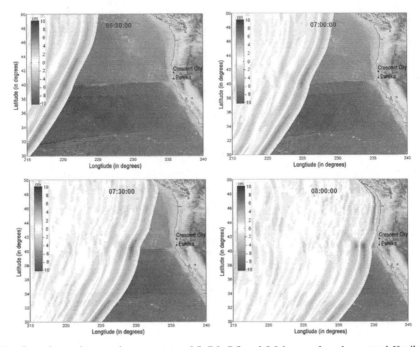

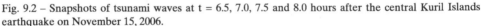

Fig. 9.2 – Snapshots of tsunami waves at t = 6.5, 7.0, 7.5 and 8.0 hours after the central Kuril Islands
earthquake on November 15, 2006.

This finding has significant impact on the tsunami warning practice in the future, since the MFZ will most likely play the same role as a wave guide for any
tsunami generated along the Kuril islands fault zone and a portion of the Aleutian
islands fault zone. The policy for cancelation of a tsunami warning advisory for the
Northern California and Oregon coasts needs to be re-examined.

7.   Concluding Remarks

In this chapter, a rigorous derivation for depth-integrated tsunami propagation
models is presented. These models are categorized according to the relative importance of the frequency dispersion and nonlinearity. As shown in figure 9.1 for most
tsunami events, linear and nondispersive wave model (*i.e.,* linear shallow water
wave equation model) is adequate for estimating the arrival time and the amplitude of the leading waves. However, tsunamis usually contain energy in high frequency range, a linear and dispersive wave model might be desirable if one is
interested in the trailing waves. On the other hand, as tsunamis propagate into
shallower water, wave amplitude will grow and the nonlinearity will transfer wave
energy to both super- and sub-harmonic components. To capture this process, the
Boussinesq-type wave equations might be necessary.

Although the temptation is strong to use the most complicated wave model so that all the processes are included, it is not always cost-effective. Before selecting a wave model for tsunami simulation, one needs to understand the range of physical parameters involved and the purpose of modeling, *i.e.* different models might be chosen for early warning system or tsunami hazard mapping.

Regardless which model is used, the accuracy and resolution of the results are frequently determined by the bathymetry data. This is particularly true in the relatively shallow water.

Acknowledgment

The author would like to acknowledge the continuing support from the National Science Foundation through grants to Cornell University.

References

Chen, Y., and Liu, P. L.-F. 1995. "Modified Boussinesq equations and associated parabolic model for water wave propagation", *J. Fluid Mech.,* **228,** 351–381.

Fujima, K., Masamura, K., and Chiaka, G. 2002. "Development of the 2D/3D hybrid model for tsunami numerical simulation." *Coastal Engrg Jour.,* **44(4),** 373–397.

Geist, E. L. 1998. "Local tsunami and earthquake source parameters", *Advances in Geophysics,* **39,** 117–209.

Gobbi, M. F., Kirby, J. T., and Wei, G. 2000. "A fully nonlinear Boussinesq model for surface waves. II Extension to $O((kh)^4)$," *J. Fluid Mech.* **405,** 182–210.

Hammack, H. L. 1973. "A note on tsunamis: Their generation and propagation in an ocean of uniform depth," *J. Fluid Mech.* **60,** 769–799.

Imamura, F. 1996. "Review of tsunami simulation with a finite difference method". In *Long Wave Runup Models,* edited by H. Yeh, P. Liu, and C. Synolakis, World Scientific, 25–42.

Kanamori, H. 1977. "The energy release in great earthquakes", *J. Geophys. Res.,* **82,** 2981–2987.

Kennedy, A. B., Kirby, J. T., Chen, Q. and Dalrymple, R. A. 2001. "Boussinesq-type equations with improved nonlinear behaviour," *Wave Motion,* **33,** 225–243.

Lin, P. and Liu, P. L.-F. 1998a. "A numerical study of breaking waves in the surf zone", *J. Fluid Mech.,* **359,** 239–264.

Lin, P. and Liu, P. L.-F. 1998b. "Turbulence transport, vorticity dynamics, and solute mixing under plunging breaking waves in surf zone," *J. Geophys. Res.,* **103,** 15677–15694.

Lin, P., Chang, K.-A., and Liu, P. L.-F. 1999. "Runup and rundown of solitary waves on sloping beaches", *J. Waterway, Port, Coastal and Ocean Engrg.,* ASCE, **125(5),** 247–255.

Liu, P. L.-F., 1994. "Model equations for wave propagations from deep to shallow water," in *Advances in Coastal and Ocean Engineering,* Vol. **1** (ed. P. L.-F. Liu), 125–158.

Liu, P. L.-F., Cho, Y.-S., Yoon, S. B. and Seo, S. N., 1994. "Numerical simulations of the 1960 Chilean tsunami propagation and inundation at Hilo, Hawaii," in *Recent Development in Tsunami Research,* edited by M. I. El-Sabh, Kluwer Academic Publishers.

Liu, P. L.-F., Wu, T.-R., Raichlen, F., Synolakis, C., and Borrero, J. 2005. "Runup and rundown generated by three-dimensional sliding masses," *J. Fluid Mech.,* **536,** 107–144.

Lynett, P. and Liu, P. L.-F. 2002. "A numerical study of submarine landslide generated waves and runups", *Proc. Roy. Soc. London,* A, **458,** 2885–2910.

Lynett, P. J., Borrero, J., Liu, P. L.-F., and Synolakis, C. E. 2003. "A review of the Papua New Guinea tsunami", *Pure and Applied Geophysics,* **160,** 2119–2146.

Madsen, P.A., and Schäffer, H.A., 1998. "Higher-Order Boussinesq-type equations for surface gravity waves: derivation and analysis," *Phil. Trans. R. Soc. Lond.* A, **356,** 2123–3184.

Mansinha, L. and Smylie, D. E. 1971. "The displacement fields of inclined faults," *Bull. Seismol. Soc. Am.,* **61,** 1,433–1,440.

Mei, C. C. 1989. *The Applied Dynamics of Ocean Surface Waves,* World Scientific Publishing Co.

Nwogu, O. 1993. "Alternative form of Boussinesq equations for nearshore wave propagation", *J. Wtrwy, Port, Coast and Ocean Engrg.,* ASCE, **119,** 618–638.

Okada, Y. 1985. "Surface deformation due to shear and tensile faults in a half-space", *Bull. Seismol. Soc. Am.,* **75,** 1135–1154.

Priest, G.R., *et al.* 1997. *Cascadia Subduction Zone Tsunamis: Hazard Mapping at Yaquina Bay, Oregon. Final Technical Report to the National Earthquake Hazard Reduction Program,* DOGAMI Open File Report 0–97-34, 143pp.

Titov, V. V. and Synolakis, C. E. 1998. "Numerical modeling of tidal wave runup." *J. Waterway, Port, Coastal and Ocean Eng.,* ASCE, **124(4),** 157–171.

Wang, X. and Liu, P. L.-F. 2006. "An Analysis of 2004 Sumatra Earthquake Fault Plane Mechanisms and Indian Ocean Tsunami", *J. Hydraulic Research,* **44(2),** 147–154.

Woo, S-B. and Liu, P. L.-F. 2004. "A finite element model for modified Boussinesq equations. Part II: Applications to nonlinear harbor oscillations", *J. Waterway, Port, Coastal and Ocean Engrg.,* **130(1),** 17–28.

Yeh, H., Liu, P. L.-F., and Synolakis, C. (Eds) 1996. *Long-wave runup models.* Proc. 2nd Int. Workshop on Long-wave Runup Models, World Scientific Publishing Co., Singapore.

Appendix: Derivation of Two-dimensional Governing Equations

In deriving the two-dimensional governing equations, the frequency dispersion is assumed to be weak, *i.e.,*

$$O\left(\mu^2\right) \ll 1.$$

This also requires that the vertical length scale must be small in comparison with the horizontal length scale. Using μ^2 as the small parameter, we can expand the dimensionless physical variables as power series in terms of μ^2.

$$f = \sum_{n=0}^{\infty} \mu^{2n} f_n ; \quad \left(f = \zeta, p, \mathbf{u}\right), \tag{62}$$

$$w = \sum_{n=1}^{\infty} \mu^{2n} w_n . \tag{63}$$

Furthermore, we will adopt the following assumption in the vorticity field. We assume that the vertical vorticity component, $(u_y - v_x)$, is of $O(1)$, while the horizontal vorticity components are weaker and satisfy the following conditions

$$\frac{\partial}{\partial z}\mathbf{u}_0 = 0 , \quad \frac{\partial}{\partial z}\mathbf{u}_1 = \nabla w_1 \tag{64}$$

Consequently, from (3), the leading order horizontal velocity components are independent of the vertical coordinate, *i.e.,*

$$\mathbf{u}_0 = \mathbf{u}_0(x, y, t). \tag{65}$$

Substituting (62) and (63) into the continuity equation (3) and the boundary condition (8), we collect the leading order terms as

$$\nabla \cdot \mathbf{u}_0 + w_{1z} = 0, \quad -h < z < \varepsilon\zeta, \tag{66}$$

$$w_1 + \mathbf{u}_0 \cdot \nabla h + \frac{h_t}{\varepsilon} = 0, \quad \text{on } z = -h. \tag{67}$$

Integrating (66) with respect to z and using (67) to determine the integration constant, we obtain the vertical profile of the vertical velocity components:

$$w_1 = -z\nabla \cdot \mathbf{u}_0 - \nabla \cdot (h\mathbf{u}_0) - \frac{h_t}{\varepsilon}. \tag{68}$$

Similarly, integrating (64) with respect to z with information from (67), we can find the corresponding vertical profiles of the horizontal velocity components:

$$\mathbf{u}_1 = \frac{z^2}{2}\nabla(\nabla \cdot \mathbf{u}_0) - z\nabla\left[\nabla \cdot (h\mathbf{u}_0) + \frac{h_t}{\varepsilon}\right] + \mathbf{C}_1(x, y, t), \tag{69}$$

in which $\mathbf{C}_1$ is an unknown function to be determined. Up to $O(\mu^2)$, the horizontal velocity components can be expressed as

$$\mathbf{u} = \mathbf{u}_0(x, y, t) + \mu^2\left\{-\frac{z^2}{2}\nabla(\nabla \cdot \mathbf{u}_0) - z\nabla\left[\nabla \cdot (h\mathbf{u}_0) + \frac{h_t}{\varepsilon}\right] + \mathbf{C}_1(x, y, t)\right\},$$
$$+ O(\mu^4), \quad -h < z < \varepsilon\zeta. \tag{70}$$

Now, we can define the horizontal velocity vector, $\mathbf{u}_\alpha(x, y, z_\alpha(x, y, t), t)$ evaluated at $z = z_\alpha(x, y, t)$ as

$$\mathbf{u}_\alpha = \mathbf{u}_0 + \mu^2\left\{\frac{z_\alpha^2}{2}\nabla(\nabla \cdot \mathbf{u}_0) - z_\alpha\nabla\left[\nabla \cdot (h\mathbf{u}_0) + \frac{h_t}{\varepsilon}\right] + \mathbf{C}_1(x, y, t)\right\} + O(\mu^4). \tag{71}$$

Subtracting (71) from (70) we can express $\mathbf{u}$ in terms of $\mathbf{u}_\alpha$ as

$$\mathbf{u} = \mathbf{u}_\alpha - \mu^2\left\{\frac{z^2 - z_\alpha^2}{2}\nabla(\nabla \cdot \mathbf{u}_\alpha) + (z - z_\alpha)\nabla\left[\nabla \cdot (h\mathbf{u}_\alpha) + \frac{h_t}{\varepsilon}\right]\right\} + O(\mu^4). \tag{72}$$

Note that $\mathbf{u}_\alpha = \mathbf{u}_0 + O(\mu^2)$ has been used in (72).

We can define the depth-averaged horizontal velocity vectors as

$$\bar{\mathbf{u}} = \frac{1}{h + \varepsilon\zeta} \int_{-h}^{\varepsilon\zeta} \mathbf{u}\, dz = \mathbf{u}_0 + \mu^2 \left\{ -\frac{1}{6}\left(h^2 - \varepsilon\zeta h + \varepsilon^2\zeta^2\right)\nabla(\nabla \cdot \mathbf{u}_0) \right.$$

$$\left. -\frac{1}{2}(\varepsilon\zeta - h)\nabla\nabla \cdot (h\mathbf{u}_0) + \mathbf{C}_1 \right\} + O(\mu^4). \tag{73}$$

We can further construct the relationships among $\bar{\mathbf{u}}$ and $\mathbf{u}_\alpha$ from (70), (71), (72) and (73). Thus

$$\mathbf{u}_\alpha = \bar{\mathbf{u}} + \mu^2 \left\{ \left[\frac{1}{6}\left(\varepsilon^2\zeta^2 - \varepsilon\zeta h + h^2\right) - \frac{1}{2}z_\alpha^2\right]\nabla(\nabla \cdot \bar{\mathbf{u}}) \right.$$

$$\left. + \left[\frac{1}{2}(\varepsilon\zeta - h) - z_\alpha\right]\nabla[\nabla \cdot (h\bar{\mathbf{u}})] \right\} + O(\mu^4). \tag{74}$$

The exact continuity equation (9) can be rewritten approximately in terms of ζ and $\mathbf{u}_\alpha$. Substituting (72) into (9), we obtain

$$\frac{1}{\varepsilon}H_t + \nabla \cdot (H\mathbf{u}_\alpha) - \mu^2 \nabla \cdot \left\{ H\left[\left(\frac{1}{6}\left(\varepsilon^2\zeta^2 - \varepsilon\zeta h + h^2\right) - \frac{1}{2}z_\alpha^2\right)\nabla(\nabla \cdot \mathbf{u}_\alpha)\right.\right.$$

$$\left.\left. + \left(\frac{1}{2}(\varepsilon\zeta - h) - z_\alpha\right)\nabla\left(\nabla \cdot (h\mathbf{u}_\alpha) + \frac{h_t}{\varepsilon}\right)\right]\right\} = O(\mu^4), \tag{75}$$

in which $H = h + \varepsilon\zeta$.

Equation (75) is one of three governing equations for ζ and $\mathbf{u}_\alpha$. The other two equations come from the horizontal momentum equation, (4). However, we must find the pressure field first. This can be accomplished by approximating the vertical momentum equation (5) as

$$\varepsilon p_z = -1 - \mu^2\left(\varepsilon w_{1t} + \varepsilon^2\mathbf{u}_0 \cdot \nabla w_1 + \varepsilon^2 w_1 w_{1z}\right) + O(\mu^4), \quad -h < z < \varepsilon\zeta. \tag{76}$$

We can integrate the equation above with respect to z to find the pressure field as

$$p = \left(\zeta - \frac{z}{\varepsilon}\right) + \mu^2\left\{\frac{1}{2}\left(z^2 - \varepsilon^2\zeta^2\right)\nabla \cdot \mathbf{u}_{0t} + (z - \varepsilon\zeta)\left[\nabla \cdot (h\mathbf{u})_{0t} + \frac{h_{tt}}{\varepsilon}\right]\right.$$

$$+ \frac{\varepsilon}{2}\left(z^2 - \varepsilon^2\zeta^2\right)\mathbf{u}_0 \cdot \nabla(\nabla \cdot \mathbf{u}_0) + \varepsilon(z - \varepsilon\zeta)\mathbf{u}_0 \cdot \nabla\left[\nabla \cdot (h\mathbf{u}_0) + \frac{h_t}{\varepsilon}\right] + \frac{\varepsilon}{2}\left(\varepsilon^2\zeta^2 - z^2\right)(\nabla \cdot \mathbf{u}_0)^2 \tag{77}$$

$$\left. + \varepsilon(\varepsilon\zeta - z)\left[\nabla \cdot (h\mathbf{u}_0) + \frac{h_t}{\varepsilon}\right]\nabla \cdot \mathbf{u}_0\right\} + O(\mu^4), \quad -h < z < \varepsilon\zeta.$$

We remark here that (70) has been used in deriving (77). To obtain the governing equations for $\mathbf{u}_\alpha$ we first substitute (72) and (77) into (4) and obtain the following equation, up to $O(\mu^2)$,

$$
\mathbf{u}_{\alpha t} + \varepsilon \mathbf{u}_\alpha \cdot \nabla \mathbf{u}_\alpha + \nabla \zeta + \mu^2 \left\{ \frac{1}{2} z_\alpha^2 \nabla \left(\nabla \cdot \mathbf{u}_{\alpha t} \right) + z_\alpha \nabla \left[\nabla \cdot (h\mathbf{u}_\alpha)_t + \frac{h_{tt}}{\varepsilon} \right] \right\}
$$

$$
+ \mu^2 z_{\alpha t} \left\{ z_\alpha \nabla \left(\nabla \cdot \mathbf{u}_\alpha \right) + \nabla \left[\nabla \cdot (h\mathbf{u}_\alpha) + \frac{h_t}{\varepsilon} \right] \right\}
$$

$$
+ \varepsilon \mu^2 \left\{ \left[\nabla \cdot (h\mathbf{u}_\alpha) + \frac{h_t}{\varepsilon} \right] \nabla \left[\nabla \cdot (h\mathbf{u}_\alpha) + \frac{h_t}{\varepsilon} \right] \right.
$$

$$
- \nabla \left[\zeta \left(\nabla \cdot (h\mathbf{u}_\alpha)_t + \frac{h_{tt}}{\varepsilon} \right) \right] + (\mathbf{u}_\alpha \cdot \nabla z_\alpha) \nabla \left[\nabla \cdot (h\mathbf{u}_\alpha) + \frac{h_t}{\varepsilon} \right] \tag{78}
$$

$$
+ z_\alpha \nabla \left[\mathbf{u}_\alpha \cdot \nabla \left(\nabla \cdot (h\mathbf{u}_\alpha) + \frac{h_t}{\varepsilon} \right) \right] + z_\alpha (\mathbf{u}_\alpha \cdot \nabla z_\alpha) \nabla \left(\nabla \cdot \mathbf{u}_\alpha \right) + \frac{z_\alpha^2}{2} \nabla [\mathbf{u}_\alpha \cdot \nabla (\nabla \cdot \mathbf{u}_\alpha)] \right\}
$$

$$
+ \varepsilon^2 \mu^2 \nabla \left\{ -\frac{\zeta^2}{2} \nabla \cdot \mathbf{u}_{\alpha t} - \zeta \mathbf{u}_\alpha \cdot \nabla \left[\nabla \cdot (h\mathbf{u}_\alpha) + \frac{h_t}{\varepsilon} \right] + \zeta \left[\nabla \cdot (h\mathbf{u}_\alpha) + \frac{h_t}{\varepsilon} \right] \nabla \cdot \mathbf{u}_\alpha \right\}
$$

$$
+ \varepsilon^3 \mu^2 \nabla \left\{ \frac{\zeta^2}{2} \left[(\nabla \cdot \mathbf{u}_\alpha)^2 - \mathbf{u}_\alpha \cdot \nabla (\nabla \cdot \mathbf{u}_\alpha) \right] \right\} = O(\mu^4).
$$

Equations (75) and (78) are the coupled governing equations, written in terms of $\mathbf{u}_\alpha$ and ζ, for fully nonlinear, weakly dispersive waves generated by a submarine landslide.

Chapter 10.   Tsunami Modeling: Calculating Inundation and Hazard Maps

FUMIHIKO IMAMURA

Tohoku University

Contents

1.   Tsunamis on land
2.   Resistance formula in the simulation
3.   Inundation and Hazard maps

1. Tsunamis on land

A tsunami amplified in the shallow sea, with increased steepness of the wavefront, would inevitably break down, reducing the energy at the wave front. However, a tsunami could cause flooding, due to the formation of a turbulent bore and finally reach inland, inundating the coastal area. Its wave-length is so large that a huge mass of water follows behind the wave front as shown in Fig.10.1. The run-up heights of a tsunami depend upon the configuration of the shore and land, diffraction, standing wave resonance, and the generation of the edge waves. A narrow valley on land could focus the kinetic energy so that a tsunami could flood the top of the valley, and run-down. A strong back current, accelerated on the slope could cause erosion on land and shore. In addition, a tsunami could also propagate many times through a river mouth, forming an edge bore or solitary waves. With time, the friction at the bottom, the existence of structures and vegetation on land would dissipate the energy transmitted by the waves.

Estimation of tsunami behavior on land is essential for tsunami mitigation and also for designing an optimal community evacuation system along the coast, in the event of such a catastrophe. The inundation area marked on local community maps could be used as a source for public education. Such maps, called hazard maps, could be updated when more information regarding the above mentioned processes becomes better understood. Computer animations of tsunami propagation and inundation are also an effective tool for the general public awareness as well as research for experts to understand the behavior of tsunamis.

The Sea, Volume 15, edited by Eddie N. Bernard and Allan R. Robinson
ISBN 978–0–674–03173–9 ©2009 by the President and Fellows of Harvard College

Inundation area

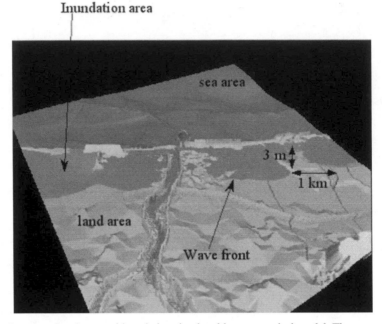

Fig. 10.1 – An example of tsunami inundation simulated by a numerical model. The ocean is towards the top. One piece of valuable information for tsunami mitigation is an inundation map which identifies the tsunami run-up area. The behavior of a tsunami on land is non-linear and the wave front on land is one of most difficult problems to simulate.

1.1 Model of the run-up on land and associated difficulties

Non-liner shallow wave theory can simulate the behavior of tsunamis in shallow sea as well as on land. The convection and bottom friction terms become significant, causing physical as well as numerical difficulties on land in the numerical simulation of tsunamis. The Manning and resistance laws are widely used for bottom friction, the coefficients of which are experimentally selected according to the surface condition of the land and type of land use. A rational and reasonable method for their determination is required. The denominator of the terms involves the total water depth, which is very small at the front, thus causing numerical instability. Therefore, the condition of the wavefront is also critical and a difficult one for simulating the tsunami on the land. Tsunami simulations using Finite Difference Methods and Finite Element Methods are based on the Eulerian reference so that a boundary condition for the front is necessary for the inland run-up.

1.2 Condition of the wavefront on land

Run-up is taken into consideration only in nonlinear computations due to the break down of the linear theory in shallow regions, including the land area. Whether a cell is dry or submerged can be judged as follows:

$$D = \begin{cases} h + \eta > 0, \_then\_cell\_is\_submerged \\ h + \eta \leq 0, \_then\_cell\_is\_dry \end{cases} \quad (10.1)$$

where h is the still water depth, and η, the water surface elevation. Figure 10.2 shows an example of the case of a staggered scheme in space and time. The cell at *j+1* is dry while that at *j* is submerged or wet. The wave front should be located between the "dry" and the "submerged" cells. In general, the discharge across the boundary between the two cells can be calculated, if the ground height at the dry cell is lower than the water level at the submerged cell (case A in Fig.10.2). In other cases, the discharge considered to be zero (case B in Fig.10.2).

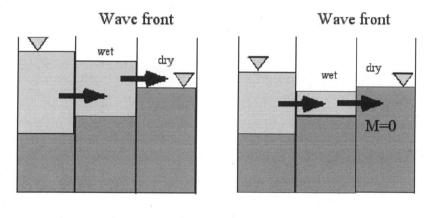

Fig. 10.2 – The condition of the wave front on land as viewed in model cells. The discharge across the boundary between the two cells can be calculated, if the ground height at the dry cell is lower than the water level at the submerged cell as shown in Case-A.

For the run-up condition, it is a unique approach to introduce a "slot" at every grid point and into the equation of continuity (Bundgaard and Warren, 1991). A "slot" is a very narrow channel as shown in Fig.10.3, and sets a grid point to be a "wet" point, any time during a simulation. Therefore, the concept of a wave front between a dry and a wet cell is not necessary. When a water surface reaches a certain level, it is considered to be submerged or wet. However, the problem is to determine the proper slot size. Previous studies regarding the determination of the wave front or a shoreline boundary can be summarized as follows (cf. Fig.10.4). Titov and Synolakis (1993) determine the wave front to be the intersection of the beach with the horizontal projection of the last "wet" point. Other methods use the equations of continuity or motion. Sielecki and Wurtele (1970) proposed the method based on the continuity equation, in which the sea level was extrapolated at the first dry point. On the other hand, Kim and Shimazu (1982) applied the equation of motion, neglecting friction and inertia, which is unreasonable because Matsumoti (1983) has theoretically shown the importance of these factors at the wave front. Matsumoti (1983) determines the wave front by making use of

Whitham's assumptions, which forms a good first approximation at the tip of the flow on a dry bed. Kirkgoz (1983) uses the equation of motion, including the shear stress on the bottom and the inertia, thus determining the height of the wave front in the up-rush zone.

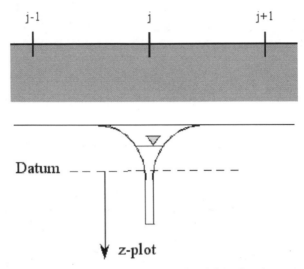

Fig. 10.3 – A schematic description of the slot model. A "slot" is introduced at every grid point and into the equation of continuity as introduced by (Bundgaard and Warren, 1991). A "slot" is a very narrow channel and sets a grid point as a "wet" point, any time during a simulation.

1.3   Calculation of discharge at the wave front

In the case of the staggered scheme in space and time, the need for determining the exact location of a wave front, between the two cell points ("dry" and "wet"), can be avoided while the computation of the equations of motion and continuity proceed alternatively proceeded. However, there is problem of calculating the discharge between the two cells, which strongly affects the water level at the next time step. The following is a summary of methods for estimating the inflow discharge near a wave front as shown in Fig.10.4.

Iwasaki and Mano (1979) assumed that the line connecting the water level and the bottom height gives the surface slope to a first-order approximation. Then the equation of momentum, in the absence of the convection terms can be used directly to calculate the discharge. Hibberd and Peregrine (1979) gave the water level in the dry cell provisionally on a linearly extrapolated water surface. The discharge calculated using this provisional water level gives the total amount of water into the dry cell and hence the depth of water in the cell. If necessary, the computation can be repeated with the modified water level. Aida (1977) and Houston & Butler (1979) evaluated the discharge into the dry cell with broad-crested weir formulas wherein the depth of water in the dry cell was substituted. The coefficient of the ratio of the discharge to sqrt(gh) should be determined by a flow condition such as the Froud number (=u/sqrt(gh)). Imamura (1996) evaluated the discharge directly by applying the equation of momentum, whilst keeping the total depth on the first "dry" cell at zero. In the Imamura model, the total depth at the point of discharge was given by the difference between the ground level on a dry cell and the water level on a wet one. This could overcome the problem in Iwasaki and Mano (1979) that the discharge was evaluated even when level in the submerged cell was lower than the ground level in a dry cell as shown for case-B in Fig.10.2.

These approximations are convenient, but introduce numerical errors in the estimation of the wave front, which was studied by Goto and Shuto (1983). The run-up height as computed with the Iwasaki-Mano method was found to agree with the theoretical solution within an error range of 5%, provided the following condition was satisfied;

$$\Delta x / \alpha g T^2 < 4 \times 10^{-4} \tag{10.2}$$

in which α angle of slope, g is the gravitational acceleration, x is the spatial grid length, and T is the wave period.

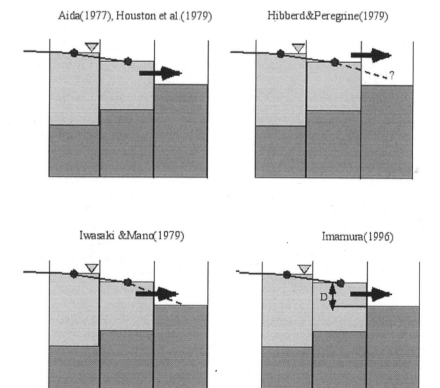

Fig. 10.4 – Numerical methods to estimate the discharge at the wave front are schematized from various models.

1.4 Scheme for the bottom friction

The frictional term also becomes a source of instability when it is discretized using an explicit scheme (Ramming and Kawalik, 1980). This is a good example for understanding the mechanism of numerical instability. Let us look at the following linearized momentum equation without the convection terms:

$$\frac{\partial M}{\partial t} + gD\frac{\partial \eta}{\partial x} + \frac{gn^2}{D^{7/3}} M|M| = 0 \tag{10.3}$$

The explicit form of the equation is

$$M^{n+1} = \left[1 - \frac{\Delta t g n^2 |M|}{D^{7/3}}\right] M^n - gD\Delta t \frac{\partial \eta}{\partial x} \tag{10.4}$$

where M is the discharge in the x-direction, n the Manning coefficients, g the gravitational acceleration, and Δt the time step.

When the velocity becomes large or the total depth is small in a shallow region or on land, the absolute value of the coefficient (amplification factor) of the first term on the right-hand side of Eq. (10.4) become larger than unity, thus amplifying the velocities afterwards, leading to numerical instability. In order to overcome this problem, basically, an implicit velocity via the frictional term can be introduced. For example, a simple implicit form can be

$$M^{n+1} = \frac{\left[M^n - gD\Delta t \frac{\partial \eta}{\partial x}\right]}{\left[1 + \frac{\Delta t g n^2 |M|}{D^{7/3}}\right]} \tag{10.5}$$

This ensures numerical stability, since the amplification factor on the right-hand side of Eq. (10.5) is always less than unity. However, the effect of friction in shallow water becomes so large that numerical results would be inaccurate. Another implicit form, which is a combined implicit one for the frictional term is given by

$$M^{n+1} = M^n \frac{\left[1 - \frac{\Delta t g n^2 |M|}{2D^{7/3}}\right]}{\left[1 + \frac{\Delta t g n^2 |M|}{2D^{7/3}}\right]} - gD\Delta t \frac{\partial \eta}{\partial x} \frac{1}{\left[1 + \frac{\Delta t g n^2 |M|}{2D^{7/3}}\right]} \tag{10.6}$$

This scheme also provides a stable result. The amplification factor would be kept at less than unity and was found to be larger than that in Eq. (10.5). This means that the numerical dissipation would be smaller. The above scheme causes numerical oscillation at the wave front because the amplification factor could be negative in the shallow region having a small total depth, suggesting that the water mass was pulled backwards by unrealistically large friction. We should select the best scheme from among the various implicit schemes in order to apply the bottom frictional term with Manning's roughness.

2. Resistance formula in the simulation

2.1 Bottom friction under the sea

As long as the Manning formula is applicable for tsunami propagation, only the bottom friction needs to be considered for the resistance model. In general, a coefficient of 0.025 in the formula is widely used. Masamura et al.(2000) proposed the following relationship between the Manning coefficient, n, and the equivalent sand diameter, K_s:

$$n=0.15K_s^{1/6}/g \quad (10.6)$$

This suggests that an n=0.025 is equivalent to be sand diameter of 2 cm.

2.2 Resistance model having structures on land

Usually, land would be covered with sand, soil, and land use: vegetation, field, houses, buildings and roads so a model should be selected carefully. There are two models proposed for the run-up on land structure; the first one is a topography model that approximates a house and building by using a special grid, and the second, a model with the Manning formula having equivalent coefficients for resistance of the structures.

1) Topography model. When the spatial grid size in the simulation is smaller than the scale length of the structures (houses and buildings), they could then be approximated by the grid data along with the elevation. The convergence of flow between them or divergence after passing through them can be simulated well, which then enables us to include the forces of resistance via the existence of structures. The grid size should be small enough to approximate them with a square or rectangular mesh. The laser Doppler technique can provide topographical data with a special resolution of 1-2m. Structural destruction or disappearance as a result of the interaction with tsunami waves would then be taken into account for realistic simulations.

2) Model of Manning formula with the equivalent coefficient. The Manning formula with a coefficient selected by the condition of the surface and the land use is well documented (Kotani et al., 1998). If the spatial grid size is larger than the scales of structures (houses and buildings), then these coefficients should be applied. Aida (1969) proposed the conventional way for selecting the value experimentally, based on land use at areas that are tsunami prone in Japan, which should be replaced by a reasonable method.

TABLE 10.1
Values of Coefficient of Bottom Friction n (after Kotani et al., 1998)

Condition of land	n	Condition of land	n
Farming land or waste land	0.020	Residential area or urban district with low density (1–20% occupation ratio of building/house)	0.040
Natural coast and channels and river	0.025	Residential area or urban district with middle density (20–50% occupation ratio of building/house)	0.060
Coastal forest	0.030	Residential area or urban district with high density (50–80% occupation ratio of building/house)	0.080

Aburaya et al. (2000) proposed a method comprising of the bottom roughness, which also included the structures and their location on the grid;

$$R_1 = \rho g D \frac{n_o u^2}{D^{4/3}} dxdy \left(1 - \frac{\theta}{100}\right)$$

$$R_2 = \frac{\frac{1}{2} C_D \rho u^2 (kD) \frac{\theta}{100} dxdy}{k^2}$$

(10.7)

The sum of R_1 and R_2 should equal the force of resistance, and thus equivalently, with the coefficient of Manning. C_D is the drag coefficient and D is the total depth. This leads to the following formula;

$$n = \sqrt{n_o + \frac{C_D}{2gk} \frac{\theta}{100 - \theta} D^{4/3}}$$

(10.8)

where R_1, is the total force of resistance as a result of the roughness at the surface of bottom, R_2 the total force of resistance due to the structures (houses and buildings), u is the average velocity, n_o the Manning coefficient of the bottom , θ the percentage of structures in the spatial grid, and k the equivalent coefficient of Manning.

3.   Inundation and Hazard maps

Tsunamis cause high velocity flooding destroying most structures in its path and quickly killing a lot of coastal residents. History has taught that the best human survival strategy is to avoid the flooding area. Three effective steps to create a tsunami-resistant community are to 1) produce tsunami hazard maps to identify areas susceptible to flooding, 2) implement and maintain an awareness/educational program on indicators of tsunami dangers, and 3) develop early warning systems to alert coastal residents that danger is present (Bernard, 1999).

3.1 Design Tsunami for mitigation plan

Tsunami prevention works as countermeasures can be divided into three categories; (1) permanent structures, (2) emergency management, (3) re-construction and/or re-settlement after the damage. In (1), a permanent structure includes re-settlement to a higher/elevated place, construction of sea wall, break water and tide gate along coastal area with the purpose of prevention for a tsunami disaster, which are called hard countermeasures. In (2), an emergency management program covering tsunami warning, evacuation, the activities of saving lives, relief and rescue, is planned for reducing the damage when a tsunami is generated and threatens the coastal area, which is called a soft type countermeasure. And, in (3), the re-construction and/or re-settlement recovery program, using insurance and land-use regulation, is called regional planning.

In order to make a plan and design a tsunami prevention work, four types of tsunami heights are proposed and used at each stage for design of the structure. The first is a historically maximum height recorded or reported in the past, obtained from the historical documentation and database. The second is a height estimated by the numerical simulations with the bathymetric data in ocean and coast by using data of the earthquake in the past and at seismic gap and potential zones in the future. The third is a tsunami height for planning, which is selected one from the historically maximum and estimated values. The fourth is a tsunami height for design, which is the final one selected among three, taking the total effects such as environment, scenery, utilization and ecology in each region into consideration. The following are the summary of the tsunami heights;

- Historically maximum tsunami height
- Tsunami height estimated by the simulation in the past and future
- Tsunami height for the planning
- Tsunami height for the design

Sea wall, break-water, tsunami bay-mouth breakwater and tide gates are typical structures for the tsunami hard countermeasure, which should be properly designed for preventing the tsunami impact along the coast. Tsunami risk analysis by estimating the tsunami damage at a target area is also important for selecting type and location of structures for the mitigation.

3.2 Making Tsunami Hazards Map with the People

The first step in mitigation is to identify areas that are susceptible to flooding before the tsunami occurs. The ideal way to identify those areas is to use historical information as a guide. Due to the short history available to most coastal communities and the lack of careful surveys, historical data on tsunamis are rare. Scientists have developed numerical models to simulate the behavior of tsunamis to estimate the areas that could be flooded.

In 1989, the IUGG Tsunami Commission and the United Nations Intergovernmental Oceanographic Commission (IOC) formed a partnership to develop an internationally accepted methodology to produce tsunami inundation maps as a contribution to the International Decade of Natural Disaster Reduction

(IDNDR). Professor Nobuo Shuto (Tohoku University) of the Tsunami Commission with support from Japan and the IOC, established the Tsunami Inundation Modeling Exchange Program to transfer tsunami inundation mapping technology to other countries through a comprehensive training program. The technology and training exists to produce tsunami hazard maps for any tsunami threatened community.

The results from the simulation of the tsunami and the estimated inundation/damage can be added the hazard map, which is an effective way to let people understand and be aware of tsunami. First, the earthquake and tsunami scenario is required to define targets on the map. Any single scenario would fail to the necessary awareness amongst the general public, as a result of biased and rigid knowledge base. Secondly, tsunami simulations should be carried out for estimating the arrival time, heights of the first waves, and inundation at the target areas as shown in Fig.10.5. These results could then be compiled on the map. A novel idea of integrating the local and public hazard maps is proposed, which would also serve as a venue for education and outreach (Komura & Hirano, 1997). This would help the general public to understand the mechanism of these natural hazards and discuss concrete ways for reducing damage to their community through the experience from the hazard maps. The picture (Fig.10.6) shows such a workshop held in the city for the discussion of the tsunami hazard map. Such activities help to foster an understanding and also to reach an agreement between the people and the government.

Once the areas of tsunami flooding hazard have been identified, a community-wide effort of tsunami hazard awareness is essential to educate the residents as to appropriate actions to take in the event of a tsunami. Awareness education must include at a minimum: the creation of tsunami evacuation procedures to remove residents from the tsunami hazard zones; the implementation of an education program for schools to prepare students at all age levels; the conduct of periodic practice drills to maintain the preparedness level; and involvement of community organizations to educate all sectors of the population at risk. The IOC has a program to assist countries in implementing tsunami awareness. Written educational materials in numerous languages, educational curriculums, videos, and reports from communities with comprehensive awareness programs are available through the International Tsunami Information Center (e-mail: itic@noaa.gov). A workshop making hazard map with the residents, government member and experts in the community are important for awareness, sharing experiences, and making the plan for the evacuation.

Fig. 10.5 – An example of a hazard map, showing the inundation area on land with the location of evacuation area.

Fig. 10.6 – This photo depicts the process of generating a hazards map through a workshop involving the participation of local people as well as experts.

References

Aburaya, T. and F. Imamura (2002), Proposal of a tsunami run-up simulation using the combined equivalent roughness model, *Proc. of Coastal Eng. in Japan, JSCE,* Vol.49, pp. 276–280, (in Japanese).

Aida, I. (1969), Numerical experiments for the tsunami propagation -the 1964 Niigata tsunami and the 1968 Tokachi-oki tsunami, *Bull. Earth. Res. Inst.,* Vol.47, pp. 673–700.

Aida, I. (1977), Numerical experiments for inundation of tsunamis, Susaki and Usa, in the Kochi prefecture, *Bull. Earth. Res. Inst.,* Vol.52, pp. 441–460 (in Japanese).

Bernard, E. (1999), Tsunami, *Geological Hazards,* Natural Disaster Management, United Nation, Tutor Rose Pub., pp. 58–60.

Bundgaard, H. I. and I. R. Warren (1991), Modeling of tsunami generation and run-up, *Sci. Tsunami Hazard,* Vol.10, pp. 23–29.

Goto, C. and N. Shuto (1983), Numerical simulation of tsunami propagation and run-up, in K. Iida and T. Iwasaki (eds.), *Tsunamis: Their Science and Engineering,* Terra Science Pub.Co., pp. 439–451.

Hibberd, S. and D. H. Peregrine (1979), Surf and run-up on a beach: uniform bore, *J.Fluid Mech.,* Vol.95, pp. 322–345.

Houston, J. R. and H. L. Boutler (1979), A numerical model for tsunami inundation, *WES Rech. Rep.* HL-79-2.

Imamura, F. (1996), Review of tsunami simulation with a finite difference method, *Long-wave Run-up Models* edited by H.Yeh, P.Liu and C.Synolakis, World Scientific (ISBN981-02-2909-7), pp. 231–241.

Iwasaki, T. and A. Mano (1979), Two-dimensional numerical simulation of tsunami run-ups in the Eulerian description, *Proc. 26th Conf.Coastal Eng., JSCE,* pp. 70–72 (in Japanese).

Kim, S. K. and Y. Shimazu (1982), Simulation of tsunami run-up and estimate of tsunami disaster by the expected great earthquake in the Tokai District, Central Japan, *J.Earth Sci., Nagoya Univ.,* Vol.30, pp. 1–30.

Kirkgz, M. S. (1983), Breaking and run-up of long waves, In K.Iida and T.Iwasaki, eds, *Tsunamis: their science and engineering,* Terra Sci., Tokyo, pp. 467–478.

Komura, T. and M.Hirano (1997), Map exercise of DIG (Disaster Imagination Game), *Proc. of Region Safety Assoc.,* Vol.7, pp. 136–139. (in Japanese)

Kotani, M., F. Imamura and N. Shuto (1998), New method of tsunami run-up and estimation of damage using GIS data, *Proc. of Coastal Eng. in Japan, JSCE,* Vol.45, pp. 356–360 (in Japanese).

Matsutomi, H. (1983), Numerical analysis of the run-up of tsunamis on dry bed, In K. Ida and T. Iwasaki, eds, *Tsunamis: their science and engineering,* Terra Sci., Tokyo, pp. 479–493.

Masamura, K., K. Fujima, C. Goto, K. Iida and T. Shigemura (2000),Theoretical Solution of Long Wave Considering the Structure of Bottom Boundary layer and Examinations on Wave Decay due to Sea Bottom Friction, Journal of Hydraulic, *Coastal and Environmental Engineering, JSCE,* No.663/II-53, pp. 69–78, 2000.

Ramming, H. G. and Z. Kowalik (1980), Numerical modeling of marine hydrodynamics - Applications to dynamic physical processes, *Elsevier Oceanography Series 26,* Elsevier Scientific Pub. Co.Ltd., 360p.

Sielecki, A. and M. G. Wurtele (1970), The numerical integration of the nonlinear shallow-water equations with sloping boundaries, *J.Comput.Phys.,* Vol.6, pp. 219–236.

Titov, V. V. and C. E. Synolakis (1993), A numerical study of wave run-up of the September 1, 1992 Nicaraguan tsunami., *Proc. of IUGG/IOC Int. Tsunami Symp.,* Wakayama, pp. 627–635.

Chapter 11.   TSUNAMI IMPACTS ON COASTLINES

Oregon State University

Contents

1. Introduction

Onshore tsunami behaviors and characteristics are quite distinct from other coastal hazards. The effects cannot be inferred from common knowledge or intuition. The primary reason for the difference can be attributed to the unique spatial and time scales associated with tsunami phenomena. For a typical tsunami, the water surface near the shore fluctuates with amplitude of several meters during a period of tens of minutes. This timescale is intermediate between hours to days typical of river-floods, and tens of seconds or less associated with wind waves. It is also difficult to predict hydrodynamic characteristics of tsunamis because they are influenced by the tsunami waveform and the surrounding topography and bathymetry. Nonetheless, based on previous research—including field observations, tsunamis have the following general characteristics in terms of their coastal effects, although there are many exceptions.

At the source, tsunami waveform contains a wide range of wave components—from short to long wavelengths. Long gravity wave components propagate faster than the shorter wave components. Therefore, a transoceanic tsunami is usually characterized by a long-period wave (several to tens of minutes); shorter wave components are left behind and attenuated by radiation and dispersion. However, when the source area is sufficiently large, shorter waves can still propagate in a long distance just as swells generated by storms. For distant tsunamis, such swell-like waves could arrive the shore a long time after the main tsunami attack, and could cause problems in coastal activities, for example, port and harbor operations.

The Sea, Volume 15, edited by Eddie N. Bernard and Allan R. Robinson
ISBN 978–0–674–03173–9 ©2009 by the President and Fellows of Harvard College

For a locally generated tsunami, the leading wave is often a receding water level followed by an advancing positive heave (an elevation wave). This may not be the case if the coastal ground itself subsides by co-seismic displacement; in this case, the leading elevation wave may result. On the other hand, the leading wave of a far-source-generated tsunami is often elevation. These trends are related to the pattern of seafloor displacement resulting from a subduction-type earthquake, which causes ground-surface subsidence followed by uplift in the direction of subduction as depicted in Fig. 11.1. Figure 11.2 shows the leading depression wave measured at the tide gage station in Thailand during the 2004 Indian Ocean Tsunami, in contrast with the leading elevation wave measured at the southern end of India.

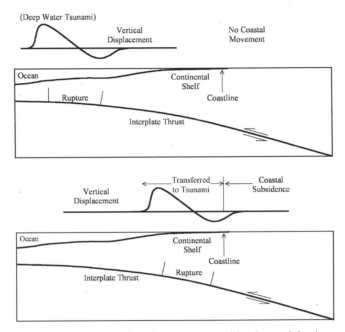

Fig. 11.1 – Schematic diagrams of the vertical displacement resulting from subduction-type fault dislocation: a) rupture zone located far offshore and b) rupture zone adjacent to coastline with coastal subsidence. (after Geist, 1998)

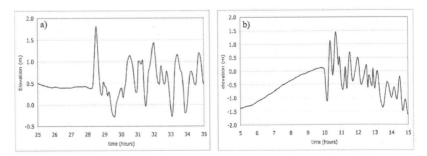

Fig. 11.2 – Tide gage records for the 2004 Indian Ocean tsunami: a) at Ta Phao Noi, Thailand, showing the leading depression wave; b) at Titicorin, India, showing the leading elevation wave. (Digitized by Robertson, 2007)

Tsunami runup heights deviate significantly within neighboring areas. This characteristic may be caused by tsunami's reflective behavior as well as the effects of local bathymetry and coastal topography (Yeh, 1998; SCOR, 2001). Figure 11.3 shows significant variation in runup heights measured along 4 km of the northwest coastline of Okushiri Island by the 1993 Okushiri Tsunami.

One of the important coastal effects to evaluate is tsunami force, which causes building and infrastructural damage. Shuto (1994) summarized the degrees of building damage as shown in Fig. 11.4. Although no detailed information is given and the tsunami heights in the figure are the values of the maximum runup heights in the vicinity of the buildings, his data clearly indicates that reinforced concrete buildings can withstand a majority of tsunami attacks: the exception shown in the figure is the total destruction of Scotch Cap Lighthouse destroyed by the 1946 Aleutian tsunami; the lighthouse was located right at the shoreline as shown in Fig. 11.5.

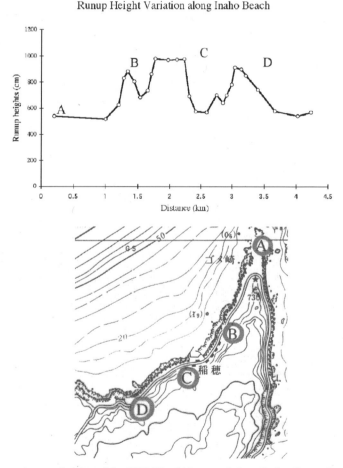

Fig. 11.3 – Measured runup heights of the 1993 Okushiri tsunami along Inaho Coast, demonstrating that the runup height varies significantly within the neighboring area.

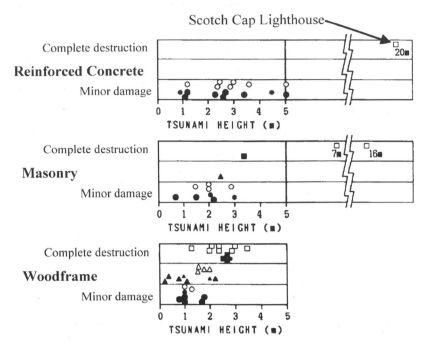

Fig. 11.4 – Degrees of building damage v.s. tsunami runup height. The marks filled in black are the data from the 1993 Okushiri tsunami; the hollow marks are the data from the previous tsunami events. (after Shuto, 1994).

Fig. 11.5 – Scotch Cap Lighthouse destroyed by the 1946 Aleutian Tsunami.

Structural damage by tsunami can be caused primarily by direct water forces, impact forces by water-borne missiles, fire spread by floating materials (including burning oil), scour and slope/foundation failure, and winds induced by the wave motion. The last cause, tsunami-induced winds, is localized and minor, but recorded as the cause of partial damage of a wood frame house by the 1993 Okushiri Tsunami (Shuto, 1994).

The 1993 Okushiri Tsunami completely destroyed the entire town of Aonae. Figure 11.6 shows remains of wood-frame residential houses. Note that the scene is typical that we often observed during the tsunami reconnaissance surveys. Unlike earthquake damage, the debris had been swept away and none left by the tsunami; the only remaining sign of the houses is the concrete-slab foundations. The 1992 Nicaragua Tsunami event provided other examples of damaged wood-frame and masonry houses as shown in Fig. 11.7. The complete destruction of the house with severe scour around the foundation contrasts with the survival of the relatively rigid masonry house and the weak wooden building but with the elevated floor. All three houses shown in Fig. 11.7 are located on a beach berm in the same vicinity, less than 200 m apart. Building failures also occur when waterborne missiles traveling at significant speeds impact buildings. An example of the destruction caused by the impact of a water-borne missile (fishing boats) is shown in Fig. 11.8.

Fig. 11.6 – Total destruction of a group of wood-frame houses in Aonae Village, Okushiri, Island, Japan, by the 1993 Okushiri Tsunami.

Fig. 11.7 – Damaged beach houses in El Popoyo, Nicaragua (1992 Nicaragua Tsunami). All three houses are at the same vicinity.

Fig. 11.8 – Damage cause by a water-borne missile (a boat) in Nagappattinam, India, by the 2004 Indian Ocean Tsunami.

It is evident that tsunami behaviors on coastlines are highly variable and cannot be characterized with a single pattern. Knowing the complexities, attempts are made in this chapter to present the available prediction methodologies to describe tsunami runup behaviors. In the next section, we describe typical waveforms of tsunamis when they approach the shore. With the consideration of the typical waveforms, the prediction models for tsunami hydrodynamics in the runup zone are presented in Section 3. Here we focus on analytic models based on the shallow-water-wave theory under simple geometry and tsunami conditions, i.e. normal incidence of tsunami to a straight shoreline with a uniform sloping beach; hence, the proper physics is retained unambiguously in the analyses. Based on the fundamental behaviors in tsunami runup, practical prediction guidelines to estimate tsunami forces are presented in Section 4. It is emphasized at the outset that no attempt is made in this chapter to incorporate the foregoing complexities associated with site-specific situations.

2.   Typical Processes of Tsunami Runup onto the Shore

The majority of eyewitness accounts and visual records (videos and photos) indicate that an incident tsunami will break offshore forming a 'bore' or a series of bores as it approaches the shore, see Fig. 11.9. A bore is defined as a broken wave having a steep and turbulent wave front, propagating over still water of a finite depth. These broken waves (or bores) are considered relatively short waveforms (although still much longer than wind-generated waves) riding on a much longer main heave of the tsunami. After a bore reaches the shore, the tsunami rushes up on dry land in the formation of a 'surge', as shown in Fig. 11.10.

In some cases, especially when a long-wavelength, leading-elevation, and far-source-generated tsunami attacks land on a steep slope, the runup can be characterized as a gradual rise and fall of water as shown in Fig. 11.11. The impacts of the 1960 Chilean tsunami at some Japanese localities and the 1964 Alaska tsunami at the town of Port Alberni, Canada, are classic examples of this type.

In rare situations, tsunamis could break right at the shoreline (this is called a "collapsing" breaker) under the condition of very steep terrain. The destruction of the Scotch Cap Lighthouse in Unimak Island by the 1946 Aleutian Tsunami could have resulted from a wave breaking directly onto the structure, considering the fact that the lighthouse was located right at the shoreline (see Fig. 11.5).

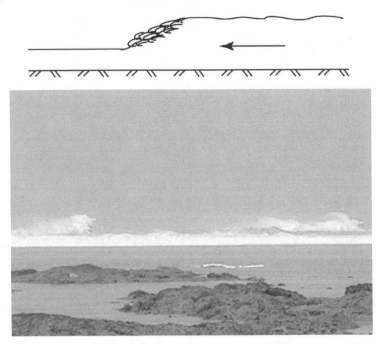

Fig. 11.9 – A sketch of a bore and a photo of the 2004 Indian Ocean Tsunami at Khao Lak, Thailand, showing the formation of a bore (photo by Knill).

Fig. 11.10 – A sketch of a surge and a photo of the 1983 Nihonkai-Chubu Tsunami showing the formation of a surge (photo by Nara).

Fig. 11.11 – A sequence of photos of the 1983 Nihonkai-Chubu Tsunami showing the gradual flooding of tsunami runup (photo by Sato).

A single uniform bore and the subsequent runup formation of surge were simulated in the laboratory by Yeh, et al. (1989). They used the laser-induced fluorescent technique for the visualization. (A 4-watt Argon-ion laser beam is converted to a thin sheet of laser light through a resonant scanner. The generated laser sheet is projected from above in the cross-shore direction along the centerline of the tank, which illuminates the vertical longitudinal plane of the water dyed with fluorescein. Because the laser sheet is approximately 1 mm thick, the method can provide the flow visualization in a virtually two-dimensional plane.) A typical image of the bore is shown in Fig. 11.12. The profile in the figure clearly indicates that the bore is fully developed, i.e. the entire front face is turbulent. The transition process from the incident bore to the runup surge is found to involve the "momentum exchange" between the bore and the small wedge-shaped water body along the shore. As shown in Fig. 11.12, the bore front itself does not reach the shoreline directly, but the large bore mass pushes the small, initially quiescent water in front of it. The term "momentum exchange" is used to describe this transition since the process is analogous to the collision of two bodies; a fast-moving large mass (i.e. bore) collides with a small stationary mass (the wedge-shaped water mass along the shoreline).

The transition process of undular bore to runup is different from that of a fully developed bore as shown in Fig. 11.13. A bore becomes undular when a broken wave front does not form and there is instead a train of unbroken short waves riding behind the front. Instead of the momentum exchange, which occurs for a fully developed bore, the front of undular bore overturns directly onto the dry beach surface, i.e. wave breaking at the shoreline that is the characteristic of the collapsing breaker. The runup motion then commences with the formation of a thin layer of splash-up water.

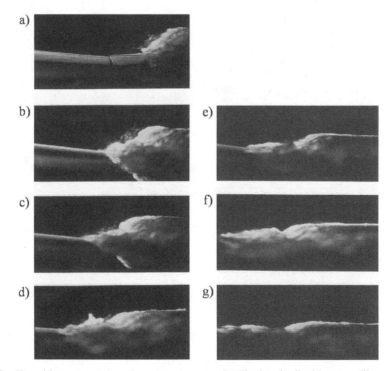

Fig. 11.12 – Transition process from bore to run-up mode. The longitudinal bore profiles were illuminated by the laser sheet. Initial Froude number, F = 1.43. a) bore approaching the shore, b)–e) transition, f)–g) run-up. (after Yeh & Ghazali, 1988).

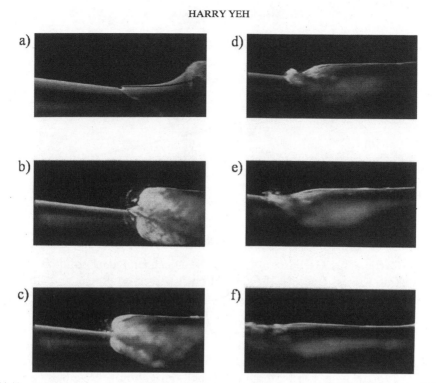

Fig. 11.13 – Transition process from undular bore to run-up mode. Initial Froude number, F = 1.18. a) bore approaching the shore, b)–e) transition, f) run-up. (after Yeh & Ghazali, 1988).

3. Mathematical Analyses

To understand the basic behavior of tsunami effects on coastal areas, we consider the simplest and idealized condition, viz. tsunami runup onto a plane beach as shown in Fig. 11.14. Even in this simple situation, the problem of tsunami runup is nonlinear, the flow is turbulent, and the boundary (beach surface) plays a role in the flow. Nonetheless, as for the approximation, it is customary to formulate the problem with the shallow-water-wave equations, neglecting turbulence and boundary layer effects but retaining the nonlinearity of the wave. Assuming that the beach slope is mild, the pressure field is hydrostatic, and the horizontal water-particle velocity u is uniform over the depth, the depth-integrated conservation equations of mass and momentum can be written as

$$\frac{\partial \eta}{\partial t} + \frac{\partial}{\partial x}\{u(x\alpha + \eta)\} = 0,$$
$$\frac{\partial u}{\partial t} + u\frac{\partial u}{\partial x} + g\frac{\partial \eta}{\partial x} = 0,$$

(1)

respectively, where the x-coordinate points in the offshore direction from the shoreline, $\eta(x, t)$ is the departure of the water surface from the quiescent water depth $h_0(x) = \alpha x$, α is tangent of the beach slope, and g is the acceleration of gravity. Equation (1) is often called the fully nonlinear shallow-water equations.

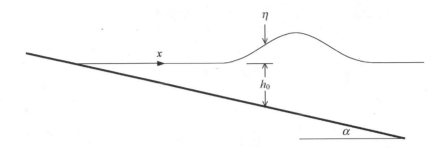

Fig. 11.14 – Definition sketch.

3.1   Non-Breaking Front

In 1958, Carrier and Greenspan derived the exact analytical solution of (1) for non-breaking waves. Because the derivation involves nonlinear and hodograph-type transformation, the Carrier-Greenspan solution is not in a convenient form for the inversion to presentations in real time and space domains. Because of the difficulty, only a limited number of the examples have been reported in the past. Recently, Carrier, Wu, and Yeh (2003) developed a more convenient solution algorithm for arbitrary initial conditions: hereinafter this is called the CWY algorithm. The following is a concise sketch of their derivation—the details are given in Appendix.

Using the following scaling parameters:

$$\hat{u} = \frac{u}{\sqrt{g\alpha L}} \; ; \quad \hat{\eta} = \frac{\eta}{\alpha L} \; ; \quad \hat{x} = \frac{x}{L} \; ; \quad \hat{t} = t\sqrt{\frac{\alpha g}{L}} \; , \tag{2}$$

where L is any convenient horizontal length scale, the shallow-water-wave equations (1) can be expressed in the following dimensionless forms:

$$\frac{\partial \hat{\eta}}{\partial \hat{t}} + \frac{\partial}{\partial \hat{x}}\left\{\hat{u}\left(\hat{x} + \hat{\eta}\right)\right\} = 0,$$
$$\frac{\partial \hat{u}}{\partial \hat{t}} + \hat{u}\frac{\partial \hat{u}}{\partial \hat{x}} + \frac{\partial \hat{\eta}}{\partial \hat{x}} = 0, \tag{3}$$

Note that the beach slope α and the gravity g no longer appear in the governing equations. After a few stages of nonlinear transformation (see Appendix), (3) can be reduced to the form of linear cylindrical-wave equation:

$$\frac{\partial^2 \varphi}{\partial \tau^2} - \frac{1}{4\sigma}\frac{\partial}{\partial \sigma}\left(\sigma\frac{\partial \varphi}{\partial \sigma}\right) = 0 \; , \tag{4}$$

where $\sigma = \sqrt{h} \equiv \sqrt{\hat{x} + \hat{\eta}}$, $\tau = \hat{t} - \hat{u}$, and the function φ is defined as $\partial \varphi / \partial \tau = \hat{\eta} + \hat{u}^2 / 2$. Carrier, Wu, and Yeh (2003) solved this cylindrical wave equation with general initial conditions applying the Fourier-Bessel transform.

Once $\varphi\,(\sigma,\,\tau)$ is computed, the physical (yet non-dimensionalized) variables in the $\hat{x} - \hat{t}$ space are obtained. Unlike the direct numerical simulations, this method can obtain the very accurate solution for arbitrarily selected time and location without computing the rest of the computational domain. Figure 11.15 shows the incoming wave and the outgoing wave generated from the initial static water-surface deformation of the Gaussian shape. Note that the incident wave of single elevation causes the reflected wave in the dipole formation.

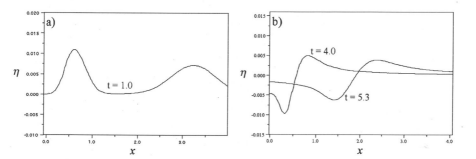

Fig. 11.15 – Computed water-surface profile for a) the incoming wave at t = 1.0 and b) the reflected waves at t = 4.0 and 5.3. (after Carrier, Wu & Yeh, 2003).

Using the developed algorithm, the detailed runup and drawdown motions can be evaluated for arbitrary initial conditions. Figure 11.16 depicts the four initial static water-surface deformations: a) Gaussian shape, b) negative Gaussian shape, c) leading depression N-wave shape, typically caused by a seismic fault dislocation by subduction earthquake, and d) leading depression N-wave shape, typically caused by an offshore submarine landslide. Figure 11.17 shows the temporal and spatial variations of the inundation depth. Temporal variations of the shoreline locations show that the initial wave forms of the predominantly positive displacement, Cases a and c, result in higher runup heights than the heights caused by the initial wave forms of the predominantly negative displacement, Cases b and d. The converse can be observed for the withdrawal distances; those in Cases b and d are greater than Cases a and c. In spite of their equal positive displacements in Cases a and c, the maximum runup height is substantially greater in Case c (the leading depression N-wave) than in Case a (the single positive displacement of the Gaussian shape). This higher runup height of the leading depression N-wave is consistent with the findings of Mazova and Pelinovsky (1991) and Tadepalli and Synolakis (1994).

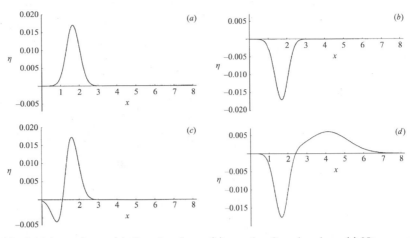

Fig. 11.16 – Initial waveforms. (a) Gaussian shape, (b) nagative Gaussian shape,(c) N-wave caused by co-seismic seabed dislocation, and (d) N-wave caused by submarine landslide.

The magnitude of the maximum runup height of Case a (positive Gaussian wave) is identical to that of the minimum shoreline elevation of Case b (negative Gaussian wave), and the maximum runup height of Case b has an equal magnitude to the minimum of Case a. This coincidence can be interpreted as that the extreme shoreline location computed by fully nonlinear theory is identical to those predicted by linear theory, while the shoreline trajectories and waveforms are different for the two cases as observed in Fig. 11.17a and b. This characteristic was pointed out by Carrier (1971) and also discussed by Synolakis (1987).

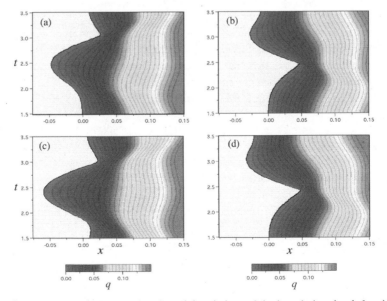

Fig. 11.17 – (See color insert) Temporal and spatial variations of the inundation depth for a) the initial waveforms of the Gaussian shape (Case a), b) the negative Gaussian shape (Case b), c) the leading depression N-wave (Case c), and d) the waveform caused by the submarine landslide (Case d). (after Carrier, Wu & Yeh, 2003).

Temporal and spatial variations of the runup flow velocities in the $\hat{x} - \hat{t}$ plane are shown in Fig. 11.18. Note that the maximum and minimum flow velocities occur at the shoreline, and the maximum flow speeds occur near the minimum drawdown position. For Cases a and c (predominantly positive displacement in the initial wave form), the maximum shoreline speed occurs during the drawdown process, i.e., in the offshore direction. On the other hand, the maximum speed occurs during the runup process, i.e., in the inshore direction, for Cases b and d, which are generated by predominantly negative initial water-surface displacement. This shoreline velocity behavior implies that objects (e.g. sediments, boulders, etc.) will be likely carried inshore in the event of predominantly negative initial waves, while they will be carried offshore in the case of positive initial waves.

The linear-momentum flux is also an important parameter to examine the effects of tsunami runup and rundown. The momentum flux per unit breadth for the quasi-steady flows can be expressed as

$$f = \hat{h}\hat{u}^2, \tag{5}$$

which can be interpreted as the (non-dimensionalized) drag force per unit breadth for a surface-piercing stationary object being placed vertically over the flow depth. Hence the quantity evaluated by (5) provides a measure for the net force exerted on such an object.

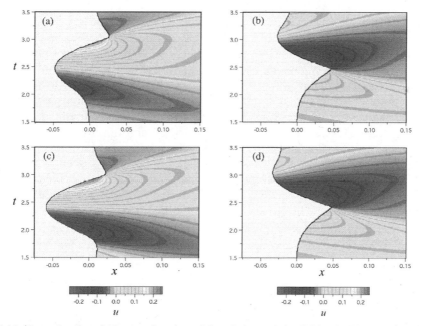

Fig. 11.18 (See color insert) Temporal and spatial variations of the fluid velocities for a) the initial waveforms of the Gaussian shape (Case a), b) the negative Gaussian shape (Case b), c) the leading depression N-wave (Case c), and d) the waveform caused by the submarine landslide (Case d). (after Carrier, Wu & Yeh, 2003).

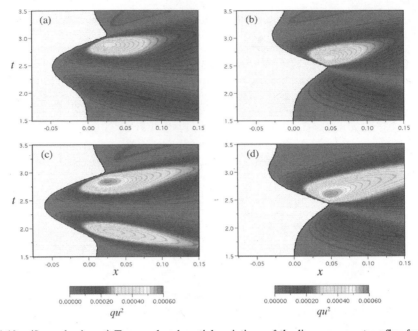

Fig. 11.19 – (See color insert) Temporal and spatial variations of the linear-momentum flux for a) the initial waveforms of the Gaussian shape (Case a), b) the negative Gaussian shape (Case b), c) the leading depression N-wave (Case c), and d) the waveform caused by the submarine landslide (Case d). (after Carrier, Wu & Yeh, 2003).

The computed momentum flux f shown in Fig. 11.19 indicates that the maximum momentum flux occurs near the extreme drawdown location, $x > 0$, immediately prior to or after the flow reversal. For predominantly positive initial waves (Cases a and c), the maximum momentum flux results during the drawdown phase and points offshore. On the other hand, for the cases of predominantly negative initial waveform (Cases b and d), the maximum momentum flux occurs during the runup immediately after the extreme drawdown, acting in the inshore direction. Also note that for the incident wave of leading depression N-wave (Case c), both inshore and offshore forces are significant, although the drawdown force is still greater.

Using this CWY algorithm, we examined four initial wave conditions, which represent typical tsunami generations. It was found that the gross characteristics in the initial wave form are important for the behaviors in runup and rundown motion, i.e. the motions in Cases a and c are similar, and Cases b and d are also similar.

The CWY algorithm assumes no wave breaking, i.e. no singularity for the nonlinear transformation. Hence, once the value of the Jacobian of the transformation (see Appendix) becomes zero, physical interpretations of the solution become uncertain. Fortunately, as pointed out by Synolakis (1987) with his laboratory experimental results, even if the Jacobian becomes zero at the tip of the shoreline, the solution of nonlinear shallow-water wave theory recovers immediately and yields extremely accurate predictions for later times.

3.2 Bore Front

When a tsunami breaks farther offshore (see Fig. 11.9), the waveform becomes analogous to a bore. Then, it is uncertain if the CWY algorithm be valid for the description of such a prolonged broken wave formation. The behavior of bore propagation is often modeled as a discontinuity in water depth and velocity in the shallow-water wave theory described in (1).

For a general conservation equation of the form:

$$\frac{D\Phi}{Dt} + \Phi\nabla\cdot\mathbf{u} + \nabla\cdot G + H = 0, \tag{6}$$

the corresponding jump condition was derived by Yeh (1995) such that:

$$\left[\!\left[\Phi(\mathbf{u}\cdot\mathbf{n}-\mathbf{U}\cdot\mathbf{n})+G\cdot\mathbf{n}\right]\!\right]=0. \tag{7}$$

Note that in (6) and (7), $\Phi(\mathbf{x}, t)$ is any integrable scalar or vector function of position x and time t, $\mathbf{n}$ is the unit vector normal to the jump surface, $\mathbf{u}$ is the fluid particle velocity, $\mathbf{U}$ is the velocity of the discontinuity, and the double bracket denotes the jump of the quantity across the discontinuity, e.g.

$$\left[\!\left[\Phi\right]\!\right]= \lim_{\substack{x\to\xi\in\Sigma \\ x\in D_2}} \Phi - \mathrm{Lim}_{\substack{x\to\xi\in\Sigma \\ x\in D_1}} \Phi, \tag{8}$$

where ξ is the location of the discontinuity surface Σ between domains D_1 and D_2. Using (6) and (7) for shallow-water-wave equations (1) yields the jump conditions of the depth-integrated conservation equations of mass and momentum, respectively:

$$\begin{aligned}
\left[\!\left[uh\right]\!\right]-U\left[\!\left[\eta\right]\!\right]&=0, \\
\left[\!\left[u^2h-\tfrac{1}{2}g\eta^2+g\eta h\right]\!\right]-U\left[\!\left[uh\right]\!\right]&=0
\end{aligned} \tag{9}$$

where $h = x\alpha + \eta$, η is the bore height from the quiescent water level, and U is the propagation speed of the bore front. It is emphasized that (9) is the jump conditions for the condition with a uniformly sloping plane beach. Solving the pair of equations of (9) yields

$$U = \sqrt{\frac{gh(2h-\eta)}{2(h-\eta)}}; \quad u = \frac{\eta}{h}U \tag{10}$$

Now, following Whitham (1958), we consider the 'positive' characteristic of (1), $dx/dt = u + c$, to yield the characteristic relation:

$$du + 2\,dc = g\,\alpha\,dt = \frac{g\,\alpha\,dx}{u+c} \tag{11}$$

where $c = \sqrt{gh}$. Then using (10), the bore height is given by

$$\eta = 2h_0\left(F^2 - 1\right) \tag{12}$$

in which $h_0 = \alpha x$ is the quiescent water depth, and we introduced the Froude parameter, $F = U/c$. And the bore velocity can be expressed as:

$$\frac{U}{\sqrt{gh_0}} = F\sqrt{2F^2 - 1}. \tag{13}$$

Furthermore, from (11) and (12),

$$\frac{c}{\sqrt{gh_0}} = \sqrt{2F^2 - 1}; \quad \frac{u}{\sqrt{gh_0}} = \frac{2F\left(F^2 - 1\right)}{\sqrt{2F^2 - 1}} \tag{14}$$

Substituting (14) in (11) yields

$$\frac{1}{h_0}\frac{dh_0}{dF} = \frac{1}{x}\frac{dx}{dF} = -\frac{2\left(2F^3 + 2F^2 - 2F - 1\right)\left(F+1\right)\left(2F-1\right)^2}{\left(2F^2 - 1\right)\left(2F^5 + 4F^4 - 4F^3 - 5F^2 + F + 2\right)} \tag{15}$$

which is a slightly different presentation (but equivalent) form from Whitham (1958). Note that the value of F must be greater than unity for bore propagation. Since the right-hand-side of (15) is negative, F always increases as h_0 (or x) decreases and vice versa. For a weak bore ($F \approx 1$), the asymptotic form of (15) is

$$\frac{1}{h_0}\frac{dh_0}{dF} = \frac{1}{x}\frac{dx}{dF} = -\frac{4}{5}\frac{1}{F-1}. \tag{16}$$

Therefore, $F - 1 \propto h_0^{-5/4}$, which means

$$\eta \propto h_0^{-1/4}. \tag{17}$$

This is the Green Law: the result of the linear theory.

For $F \gg 1$ (strong bores), the asymptotic form of (15) becomes

$$\frac{1}{h_0}\frac{dh_0}{dF} = \frac{1}{x}\frac{dx}{dF} = -\frac{4}{F} \tag{18}$$

which leads to

$$\eta \propto h_0^{\frac{1}{2}} \propto x^{\frac{1}{2}}; \quad u \approx U \propto h_0^{-\frac{1}{2}} \propto x^{-\frac{1}{2}}. \tag{19}$$

Hence, the shallow-water wave theory predicts that the height of a strong bore tends to vanish as it approaches the shoreline. At the shoreline, the fluid velocity, u, and bore front velocity, U, increase as $x^{-1/2}$ and approach their common finite value U^* (Ho and Meyer, 1962), whereas their accelerations become singular at the shoreline. This behavior at a shoreline involving the rapid conversion of potential to kinetic energy is often called "bore collapse".

The experimental results of Yeh and Ghazali (1986, 1988) demonstrate the detailed transition process at the shoreline as shown in Fig. 11.12. As discussed earlier, the transition process is not exactly the "bore collapse" predicted by (19), but involves the "momentum exchange" between the bore and the small wedge-shaped water body along the shore. As shown in Fig. 11.12, the bore front itself does not reach the shoreline directly, but the large bore mass pushes the small, initially quiescent water in front of it. The term "momentum exchange" is used to describe this transition since the process is analogous to the collision of two bodies; a fast-moving large mass (i.e. bore) collides with a small stationary mass (the wedge-shaped water mass along the shoreline). Yeh and Ghazali (1986, 1988) also found that, because $u \to U$ near the shoreline, the turbulence on the front face of a bore as well as that generated at the transition process is advected forward onto the dry beach with the runup motion instead of leaving it behind the wave front.

The transition process could be related to turbulence nearshore that penetrates to the beach bottom, which causes the "splash-up" of the water along the shoreline as shown in Fig. 11.12. Nonetheless, turbulence observed by Yeh and Ghazali (1986, 1988) seems too weak for that explanation to describe the process. Instead, the following conjecture was presented. If the bore front were a discontinuity of flow as modeled in the shallow-water wave theory, conservation of momentum predicts that a sudden acceleration of propagation would occur near the shoreline since a mass of water diminishes in front of the bore as it approaches the shore. However, a real bore front has a finite length (not a discontinuity) and the local flow at the front cannot be described by the shallow-water wave theory (not a hydrostatic pressure field).

Suppose that there is a smooth precursor of a bore moving with the same speed as the bore propagation, U, and it is small enough to be considered to be a linear wave. In water of uniform depth, such a leading wave precursor has the velocity potential ϕ of the form $\exp(kx + \omega t)\cos(k(y + h_0))$ with the dispersion relation $\omega^2 = gk\tan(kh_0)$, just like the representation of the outskirts of a solitary wave as pointed out by Lamb (Art. 252, 1932). (Note that the y-coordinate points vertically upward from the quiescent water surface.) A similar wave associated with a bore on a uniformly sloping beach with $\alpha = \pi/4n$, where n is a positive integer, is inferred from the oscillatory wave solution provided by Hanson (1926) and can be found to be

$$\phi = \Re\left[e^{-\omega t}\left\{ \begin{array}{l} C_{0,2}\, e^{-ikx-ky} \\[2mm] + \displaystyle\sum_{j=1}^{m-1}\sum_{p=1}^{2} C_{j,p}\, e^{\left\{(-1)^{p+1}ky\cos(2j\alpha)-kx\sin(2j\alpha)+i\left(-kx\cos(2j\alpha)+(-1)^{p}ky\sin(2j\alpha)\right)\right\}} \\[4mm] + C_{m,1}\, e^{-kx-iky} \end{array}\right\}\right] \tag{20}$$

where $\Re e$ is the real part of the function, $i^2 = -1$, and

$$\omega^2 = gk; \quad C_{m,1} = (i-1)A = C_{m-1,2}; \quad C_{j,1} = -iC_{j,2}\tan j\alpha; \quad C_{j-1,2} = C_{j,1}$$

and A is a constant. Note that (20) exactly satisfies the Laplace equation in the fluid domain, the boundary condition along the uniformly sloping beach, and the linearized free-surface boundary condition:

$$\nabla^2\phi = 0,$$

$$\frac{\partial\phi}{\partial x}\sin\alpha = \frac{\partial\phi}{\partial y}\cos\alpha \quad on\ x\sin\alpha = y\cos\alpha \tag{21}$$

$$\frac{\partial^2\phi}{\partial t^2} + g\frac{\partial\phi}{\partial y} = 0 \quad on\ y = 0$$

Under the beach condition with $\alpha = \pi/24$ (the identical beach slope for the laboratory results presented in Fig. 11.12), the water surface variation predicted by (20) is shown in Fig. 11.20. Note that the water surface η monotonically increases in the offshore direction; hence the solution is physically viable only if it describes a local flow near the shoreline and the offshore boundary condition is provided, such as a bore advancing toward the shore. The horizontal component of fluid particle velocity, u, is computed from (20) at various offshore locations kx and presented in Fig. 11.21. Instead of a uniform velocity profile over the depth, fluid particles at the leading bore front can have increasing velocities with depth. Particle trajectories created by this wave are shown in Fig. 11.22. While a particle initially at the bottom travels along the beach surface with an increasing velocity, a particle near the surface initially moves horizontally but then gradually changes its course upward and eventually moves offshore with a significant vertical velocity component. At this stage, our linearized wave approximation, of course, breaks down; hence the final trajectories shown in Fig. 11.22 are not realistic. Nonetheless, both Figs. 11.21 and 11.22 clearly demonstrate that, near the shoreline, this non-uniform velocity effect manifests itself as "pushing" the wedge of water along the shore just like the "momentum exchange" process as described earlier.

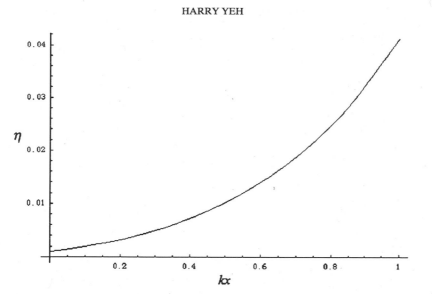

Fig. 11.20 – The water surface profile based on (20). The shoreline is at $kx = 0$, and the x-coordinate points offshore. $\alpha = \pi/24$, $A = 0.1$, $k = 1.0$.

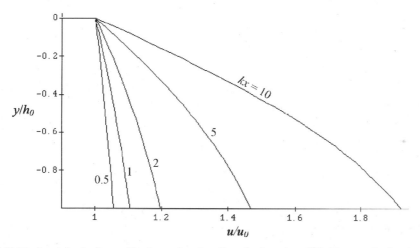

Fig. 11.21 Horizontal velocity profiles for various locations kx, based on (20). The magnitude of velocity is normalized by that at the surface. $\alpha = \pi/24$.

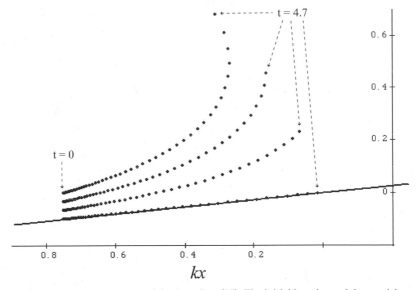

Fig. 11.22 The trajectories of water particles based on (20). The initial locations of the particles are at kx = 0.75, ω = 1.0. The time t is varied from 0.0 to 4.7 by the increment of 0.1. The solid line represents the beach surface. $\alpha = \pi/24$.

3.2.1   Runup of Bore

Based on the shallow-water wave theory, Shen and Meyer (1963) analyzed the wave runup, which subsequently occurs after the bore collapse. They found that the motion of a runup front is totally governed by the gravity force. Hence, the runup front velocity can be expressed as

$$u_s = \sqrt{2 g \alpha |x|} \,, \tag{22}$$

where u_s is the velocity of the runup front tip: note that the runup position is always negative in x (recall that the x-coordinate points offshore from the shoreline at quiescent state). The maximum runup height, R, is simply,

$$R = \frac{U^{*2}}{2g} \tag{23}$$

which is a total conversion of kinetic energy at the bore collapse to potential energy: recall that U^* is the velocity of the bore front (as well as the fluid) at the shoreline. Shen and Meyer (1963) found that the water depth close to the front could be approximated by

$$h(x,t) \rightarrow \frac{\left(x_s(t) - x\right)^2}{9 t^2} \tag{24}$$

as $(x_s - x) \to 0$. According to (24), the water surface is tangential to the beach surface at the front and the sheet of runup becomes thinner as time increases. Note that this result is for inviscid fluids, and in reality, there forms the runup-tongue where the entire depth is saturated by the boundary layer as discussed by Whitham (1955).

Following Shen and Meyer (1963), Peregrine and Williams (2001) provided the formulae for the temporal and spatial variations in fluid velocity and flow depth of the bore runup near the runup tongue. With slightly different scaling, Peregrine and Williams' formulae for the flow depth and velocity can be expressed, respectively:

$$\delta = \frac{1}{36\,\tau^2}\left(2\sqrt{2}\,\tau - \tau^2 - 2\zeta\right)^2, \tag{25}$$

and

$$v = \frac{1}{3\tau}\left(\tau - \sqrt{2}\,\tau^2 + \sqrt{2}\,\zeta\right). \tag{26}$$

where:

$$\delta = {d}/{R}\,; \ v = \frac{u}{\sqrt{2gR}}\,; \ \tau = t\tan\alpha\sqrt{{g}/{R}}\,; \ \zeta = {z}/{R}\,, \tag{27}$$

and d is the water depth, R is the ground elevation at the maximum penetration of tsunami runup measured from the initial shoreline, t is the time ($t = 0$ when the bore passes at the initial shoreline), and z is the ground elevation of the location of interest, measured from the initial shoreline.

Laboratory data of the bore propagation and runup onto a uniformly sloping beach from Yeh et al. (1989) were shown in Fig. 11.23 together with the theoretical predictions made by the shallow-water-wave theory. The scatter in the measured data is not measurement errors or a repeatability problem but is due to the irregularities associated with the bore propagation itself, i.e. a bore front is not smooth and the propagation of the front widely fluctuates. Figure 11.23 also indicates that the bore velocity offshore, U, decelerates faster than the prediction by (15). A similar discrepancy was found in Miller's (1968) experimental results; the value of U decelerates to 68 % of the velocity measured at the beach toe, which is equivalent to $U/U^* \approx 0.39$ in Fig. 11.23. Besides the frictional effects and the effects of frequency dispersion, the reason for this discrepancy is not clear.

In Fig. 11.23, the measured results support qualitatively the occurrence of bore collapse, i.e. a sudden acceleration of the bore propagation is evident from the results. It is emphasized, however, that the photographic results shown in Fig. 11.12 indicate that the actual transition involves the "momentum exchange" between the incident bore and the small wedge-shaped water along the shoreline, but not a genuine bore collapse. The widely scattered data near the shoreline are due to this momentum exchange process. The acceleration caused by the collided water mass appears to commence earlier than the theoretical bore collapse prediction.

During the runup process, the leading-tongue velocity is always smaller than the prediction. The runup front velocities predicted by (22), are modified by using the measured maximum velocity (0.82 U^*) at the shoreline and presented by the broken line in Fig. 11.23. The measured values seem to be in fairly good agreement with the modified prediction. Considering the transition process at the shoreline, the runup motion is initiated by "pushing" the water mass. This momentum exchange process is not instantaneous, and the runup motion forms a thick layer of flow. Hence, the driving force for the runup motion appears to be the pressure gradient existing during the transition (Note that an infinitesimally thin layer of flow could not be influenced by the pressure gradient, as discussed by Shen and Meyer in 1963).

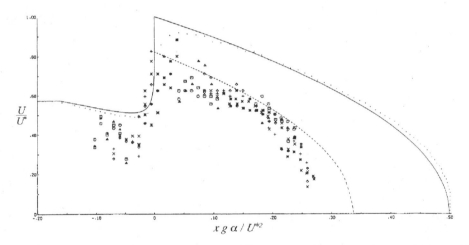

Fig. 11.23 – Variation of bore and runup front velocities. Initial Froude number = 1.43; U^* = 243 cm/sec.; the solid line, theoretical velocities by (15) and (22); the dotted line, numerical results; the broken line, theoretical velocities modified by 0.82 U^*. Different symbols denote measured velocity data taken from the repeated experiments. (after Yeh & Ghazali, 1988).

3.3   Summary

Foregoing theoretical predictions provide tsunamis' near-shore and runup characteristics. It is emphasized that the results are drawn from the fully nonlinear shallow-water theory that assumes 1) hydrostatic pressure field, 2) uniform horizontal velocity over the depth, which implies inviscid fluids, and 3) the uniformly sloping plane beach configurations. The CWY algorithm solves a long-wave runup from an arbitrary initial condition. This method is robust and general, for it is capable of computing both incoming and outgoing waves from arbitrary initial wave condtions. However, because of the nonlinear transformation involved, the results are limited for non-broken tsunamis in theory, although, even if the Jacobian becomes zero at the tip of the shoreline, the solution of nonlinear shallow-water wave theory recovers immediately and yields accurate predictions for later times. Based on this theoretical work, we found that the maximum flow velocity occurs at the moving shoreline during the drawdown phase for the predominantly positive initial waves, while the maximum velocity for the predominantly negative initial waves also occurs at the shoreline but during the runup. Since flow stresses are proportional to the square of velocity, submerged objects such as sediments, boulders, and alike tend to move shoreward when the initial waveform is a depression. On the other hand, they tend to move offshore when the initial wave is an elevation. For the initial positive waveform, the maximum momentum flux (force) over the flow depth occurs after the maximum penetration, at the vicinity of the extreme drawdown location prior to the subsequent flow reversal; hence the maximum momentum flux is in the offshore direction. On the other hand, the maximum momentum flux occurs in the inshore direction prior to the maximum runup penetration for the initial negative waveform, and it occurs immediately after the flow reversal from the initial withdrawal caused by the leading depression wave. This indicates that the dominant momentum flux is shoreward for the predominantly negative initial waveform. Regardless of the initial waveform, the momentum flux, that is a measure of the net force exerted on a surface-piercing object over the depth, becomes the maximum near the extreme drawdown location. This indicates that offshore structures placed close to the shore are vulnerable to tsunami attacks: for example, breakwaters, piers, LNG and oil loading terminals, ships moored offshore, etc.

Behaviors of a single bore propagating onto quiescent water on a plane beach and their ensuing runup motions were described based on the shallow-water theory. The speed of bore propagation decelerates when it approaches the shore. Although the qualitative behavior of the propagation during the transition from bore to runup mode resembles the "bore collapse" predicted by the theory, the acceleration is resulted from the "pushed-up" water by the momentum exchange process but not due to the speed of the bore front itself. The momentum exchange phenomenon is a consequence of non-uniform velocity and acceleration fields over the water depth caused by a leading wave of the form (20), which propagates as the precursor of the bore front. The subsequent runup motion is governed by the gravity. With foregoing theoretical predictions, we now develop tools to evaluate the coastal effects of tsunamis.

4. Applications

In this section, we introduce simple and useful methodologies to estimate tsunami forces on inshore structures ($x < 0$) by utilizing the theoretical understandings of tsunami runup characteristics described in Sec. 3. It is emphasized that our analyses presented in this section are not applicable in the offshore region ($x > 0$).

When water flows around a structure, hydrodynamic forces are exerted on the structure. These forces are induced by quasi-steady flows, and are a function of fluid density ρ, flow velocity u and structure geometry. They are often called drag forces. The maximum hydrodynamic forces can be computed by

$$F_d = \frac{1}{2} \rho \, C_d \, B \left(h u^2 \right)_{max} \qquad (28)$$

where h is the flow depth at the location of interest when there is no flow obstruction (i.e. no structure); the combination $h \, u^2$ represents the momentum flux per unit mass per unit breadth; B is the breadth of the structure in the plane normal to the flow direction. The drag coefficient takes a value of approximately $1.0 < C_d < 2.0$ depending on the shape of the objects (FEMA, 2005; Arnason, 2005). At a given location, the maximum flow depth, h_{max}, and maximum flow velocity, u_{max}, may not occur at the same time: $(h \, u^2)_{max} \neq h_{max} \, u^2_{max}$. The hydrodynamic forces must be based on the maximum momentum flux, $(h \, u^2)_{max}$, occurring at the site. In the area of coastal engineering, hydrodynamic forces are evaluated in combination of the drag force (28) and the inertial force. Because tsunami wave periods are very long, the inertial force is considered unimportant except at the impact of tsunami's leading edge, which is called surge forces.

Surge forces are caused by the leading edge of surge of water impinging on a structure. Ramsden (1993) performed comprehensive experiments on surging forces. His laboratory data shown in Fig. 11.24 demonstrate no initial impact force (surging force) in dry-bed surges, but an overshoot in force was observed in bores. (Note that there is water (not dry) in front of the bore propagation.) The maximum overshoot for bore impact is approximately 1.5 times the subsequent hydrodynamic force, which is consistent with independent laboratory data obtained by Arnason (2005). The lack of overshoot in dry-bed surge can be attributed to the relatively mild slope of the front profile of the water surface, while the impact momentum increases with the sudden slam of the steep front of bores (see, Fig. 11.24b). This suggests that the initial impingement of a 'dry-bed' surge, i.e. inertia force, is not important for the evaluation of tsunami forces in inshore locations.

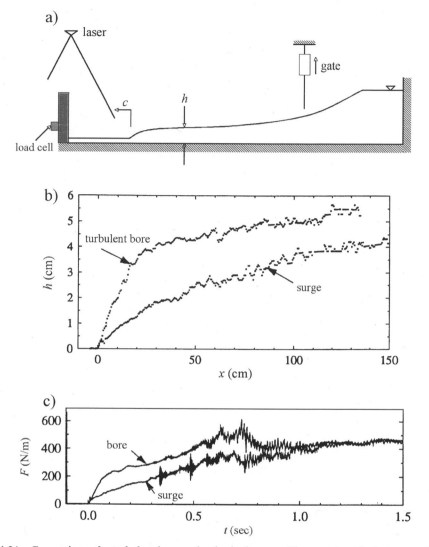

Fig. 11.24 Comparison of a turbulent bore and a dry-bed surge with approximately the same celerity (Ramsden, 1993): a) experimental set up, b) water-surface profiles, c) forces measured by the load cell.

The hydrodynamic and surge forces can be computed by (28). A problem is how to estimate the maximum value of $(h\,u^2)_{max}$. In fact this is the reason why many existing force formulae are expressed with a function of depth only (but not flow velocity). Ideally, the maximum value of hu^2 should be obtained by running a detailed numerical simulation model. However the numerical model in the runup zone must be run with a very fine grid size (say, less than 5 m) to ensure adequate accuracy in the prediction of $h\,u^2$.

Using the CWY algorithm discussed in Sec. 3, the maximum fluid-force distribution in near-shore region can be computed and the results are shown in Fig. 11.25. As we mentioned earlier, the CWY algorithm assumes no wave breaking, i.e. the

Jacobian of the nonlinear transformation cannot vanish in theory for the validity of the solution. As an appropriate comparison, the distribution curve computed for a climb-up of a uniform bore is included in Fig. 11.25.

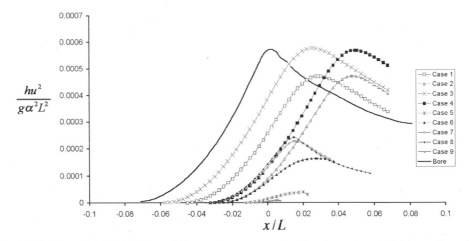

Fig. 11.25 – The spatial distribution of the maximum momentum flux for various initial tsunami conditions. The spatial origin x is at the initial shoreline location, and the x points offshore. (after Yeh, 2006).

Because the scaling parameter L in (2) can be 'any' horizontal length scale, the results in Fig. 11.25 can be rescaled. It is more convenient to represent the tsunami force distribution with the runup (inundation) distance ℓ: the distance between the initial shoreline and the maximum runup location. In Fig. 11.26, the origin of x is now shifted to the maximum runup location. In spite of a variety of the initial waveforms (including a bore), amplitudes, and the source locations, the resulting distribution curves in the runup zone appear to form the envelope. Note that three curves falling under the others are for the cases with much weaker initial conditions, i.e. smaller displacement amplitude with broader displacement breadth. With a curve fit, the envelope of the maximum value of $h\,u^2$ can be expressed by

$$\frac{\left(h u^2\right)_{\max}}{g\,\alpha^2\,\ell^2} = 0.11\left(x/\ell\right)^2 + 0.015\left(x/\ell\right) \tag{29}$$

It must be emphasized that this solution is for a uniform beach slope; therefore, some adjustments need to be made for a real situation. It is further convenient to express (29) as a function of the ground elevation, instead of the horizontal location—this maneuver is motivated by the fact that tsunami runup motion may be modeled as a conversion of kinetic energy to potential energy as demonstrated in (22) and (23). Transforming (29) yields

$$\frac{\left(h u^2\right)_{\max}}{g\,R^2} = 0.125 - 0.235\frac{z}{R} + 0.11\left(\frac{z}{R}\right)^2 \tag{30}$$

where R is the ground elevation at the maximum penetration of tsunami runup, measured from the initial shoreline, and z is the ground elevation of the location of interest. The values of R and z can be obtained if the tsunami inundation map is available.

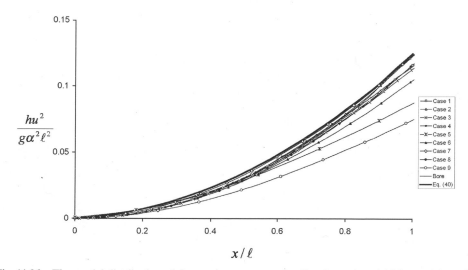

Fig. 11.26 – The spatial distribution of the maximum momentum flux for various initial tsunami conditions, scaled with the inundation distance ℓ. The spatial origin x is at the maximum runup location.

Impact forces from waterborne missiles (e.g., floating driftwood, boats, shipping box containers, automobiles, buildings) are another dominant cause of structural destruction, and, unfortunately, it is difficult to estimate this force accurately. Most available models are based on the impulse-momentum concept, in which the impulse I of the resultant force F acting for an infinitesimal time t^* is equal to the change in linear momentum:

$$I = \int_0^{t^*} F \, dt = d\left(m u_b\right); \quad t^* \to 0, \tag{31}$$

where m is the mass of water-borne missile and u_b is the velocity of the missile.

Another approach is to apply the dynamic model. Haehnel and Daly (2002) used the linear dynamic model with one degree of freedom. Since the collision occurs over a short duration, damping effects are neglected; hence the model can be formulated by:

$$m\frac{d^2\xi}{dt^2} + k\xi = 0 \tag{32}$$

where ξ is the summation of the compression of the structure and the floating debris during impact and rebound, and k is the effective constant stiffness associated with both the debris and the structure. Solving (32) yields the maximum force:

$$F_I = Max.\langle k\,\xi \rangle = u_b \sqrt{k\,m}\,.$$

(33)

Review of the previous works demonstrates the immaturity and uncertainty of our present understanding of missile-impact forces. Nonetheless, the evaluation of impact forces requires estimating the maximum flow velocity, regardless of the method to be chosen. As we discussed earlier, for prediction of flow velocities and depths at a site of interest for a given design tsunami, the best practice is to run a detailed numerical simulation model with a very fine grid size (say, less than 5 m) in the runup zone. The numerical simulation can provide the complete time history of flow velocity and depth at the site of interest.

Alternatively, just as we presented earlier for the value of $\left(hu^2\right)_{max}$, the use of analytical solution can be considered to compute the maximum velocity at a given runup location. The maximum fluid velocity of a bore occurs at the leading runup tip and their solution can be expressed as a function of the ground elevations z and R by rewriting (22):

$$u_{max} = \sqrt{2\,g\,R\left(1-\frac{z}{R}\right)}.$$

(34)

Equation (34) indicates that the flow of the leading tip moves up the beach under gravity just like a particle: simple energy exchange between its kinetic and potential energies. The solutions by the CWY algorithm showed that (22) (or (34)) provides the upper-limit envelope of the flow velocity for all incident tsunami forms (Yeh, 2006). This flow velocity occurs at the leading tongue of the flow where the flow depth is nil. Hence, excessive overestimation of the impact load may result.

For a given tsunami penetration, the incident-bore formation should yield the maximum flow velocity not only at the leading tongue, but also behind the tongue in a shallow-but-finite flow depth; it is because the gradual flooding of non-breaking tsunamis should result slower flow velocities. Therefore, (25) and (26) can be used to determine the maximum flow velocity at a given location for a given flow depth. Combining the two equations and eliminating τ derives Fig. 11.27. Each curve in the figure represents the flow velocity vs. the location in terms of its ground elevation z for a given local flow depth h. This figure can be used to evaluate the maximum flow velocity that can carry floating missile with a finite draft, because draft of the missile must be smaller than the flow depth to make the missile afloat. The lower curve in Fig. 11.27 represents the maximum flow velocity at the maximum runup penetration of flow depth. This curve can be considered as the limit of the maximum flow velocity with a given flow depth. Note that the results in Fig. 11.27 are for incident bores. Local inundation depth of other tsunami forms may exceed that of a bore runup; however, the maximum flow velocity is lower than the limit curve in Fig. 11.27. Hence when a floating missile has its draft that exceeds the flow depth of the bore runup, the design velocity u_{max} can be estimated conservatively with the lower limit curve.

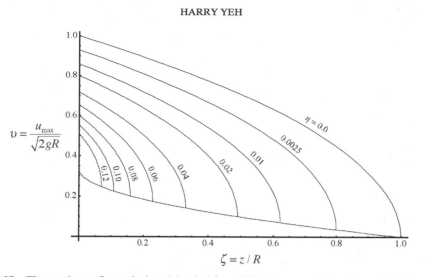

Fig. 11.27 – The maximum flow velocity of depth d ($\eta = d/R$) at the ground elevation z. R is the maximum runup elevation. The bottom curve represents the lower limit of maximum flow velocity. (after Yeh, 2007).

5. Concluding Remarks

Coastal impacts of tsunami are complex due to tsunami's sensitive behaviors at the coastline. Tsunamis are much more reflective than wind waves, and highly sensitive to the near-shore bathymetry and topography. In spite of such apparent complexities, we focused on the analytic models for one-dimensional tsunami runup on a uniformly sloping plane beach. Both incident tsunamis with no wave breaking and in the formation of bores are discussed in detail. The physically relevant findings are summarized in Sec. 3.3.

Tsunami forces in the runup zone are then evaluated using the CWY algorithm, as well as the solution of a bore propagation and runup. The assumptions involved in the algorithm make the force evaluation valid only for the ideal and simplistic situations: i.e. 1-D spatial variation on a uniformly sloping beach. The theory also assumes inviscid-fluid motions in spite of the runup/rundown forming a thin layer of the boundary-dominated flow. Although the simplified conditions do not accurately represent a real coastal environment, the force formula (28) with (30) should provide a convenient estimate of the tsunami-force envelope from the initial shoreline to the maximum runup location. Since the estimation is theoretically based on unambiguous assumptions, the results can be used, for example, as a basis for the design guideline for tsunami resistant structures located in the runup zone and/or to assess ecological and environmental impacts caused by tsunamis.

The evaluation of missile-impact forces is difficult, although we know the evaluation requires estimating the maximum flow velocity. As for the maximum flow velocity, the analytical prediction of a bore runup by (34) yields a conservative envelope. Nonetheless, a floating missile is unlikely carried with this critical speed at the runup tip where the water depth vanishes; the actual transport speed must be less than the critical. For a floating missile with finite draft, the best—yet still conservative—estimate for the maximum flow velocity is evaluated with Fig.

11.27. Again, since the result shown in Fig. 11.27 is theoretically based, it can be used as a basis for the design guideline and other damage assessments.

Foregoing conclusions are valid for tsunami effects in inshore regions. In offshore regions—no matter how close to the initial shoreline is—the incident bore formation may or may not be the critical condition for the flow velocities and the fluid forces as demonstrated in Fig. 11.25. The maximum fluid forces (or linear momentum) occur at the offshore location: near the extreme drawdown location.

Acknowledgments

This research was supported in part by National Science Foundation (CMS-0245206 and SBE- 0527520) and the Edwards endowment through Oregon State University.

References

Arnason, H. 2005. Interactions between and Incident Bore and a Free-Standing Coastal Structure, Ph.D. Thesis, University of Washington, Seattle, 172pp

Carrier, G. F., 1971. The dynamics of tsunamis. In Mathematical Problems in the Geophysical Sciences: 1. Geophysical Fluid Dynamics, (ed. Reid, W. H.). *Amer. Math. Soc.,* 157–187.

Carrier, G. F. and Greenspan, H. 1957. Water waves of finite amplitude on a sloping beach. *J. Fluid Mech.,* 62, 97–109.

Carrier, G. F., Wu, T. T. and Yeh, H. 2003. Tsunami Runup and Drawdown on a Plane Beach. *J. Fluid Mech.,* 475, 79–99.

FEMA, 2005. Coastal Construction Manual, FEMA 55 Report, Edition 3, Federal Emergency Management Agency, Washington, D.C

Geist, E. L. 1998. Local tsunami and earthquake source parameters. Advances in Geophysics, vol. 39, 117–209.

Haehnel, R. B., and Daly, S. F. 2002. Maximum impact force of woody debris on floodplain structures. Technical Report: ERDC/CRREL TR-02-2, US Army Corps of Engineers, 40 pp.

Hanson, E. T., 1926. The theory of ship waves. *Proc. R. Soc. London Ser. A.* 111, 491–529.

Ho, D. V. and Meyer, R. E. 1962. Climb of a bore on a beach. Part 1: Uniform beach slope. *J. Fluid Mech.* 14. 305–318.

Lamb, H., 1932. Hydrodynamics, 6th ed. Cambridge University Press.

Mazova, R.KH. and Pelinovsky, E. N., 1991. The increasing of tsunami runup height with negative leading wave. Proc. IV Int. Symp. on Geophys. Hazards, Perugia, Italy, p. 118.

Miller, R. L., 1968. Experimental determination of run-up of undular and fully developed bores. *J. Geophys. Res.* 73, 4497–4510.

Peregrine, D. H. and Williams, S. M. 2001. Swash overtopping a truncated plane beach. *J. Fluid Mech.,* 440, 391–399.

Ramsden, J. D. 1993. Tsunamis: Forces on a vertical wall caused by long waves, bores, and surges on a dry bed. Report No. KH-R-54, W. M. Keck Laboratory, California Institute of Technology, Pasadena, Calif.

Robertson, I. 2007. personal communication.

SCOR. 2001. Improved Global Bathymetry: Final Report of SCOR Working Group 107. Technical Series 63, Intergovernmental Oceanographic Commission, UNESCO, 122pp.

Shen, M. C. & Meyer, R. E. 1963. Climb of a bore on a beach. Part 3. Run-up. *J. Fluid Mech.* 16, 113–125.

Shuto, N. 1994. 北海道南西沖地震津波による家屋の被害 (Damage of buildings resulted from the 1993 Hokkaido Nansei-oki tsunami). Tsunami Engineering Technical Report No. 11, Tohoku University, 11–28.

Synolakis, C. E. 1987. The runup of solitary waves. *J. Fluid Mech.*, 185, 523–545.

Synolakis, C. E. and Bernard, E. N. 2006. Tsunami science before and beyond. *Phil. Trans. R. Soc. A,* 364, 2231–2265.

Tadepalli, S. and Synolakis, C. E., 1994. The run-up of N-waves on sloping beaches, *Proc. R. Soc. Lond.* A, 445, 99–112.

Whitham, G. B., 1958. On the propagation of shock waves through regions of non-uniform area of flow. *J. Fluid Mech.* 4, 337–360.

Whitham, G. B., 1955. The effects of hydraulic resistance in the dam-break problem. *Proc. Roy. Soc. Lond.* A. 227, 399–407.

Yeh, H. H. 1995. Free-surface dynamics. In: Advances in Coastal and Ocean Engineering (Ed: P. L.-F. Liu), World Scientific Publishing Co., 1–75.

Yeh, H. H. 1998. Tsunami Researchers Outline Steps for Better Data. *EOS, Trans. Amer. Geophys. Union*, 79, 480 & 484.

Yeh, H. 2006. Maximum fluid forces in the tsunami runup zone. *J. Waterw., Port, Coastal, Ocean Eng.,* 132, 496-500.

Yeh, H. 2007. Design tsunami forces for onshore structures. *J. Disaster Research*, 2, 531–536.

Yeh, H. H. & Ghazali, A., 1986. Nearshore behavior of bore on a uniformly sloping beach. *Proc. Conf. Coastal Eng.* 20th, 877–888.

Yeh, H. H. & Ghazali, A., 1988. On bore collapse. *J. Geophys. Res.* 93, 6930–6936

Yeh, H., Ghazali, A., and Marton, I. 1989. Experimental Study of Bore Runup. *J. Fluid Mech.*, 206, 563–578.

Appendix – the CWY algorithm

The fully nonlinear shallow-water-wave equation of the form (1) can be written in the non-dimensionalized form with the scaling parameters presented in (2):

$$\frac{\partial \eta}{\partial t} + \frac{\partial}{\partial x}\{u(x+\eta)\} = 0,$$
$$\frac{\partial u}{\partial t} + u\frac{\partial u}{\partial x} + \frac{\partial \eta}{\partial x} = 0, \tag{A1}$$

Note that the beach slope α and the gravity g in (1) no longer appear in these governing equations. Also note that ^ indicated for the non-dimensionalized parameters in (3) is omitted in Appendix for brevity. Then, (A1) is transformed by introducing the distorted coordinates h and λ such that

$$\tau = t - u; \quad h = x + \eta. \tag{A2}$$

Note that

$$\frac{\partial x}{\partial h} = 1 - \frac{\partial \eta}{\partial h}, \quad \frac{\partial x}{\partial \tau} = -\frac{\partial \eta}{\partial \tau}, \quad \frac{\partial t}{\partial h} = \frac{\partial u}{\partial h}, \quad \frac{\partial t}{\partial \tau} = 1 + \frac{\partial u}{\partial \tau}. \tag{A3}$$

The Jacobian in the transformation is $J = \dfrac{\partial x}{\partial h}\dfrac{\partial t}{\partial \tau} - \dfrac{\partial x}{\partial \tau}\dfrac{\partial t}{\partial h}$, and unless J (or $1/J$)

vanishes in the x–t domain, the transformation is single-valued. A physical implication of $J = 0$ is the occurrence of wave breaking. It follows from (A3) that

$$\frac{\partial h}{\partial x} = \frac{1}{J}\frac{\partial t}{\partial \tau}, \quad \frac{\partial h}{\partial t} = -\frac{1}{J}\frac{\partial x}{\partial \tau}, \quad \frac{\partial \tau}{\partial x} = -\frac{1}{J}\frac{\partial t}{\partial h}, \quad \frac{\partial \tau}{\partial t} = \frac{1}{J}\frac{\partial x}{\partial h}. \tag{A4}$$

hence, the transformation of (A2) leads (A1) to become

$$\frac{\partial}{\partial h}(h\,u) + \frac{\partial}{\partial \tau}\left(\eta + \frac{u^2}{2}\right) = 0,$$
$$\frac{\partial u}{\partial \tau} + \frac{\partial}{\partial h}\left(\eta + \frac{u^2}{2}\right) = 0. \tag{A5}$$

Defining $\psi = \eta + \dfrac{u^2}{2}$ and replacing h by σ^2 such that $\sigma = \sqrt{h} = \sqrt{x + \eta}$, (A5) can be expressed by

$$\frac{\partial}{\partial \sigma}\left(\sigma^2 u\right) + 2\sigma\frac{\partial \psi}{\partial \tau} = 0,$$
$$\frac{\partial u}{\partial \tau} + \frac{1}{2\sigma}\frac{\partial \psi}{\partial \sigma} = 0. \tag{A6}$$

and the elimination of u yields

$$\frac{\partial^2 \psi}{\partial \tau^2} - \frac{1}{4\sigma}\frac{\partial}{\partial \sigma}\left(\sigma\frac{\partial \psi}{\partial \sigma}\right) = 0. \tag{A7}$$

Equation (A7) is not convenient because the solution for ψ will not provide the solutions for η and u, separately. To circumvent this difficulty, we introduce the function φ such that

$$\psi = \frac{\partial \varphi}{\partial \tau} = \eta + \frac{u^2}{2}. \tag{A8}$$

From (A6), it is readily seen that

$$u = -\frac{1}{2\sigma}\frac{\partial \varphi}{\partial \sigma} \quad \text{and} \quad \eta = \frac{\partial \varphi}{\partial \tau} - \frac{1}{8\sigma^2}\left(\frac{\partial \varphi}{\partial \sigma}\right)^2. \tag{A9}$$

Then, (A7) can be replaced by

$$\frac{\partial^2 \varphi}{\partial \tau^2} - \frac{1}{4\sigma}\frac{\partial}{\partial \sigma}\left(\sigma\frac{\partial \varphi}{\partial \sigma}\right) = 0 \qquad (A10)$$

Note that $\sigma = 0$ represents the moving shoreline, and, since $h \geq 0$, σ is always real in the fluid domain. Equation (A10) is the linear cylindrical-wave equation, which is the same form as that derived by Carrier and Greenspan (1958) but with different and more convenient non-dimensionalization. Carrier, Wu, and Yeh (2003) solved this cylindrical wave equation with the general initial conditions (at $\tau = 0$) applying the Fourier-Bessel transform (also called the Hankel transform):

$$\bar{\varphi}(\rho,\tau) = \int_0^\infty \sigma J_0(\rho\sigma)\varphi(\sigma,\tau)\,d\sigma, \quad \varphi(\sigma,\tau) = \int_0^\infty \rho J_0(\rho\sigma)\bar{\varphi}(\rho,\tau)\,d\rho, \qquad (A11)$$

with the general initial conditions:

$$\varphi(\sigma, 0) = P(\sigma), \quad \frac{\partial}{\partial \tau}\varphi(\sigma, 0) = F(\sigma), \qquad (A12)$$

in which

$$P(\sigma) = -\int_0^\sigma 2\sigma'u(\sigma',0)\,d\sigma', \quad and \quad F(\sigma) = \eta\,(\sigma,0) + \frac{u^2(\sigma, 0)}{2}. \qquad (A13)$$

Then, the solution of (A10) in the transformed space can be readily found to be

$$\bar{\varphi}(\rho,\tau) = \bar{P}(\rho)\,\cos\left(\tfrac{1}{2}\rho\tau\right) + \frac{2}{\rho}\bar{F}(\rho)\,\sin\left(\tfrac{1}{2}\rho\tau\right). \qquad (A14)$$

The inversion can be obtained as

$$\varphi(\sigma,\tau) = 2\left\{\int_0^\infty F(b)\,G(b,\sigma,\tau)\,db + \int_0^\infty P(b)\,\frac{\partial}{\partial \tau}G(b,\sigma,\tau)\,db\right\}, \qquad (A15)$$

where the Green function is

$$G(b, \sigma, \tau) = b\int_0^\infty J_0(\rho\sigma)\,\sin\left(\tfrac{1}{2}\rho\tau\right)J_0(\rho b)\,d\rho \qquad (A16)$$

Although the Green function $G(b, \sigma, \tau)$ appears an innocent looking, it is not easy to evaluated numerically, as demonstrated in Fig. 11.A1.

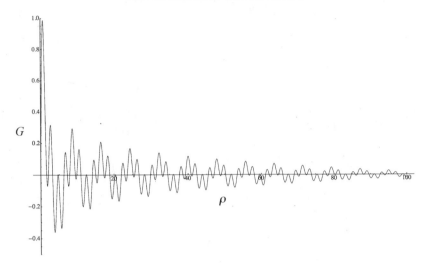

Fig. 11.A1 – The integrand of G in (A16) with $b = 1.2$, $\sigma = 0.02$, $\tau = 4.0$. The larger the value of σ, the slower the amplitude decay. The larger the value of τ, the faster the oscillation.

To circumvent this difficulty, Carrier, Wu and Yeh (2003) evaluate the Green function explicitly as

$$G(b,\sigma,\tau) = \begin{cases} 0 & for \ \dfrac{\tau}{2} < |\sigma - b| \\[4mm] \dfrac{1}{\pi}\sqrt{\dfrac{b}{\sigma}}\, K\!\left(\dfrac{\tau^2 - 4(\sigma-b)^2}{16\sigma b}\right) & for \ |\sigma - b| < \dfrac{\tau}{2} < |\sigma + b| \\[4mm] \dfrac{4}{\pi}\dfrac{b}{\sqrt{\tau^2 - 4(\sigma-b)^2}}\, K\!\left(\dfrac{16\sigma b}{\tau^2 - 4(\sigma-b)^2}\right) & for \ \dfrac{\tau}{2} > |\sigma + b| \end{cases} \tag{A17}$$

where $K(k) = \displaystyle\int_0^{\pi/2} \dfrac{dv}{\sqrt{1 - k\sin^2 v}}$ is the Complete Elliptic Integral of the first kind.

The Green function has a singularity at $b = \tau/2 - \sigma$ as shown in Fig. 11.A2; however, (A15) can be integrated numerically with standard integration software for any regular initial conditions.

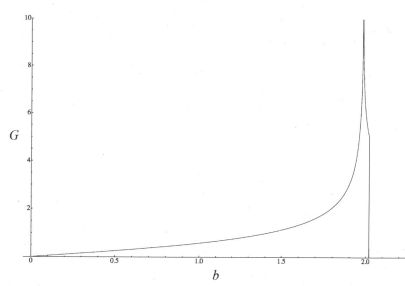

Fig. 11.A2 – G (b, $\sigma = 0.02$, $\tau = 4.0$).

Once $\varphi\,(\sigma,\ \tau)$ is computed with (A15 and A17), the physical variables in the $x - t$ space can be obtained using (A9), then (A2). It is emphasized that this method can obtain the very accurate solution for arbitrarily selected time and location without computing the rest of the computational domain: this is the primary advantage over the direct numerical simulations.

Chapter 12.   Tsunami Forecasting

VASILY V. TITOV

NOAA/PMEL, University of Washington/JISAO

Contents

1.   Introduction

Tsunami forecast tools provide real-time guidance for rapid, critical decisions in which lives and property are at stake. Tsunami Warning Centers are tasked with issuing tsunami warnings that lead to immediate actions by local authorities to mitigate potentially deadly tsunami inundation at a coastal community. The more timely and precise the warnings, the more effective actions local emergency managers can take and more lives and property can be saved. At present, the Tsunami Warning Centers' personnel face a difficult challenge: to issue tsunami warnings based on incomplete and often ambiguous data. The initial warning decisions are based on analysis of seismic waves produced by an earthquake. Confirmation that a tsunami exists is provided by coastal tide gages, but may be too late for timely evacuation measures. This inaccurate and untimely information can lead to high false alarm rates and ineffective responses to the tsunami warning. Tsunami forecasts based on new tsunami measurement technologies and modeling techniques can provide crucial, accurate information to guide tsunami mitigation measures before the tsunami arrives.

A tsunami forecast should provide site- and event-specific information about tsunamis well before the first wave arrives at a threatened community. Historically, the only forecasts have been tsunami arrival times, which are based on indirect

The Sea, Volume 15, edited by Eddie N. Bernard and Allan R. Robinson
ISBN 978–0–674–03173–9 ©2009 by the President and Fellows of Harvard College

seismic information about the earthquake source. The next generation tsunami forecast will provide estimates of all critical tsunami parameters (amplitudes, inundation distances, current velocities, etc.) based on direct tsunami observation. The technical obstacles to achieving this goal are many, but three primary requirements are *accuracy, speed,* and *robustness.*

Tsunami measurement and numerical modeling technology must be integrated to create an effective tsunami forecasting system. Neither technology can provide satisfactory forecasts by itself. Observational networks will never be adequate because the ocean is vast. Establishing and maintaining monitoring stations is costly and difficult, especially in deep water. Numerical model accuracy is inherently limited by errors in bathymetry and topography and uncertainties in the generating mechanism. But combined, these techniques can provide useful tsunami forecasts.

Critical components of tsunami forecasting technology exist now that could provide rapid, usably accurate forecasts of at least the first few waves of an earthquake-generated tsunami. In particular, it is feasible to develop a forecast system that combines real- time seismic and tsunami data with a forecast database of pre-computed scenarios that have been thoroughly tested and scrutinized for reasonableness and sensitivity to errors. Later waves could also be forecasted by processing more real-time tsunami data with forecast models. Implementation of this technology requires integration of these components into a unified, robust system.

2. Background

Since 1949, the Pacific tsunami warning system has provided warnings of potential danger in the Pacific basin by monitoring earthquake activity and the passage of tsunamis at coastal tide gauges. However, neither seismometers, used to detect earthquakes, nor tide gauges, used to monitor tidal fluctuations, provide data that allow accurate prediction of the impact of a tsunami at a particular coastal location. Monitoring earthquakes gives a good estimate of the potential for tsunami generation, based on earthquake size and location, but gives no direct information about the tsunami itself. The variation in local bathymetry and harbor shapes severely limits the effectiveness of harbor tide gauges in providing useful measurements for tsunami forecasting (Synolakis and Bernard, 2006). Tide gauges, however, do provide an independent verification of tsunami forecast skill. Partly because of these data limitations, 15 of 20 tsunami warnings issued since 1946 were considered false alarms because the waves that arrived were too weak to cause damage. Recently developed real-time, deep-ocean tsunami detectors (tsunameters) provide the data necessary for models to make forecasts (Gonzalez et al., 2005). Recent distant tsunamis in 2003 and local tsunamis in 2005 have underscored the accuracy of this observation/model technology (Bernard et al., 2006). No tsunameters existed in the Indian Ocean in 2004, but later simulations suggest that had they been in place, they would have provided adequate time at least for warning the more distant shores in India, Sri Lanka, Oman, and Africa, and many locations in Southeast Asia.

3. Goals and Challenges of Tsunami Forecasting

3.1 Long-term forecasting (tsunami inundation mapping)

In principle, tsunami inundation studies to produce tsunami inundation maps for coastal locations can also be considered as tsunami forecasts. Such studies forecast tsunami inundation far in advance for hazard mitigation and planning. The common steps of the long-term forecast procedures are generally the same as for the short-term forecast, however the time constraints are much more forgiving.

An inundation modeling study attempts to recreate the tsunami generation in deep or coastal waters, wave propagation to the impact zone, and inundation along the study area. To reproduce the correct wave dynamics during the inundation computations high-resolution bathymetric and topographic grids are used in this type of study. The high quality bathymetric and topographic data sets needed for development of inundation maps require maintenance and upgrades as better data become available and coastal changes occur.

Inundation studies can be conducted taking a probabilistic approach in which multiple tsunami scenarios are considered, and an assessment of the vulnerability of the coast to tsunami hazard is evaluated; or they may focus on the effect of a particular "worst case scenario" and assess the impact of a particular high-impact event on the areas under investigation.

The results of a tsunami inundation study should include information about the maximum wave height and maximum current speed as a function of location, maximum inundation line, as well as time series of wave height at different locations, indicating wave arrival time. This information can be used by emergency managers and urban planners primarily to establish evacuation routes and to evaluate the resilience of vital infrastructure.

The focus of this paper is the short-term inundation forecast, which is performed in real time. However, all tools and principles described are the same for both time scales and can be applied for the long-term forecast. Since the time constraint is not applicable to the mapping studies, accuracy becomes the dominant factor in the long-term forecast.

3.2 Short-term forecast (real-time)

The main objective of a forecast model is to provide an estimate of wave arrival time, wave height, and inundation area immediately after a tsunami event. Tsunami forecast models are run in real time while a tsunami is propagating in the open ocean. Consequently they are designed to perform under very stringent time limitations. Three primary technical requirements for producing a reliable tsunami forecast are *accuracy, speed,* and *robustness.*

Accuracy

Errors and uncertainties will always be present in any forecast. A practical forecast, however, minimizes the uncertainties by recognizing and reducing possible errors. In the tsunami forecast, measurement and modeling errors present formidable challenges; but advancements in the science and engineering of tsunamis have identified and studied most of them.

1) *Measurement Error.* Tsunami measurements are always masked by noise from a number of sources: tides, harbor resonance, instrument response, to name a few. Most of the noise can be eliminated from the record by careful consideration of its sources. However, eliminating it in automatic mode during real-time assessment presents a serious challenge.

2) *Model Approximation Error.* The physics of tsunami propagation is better understood than that of generation and inundation. For example, landslide generation physics is currently a very active research subject, and comparative studies have demonstrated significant differences in the ability of inundation models to reproduce idealized test cases and/or field observations.

3) *Model Input Error.* Model accuracy can be degraded by errors in (a) the initial conditions set for the sea surface and water velocity, due to inadequate physics and/or observational information, and (b) the bathymetry/topography computational grid, due to inadequate spatial coverage, resolution, and accuracy, including the difficult issues encountered when merging data from different sources.

Speed

We refer here to *forecast speed,* relating to the time taken to make the first forecast product available to an emergency manager for interpretation and guidance. This process involves at least two important and potentially time-consuming, steps:

1) *Data stream to TWC.* Seismic wave data are generally available first, but finite time is required to interpret these signals in terms of descriptive parameters for earthquakes, landslides, and other potential source mechanisms. Tsunami waves travel much slower. In addition, the time of at least one quarter wave period (when the leading tsunami wave crests) will be needed from deep ocean tsunameters to incorporate these data into a forecast. Seismic networks are much more dense than tsunami monitoring networks (for example, NOAA's tsunami warning centers use data from over 250 seismic stations to determine earthquake parameters, but data from only 44 tsunameter stations to determine tsunami parameters), but inversion algorithms for both are needed to provide source details.

2) *Model simulation speed.* Currently available computational power can provide useful, far-field, real-time forecasts before the first tsunami strikes a threatened community. If the time available for forecasting is sufficiently large, the source can be quickly specified and an accurate computational grid made available. In fact, if powerful parallel computers and/or precomputed model results are exploited, model execution time can be reduced almost to zero, at least in principle. In practice, of course, there will always be situations for which the source proximity would make it impossible to provide timely forecasts for nearby coasts. But even a late forecast will still provide valuable assessment guidance to emergency managers responsible for critical decisions regarding response, recovery, and search-and-rescue.

Robustness

With lives and property at stake, reliability standards for a real-time forecasting system are understandably high; and the development of such a system is a difficult challenge. It is one thing for an experienced modeler to perform a hindcast study and obtain reasonable, reliable results. Such exercises typically take months to complete, during which multiple runs can be made with variations in the model input and/or the computational grid that are suggested by improved observations. The results are then examined for errors and reasonableness. It is quite another matter to design and develop a system that will provide reliable results in real time, without the oversight of an experienced modeler.

4. Technology of tsunami forecasting

4.1. Measurement methods

Several real-time data sources are traditionally used for tsunami warning and forecast. They are (1) seismic data to determine source location and source parameters, (2) coastal tide gage data used for direct tsunami confirmation and for tsunami source inversion studies (mostly research studies not in real-time mode), and (3) new real-time deep-ocean data from the tsunameter network (Gonzalez et al., 2003).

Seismic

Most tsunamis are induced by seismic events. Therefore, seismic data is traditionally the primary data for tsunami studies. Seismic waves propagate much faster than tsunami waves, therefore use of seismic measurements provide a time advantage for tsunami forecast applications. The difficulty of using seismic data for predicting tsunamis is that these are indirect measurements of tsunamis. The problem of inferring tsunami characteristics from seismic measurements is notoriously difficult, particularly if the goal is to provide quantitative tsunami predictions. For example, the uncertainty of one tenth of the earthquake magnitude scale (for example, Mw = 8.1 ± 0.1—a typical magnitude uncertainty few minutes after an earthquake) can lead to more than 100% uncertainty of the tsunami amplitude prediction (an increase of 0.2 in the logarithmic scale of the Mw produces twice the fault slip and, therefore, twice the tsunami amplitudes, if all other source parameters are the same).

Water level

A proven effective strategy for real-time forecasting is to use deep-ocean measurement as a primary data source for making the tsunami forecast. There are several key features of deep-ocean data that make it indispensable for the forecast model input:

1) *Rapid tsunami observation.* Since tsunamis propagate with much greater speed in deeper water, the wave will reach a deep-ocean tsunameter much sooner than an equally distant coastal gage. Therefore, a limited number of strategically placed deep-ocean gages can provide advanced tsunami observation for a large portion of the coastline.

2) *Harbor response.* Tsunameters are placed in deep water in the open ocean where a tsunami signal is not contaminated by local coastal effects. Coastal tide-gages, on the other hand, are usually located inside harbors where measurements are subjected to the harbor response. As a result, only part of the tsunami frequency spectrum is accurately measured by coastal gages. In contrast, the tsunameter recording provides "unfiltered" time series with the full spectrum of tsunami energy.

3) *Instrument response.* The Bottom Pressure Recorder (BPR) of the tsunameter has a very flat and constant frequency response in the tsunami frequency range (e.g. Milburn et al., 1996). Many coastal gages, on the contrary, have complicated and changing response characteristics. Since most of the tide gages are designed to measure tides, they often do not perform well in the tsunami frequency band.

4) *Linear process.* The dynamics of tsunami propagation in the deep ocean is mostly linear, since amplitudes are very small compared with the wavelength. This process is relatively well understood, and numerical models of wave propagation are very well developed. The linearity of wave dynamics allows for the application of efficient inversion schemes.

4.2   Modeling methods

Models for tsunami forecast

The numerical modeling of tsunami dynamics has become a standard research tool in tsunami studies. Modeling methods have matured into a robust technology that has proven to be capable of accurate simulations of past tsunamis, after careful consideration of field and instrumental historical data. Imamura (this volume) presents and excellent overview of the of the numerical techniques used in tsunami modeling. The majority of the results in this chapter are computed using the MOST (Method Of Splitting Tsunamis – Titov and Synolakis, 1995,1998, Titov and Gonzalez, 1997) model.

The MOST propagation code uses the non-linear shallow water equation in spherical coordinates with Coriolis force and a numerical dispersion scheme to take into account the different propagation wave speeds with different frequencies. The equations, shown below, are numerically solved using the splitting method (Titov and Synolakis, 1998):

$$h_t + \frac{(uh)_\lambda + (vh\cos\phi)}{R\cos\phi} = 0 \tag{3}$$

$$u_t + \frac{uu_\lambda}{R\cos\phi} + \frac{vu_\phi}{R} + \frac{gh_\lambda}{R\cos\phi} - \frac{uv\tan\varphi}{R} = \frac{gd_\lambda}{R\cos\phi} - \frac{C_f u|u|}{d} + fv \tag{4}$$

$$v_t + \frac{uv_\lambda}{R\cos\phi} + \frac{vv_\phi}{R} + \frac{gh_\lambda}{R} + \frac{u^2\tan\phi}{R} = \frac{gd_\phi}{R} - \frac{C_f v|u|}{d} - fu \tag{5}$$

where:

λ = longitude
φ = latitude
$h = h(\lambda, \phi, t) + d(\lambda, \phi, t)$
$h(\lambda, \phi, t)$ = amplitude
$d(\lambda, \phi, t)$ = undisturbed water depth
$u(\lambda, \phi, t)$ = depth averaged velocity in longitude direction
$v(\lambda, \phi, t)$ = depth averaged velocity in latitude direction
g = gravity
R = radius of the earth
$f = 2\omega\sin\varphi$, a coriolis parameter
$C_f = gn^2/h^{1/3}$, n is Manning coefficient

Model use

The forecast scheme, in contrast to hindcast studies, is a two-step process where numerical models operate in different modes:

1) *Data assimilation mode.* The model is a part of the data assimilation scheme where it is adjusted "on-the-fly" by a real-time data stream. The model requirement in this case is similar to hindcast studies: the solution must provide the best fit to the observations.
2) *Forecast mode.* The model uses the simulation scenario obtained at the first step to extend the simulation to locations where measured data is not available – providing the forecast. It is difficult to fully assess the forecast potential of a particular model, since the quality and accuracy of the prediction will always depend on the scenario chosen by the data assimilation step. Accurate simulation of the near-shore tsunami dynamics and inundation are especially important.

Model standards

Model standards are needed to ensure a minimum level of quality and reliability for forecasting and inundation products. Incorrectly assessing possible inundation can be costly both in terms of lives lost, or in unnecessary evacuations in areas larger than warranted that may put lives at risk and reduce the credibility of the system, even in areas that were not directly affected.

To forecast tsunami currents, forces and runup on coastal structures, and inundation of coastlines one must calculate numerically the evolution of the tsunami wave from the deep ocean to its coastal community. Numerical models require validation (the process of ensuring that the model solves the parent equations of motion accurately) and verification (the process of ensuring that the model used represents geophysical reality appropriately), as both are essential parts of the model development. Validation ensures that the model performs well in a wide range of circumstances and is accomplished through comparison with analytical solutions. Verification ensures that the computational code performs well over a range of geophysical problems. Many analytic solutions have been validated with laboratory data. Very few existing numerical models have been both validated with the analytical solutions and verified with laboratory measurements and field measurements, thus establishing a gold standard for numerical codes for inundation mapping. Further, the operational experience in tsunami forecasting with one of the models provides a blueprint for testing codes in the future (Kânoğlu and Synolakis, this volume).

4.3 Data Inversion techniques

An effective tsunami forecast scheme would automatically interpret incoming real-time data to develop the best model scenario that fits this data. This is a classical inversion problem, where initial conditions are determined from an approximated solution. Such problems can be successfully solved only if proper parameters of the initial conditions are established. These parameters must effectively define the solution; otherwise the inversion problem is ill-posed.

There are several parameters that describe a tsunami source commonly used for tsunami propagation simulations. Choosing the subset of those parameters that control the deep-ocean tsunami signal is the key to developing a useful inversion scheme for tsunameter data. In a study designed specifically to explore this problem, Titov et al. (1999, 2001) have investigated the sensitivity of far-field data to different parameters of commonly used tsunami sources. The results show that source *magnitude* and *location* essentially define far-field tsunami signals for a wide range of subduction zone earthquakes. Other source parameters have secondary influence and can be ignored during the inversion. This result substantially reduces the size of the inversion problem for the deep-ocean data.

An effective implementation of the inversion can be achieved by using a discrete set of Green's functions to form a model source. An algorithm can be developed to choose the best fit to a given tsunameter data among a limited number of unit solution combinations by minimizing a misfit function. The misfit function can be RMS or any other norm. The brief description of this mathematical formulation is as follows.

Let $w(t)$ represent the wave heights arriving at one particular DART buoy. In practice we do not observe $w(t)$ over a continuum of times t, but rather only at a discrete set of times, say, t_n, $n = 0, 1, \ldots, N_D - 1$, where N_D is the total number of recorded wave heights. Typically $t_n = t_0 + n\Delta_D$, where t_0 is the time at which the first wave height is observed, and Δ_D is the elapsed time between adjacent recorded heights (i.e., the sampling time); however, in what follows, it is not necessary that the t_n's be equally spaced (this allows us to handle arbitrary mixtures of data sampling rates). The available data thus take the form $w_n = w(t_n)$, $n = 0, 1, \ldots, N_D - 1$. Let $g(t)$ represent modeled wave heights produced by an earthquake at one particular source location. These modeled heights are in practice only computed over a grid of equally spaced times, so what is available is $g_n = g(n\Delta_M)$, $n = 0, 1, \ldots, N_M - 1$, where Δ_M is the sampling time between the modeled heights. We assume that $g(t)$ is smooth enough that it can be computed to sufficient accuracy at any given t based upon, say, a Fourier series or spline fit to the available g_n. We assume that this fit also allows us to compute the derivative $g(t)$ of $g(t)$ accurately at any given t. We want to fit a selected subset of the observed w_n series to a possibly shifted and stretched version of the modeled wave heights, which we can denote by $g_{a,b}(t) = g([t-a]/b)$, where we assume that $a_L \le a \le a_U$ and $b_L \le b \le b_U$; i.e., by physical considerations, both a and b are constrained to be within certain predefined limits (presumably the interval $[a_L, a_U]$ would include zero, whereas the interval $[b_L, b_U]$ would include unity). The idea is to determine A, a and b such that

$$S(A,a,b) \equiv \sum_{n=n_L}^{n_U} [w_n - A g_{a,b}(t_n)]^2 = \sum_{n=n_L}^{n_U} [w_n - h_n(A,a,b)]^2$$

is minimized, where $h_n(A, a, b) \equiv A g_{a,b}(t_n) = A g([t_n - a]/b)$, and the indices n_L and n_U specify the subset of observed data to be used to determine the fit. As it stands, finding the values of A, a and b that minimize $S(A, a, b)$ is a constrained nonlinear optimization problem. The minimizing values can be found efficiently using an iterative scheme that starts with initial approximations A_0, a_0 and b_0 to the desired solutions and then computes a series of improved approximations A_m, a_m and b_m, $m = 1, 2, \ldots$, that must in theory converge to the desired least squares solutions. In practice, the iterative scheme is carried out M times beyond the initial approximations, and the true least squares solutions are taken to be sufficiently approximated by a_M, b_M and A_M, where M is selected by some stopping rule criteria (e.g., $|A_M - A_{M-1}|$, $|a_M - a_{M-1}|$, $|b_M - b_{M-1}|$ and $|S(A_M, a_M, b_M) - S(A_{M-1}, a_{M-1}, b_{M-1})|$ are all small).

5. Existing Tsunami Forecast Systems

To date, three different tsunami forecast systems have been developed by Australia, Japan and the U.S. They employ different methodologies, different tools, and are currently at different stages of implementation. All three systems use a database of pre-computed propagation models to reduce forecast time. Two systems are based solely on seismic data and propagation modeling (Japan and Australia), while the U.S. system uses a combination of seismic and direct tsunami measurements with inundation modeling for coastal predictions. Here, the U.S. system will be more thoroughly described, since the author is involved in this system development. Please refer to the supplied references for the detailed description and tests of the Japanese and the Australian systems.

5.1 Quantitative Tsunami Forecast System for Japan
(Japan Meteorological Agency – JMA)

The JMA forecast system is the only one that is fully implemented and operational. The system uses seismic data to produce coastal tsunami amplitude forecasts. It consists of a model database, interpolation scheme, and coastal amplitude estimate.

The pre-computed database combines individual propagation scenarios computed with a linear long wave propagation model (Satake, 2002) using the Okada (1985) model of the tsunami source. The database consists of more than 100,000 scenarios (approximately 4,000 epicenter locations x 4 magnitudes x 6 hypocenter depths). The model source parameters are inferred from historical sources.

The interpolation scheme is applied to forecast a particular tsunami. The earthquake magnitude, location, and depth are used as input parameters. The database entries closest to the determined epicenter, depth, and magnitude are all interpolated to produce a forecast of tsunami propagation.

The coastal amplitude forecast is produced from the propagation model estimates using linear Green's law for plane waves, which estimates the wave amplitude H at depth h from a given amplitude H_1 at depth h_1 as follows:

$$H = \sqrt[4]{\frac{h_1}{h}} H_1$$

This formula assumes simplified linear one-dimensional wave dynamics between depth h_1 and h.

Details of the method are described by Kuwayama (2006) and Tatehata (1997).

5.2 First-generation real-time tsunami forecasting system for the Australian Region (Bureau of Meteorology, Australia)

A 1st-generation operational model-based tsunami prediction system, a component of the Australian Tsunami Warning System, is developed and implemented by the Bureau of Meteorology. The system consists of a tsunami scenario database, i.e., a number of tsunami model runs that are calculated ahead of time and stored. When an earthquake event occurs, the closest scenario can be extracted from the database and used as forecast guidance. The MOST model is used to generate the scenarios, along with a sub-sampled version of the Bluelink bathymetry dataset. The source locations for the scenarios are at 100 km intervals along the subduction zones within the Australian region. Each source location has four scenarios associated with it, with moment magnitudes of 7.5, 8, 8.5, and 9. The definition of the assumed rupture details for each scenario is prescribed. Each scenario has forecast guidance products associated with it. These model guidance products are a map of expected maximum wave amplitude at each model gridpoint and the expected maximum wave amplitude for specific locations. The database contains a total of 741 scenarios.

Once the earthquake location and magnitude are determined, the closest scenario is chosen from the database to produce the forecast. Future development of the system plan includes the use of sea level data in the forecast process. Several DART systems are planned around Australia with one currently deployed south of New Zealand.

Details of the forecast method are described by Greenslade et al. (2007).

5.3 Short-term Inundation Forecast for Tsunami for the U.S coastlines (NOAA, U.S.)

NOAA's TWCs have been using a 1st-generation model forecast tool for several years in test mode (Whitmore, 2003). The tool uses pre-computed propagation scenarios to predict amplitudes at tide gages. The predictions can be adjusted by scaling the scenario with measured tsunami wave height (Whitmore, this volume).

NOAA's Pacific Marine Environmental Laboratory has developed a next-generation methodology that combines the real-time deep ocean measurements with tested and verified model estimates to produce a real-time tsunami forecast for coastal communities. DART™ technology is combined with NOAA's MOST numerical model (Titov and Gonzalez, 1997, Titov and Synolakis, 1995, 1998) for the development of the tsunami forecasting scheme called Short-term Inundation Forecast for Tsunamis (SIFT).

To forecast tsunami inundation and other critical local tsunami impact parameters (amplitudes at tide gages, flow velocities, wave impact indices), seismic parameter estimates and tsunami measurements are used in combination with model results. The system sifts through a pre-computed generation/propagation forecast database and selects an appropriate (linear) combination of scenarios that most closely matches the observational data. This produces estimates of tsunami characteristics in deep water which can then be used as initial conditions for a site-specific (non-linear) inundation algorithm. The inundation model can provide a high-resolution tsunami forecast scenario showing predicted tsunami dynamics at a specific local community. The results are made available in real time to TWCs and, potentially, to local emergency managers to aid in hazard assessment and decision-making in real time, before the tsunami reaches the community. Figure 12.1 (see color insert) shows the components of the SIFT system.

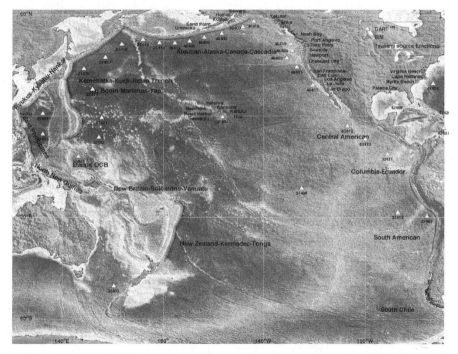

Fig. 12.1 – (see color insert) Components of the Short-term Inundation Forecast for Tsunamis in the Pacific. △ are DART locations; ● are U.S. forecast sites; ⊞ are tsunami source functions.

Linear propagation model database for unit sources

The source sensitivity study (Titov et al., 1999) has established that only a few source parameters are critical for the far-field tsunami characteristics, namely the location and the magnitude (assuming some typical mechanism for the displacement). Therefore, a discrete set of unit sources (Green's functions) can provide the basis for constructing a tsunami scenario that would simulate a given tsunameter data set. Numerical solutions of tsunami propagation from these unit sources, when linearly combined, provide arbitrary tsunami simulation for the data assimilation step of the forecast scheme.

This principle is used to construct a tsunami forecast database of pre-computed propagation solutions for unit sources around the North Pacific (Fig. 12.1, see color insert). All unit scenarios are thoroughly tested and scrutinized for reasonableness and sensitivity to errors. Presently, the database contains 1299 unit source scenarios that cover potentially tsunamigenic areas around the Pacific, Caribbean in the Atlantic, and the Indian Ocean. The database stores all simulation data for each unit solution, including amplitudes and velocities for each offshore location around the North Pacific. This approach allows completing data assimilation without additional time-consuming model runs. The methodology also provides the offshore forecast of tsunami amplitudes and all other wave parameters around North Pacific immediately after the data assimilation is complete.

Source correction using tsunameter

The previously described inversion algorithm is implemented to work with the forecast database. It combines real-time tsunameter data of offshore amplitude with the simulation database to improve accuracy of an initial offshore tsunami scenario. At present, 44 tsunameters using DART™ technology are deployed in the Pacific, Atlantic, and Indian Ocean (Fig. 12.1, see color insert).

Inundation estimates with non-linear model

Once the offshore scenario is obtained, the results of the propagation run are used for the site-specific inundation forecast. Tsunami inundation is a highly nonlinear process. Therefore, linear combinations of different inundation runs cannot be combined to obtain a valid solution. A high-resolution 2-D inundation model is run to obtain a local inundation forecast. The data input for the inundation computations are the results of the offshore forecast—tsunami parameters along the perimeter of the inundation computation area. The forecast inundation model can be optimized to obtain local forecasts within minutes on modern computers. These Stand-by Inundation Models (SIMs) will be run in real time for most vulnerable coastal communities. Three levels of telescoping grids with increasing resolution are employed to model tsunami dynamics and inundation onto dry land. Figure 12.2 illustrates grid setup examples for four SIMs in Hawaii. Each SIM is implemented and optimized for speed and accuracy. Each developed SIM has been validated thoroughly with historical tsunamis as well as results from a corresponding reference inundation model of higher resolution (Tang et al., 2007). SIMs are designed to simulate 4 hours of coastal tsunami dynamics in less than 10 minutes, meaning that a forecast of the first few waves will be obtained within 2–3 minutes. NOAA plans to develop at least 75 SIMs along the U.S. coasts in the next few years. 26 SIMs have already been developed and are now available for forecasting (Fig. 12.1).

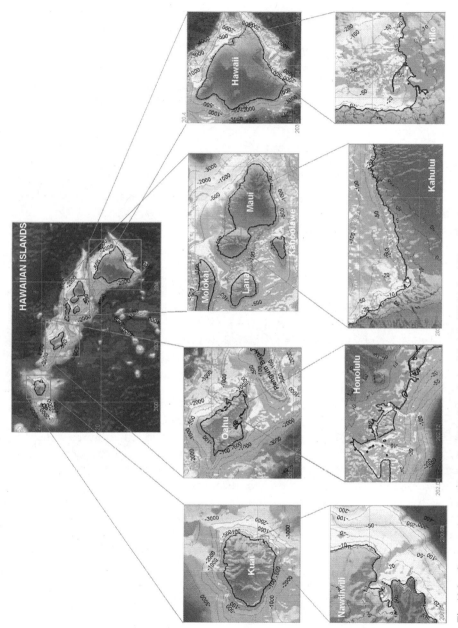

Fig. 12.2 – Computational grids for Stand-by Inundation Models (SIMs) for 4 Hawaiian locations.

Simplified methods of inundation estimation are also being considered for fast preliminary estimates of coastal amplitudes, such as, one-dimensional runup estimates (one spatial dimension), analytical extrapolation of the offshore values to the coast and some others.

6. Testing of the Tsunami Forecast Method

Some preliminary tests of the newly developed BoM forecast are described by Greenslade et al. (2007). However, the test of the Australian system performance with recent events has not been published at the time of this publication. JMA system performance is discussed briefly by Kuwayama (2006) and Tatehata (1997). Here, several important tests of NOAA's forecast system are presented. The system is currently being implemented at TWCs and is not fully operational. However, the described methods have been tested with data from 15 different tsunamis that were measured in deep waters to provide deep-ocean amplitude constraints for the model forecast. Eight of these events have been modeled in true forecast mode, when model results were obtained before the tsunami reached the modeled locations. Some of these forecast results are illustrated in this chapter.

6.1   November 17, 2003 Rat Islands – single site forecast

The 17 November 2003 Rat Islands tsunami in Alaska was the first test and the proof of concept for the SIFT methodology for distant tsunamis. The Mw 7.8 earthquake on the shelf near Rat Islands, Alaska, generated a tsunami that was detected by two tsunameters located along the Aleutian Trench—the first such detection by then newly developed real-time tsunameter system. These real-time data, combined with a model database, were then used to produce the real-time model tsunami forecast. For the first time, tsunami model predictions were obtained during the tsunami propagation before the waves had reached many coastlines. The initial offshore forecast was obtained immediately after preliminary earthquake parameters (location and magnitude Ms 7.5) became available from the West Coast/Alaska Tsunami Warning Center (about 15–20 minutes after the earthquake). The model estimates provided the expected tsunami time series at tsunameter locations. When the closest tsunameter (D171) recorded the first tsunami wave about 80 minutes after generation, the model predictions were compared with the deep-ocean data to refine the model offshore prediction. These offshore model predictions were then used as input for the high-resolution inundation model for Hilo Bay (Titov et al. 2005). The model computed tsunami dynamics on several telescoping grids, with the highest spatial resolution of 30 m inside Hilo Bay. The tsunami did not produce inundation at Hilo, but a nearly 0.5 m (peak-to-trough) tsunami was recorded at the Hilo gauge. A comparison of the model forecast with the tide gauge observations (Fig. 12.3) demonstrated that amplitudes, arrival time, and periods of several initial waves of the tsunami wave train were accurately forecasted (Titov et al. 2005).

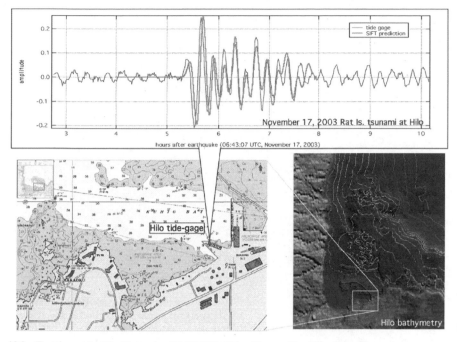

Fig. 12.3 – Test forecast of the November 17, 2003 Rat Island tsunami for Hilo, Hawaii.

6.2   May 3, 2006 Tonga tsunami – multiple sites forecast test

The Tonga tsunami provided a good test of the inversion component of the system and the first test of multiple SIMs that had been developed by that time.

At local time 04:26:39 on 4 May (15:26:39 on 3 May UTC) 2006, a seismic moment magnitude (Mw) 7.9 earthquake occurred on the subduction zone about 160 km (100 miles) NE of Nuku'alofa, Tonga in the Southern Pacific Ocean. The earthquake generated tsunami waves that propagated throughout the entire Pacific basin. The tsunami provided a test of NOAA's next generation tsunami forecast system, which includes DART (Deep-ocean Assessment and Reporting of Tsunamis) detection, inversion and site-specific inundation estimates for eight coastal communities (Fig. 12.4).

Two DART systems near Hawaii, DART-II 51407 located 4818 km from the epicenter and an experimental DART-ETD (Easy-To-Deploy) system 193 km to the West of the 51407 recorded the tsunami generated in Tonga. The data showed a 0.5 cm high first wave arrived offshore of the Hawaiian Islands about 6 hours after the earthquake, followed by a train of oscillations with maximum peak-to-trough of 2.6 cm (Fig. 12.5b and 12.5c). These DART data were not useful for a real-time forecast in Hawaii, since the detectors were so close to the coastline, but it provided a hard test for the inversion scheme of the forecast system.

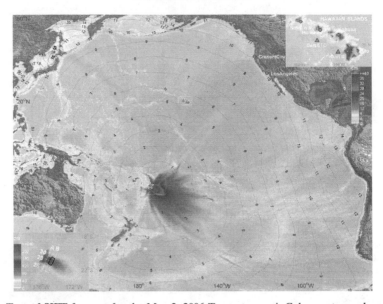

Fig. 12.4 – Test of SIFT forecast for the May 3, 2006 Tonga tsunami. Color contours show maximum forecast tsunami amplitudes, yellow triangles show DARTs deployed at the time of the earthquake, Red circles show locations of forecast sites, white rectangles indicate pre-computed sources of tsunami forecast database.

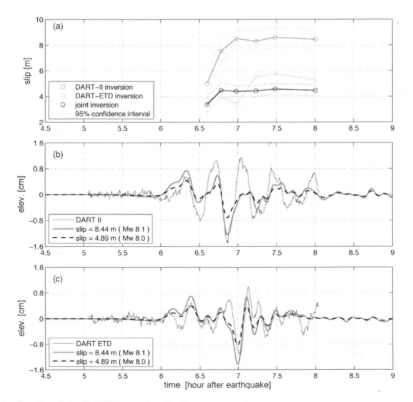

Fig. 12.5 – Results of the DART inversion for Tonga tsunami.

A set of model tsunami source functions from six pre-computed model scenarios in the database, A28 to A30 and B28 to B30 near the epicenter at the Tonga subduction zone, were chosen for inversion (Fig. 12.4). A linear least square scheme was then tested with both single and combined sets of DART data of different length of time series, respectively (Fig. 12.5). All the results indicate the tsunami source was most likely originated from source B29. As more data of longer time series are used, the slip converges quickly. The inverted source corresponded to a moment magnitude of 8.0. Better comparisons of wave amplitudes were obtained at DART-ETD (Fig. 12.2b). DART-ETD is located in a more open offshore position and thus has less interference from the later waves from the Hawaiian Islands (Fig. 12.4).

The propagation scenario obtained from the inversion was used to force inundation computations for eight SIMs. Figure 12.6 presents the comparisons of observed and modeled waveforms at 8 SIMs, four in Hawaii, two at the West Coast and two in Alaska. The comparison shows excellent model-observation agreement of first arrivals, wave amplitudes, periods, and amplitude decay for up to 24 hr after the earthquake. Further investigation of the model animation revealed that the model reproduced wave reflections from various bathymetric features of the South Pacific (Fig. 12.5).

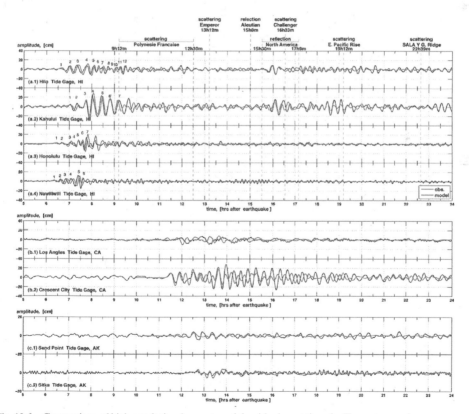

Fig. 12.6 – Comparison of high-resolution forecast models with observations for Tonga tsunami.

The approximate arrivals of major wave trains in Hawaiian Islands are also indicated in Figure 12.6. Hilo and Kahului SIMs accurately reproduced arrivals, amplitudes, and periods of the waves reflected from North America, which arrived approximately 16 hours after the earthquake.

6.3   November 15, 2006, Kuril Island – real-time multiple-site forecast

The tsunami generated by a Mw 8.3 (NEIC: Mw = 7.9, Harvard: Mw = 8.3) earthquake offshore the central Kuril Islands offered another test for the NOAA forecast software. This perfect research tsunami was strong enough to generate a rich set of measurement data from around the Pacific; at the same time, the remote earthquake location and specific patterns of tsunami energy distribution spared population centers from significant damage and loss of life.

The strong thrust earthquake occurred on the boundary between the Pacific plate and the Okhotsk plate at 11:14:16 UTC (23:14:16 local time at epicenter). This was the largest earthquake to have occurred in the central Kuril Islands since 1915, when an earthquake occurred with estimated magnitude of about 8. No reports of local impact from the 2006 Kuril earthquake and tsunami are available at the time of publication, since the closest islands were not populated at the time of the earthquake while there was no easy connection with this remote location. On the other hand, far-field tsunami measurements of this event are plentiful. In the open ocean, 16 DART stations clearly recorded in real time the tsunami propagating offshore the Aleutian Island chain; several JMA cable systems with tsunameters have also recorded the tsunami offshore of Japan (Fig. 12.7, see color insert). At shorelines, more than 50 tide gage stations recorded the wave while it was reaching coastlines around the Pacific.

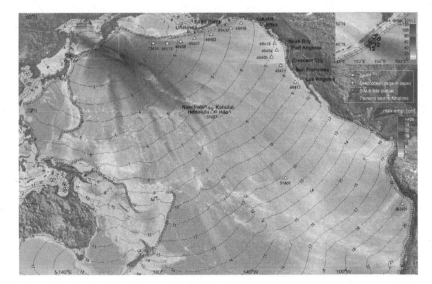

Fig. 12.7 – (see color insert) Test forecast for the November 15, 2006 Kuril Island tsunami. The symbols are the same as in Figure 4.

This wealth of data is invaluable for studying this event and comparing models against the measurements. More importantly, the abundance of real-time data allowed testing of the NOAA tsunami forecast system (SIFT) and verifying the forecast skills by comparing model predictions with tide data. Among the coastal locations with tsunami measurements, there are 12 communities in Hawaii, Alaska, and the West Coast of the U.S. where high-resolution forecast models (SIMs) had already been developed. While this high-resolution coastal forecast capability was not operationally installed at TWCs at the time of the tsunami, the forecast models were set up and ready for test runs in real time. The staff of the NOAA Center for Tsunami Research (NCTR), in coordination with the U.S. TWCs, performed the real-time test of the SIFT system during the event and analysis of the forecast results with the measurement data. The following time line illustrates the potential speed and accuracy of the components of the coastal tsunami forecast, but does not show the actual performance of the integrated system. The fully installed SIFT will automate data-model connections and many operations that were performed manually during this test, thus increasing the speed and efficiency of the forecast.

Tsunami warning and forecast time line

- *11.30 UTC:* A tsunami warning was issued by PTWC about 15 minutes after the earthquake. At that time, preliminary earthquake location and magnitude became available and could have been used for a preliminary seismic-based forecast, if it was operational at TWCs.
- *12.30 UTC:* The first test tsunami forecast was obtained by NCTR about 1 hour after the earthquake using the updated magnitude Mw = 8.1 and epicenter location. The offshore forecast provided an estimate of the tsunami amplitude and arrival at DART locations, when the tsunami was about 1 hour away from the closest DART (Fig. 12.7).
- *13.45 UTC:* The tsunami reached the closest DART 21414 about 2 hours after the earthquake. The DART recorded the first wave period of the tsunami. At that point, the inversion of the DART data had been performed (Fig. 12.8), an updated offshore forecast was obtained, and the first coastal forecasts for Hawaiian locations were run, showing maximum amplitude of slightly above 0.5 m at Kahului and Hilo, while the tsunami is about 4 hours away from those Hawaiian locations. During the next few hours, the coastal forecasts were performed for all 12 SIMs and were updated several times with new DART data, which did not substantially change coastal predictions. The coastal forecast showed borderline tsunami amplitudes for several Hawaii locations and for Crescent City, CA.
- *16:45 UTC:* The coastal forecast is finalized for 12 U.S. locations where SIMs are available. The tsunami is about 1 hour away from the Hawaii and about 3 hours away from major West Coast population centers (Fig. 12.9).

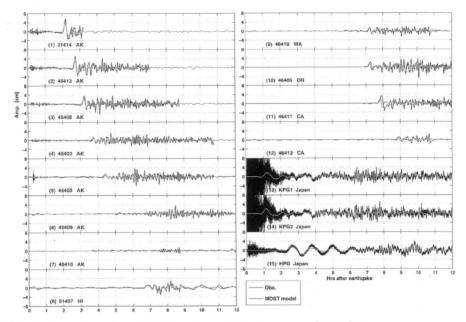

Fig. 12.8 – Comparison of model offshore forecast with deep-ocean observations.

The forecast pinpointed the main potential problem spots for the U.S coastlines early on. Predicted offshore tsunami amplitude distribution (Fig. 12.7, see color insert) shows that Alaska and most of the West Coast population centers would not experience substantial waves. At the same time, most of Hawaii would be approached by a higher amplitude tsunami. The propagation model also showed some focusing of tsunami energy toward Crescent City. Beyond the U.S. coastlines, the offshore forecast showed potential hazards along the Kuril Island chain, south Kamchatka, inside Okhotsk Sea, at Northeast Hokkaido. The propagation model also showed some focusing at South Chile and at Galapagos Islands. To quantify these preliminary qualitative assessments, the high-resolution coastal forecast models were run. The coastal forecast results showed no substantial inundation for any of the 12 SIMs (Fig. 12.9). Nevertheless, the amplitudes for several Hawaii locations approached critical levels. Figure 12.10 (see color insert) illustrates the coastal forecast at Kahului showing amplitudes above 0.5 m at a large portion of the coastline inside and outside the harbor. The predictions of the tide gage at Crescent City were especially interesting. While the model did not predict any inundation (Fig. 12.10, see color insert), the tsunami amplitudes were predicted to grow gradually at the tide gage, reaching a maximum about 2 hours after the first tsunami arrival.

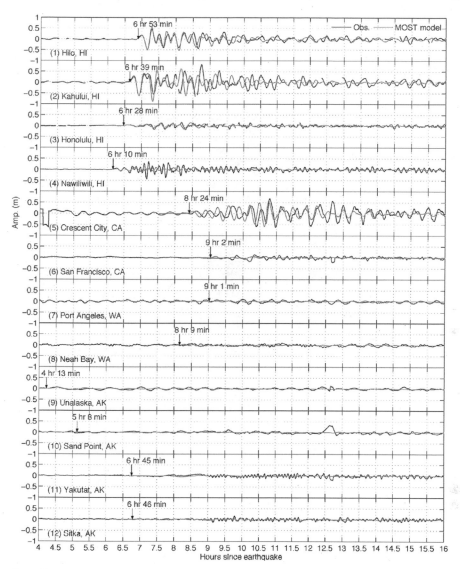

Fig. 12.9 – Comparison of SIM forecast with tide gage observations.

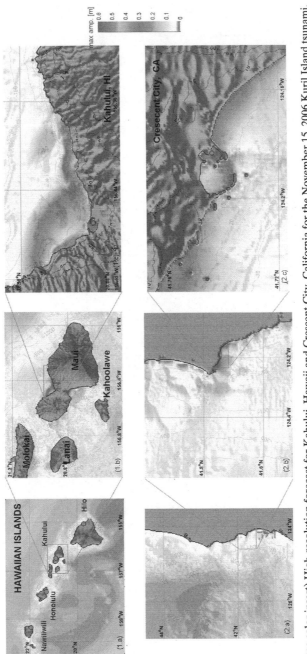

Fig. 12.10 – (see color insert) High-resolution forecast for Kahului, Hawaii and Crescent City, California for the November 15, 2006 Kuril Island tsunami.

The aftermath of the tsunami and comparisons of the forecast predictions with coastal measurements confirmed good accuracy of the coastal forecast. Figure 12.9 shows comparisons of tide gage forecasts with measurements for 12 SIM locations. An additional test of the offshore forecast quality came after comparison with the Japanese deep-ocean recording of the tsunami by the Japan Marine Science and Technology Center cable system offshore Japan. Figure 12.8 shows excellent comparison between the model and this independent data (not used in the inversion) for several first waves.

6.4   August 15, 2007, Peru

On 15 August 2007, at 23:41 UTC, a massive earthquake of moment magnitude 8.0 struck off the Pacific coast of Central Peru, i.e., Pisco, Ica. The earthquake (76.509 W, 13.354 S) was offshore about 150 km southeast of Lima at a focal depth of 39 km (USGS). This earthquake caused severe shaking and damage in nearby towns, especially in Pisco, Ica. The death toll throughout Peru is reported to be as high as 650 (Historical Tsunami Database at NOAA/National Geophysical Data Center) due to collapsing houses and infrastructures, forcing the government to declare a state of emergency. Coastal flooding induced by tsunami waves as high as 5 m was observed in harbors of Patacas (Tubbesing, personal communication, 2007). Pictures taken 4 hours after the earthquake showed that the tsunami waves resulted in minor flooding in La Punta, Callao (Power, personal communication, 2007). Fritz et al. (2007) surveyed the area and measured 2 km inundation at Lagunilla, where three bodies were found approximately 1.3 km inland. They also measured 10 m maximum runup at Playa Yumaque, with representative runup on the order of 7 m in this area.

Tsunami warning time line
The time line of the experimental forecast for the August 15 tsunami is illustrated in Figure 12.11 (from Wei et al., 2008)

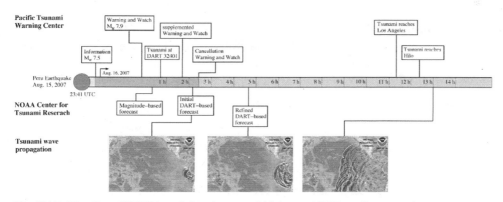

Fig. 12.11 – Timeline of NCTR's real-time forecast of 15 August 2007 Peruvian tsunami.

On 15 August, at 23:53 UTC, 12 minutes after the earthquake, PTWC disseminated its first information bulletin reporting the earthquake location with a preliminary magnitude Mw 7.5. At 00:19 UTC on 16 August, 26 minutes later, PTWC upgraded the earthquake magnitude to Mw 7.9 and issued a second bulletin for a regional tsunami warning to the entire Pacific coast of South America, as well as a tsunami watch to the countries of Central America. Based upon the updated seismic magnitude, a preliminary offshore forecast was generated, similar to Figure 12.12 (not yet constrained by DART data, as shown in the figure) at 00:27 UTC using pre-computed propagation scenarios. This map reflects the forecast maximum wave amplitude at each computational grid and is an indicator of the distribution of tsunami energy (Wei et al., 2008)

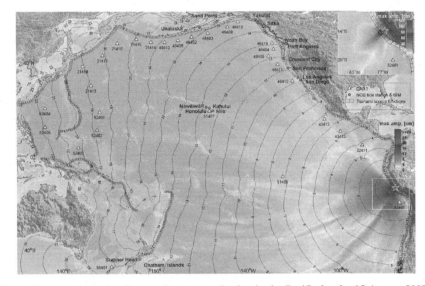

Fig. 12.12 – Forecast scenario of tsunami energy projection in the Pacific for the 15 August 2007 Peruvian Tsunami. ☆ is earthquake hypocenter; △ are DART buoys; • are U.S. forecast sites; ⊞ are tsunami source functions.

The Chilean-owned DART 32401, located approximately 700 km southeast of the epicenter, recorded the first tsunami pulse of a 6-cm-high peak 56 minutes after the main shock (Fig. 12.13a). In response to this clear tsunami signal, PTWC issued the third bulletin at 01:21 UTC to supplement the tsunami warning and watch published earlier, and to advise for potential tsunami risks at distant Pacific coasts. The 01:21 UTC bulletin explicitly stated that this supplementary warning was based on the observation of a "tsunami signal on the deep ocean gauge off northern Chile." The real-time deep ocean measurements free of coastal contamination are required as an input to the NOAA forecast system (Titov et al., 2005). The DART measurement provided the arrival time, amplitude, and half period of the wave, which are the minimum information required for the data assimilation and inversion to constrain an initial tsunami source. At 01:50 UTC, 2 hours 9 minutes after tsunami generation, NCTR obtained the initial DART-constrained tsunami source inverted from measurements of DART 32401 and produced a new tsunami energy projection map. As the tsunami spread away from South America, the refined tsunami energy projection indicated large tsunami amplitudes off the

Hawaiian Islands. Harbors in California were also of concern because they were the nearest coast of the U.S. to be impacted by the tsunami. Note that a tsunami, such as the November 15, 2006 Kuril Island tsunami, with small far-field effects could cause significant damage at the harbor of Crescent City due to the harbor resonances (Kelly et al., 2006). The final stage of NCTR's tsunami forecast was to provide an assessment of coastal tsunami impact for these U.S. forecast sites. The forecasts based on the initial tsunami source for U.S. coastal communities in Hawaii and on the U.S. West Coast, indicated no tsunami flooding for any of the sites. The propagation snapshots in Figure 12.11 show that, at the time of the forecast, the influence of the tsunami was still limited to the Pacific coast of Peru. At 02:09 UTC, 2 hours 28 minutes after the initial quake, PTWC sent out the final statement to cancel the tsunami warning and watch for both South America and Central America.

The main cycle of the tsunami waves would take at least another 8 hours to arrive at distant Pacific coasts thousands of miles away from the tsunami source. This time frame provided a valuable opportunity for NCTR scientists to exercise the tsunami forecast system, and further improve the accuracy of the forecasts. Four hours and 45 minutes after the tsunami generation, NCTR computed a refined tsunami source based on data inversion over longer time series at DART 32401. Although DART buoys 32411, 46412, and 51406 (Fig. 12.12) also recorded the passing tsunami (Fig. 12.13a), they were not used to constrain the tsunami source during the forecast because of the late tsunami arrival at these buoys. Using the updated tsunami source, the final forecast results were made public by NCTR for 14 U.S. coastal communities in Hawaii, the U.S. West Coast, and Alaska (Fig. 12.13b), including the final tsunami energy projection in the Pacific (Fig. 12.12), while the tsunami was 6 hours away from arriving at the nearest U.S. coastline.

Tsunami Forecast Results
The comparison of the wave amplitude in Figure 12.13a indicates that the forecast based on the refined tsunami source provided an excellent fit with observation at DART 32401 before the tsunami waves were reflected and scattered by the seamounts off the Pacific coast of South America. The excellent agreement of model results with wave heights at the other three DARTs verified that the refined forecast correctly estimated the tsunami source. One noticeable result in Fig. 12.13a is the time discrepancy of the arrival of the first wave at DARTs 32411, 46412, and 51406. This time difference is mainly due to the location of the pre-computed tsunami source functions, which are offshore of the earthquake epicenter (Fig. 12.12). If they had been located closer to the coast, most of the time difference would be eliminated. A secondary cause is the imprecision of ocean bathymetry, which results in error accumulation over extensive computation.

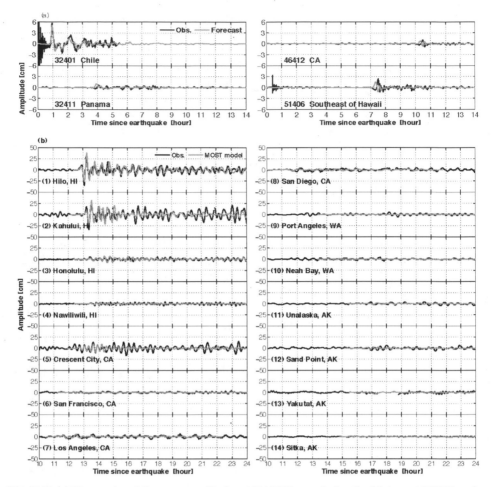

Fig. 12.13 (a) Time series of the wave amplitudes at DART buoys during the 15 August 2007 Peruvian tsunami: — observations; — forecast. (b) Time series of the wave amplitudes with +12-minute adjustment at 14 U.S. coastal communities during the 15 August 2007 Peruvian tsunami: — observations; — MOST model.

Figure 12.12 shows the basin-wide tsunami energy projection and the contours of travel time computed from the refined DART-constrained tsunami source. It is clear that the tsunami energy was mostly directed to the south and west of the Pacific coast of South America. Figure 12.12 indicates that the coasts of Ecuador, Panama, and countries in Central America were in the shadow of tsunami energy and received much less impact than Peru and Chile. The islands in the south Pacific, however, faced more serious threats. Figure 12.12 predicted high tsunami waves in the Chatham Islands, east offshore of New Zealand, and was later confirmed by the 53 cm and 46 cm peak-to-trough heights observed at tide gages in Waitangi (Weinstein, personal communication, 2007) and Kaingaroa (Bell, personal communication, 2007), respectively. Figure 12.12 also indicated higher tsunami energy at several locations on the east and north coasts of New Zealand that were verified by tide-gage measurements, including the 54 cm peak-to-trough height at Sumner Head (Bell, personal communication, 2007). The energy distribution shows minor tsunami impact along the east and north coasts of Australia, which were protected by New Zealand and the island chain to its north. The observed peak-to-trough wave height there was about 20 cm. The tsunami brought 20–25 cm high waves to Japanese coasts after a more than 20-hour propagation over thousands of kilometers, causing Japanese emergency agencies to issue tsunami warnings to coastal communities.

Inundation computations using SIMs were conducted during the event for four harbors in Hawaii, six on the U.S. West Coast, and four in Alaska. Figure 12.13b shows the refined forecasts compared with observations for the 14 tide gages. Since all the modeling forecasts were completed before the tsunami arrived, Figure 12.13b provides additional evidence of the accuracy and efficiency of the forecast system. Among the 14 harbors, Hilo, HI recorded the highest maximum wave height, 67 cm, Kahului, HI recorded 56 cm, and Honolulu, HI recorded 10.5 cm, while Crescent City, CA recorded 30.7 cm. The others had wave heights less than 10 cm that were in the range of the background noise. As noted earlier, all 14 forecast results showed consistent 12-minute earlier arrivals due to the inaccurate location of the existing tsunami source functions and ocean bathymetry. One should note that a 12-minute difference out of 12 hours represents a less than 2% error. After the 12-minute time difference was adjusted in Figure 12.13b, the model results and observations matched very well in both wave height and period at all tide stations. The most notable are Hilo, HI and Honolulu, HI, where the modeling results and observations are in excellent agreement for up to 24 hours after tsunami generation.

6.5   Other tsunamis

Ten tsunamis have been forecasted to date in experimental mode using the NOAA system under development, including all Pacific tsunamis recorded at DART since 2003 (eight total) and two tsunamis in the Indian ocean since the first DART was deployed in 2006. In addition, deep-ocean BPR data from five pre-DART tsunamis have been used to test the components of the forecast system. The results of the real-time forecast tests are summarized in Table 1. While the level of the test forecast efforts for these events varied, the results demonstrate growing confidence in this tsunami forecast capability that uses deep-ocean data for forecast model adjustments.

TABLE 1.

Results of test forecasts during 2003–2007 tsunamis.

Tsunami	# of forecasts	Forecast accuracy at tide gages above noise*	Lead time (hours)	Impact
Alaska 11/17/2003	1	above 95%	3	Evacuation avoided**
California 06/14/2005	1	amplitudes below noise	0	Evacuation avoided**
Tonga 05/03/2006	8	above 90%	3	Warning cancellation**
Kuril Is. 11/15/2006	12	above 90%	2	Evacuation Avoided
Nicobar Is. 01/08/2007	offshore only			Was not used for warning
Kuril Is. 01/13/2007	15	above 70% ***	2	Evacuation Avoided
Solomon Is. 04/01/2007	7****	all amplitudes below noise	3	No warning issued for U.S.
Peru 08/15/2007	16	above 95%	6	Warning cancellation
Sumatra 09/12/2007	1	above 90%	1	Was not used for warning
Chile 11/14/2007	4****	all amplitudes below noise	10	Warning cancellation

* Amplitude and arrival time accuracy for the first five to six tsunami waves are assessed for tsunami signals that are significantly above noise level prior to estimated tsunami arrival at a particular tide gage (noise-to-signal ration less than 25%). The noise error is subtracted from the the comparison error.

** Evacuation and warning decision was based on the offshore DART data and offshore forecast only; high resolution forecast was not available to TWCs during this tsunami warning.

*** Forecast at Hilo and Kahului over-predicted amplitudes by about 30% (25 cm predicted vs. 15 cm measured); the rest of the forecasts are above 90% accurate, or within noise.

**** The rest of the forecasts were not computed due to small expected amplitudes

7. Summary

The ability to make accurate tsunami forecasts is linked directly to the availability of timely, tsunameter measurements of the approaching tsunami. The accuracy of the forecast is directly tied to these observations coupled with validated numerical models. More tests are required to ensure that inundation forecasts are always reliable. However, with 14 successful tests to date (including 10 real-time tests listed in Table 1, and 4 hindcast test with non-real-time data), we are gaining confidence that accurate and timely forecasts are within reach. When fully tested and fully implemented, such forecasts will provide enough lead time for potential evacuation or warning cancellation for coasts threatened by distant-source tsunamis.

Including the 15 August 2007 Peru event, NOAA's tsunami forecasting system has produced excellent experimental forecasts for far-field tsunami impact for eight Pacific tsunamis since its first real-time test in 17 November 2003 Rat Island tsunami. The essential components of the forecast system are deep ocean measurements and numerical modeling. High-quality forecasts have shown the strength of the DART implementation, even for just one node of the DART network, in obtaining the accurate tsunami source. A full set of DART buoys will produce better constraints of the tsunami source while providing more timely tsunami detection and observation. With improved DART network and automated inversion processes, a substantial decrease of the forecast time is expected when this system is implemented at the TWCs.

Acknowledgments

The author is grateful to all members of the NOAA Center for Tsunami Research for their contributions and joint efforts in developing the tsunami forecast system. The author is thankful to Rachel Tang, Yong Wei, Costas Synolakis, and Eddie Bernard for their contributions and help in preparing the manuscript. This publication is funded in part by the Joint Institute for the Study of the Atmosphere and Ocean (JISAO) under NOAA Cooperative Agreement No. NA17RJ1232, JISAO Contribution 1453, PMEL Contribution 3159.

References

Bernard, E. N., H. O. Mofjeld, V. V. Titov, C. E. Synolakis, and F. I. González. 2006. Tsunami: Scientific frontiers, mitigation, forecasting, and policy implications. *Proc. Roy. Soc. Lon. A,* **364**(1845), doi:10 .1098/rsta.2006.1809, 1989–2007.

Fritz, H. M., N. Kalligeris, E. Ortega, and P. Broncano. 2007. 15 August 2007 Peru tsunami runup and inundation, *Earthquake Engineering Research Institute Newsletter,* in press.

González, F. I., V. V. Titov, H. O. Mofjeld, A. Venturato, S. Simmons, R. Hansen, R. Combellick, R. Eisner, D. Hoirup, B. Yanagi, S. Yong, M. Darienzo, G. Priest, G. Crawford, and T. Walsh. 2005. Progress in NTHMP hazard assessment. *Nat. Hazards,* **35,** 89–110. Special Issue, U.S. National Tsunami Hazard Mitigation Program. (doi: 10.1007/s1 1069-004-2406-0).

D. J. M. Greenslade, M. A. Simanjuntak, D. Burbidge, and J. Chittleborough. 2007. A first-generation real-time tsunami forecasting system for the Australian Region. BMRC Research Report 126, 84 pp.

Kelly, A., L. Dengler, B. Uslu, A. Barberopoulou, S. Yim, and K. J. Bergen. 2006. Recent tsunami highlights need for awareness of tsunami duration. *Eos Trans. AGU,* **87**(50), 566–567.

Kuwayama, T. 2007. Quantitative Tsunami Forecast System. ICG/PTWS Tsunami Warning Centre Coordination Meeting, Honolulu, HI, 17–19 January 2007.

Milburn, H. B., A. I. Nakamura, and F. I. González. 1996. Real-time tsunami reporting from the deep ocean. In Proceedings of the Oceans 96 MTS/IEEE Conference, 23–26 September 1996, Fort Lauderdale, FL, 390–394. (http://nctr.pmel.noaa.gov/milburn1996.html)

Okada, Y. 1985. Surface deformation due to shear and tensile faults in a half-space. *Bull. Seismol. Soc. Am.,* **75**(5), 1135–1154.

Satake, K. 2002. Tsunamis. In *International Handbook of Earthquake and Engineering Seismology,* pp. 437–451, Academic Press.

Synolakis, C. E., and E. N. Bernard. 2006. Tsunami science before and beyond Boxing Day, *Proc. Roy. Soc. Lon. A,* **364**(1845), 2231–2265.

Tang, L., C. Chamberlin, V. V. Titov, and E. Tolkova. 2008. A standby inundation model of Kahului, Hawaii for NOAA Short-Term Inundation Forecast for Tsunamis (SIFT). NOAA Tech. Memo. OAR PMEL, submitted.

Tatehata, H. 1997. The new tsunami warning system of the Japan Meteorological Agency. In G. Hebenstreit (ed.), *Perspectives of Tsunami Hazard Reduction,* Kluwer, pp. 175–188.

Titov, V. V., F. I. González, E. N. Bernard, M. C. Eble, H. O. Mofjeld, J. C. Newman, and A. J. Venturato. 2005. Real-time tsunami forecasting: Challenges and solutions. In *Developing Tsunami-Resilient Communities, The National Tsunami Hazard Mitigation Program,* E. Bernard (ed), *Nat. Haz.* **35**(1), 40–58.

Titov, V. V., H. O. Mofjeld, F. I. Gonzalez, and J. C. Newman. 2001. Offshore forecasting of Alaskan tsunamis in Hawaii. In: *Tsunami Research at the End of a Critical Decade,* G. T. Hebensreit (ed.),Birmingham, 19–30 July 1999, Kluwer Acad. Pub., Netherland, 75–90.

Titov, V. V., H. O. Mofjeld, F. I. González, and J. C. Newman. 1999. Offshore forecasting of Alaska-Aleutian Subduction Zone tsunamis in Hawaii. NOAA Tech. Memo. ERL PMEL-114, 22 pp.

Titov, V. V., and F. I. González. 1997. Implementation and testing of the Method of Splitting Tsunami (MOST) model. NOAA Technical Memorandum ERL PMEL-112, 11 pp.

Titov, V. V., and C. E. Synolakis. 1998. Numerical modeling of tidal wave runup. *J. Waterw. Port Coast. Ocean Eng.,* **124**(4), 157–171.

Titov, V. V., and C. E. Synolakis. 1995. Modeling of breaking and nonbreaking long wave evolution and runup using VTCS-2. *J. Waterw. Port Coast. Ocean Eng.,* **121**(6), 308–316.

Wei, Y., E. N. Bernard, L. Tang, R. Weiss, V. V. Titov, C. Moore, M. Spillane, M. Hopkins, and U. Kânoğlu (2008), Real-time experimental forecast of the Peruvian tsunami of August 2007 for U.S. coastlines, *Geophys. Res. Lett.,* **35,** L04609, doi:10.1029/2007GL032250

Whitmore, P. M. 2003. Tsunami amplitude prediction during events: A test based on previous tsunamis. In *Science of Tsunami Hazards,* **21,** 135–143.

Chapter 13. TSUNAMI WARNING SYSTEMS

PAUL M. WHITMORE

NOAA/NWS/West Coast and Alaska Tsunami Warning Center

Contents

1. Introduction

According to the U.S. National Oceanographic and Atmospheric Administration's National Geophysical Data Center (NOAA NGDC), over 400,000 people have perished in tsunamis since 1800 (NOAA tsunami data base, 2006). To reduce this loss of life, tsunami warning systems have been established.

The first warning system was established in 1949 following the death and destruction inflicted on Hawaii during the 1946 Aleutian tsunami. The original center, known as the Honolulu Observatory, was co-located at the existing Honolulu Magnetics Observatory where it still resides today. The warning center relied on seismic and tide gage data observations relayed via teletype from several observatories located around the country. The center was established just in time to issue life-saving warnings for several major tsunamis occurring in the 1950s and early-1960s.

Following the Pacific-wide impact caused by the 1960 Chile tsunami, nations throughout the Pacific Basin coordinated under the leadership of the United Nations and established a basin-wide warning system. In 1968, the Honolulu Observatory expanded its scope to provide warnings to nations throughout the Pacific and became known as the Pacific Tsunami Warning Center (PTWC; Fig. 13.1).

The Sea, Volume 15, edited by Eddie N. Bernard and Allan R. Robinson
ISBN 978–0–674–03173–9 ©2009 by the President and Fellows of Harvard College

Fig. 13.1 – The Pacific Tsunami Warning Center located in Ewa Beach, Hawaii.

The Gulf of Alaska earthquake in 1964 spurred the development of warning centers which recorded regional seismic network data in real-time and were able to issue warnings shortly after an earthquake's origin. Warnings were issued following the 1964 event, but too late for people within a 1000km of the source (Spaeth and Berkman, 1967). In 1967 the U.S. established a network of three warning centers in Alaska at Adak, Sitka, and Palmer. As communications technology improved, these three centers combined operations in Palmer during the 1970s and became known as the Alaska Tsunami Warning Center (later renamed the West Coast/Alaska Tsunami Warning Center (WCATWC) and shown in Fig. 13.2).

Fig. 13.2 – The West Coast/Alaska Tsunami Warning Center located in Palmer, Alaska.

In addition to the warning centers in Alaska and Hawaii, centers have been established in Japan, Russia, French Polynesia, Chile, Indonesia, India, and Australia. The Japanese, Russian, French Polynesia, and Chilean systems were developed throughout the 1950s and 1960s. The Indonesian, Indian, and Australian systems were initiated after the 2004 Indian Ocean tsunami. Fig. 13.3 depicts the areas-of-responsibility of these centers. This chapter will focus mainly on the operations of U.S. tsunami warning centers due to the author's expertise with these systems.

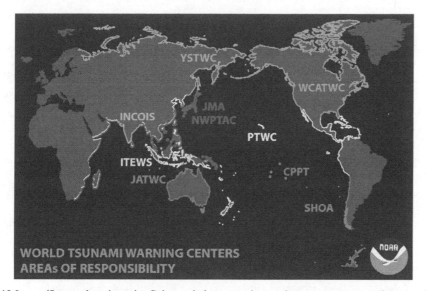

Fig. 13.3 – (See color insert) Color-coded tsunami warning center areas-of-responsibility. YSTWC=Yuzhno-Sakhalin Tsunami Warning Center; JMA=Japanese Meteorological Agency; NWPTAC Northwest Pacific Tsunami Advisory Center; INCOIS=Indian National Centre for Ocean Information Services; ITEWS=Indonesia Tsunami Early Warning System; JATWC=Joint Australian Tsunami Warning Centre; CPPT= Centre Polynésien de Prévention des Tsunamis (Polynesian Tsunami Warning Center); SHOA= Servicio Hidrográfico y Oceanográfico de la Armada (Chilean Navy Hydrographic and Oceanographic Service).

International collaboration between warning centers has been problematic due to language, geographic, and political barriers. To assist with international cooperation, the Intergovernmental Oceanographic Commission (IOC) of the United Nations Educational, Scientific and Cultural Organization (UNESCO) developed a structure for a global tsunami warning system following the impact of the 2004 Indian Ocean tsunami. This global system consists of four regional systems in the Pacific Ocean, Indian Ocean, Caribbean Sea, and the northeast Atlantic/Mediterranean regions. While the Pacific system was already mature having been initiated in 1968, development of warning systems in the other regions is still ongoing. Tsunami warning systems are reliant on a wide spectrum of information; such as that provided from seismic networks, ocean monitoring networks, processing centers, tsunami modelers for hazard definition and real-time forecasts, historical tsunami studies, warning transmission, public education, and local response. Considering this range of requirements, development of international systems takes time and commitment. The IOC emerging IOC global system is described in greater detail by Gerard (2007).

As of December 2007, three warning centers issue messages internationally: the PTWC, the Northwest Pacific Tsunami Advisory Centre (NWPTAC) of the Japanese Meteorological Agency (JMA), and the WCATWC. The PTWC issues messages to all participating warning system nations in the Pacific and Indian Oceans, as well as nations in the Caribbean Sea region. The NWPTAC issues messages to nations in the northwest Pacific region and in the Indian Ocean. The WCATWC issues messages to U.S. and Canadian states and provinces in the northeast Pacific as well as U.S. and Canadian states, provinces, and territories in the Atlantic. Tsunami warning systems consist of many components (Fig. 13.4). These include:

- Observational networks which record and transmit tsunami-related data.
- Tsunami Warning Centers (TWC) which acquire, process, and interpret the data, and issue appropriate messages. Centers strive to quickly analyze potential tsunami-generating events and produce appropriate information prior to impact.
- Robust communications networks which transmit TWC messages to coastal residents and emergency managers.
- Local emergency response organizations which respond to warnings received from a warning center, or to nature's warnings, and prepare itself and populations within its jurisdiction to respond properly to an event. Their responsibilities include pre-event planning and exercising; verifying their ability to receive warnings from a warning center; establishing and testing local warning dissemination systems, such as sirens and loudspeakers; and providing public education opportunities to prepare the public for a tsunami.

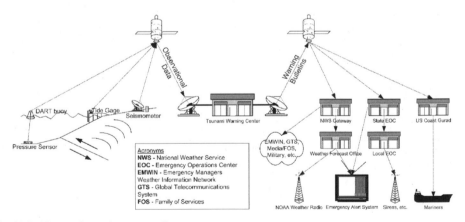

Fig. 13.4 – Tsunami warning system diagram.

Several challenges face this system. While many natural hazard warning systems are able to directly monitor the hazard for which they warn (for example, hurricanes, tornadoes, and solar storms), tsunami warning centers must provide warnings to the nearest coasts prior to observing the tsunami. Warnings are normally based on associated data such as seismic signal which is only indirectly related to tsunami intensity. This reality induces warning centers to extract as much information as possible from the seismic signal, to search for new ways to monitor tsunamis directly, and to use conservative warning protocols.

Warning communications historically have been a major problem within the tsunami warning system. However, with the modern communication technologies now available, most threatened locations can receive tsunami information through many different pathways.

Perhaps the biggest challenge to the tsunami warning system is that encountered by emergency response organizations. These organizations must be constantly prepared for an event which may only occur at their location once every 100 years or more. Preparation for tsunamis is just one of many threats for which emergency management must prepare. While tsunami impact can wreak complete destruction upon a community, preparation for these events often falls behind hazards which occur more frequently.

This chapter will discuss the tsunami warning system by looking in detail at the different components of the system. Several other general synopses have also been written on tsunami warning center operations (e.g., Sokolowski, 1999; Furumoto *et al.*, 1999; McCreery, 2005)

2.   Observational Networks

Tsunami warning centers utilize two basic types of data: seismic and sea level. The initial assessment of a tsunami's threat is normally based on seismic data. Sea level data are then used to confirm tsunami generation and, in conjunction with numerical models and historical data, forecast wave heights. As sea level data are generally not available prior to a tsunami's impact along the nearest coast, initial judgments must be made strictly on seismic data evaluation.

2.1.   Seismic Data

Earth scientists have been using seismographs and seismometers to characterize earthquakes since the late-1800s. Modern seismometers convert the kinetic energy of ground shaking to an electrical signal. Depending on design, seismometer output is proportional to ground displacement, velocity, or acceleration. Instrument response characteristics are taken into account to convert the sensor's output to an accurate representation of ground motion over time. See, for example, Wielandt (2002) for more information on seismometry.

Seismometer output contains a wealth of information which can be used to characterize both the energy source and the section of earth through which the signal passes. When these data are available in real-time to tsunami warning centers, they provide critical information regarding tsunami potential. An earthquake's primary (P) wave arrival time recorded at several stations surrounding an epicenter provides the information necessary to locate an earthquake. Ground motion amplitude is then used to determine magnitude which is related to energy release. Amplitude over time is further used to estimate an earthquake's source parameters, such as fault orientation and slip direction. Seismic analysis will be discussed further in the data processing section of the chapter.

Seismometer locations are often dictated by factors unrelated to perfect network distribution. Some of the more common factors taken into consideration are:

- Reliable source of power
- Cultural noise
- Permitting
- A site which can "see" the appropriate communications satellite
- Bedrock at or near surface
- A secure site not likely to be vandalized
- Easy accessibility for maintenance
- No duplication with other seismic networks

The final seismometer location is usually a compromise between the various factors listed above and ideal network distribution. A typical seismometer installation is shown in Fig. 13.5. For more information on seismometer installations, see McMillan (2002) or Trnkoczy *et al.* (2002).

Fig. 13.5 – WCATWC seismometer installation at Middleton Island, Alaska. Upper-left shows initial excavation and placement of the vault's outer shell. Upper-right is the outer shell with the inner insulation, barrel, and concrete base in place. The lower-left shows installation of the data communications satellite dish, and the lower right shows the final setup with the vault covered in tarp in the background.

Seismic network density and distribution are the main limiting factors which control the speed and accuracy with which a tsunami warning center can initially respond to an earthquake (Whitmore *et al.,* 2007). For example, a center can respond in five minutes with an accurate location and magnitude if the following network criteria are met (response is defined as the time of bulletin issuance minus the origin time of the earthquake):

- At least 12 evenly-distributed seismometers within 900km of an epicenter location (the P-wave travel time for 900km is 120 seconds)
- At least 80% of the seismometers are transmitting data at any given time
- Up to 30 seconds data latency (data transmission time)
- Digital, broadband seismic data (necessary to determine moment magnitudes)

If these criteria are met, a typical timeline for warning center response would be:

- 150s to record signal on at least 9 stations (120 seconds P-wave travel time + 30 seconds data latency)
- 60s more to record enough P-wave signal for magnitude computations
- 30s extra for final analyst review
- 60s to compose and transmit the appropriate message

Response timelines can be compressed further by increasing seismic network density, reducing data latency, or decreasing process time. However, response time will reach a limit due to source process times for major earthquakes which can exceed 100 seconds.

The seismic data network used at the US tsunami warning centers is shown in Fig. 13.6. Station density along the majority of U.S. and Canadian coastal regions is sufficient to support a five minute response. Warning response over the wider Pacific, Indian, and Atlantic basins is greater than 5 minutes due to sparse network coverage. Response time varies across these basins and is generally in the range of 8 to 15 minutes.

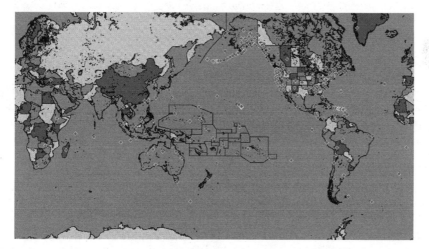

Fig. 13.6 – (See color insert) Seismometers which transmit data to U.S tsunami warning centers as of November, 2007 are shown as diamonds.

The seismic network shown in Fig. 13.6 is essentially a virtual network composed of data from many regional and global seismic networks. Seismological centers which provide data into this network are shown in Fig. 13.7. Equally important to network coverage is the ability to reliably transmit data in real-time to the warning centers. Where possible, warning centers acquire data from supporting networks via two different transmission paths. Fig. 13.7 shows that many supporting networks transmit data over both a private network known as Crestnet (Oppenheimer *et al.*, 2005) and the internet. This provides important redundancy as no transmission method is 100% reliable.

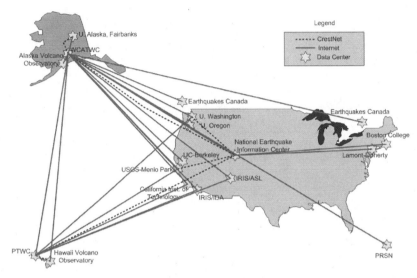

Fig. 13.7 – Seismic networks providing data connections to the U.S. Tsunami Warning Centers. IRIS/IDA and IRIS/ASL provide data from the Global Seismic Network. IRIS: Incorporated Research Institutions for Seismology. IDA: International Deployment of Accelerometers. ASL: Albuquerque Seismological Lab. PRSN: Puerto Rico Seismic Network.

2.2. Sea Level Data

Sea level variations over time are the second basic data type used at tsunami warning centers. These data provide the only direct tsunami observations available in real-time. They are critical to warning centers as seismic recordings only indicate that an earthquake has occurred which *may* generate a tsunami. Decisions to extend, restrict, or cancel warnings are based on sea level data used in conjunction with forecast models and/or historical data.

Sea level measurements are normally made either near the coast or in deep water. Measurements from both regions are important in analyzing a tsunami. Coastal observations are most useful for comparison with historical observations as historical tsunami data relates to amplitude at the coast and not in deep water. Authoritative cancellations can be declared using near source sea level data, and the data can be used to provide an all clear after a significant tsunami has occurred. Data from deep ocean sensors are advantageous when calibrating forecast models during events as these data are not influenced by harbor resonances and other near shore effects. Other advantages to the deep ocean data are that they will not be rendered inoperable during a major tsunami unlike coastal gages which may be destroyed. When located appropriately, these sensors can also help centers determine quickly whether a tsunami is confined to the near source region or if it is potentially dangerous elsewhere. One disadvantage to deep ocean data is that without corresponding forecast models, the data are difficult to interpret as there are few historical events with which the data can be compared.

Sea level gages located near the coast are often known as tide gages since most are installed with the primary purpose of recording daily tidal fluctuations. Tide gages are discussed more completely in Chapter 7 of this volume. Sea level height is normally measured within a pipe, or stilling well, suspended from a pier. Water fills the pipe from an orifice below the water surface, thus filtering out higher frequency wind waves to some extent. As tsunamis have periods ranging from several minutes to over an hour, they are well-recorded by these gages. Radar gages have recently been installed at several locations to supplement traditional tide gage networks (Urban *et al.,* 2003). These gages are also suspended from a pier, but can measure the water level directly without the use of a pipe (Fig. 13.8). Wind waves can be filtered out digitally if necessary. Examples of tsunamis recorded by sea level gages are shown in Fig. 13.9.

Fig. 13.8 – Radar gage installation at Shemya, Alaska shown without cover. This site was installed in November, 2002.

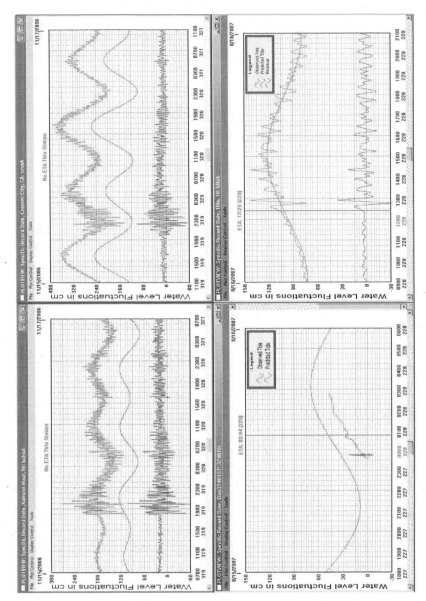

Fig. 13.9 – Tsunamis recorded on sea level gages. The upper-left graph displays the November 15, 2006 Kuril Islands tsunami recorded on the Kahului, Hawaii National Ocean Service tide gage. The upper line is the observed data, middle line is the predicted tide, and lower line is the residual between predicted and observed. The upper-right graph shows the same tsunami recorded in Crescent City, California where several million dollars damage was inflicted due to strong currents. The bottom graphs show the August 15, 2007 Peru tsunami recorded at DART gage 32401 (on the bottom-left) and in Hilo, Hawaii (on bottom-right). The DART graph shows the predicted tide on top and the high transmission rate data (one minute between samples) on the bottom. The fluctuations at the beginning of the lower trace are induced by the seismic Rayleigh wave. The tsunami starts at 0032UTC. Vertical axes on the graphs are sea level in centimeters and horizontal axes are time and day of year in UTC.

Inexpensive coastal runup gages which have recently been deployed in Hawaii are another new type of tsunami recorder (Walker and Cessaro, 2002). These devices do not transmit continuous data, but provide an indication to the warning centers when a tsunami inundates its location.

A derivative of the runup gage concept was recently developed and tested in Alaska (Burgy and Bolton, 2006). Pressure sensors were located near the low tide line of Augustine Island in lower Cook Inlet. Augustine Island is an active volcano with a history of generating tsunamis during eruptive phases. Data from these sensors were transmitted every 15 seconds via radio to an internet drop where they were made available to warning centers. As the nearest populated locations to the volcano are approximately 55 minutes tsunami travel time, the recorded splash on the volcano shores would give warning centers sufficient time to notify potentially affected residents.

Deep-ocean Assessment and Reporting of Tsunami (DART) stations provide the warning centers deep ocean sea level data (Gonzalez *et al.*, 2005a). These stations are comprised of a pressure sensor located on the ocean floor which transmits signal through an acoustic link to a surface buoy. Data from the buoys are sent via satellite to warning centers in near real-time when activated locally or remotely. The DART system is explained further in Chapter 7 of this volume, as are cabled pressure sensor systems. Ocean-bottom pressure sensors which transmit data via cable as opposed to satellite are utilized by the Japanese tsunami warning system. The planned joint Canada/U.S. Neptune project off the coasts of British Columbia and Washington will also include cabled ocean bottom pressure sensors.

Optimal sea level data coverage for a warning system is difficult to quantify. Considering an extreme case, if a warning center's requirement is to verify tsunamis prior to warning the nearest coasts and still provide warnings prior to impact, stations would have to be placed off-shore and spaced tightly. Since some tsunami sources are less than 10km wide (e.g., 5km estimate of the 1998 Papua New Guinea tsunami source by Tappin *et al.* (2001)), stations could be spaced no more than this far apart. This is clearly an impractical solution when considering the length of the world's coastlines and the cost of offshore gages. A more realistic scenario is to base tsunami warnings on seismic analysis using conservative protocols, then specify a time within which sufficient sea level data would be recorded so that the warning could be cancelled or continued. If the requirement for tsunami verification is 30 minutes following a warning for near-shore, regional sources, the proper spacing of coastal sea level gages can be computed based on a tsunami travel time analysis of the coast in question.

For distant tsunamis, the situation is different. Coastal gages near a distant source may indicate that a local tsunami was generated, but that does not necessarily indicate that a tsunami dangerous to distant shores (a tele-tsunami) was also generated. In these cases, DARTs located between the source and region of interest give the best indication of a potentially dangerous tsunami. As tele-tsunamis are almost always generated by large, subduction zone earthquakes, optimally siting these instruments is easier than siting coastal tide gages. The optimum placement of DARTs necessary to provide tsunami amplitude forecasts to distant shores is generally ocean-ward of a subduction zone, with the spacing increment related to the expected rupture length of basin-wide, tsunami-generating earthquakes. Table 1 shows a list of historic events which generated dangerous basin-wide tsunamis, and their rupture lengths.

TABLE 1.

Summary of major ocean-wide tsunami-generating earthquakes with magnitude and estimated rupture length listed. Magnitudes taken from Kanamori (1977) except as noted. Rupture lengths taken from Acharya (1979) except as noted.

Source Region	Year	Magnitude	Rupture length (km)
Ecuador	1906	8.8	277
Chile	1906	8.2	123
Chile	1922	8.5	189
Kamchatka	1923	8.3	272
Japan	1933	8.4	500
Alaska	1946	8.6	100
		Pelayo and Wiens (1991)	
Kamchatka	1952	9.0	450
Aleutian Is.	1957	8.7	850
		(Johnson et al., 1994)	(Johnson et al., 1994)
Kurile Is.	1958	8.3	160
Chile	1960	9.5	800
Kurile Is.	1963	8.5	272
Alaska	1964	9.2	800
Aleutian Is.	1965	8.7	650
Indian Ocean	2004	9.0	1200
		(USGS earthquake data base, 2006)	(Stein and Okal, 2005)

Operational constraints on sea level network design, such as maintenance requirements and costs, influence the actual network design to a greater extent than the optimal layout. Fig. 13.10 shows the present sea level network recorded by U.S. tsunami warning centers. The DART network was designed by NOAA's Pacific Marine Environmental Laboratory, and is maintained by the National Weather Service's National Data Buoy Center. Coastal tide gage networks are maintained by several agencies such as NOAA's Ocean Service, the University of Hawaii Sea Level Center, the Japanese Meteorological Agency, the Canadian Hydrographic Service, the National Tidal Facility of Australia, the Hydrographic and Oceanographic Service of the Chilean Navy, and the tsunami warning centers.

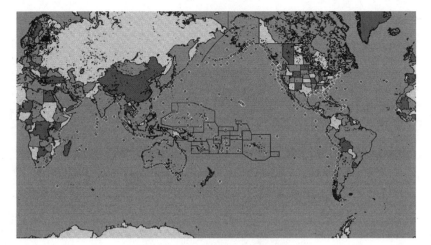

Fig. 13.10 – (See color insert) Sea level gages which transmit data to U.S. tsunami warning centers as of November, 2007 are shown as diamonds.

To be of greatest use to the tsunami warning system, data from sea level gages must be of sufficient sample rate to define tsunamis and transmission delay must be minimal. Optimum transmission consists of data sampled every 15 seconds and transmitted in real-time. This data rate and transmission latency may not always be feasible. Data sampled at one minute intervals are usually sufficient to define all but the highest frequency tsunamis. Generally, tsunami periods are in the 10 to 45 minute range. Data sampled at six minute intervals will under-sample high-frequency tsunamis.

2.3. Other Observational Data

Several other types of observational data show great promise to the tsunami warning system. Geodetic observations made with GPS receivers and transmitted in real-time are one potentially helpful data set. Recently, techniques have been devised which utilize this type of data to compute an earthquake's fault parameters shortly after an earthquake (e.g., Blewitt *et al.,* 2006). Accurate knowledge of fault parameters can then be used to determine sea floor uplift and the tsunami source. These techniques are not yet operational in warning centers, though great strides are being made to bring more raw GPS data into processing centers in real-time.

Satellite-based observations of the sea surface height are another type of new data potentially benefiting the tsunami warning system. These types of observations, if of sufficient accuracy, taken over a wide area, and provided in real-time, could revolutionize the tsunami warning system as tsunamis could be detected prior to impact along the nearest coasts. Other data types, such as hydro-acoustic and infrasound, are also being examined in relation to the tsunami warning system.

3. TWC Data Processing and Interpretation

Like most natural hazard warning centers, tsunami warning centers perform three basic functions:

- Acquire observational data related to the hazard
- Process and interpret the data
- Disseminate hazard information

These functions are described in greater detail below.

3.1. Seismic Data Processing

Many different seismic processing systems have been developed over the last 30 years by various seismic labs, government agencies, and private companies. Most of the systems have been developed for specific seismological purposes ranging from volcano monitoring to nuclear blast verification to tsunami warning response. The prime purpose of tsunami warning seismic analysis is to characterize large earthquakes very quickly. Characterization includes location, depth, magnitude, seismic moment tensor, and fault parameter determination. Some of these are easier to determine than others. In areas of dense seismic sensor coverage, earthquake location, depth, and magnitude can be computed routinely within tens of seconds to minutes. In regions of sparse seismic coverage, determining these parameters may take 8 to 15 minutes with less accuracy. Seismic moment tensor and fault parameter estimation can also be determined rapidly in some areas. Fault length and width determination based on seismic inversions are not yet computed accurately in near real-time, though many improvements are being made in that regard.

In 1998, U.S tsunami warning centers adopted the United States Geological Survey (USGS) Earthworm system (Johnson *et al.*, 1995) as the base seismic processing architecture and data transfer platform. At that time, the National Tsunami Hazards Mitigation Program provided funding and structure to improve the seismic acquisition and processing capabilities at the warning centers (Bernard, 2005; Oppenheimer *et al.*, 2005). Real-time seismic processing systems are very complex systems, especially when analyst input through a graphical interface must be re-ingested into the processing routines. Tsunami warning center systems are further complicated by the necessity to both process distant earthquakes where seismic coverage is sparse and local earthquakes where the network may be very dense. The Earthworm platform provides a base architecture which helps simplify processing. In Earthworm, the system is broken into independent modules that communicate through shared memory locations known as rings. Standard format messages are passed between processing modules using the rings.

The Earlybird seismic processing system (Whitmore and Sokolowski, 2002) was developed at the WCATWC based on previous versions of processing systems and the Earthworm architecture (Sokolowski *et al.*, 1983; Sokolowski *et al.*, 1990; and Zitek *et al.*, 1990). Earlybird provides real-time data processing modules as well as graphical user interfaces which allow analysts to interact with the data.

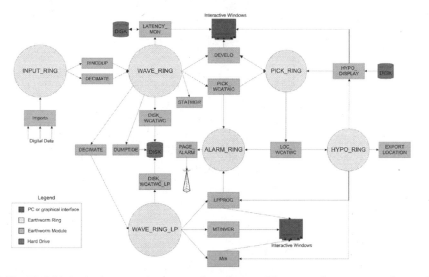

Fig. 13.11 – Earlybird seismic processing system flow diagram. The rectangles are processing modules, circles are shared memory location, and monitor icons indicate graphical user interfaces where analyst input is accepted. Main outputs from the system are alarms, graphics to accommodate analyst interaction, seismic and tide data written to disk, and earthquake source parameters used in tsunami message generation.

Fig. 13.11 displays data processing flow within Earlybird. The monitor icons indicate modules which accommodate user interaction; circles indicate shared memory rings; and rectangles indicate processing modules. Seismic data are first placed in the INPUT_RING. Here, the data are either decimated or copied into the WAVE_RING where they are further processed to determine hypocenter parameters. The modules perform the following basic functions:

- import – ingest seismic data
- export – send hypocenter and seismic data to other centers
- decimate – filter and reduce the sample rate of data for processing
- ringdup – copy messages from one ring to another
- statmgr – monitor modules attached to a ring and restart as necessary
- disk_wcatwc – log seismic data to disk
- develo – display real-time seismic data on screen and accept user input
- dumptide – log sea level data to disk
- latency_mon – track data outages and latencies for all channels
- pick_wcatwc – P-picking/magnitude determination algorithm
- loc_wcatwc – earthquake association and location module
- hypo_display – display computed hypocenter parameters and accept interactive adjustments to P data (Fig. 13.12)
- lpproc – display real-time, long period seismic data and process data for MS
- mm – process surface wave data for mantle wave magnitude (Fig. 13.13)
- mtinver – process data to determine moment tensor (Fig. 13.13)
- page_alarm – send alarms

Hypocenter Display

Force Location Mwp Display Recompute Mwp Hold Refresh

Date	O-time	Lat.	Lon.	Dep	Res	Azm	#Stn	ID	Ms	Mw	Mwp	Mb	Ml
08/02	03:21:47	51.4N	179.9W	39	1.1	148	14	0002-10			6.9-03	6.4-02	6.7-05
01/15	18:18:00	35.1N	138.7E	167	0.7	243	76	0004-50			5.8-10	5.9-69	4.1-01
01/15	18:22:17	38.6N	122.8W	5	0.9	176	6	0007-03			4.2-00?		2.5-06
06/15	03:10:05	6.0S	134.6W	20	0.9	120	5	0014-03				6.3-05	
06/15	02:50:53	41.3N	125.7W	10	1.1	160	37	0002-26			7.4-21	6.5-05	6.2-23
06/15	03:10:05	6.0S	134.6W	20	0.9	120	5	0016-03				6.3-05	
06/15	02:50:57	41.6N	125.8W	4	2.4	278	66	0006-43			7.1-27	6.3-43	6.1-03
06/15	02:50:54	41.3N	125.7W	10	1.3	181	51	0002-37			7.4-33	6.4-17	6.1-23
07/16	14:17:37	36.8N	135.1E	339	0.9	326	92	0003-50			6.3-68	6.2-113	6.0-0
07/16	14:47:51	19.5N	155.5W	13	1.1	260	6	0027-01					2.2-01

Mwp = 6.9   50 miles E of AMCHITKA, AK.

Fig. 13.12 – Screen output from the hypo_display module. The lower screen accepts input from the analyst, and adds this data into both the automatic and interactive solutions. Traces are ordered vertically by epicentral distance and are aligned horizontally by expected P-wave arrival time.

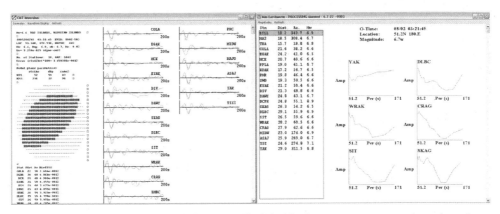

Fig. 13.13 – Earlybird processing module output. The left side shows a moment tensor inversion using a USGS technique (output from module mtinver). Both synthetic and observed traces are shown so that confidence in the inversion by comparing the two can be readily determined. The beach ball on the left indicates to an analyst the type of fault motion. The right side of the figure is the output from module mm. Surface waves are processed here to estimate moment magnitude using the technique discussed by Talandier *et al.* (1987). The graphs on the right section are surface-wave spectra for selected stations. Both graphics are based on processing of the August 2, 2007 Rat Islands earthquake.

In addition to the earthworm-based, real-time processes discussed above, several stand-alone modules are used for: event post-processing, displaying procedural prompts, and composing messages. Output from the seismic data analysis package are initial alarms to alert analysts, automated computations of hypocenter parameters, and, after analyst interaction, reviewed hypocenter parameters.

These reviewed parameters are used to initially judge an earthquake's tsunami threat.

One of the biggest challenges warning center analysts face is quickly determining an accurate magnitude during a large earthquake. Several types of magnitudes are computed at the warning centers: local magnitude M_l, body-wave magnitude m_b, surface-wave magnitude M_s, P-wave moment magnitude M_{wp} (Tsuboi *et al.*, 1995; Tsuboi *et al.*, 1999; Whitmore *et al.*, 2002), moment magnitude M_w based on seismic inversions, and M_w based on mantle-wave magnitude M_m (Talandier *et al.*, 1987; Okal and Talandier, 1987). Each of these is accurate for different sizes and types of earthquakes. Spence *et al.* (1989) provide a good review of basic earthquake magnitude formulation.

U.S. tsunami warning centers response procedures are based on moment magnitude for large, potentially tsunamigenic earthquakes. Moment magnitude is determined by several different means, though the fastest and type most often used during events is M_{wp}. Depending on the size of the earthquake, M_{wp} can be computed anywhere from 20 to 120 seconds after the P-wave arrival.

Easy access to a geographic information system (GIS) interface is critical to an analyst's perception of an event. The GIS provides access to, for example, historical earthquake and tsunami data bases, tsunami travel time and forecast model displays, cultural overlays in the epicentral region, and bathymetric data. An example of a GIS which provides these capabilities is the WCATWC EarthVu GIS. EarthVu displays are automatically driven by output from Earlybird. Some screen examples are shown in Fig. 13.14.

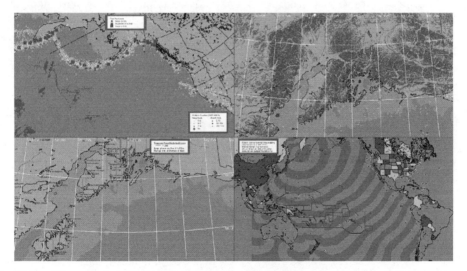

Fig. 13.14 – (See color insert) Example outputs from the EarthVu GIS. The upper-left map shows an overall location map with historic earthquake and tsunami overlain. The upper-right section shows an expanded view of the epicentral region with contours and other physical overlays. The lower-left section displays historic tsunami and the runups from the 1964 Alaska tsunami. The lower-right section is a tsunami travel time map for an event in the Gulf of Alaska. Contour intervals are one hour.

A tsunami warning center analyst must assimilate a large amount of information in a very short time. The Earlybird/EarthVu system is run on a multi-monitor platform to provide as much information as possible immediately to an analyst (Fig. 13.15). As warning centers must operate with 100% reliability, primary and secondary systems with full capabilities operate at all times.

Fig. 13.15 – WCATWC operations console.

While seismic processing capabilities have improved greatly over the years, many improvements remain to be devised and implemented. Processing improvements are necessary so that accuracy is not reduced in conjunction with decreased warning response times. Automated systems such as the Earlybird provide tools necessary for quick response. However, automated seismic processing systems are not mistake-proof. It is imperative that qualified analysts review and revise automated results prior to release as potentially millions of residents can be affected by a warning.

3.2.   Response Procedures

Based on an analyst's interpretation of the processed seismic data, the course of action is determined. As quick response is required, the initial action must be based on well-planned procedures. Procedures must be set for earthquakes of all sizes and locations. Following the initial response, analyst judgment of the situation becomes a greater part of the procedures. There are literally an infinite number of different scenarios which can play out during an event, and it is impossible to set procedures for all.

Response procedures are normally based on historic tsunami data. For example, the PTWC's Pacific-wide response protocols are based on historic data analysis which indicates that for an earthquake to trigger a dangerous tsunami outside the source region, the earthquake's moment magnitude must be magnitude 8 or over. Earthquakes below this size can also trigger tsunamis, though they have only been observed to be dangerous in the source region. Earthquakes with magnitude greater than 7 and their resulting tsunamis along the U.S west, British Columbia, and Alaska coasts are summarized on Table 2. Based on historic data like this, response criteria are set.

TABLE 2.

Summary of tsunamis impacting the U.S. west coast, British Columbia, and Alaska with an amplitude of 0.5m or greater. The second column lists the total number of earthquakes in this region in the listed magnitude range. Only further information is given for earthquakes which triggered tsunamis recorded with an amplitude of 0.5m or greater. Amplitude is defined as the height above normal sea level. Historic data obtained from the USGS earthquake data base (2006), NOAA tsunami data base (2006), and tide gage records.

Moment Magnitude	# Earthquakes	Year of tsunami	Maximum Distance where 0.5m amplitude tsunami recorded (km)	Source Zone
7.0–7.5	38	1873	80	S. Oregon
	13% triggered tsunamis	1927	20	S. California
		1946	220	British Columbia
		1989	20	C. California
		1992	150	N. California
7.6–7.8	8	1812	20	S. California
	25 % triggered tsunamis	1969	850	Kamchatka
> 7.8	13	1899 (1)	130	Gulf of Alaska
	77% triggered tsunamis	1899 (2)	360	Gulf of Alaska
		1946	Tele-tsunami	E. Aleutian Is.
		1957	Tele-tsunami	C. Aleutian Is.
		1958	230	Gulf of Alaska
		1964	Tele-tsunami	Gulf of Alaska
		1965	550	W. Aleutian Is.
		1986	120	C. Aleutian Is
		1987	200	Gulf of Alaska
		1996	80	C. Aleutians

Modeling hypothetical tsunamis can also help define procedures. For example, Knight (2006) showed by modeling potential tsunamis in the Atlantic Basin, Caribbean Sea, and Gulf of Mexico that tsunamis in the Atlantic will not pose a threat to the Gulf of Mexico and vice-versa. This type of study is particularly helpful in areas with little historical tsunami information.

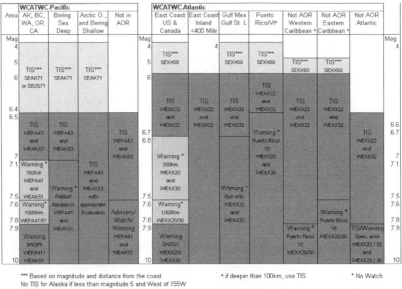

Fig. 13.16 – WCATWC procedural bar chart. Codes shown with message types are World Meteorological Organization (WMO) headers.

Based on models as described above, historic information, potential tsunami source studies, and shoreline and bathymetric configurations, a set of response procedures is defined. Procedures used by WCATWC as of December, 2007 are shown in Fig. 13.16.

3.3. Sea Level Data Processing

Once the initial analysis is complete and appropriate messages disseminated, sea level data must be examined. Data from the network shown in Fig. 13.10 are transmitted to the tsunami warning centers where they are converted to a standard format, written to disk, and displayed on a graphical user interface where they are available to an analyst for review. One of the bigger challenges faced at the warning centers is to stay current with the ever-changing formats with which sea level data are transmitted. Approximately 20 different formats must be accommodated by the centers' ingest software.

To help discern tsunami signatures in the signal, data are de-tided as part of the processing flow. Also, depending on the level of wind wave activity at a site, low pass filtering may be necessary to extract as much tsunami signal as possible.

Figs. 13.17 and 13.09 display examples of the TideView graphical user interface. Many different types of display are available with TideView. A strip-chart view is available to observe the entire network over a region. This view is convenient for monitoring network outages. Stations can be selected from here for an expanded view and analysis. Raw, de-tided, and filtered signal can be shown on the expanded view, as well as expected tsunami arrival time indicators. An analyst can interact with the data to measure amplitude and period.

Results from sea level analysis are used to determine supplemental tsunami message content. Sea level data are used in conjunction with historical data bases and forecast models to estimate tsunami severity. Post-processing of seismic data is also used to help refine the threat. Fault parameters determined by waveform inversion can indicate whether vertical or horizontal fault motion occurred during rupture. Tsunami generation probability is reduced in earthquakes with horizontal displacement versus those with vertical offset. Though, an analysis of recent, significant tsunamis by Knight (2006) showed that the lack of vertical offset in an earthquake does not preclude tsunami generation. Fourteen percent of all tsunamis over 50cm amplitude between 1977 and 2005 were generated by earthquakes with predominantly horizontal displacement. All of these tsunamis were dangerous in the source region only and not to distant shores. As local tsunami danger can not be accurately characterized by seismic data alone, sea level data are necessary to quantify threat.

3.4. Historical Data Bases

Historical tsunami data bases provide the mechanism by which a tsunami can be compared to those that have occurred in the past. This comparison supplies critical information to an analyst concerning the present event's threat. When combined with data from historical earthquake data bases, an estimate of the likelihood of a given size earthquake in a certain region to produce a tsunami can be obtained (when sufficient historical information exists).

Coastal tide gages provide the necessary data that an analyst can compare with historic information. Almost all historic tsunami records are based on coastal tide gage readings, measured runup, or visually estimated heights at the coast. The best real-time observations to compare these with are tide gage recordings. Deep ocean readings provide information critical for forecast models, but as they are a newly developed observation method, can not be used to compare directly with historic tsunamis.

Several authors have assembled historic tsunami data bases (e.g., Soloviev and Ferchev, 1961; Iida et al., 1967, see Chapter 2 this Volume). These data bases and others, along with some newly uncovered tsunami records, were compiled by the NGDC into several volumes (Lander and Lockridge, 1989; Lander et al., 1993; Lander, 1996). This NGDC data base is now kept online and is being routinely updated. An on-line data base with similar content is also maintained by the Novosibirsk Tsunami Laboratory of the Institute of Mathematics and Mathematical Geophysics of the Siberian Division of the Russian Academy of Sciences (Gusiakov, et al., 1997).

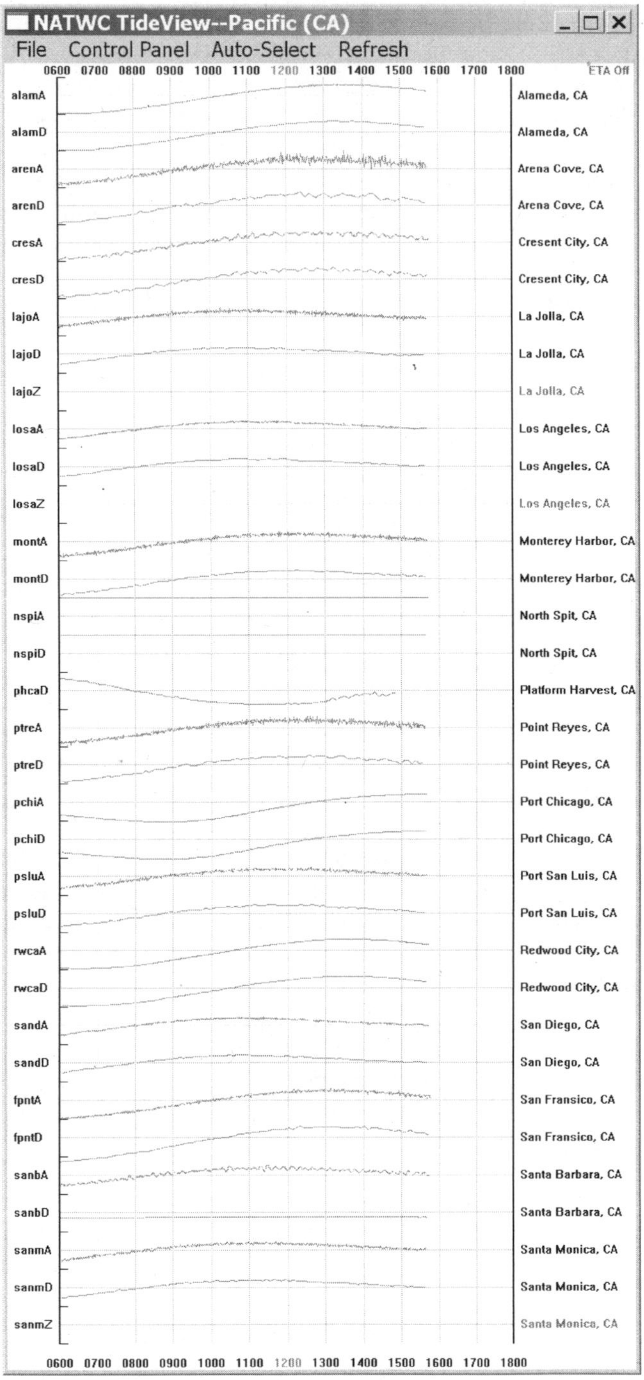

Fig. 13.17 – Sea level display showing all incoming National Ocean Service tide gage data from the California coast.

In addition to their importance in event response, historic tsunami and earthquake data bases have other uses. First, regional tsunami hazard analysis is often based on the historical record. Tsunami hazard zones in communities which have experienced their expected worst case tsunamis can be defined by examining historic inundation levels. Second, by examining historic tide gage recordings and comparing with observed damage, warning centers can determine coastal tsunami amplitudes that are likely to produce damage. Table 3 shows a comparison of recorded tide gage amplitudes and corresponding damage along the U.S. west and Alaskan coasts. Based on this table, a 50cm amplitude tsunami must be considered capable of producing damage in some coastal areas. Amplitude is defined as the height that a tsunami reaches above normal sea level.

TABLE 3.

Examples of tsunami amplitude and resulting damage (Lander *et al.,* 1993; Lander, 1996). Amplitudes measured from original tide gage records where possible.

Amp. (m)	Location – Year - Damage
0.35	Shemya, AK-1996; no damage
0.4	Santa Barbara, CA-2006; Yakutat, AK-1987; no damage
0.45	Shemya, AK-2006; no damage
0.5	Port Hueneme, CA-1957; Crescent City, CA-1994; no damage
0.5	San Francisco, CA-1960; strong current stops ferry
0.5	Crescent City, CA-1963; 1 mooring broke loose
0.5+	San Diego, CA-1957; boat/dock damage
0.51	Adak, AK-1996; no damage
0.55	Port Orford, OR-2006; no damage
0.6	Ketchikan, AK-1964; Crescent City, CA-1968; no damage
0.6	Arena Cove, CA-2006; Port San Luis, CA-2006; no damage
0.6	Los Angeles, CA-1964; $200K damage to boats
0.6	Monterrey, CA-1957; 2 almost drowned
0.6	San Diego, CA-1964; strong current, boat damage
0.7	Crescent City, CA-1957; no damage
0.7	San Diego, CA-1960; boat/pier damage (20 knot current)
0.8	Avila, CA-1927; Santa Barbara, CA-1946; no damage
0.8	Unga, AK-1946; dock washed away
0.8	Port Hueneme, CA-1946; railroad tracks flooded
0.8	San Pedro, CA-1868; wharf flooded
0.8	Santa Barbara, CA-1964; boat damage
0.8+	Los Angeles, CA-1960; $1 million damage, 1 drowning
0.9	Adak, AK-1986; Shemya, AK-1969; Crescent City, CA-1946; no damage
0.9	Anaheim, CA-1877; Santa Cruz, CA-1960; boats loose, no damage
0.9	Trinidad, CA-1992; cars stuck on beach, no damage
0.9	Yakutat, AK-1958; Mooring torn loose
0.9	Crescent City, CA-2006; dock damage >$7M
1.0	San Pedro, CA-1877; flooding, no damage
1.0	Crescent City, CA-1952; 4 boats sunk; Cape Pole, AK-1960; log boom broke
1–1.5	San Francisco Bay, CA-1964; $1 million damage
1.1	Attu, AK-1969; no damage
1.2	Annette, AK-1964; no damage
1.2	Seaside, OR-1946; boats swept away; Seldovia, AK-1964; $500K boat damage
1.2	Larsen Bay, AK-1964; warehouse flooded
1.4	Avila, CA-1952; no damage
1.4	Noyo River mouth, CA-1946; several near drownings
1.4	Santa Barbara, CA-1960; much damage

Amp. (m)	Location – Year - Damage
1.4	Ilwaco, WA-1964; Gearhart, OR-1964; streets/houses flooded
1.5	Charleston, OR-1946; Stenson Beach, CA-1960; no damage
1.5	Taholah, WA-1946; Santa Cruz, CA-1896; boats swept away, minor damage
1.5	Seaside, OR-1960; boat/pier damage; King Cove, AK-1946; cannery damage
1.5	Santa Cruz, CA-1946; 1 dead, many rescued
1.6	Attu, AK-1965; minor damage
1.7	Crescent City, CA-1960; boats sunk, pier damage, 3 injured
1.8	Surf, CA-1927; railroad station inundated
1.9	Humboldt Bay, CA-1964; Adak, AK-1957; flooding; bridge destroyed
2.0	Noyo Harbor, CA-1960; boat/dock damage; Noyo, CA-1964; 10 boats sunk
2.0	Copalis, WA-1964; some injuries, much damage
2.2	Half Moon Bay, CA-1960; 3 near drownings, flooding, boat damage
2.3	Umnak I., AK-1957; moorings destroyed; Montague I., AK-1960; damage
2.5	Pacific Beach, CA-1964; injuries, damage
2.6	Half Moon Bay, CA-1946; flooding; Drake's Bay, CA-1946; boat capsized
3.0	Santa Monica, CA-1930; boat/pier damage
3.0	Redondo Beach, CA-1930; 1 dead, many rescued
3.0	Seaside, OR-1964; 1 dead, structural damage; Cape St. Elias, AK-1964; 1 dead
3.0+	Florence, OR-1964; much damage
3.0+	Klamath River, CA-1964; 1 dead, some damage
3.4	Gaviota, CA-1812; ships run aground; Moclips, WA-1964; houses damaged
3.5	DePoe Bay, OR-1964; 4 deaths, some damage
3.7	Yakataga, AK-1964; no damage reported
4.5	Wreck Creek, WA-1964; minor damage
4.8	Crescent City, CA-1964; 10 dead, $15 million damage

3.5. Tsunami Forecasting

Tsunami forecasts provide information on the timing and amplitude of wave impact at certain locations. Initial arrival time forecasts are straightforward to compute since tsunami velocity is only dependent on depth of water and the acceleration of gravity (velocity = sqrt (gravity acc.*depth)). Applications of Huygen's principle (every point on a wave front acts as a source for a new wave) can be used with gridded bathymetric data to determine the travel time from one point in an ocean to all other points (Crowley, *et al.*, 2007). Travel times are computed in reverse using this technique. That is, potential impact locations along the coast are set as the origin point and the travel times to all other points in the ocean basin are saved. These times are saved in lookup tables to provide quick retrieval during an event.

Tsunami impact, or amplitude, forecasting is more complicated than determining tsunami travel times as hydrodynamic models must be used (see Chapter 12 of this volume). Like many other natural phenomena, tsunamis are often forecasted by assimilating observed data into numerical models. One major challenge in tsunami forecasting is that appropriate numerical models compute slower than a tsunami propagates. To accommodate this problem, forecasts can be based on observations and pre-computed models.

Tsunami arrival time forecasting has been implemented at TWCs for many decades. Maximum amplitude forecasts have been computed at U.S. and Japanese TWCs since the mid-1990s. U.S. TWCs use the forecasts to control warning decisions and can output the forecasts as experimental products on web sites. The Japanese TWC issues maximum amplitude forecasts within their official products. Wave amplitude forecasts over time are computed by the SIFT system (Chapter 12). A major developmental project is underway within NOAA to bring this capability into the U.S. TWCs.

The first tsunami forecasting routine used at U.S. tsunami warning centers was implemented in the mid-1990s (Kowalik and Whitmore, 1991; Whitmore and Sokolowski, 1996). In this method, an initial tsunami profile is computed from assumed earthquake fault parameters and the static dislocation formulae of Okada (1985). Tsunamis are propagated using shallow-water wave equations with non-linear terms and friction included in areas of fine grid resolution. The equations are computed using an explicit-in-time finite difference scheme with grid increments of approximately 9km over the deep ocean, approximately 2km over the shelf and approximately 350m where necessary to describe near-shore coastline configuration. Grids interact dynamically throughout the computations, and the ocean/land boundary is fixed.

Tsunami models were computed for several hundred hypothetical earthquakes along the coasts of northern Honshu, Kuril Is., Kamchatka, Aleutian Is., Alaska, British Columbia, the U.S. Pacific Northwest, and Chile. Hypothetical earthquake source parameters were determined by examining regional tectonic setting and past earthquakes. Moment magnitudes of the hypothetical earthquakes range from 7.5 to 9.5.

Maximum modeled amplitudes were saved at approximately 100 locations along the Pacific coasts of Alaska, British Columbia, Washington, Oregon, California, Hawaii, and at the DARTs for each model. During an event, the model closest to the epicenter with the nearest moment magnitude is chosen. The pre-computed amplitudes at all modeled sites are scaled as the tsunami is recorded on tide gages or DARTs by simple proportions. Scaling can only be performed with data from tide gages which were included within the model fine grids or from DART data. Forecasted amplitudes are the scaled model results.

The technique was tested by Whitmore (2003) on nine well-recorded tele-tsunamis within the model area. The test showed that the technique would have provided proper guidance regarding warning cancellation or expansion in each of the nine events.

The EarthVu GIS provides a platform for analysts to use the technique. Pre-computed models are chosen by interaction with the EarthVu screen. Forecasts are printed and displayed on the screen as shown in Fig. 13.18. Observed tsunami amplitudes are input for model scaling through the EarthVu interface.

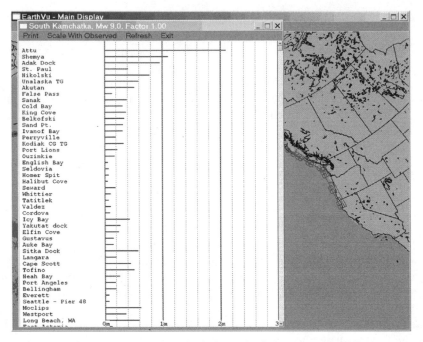

Fig. 13.18 – Forecasted amplitudes for a hypothetical magnitude 9 earthquake off the coast of southern Kamchatka. These forecasted amplitudes must be scaled with observed sea level amplitudes at DARTs or coastal tide gages which were modeled with detailed grids.

More advanced tsunami forecasting techniques are presently being developed by several modeling groups (e.g., Titov *et al.,* 2005, Kowalik *et al.,* 2005). A technique developed at the Pacific Marine Environmental Laboratory (PMEL), known as SIFT, is based on the method-of-splitting-tsunamis (Titov and Gonzalez, 1997). Tsunamis are pre-computed using sources covering the Pacific's major subduction zones. Deep ocean sea level time series are computed and saved in a data base and are correlated with observed DART data during an event. This inversion process delineates the source, and the data base is scaled accordingly. The deep-ocean heights are then used to drive inundation models at selected coastal regions (see Chapter 12 of this volume for further information). Work is ongoing to bring this capability to tsunami warning centers. Fig. 13.19 shows the SIFT graphical user interface.

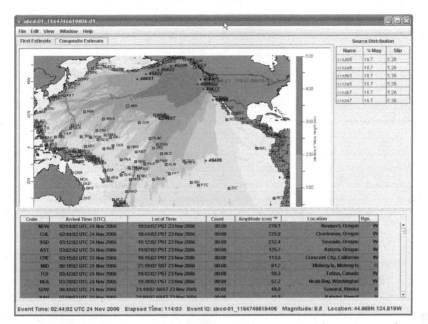

Fig. 13.19 – (See color insert) Example of the PMEL SIFT tsunami forecasting user interface. This display shows the expected coastal amplitudes and propagation pattern for a hypothetical 8.5 earthquake off the coast of Washington.

Several other methods for forecasting tsunamis have also been proposed. For example, the Japanese Meteorological Agency (JMA) tsunami warning system forecasts wave amplitudes based on earthquake location and magnitude (Tatehata, 1998). The center has computed a large data base of pre-computed tsunamis for different size earthquakes throughout the Japanese coastal regions. Based on location and magnitude, the most appropriate pre-computation is retrieved and tsunami amplitudes are forecasted directly without calibration with observed tsunamis. The JMA issues these forecasts within their messages. Other methods have been devised to forecast amplitudes at distant locations based strictly on moment magnitude (e.g., Blackford, 1984; Reymond *et al.*, 1993).

Regardless of how tsunamis are forecasted, the results are used to guide supplemental warning decisions. Forecasts indicate to the analyst that a cancellation, restriction, or expansion is warranted. Based on the historical damage summary given in Table 3, tsunamis in the range 50-150cm amplitude generally produce damage by strong currents without significant inundations. Tsunamis larger than this are more likely to produce inundations. Assuming a forecast accuracy of 50%, the WCATWC issues tsunami advisories (or low-level warnings) when the forecasted height is in the range 30 to 100cm. Warnings are issued if the forecasted amplitude is above 100cm. If the forecast is less than 30cm amplitude, damaging tsunamis are not expected.

3.6. Other TWC Functions

In addition to the basic functions of tsunami warning centers, several associated activities must be conducted to keep the warning center functioning properly. Since tsunamis are not common events, the warning centers' systems and procedures must be exercised regularly. Practice scenarios must be routinely set up and executed by analysts. In addition to preparing analysts for these rare events, each scenario will effectively test the sensibility of procedures and verify that message content is appropriate.

Outreach activities are also critical to warning system success. These efforts enable primary warning system contacts and the general public to remain in touch with warning center procedures, and vice-versa. Within the NWS, Weather Forecast Office Warning Coordination Meteorologists (WCMs) are the primary individuals responsible for public outreach, emergency management interaction, and promoting the NWS TsunamiReady program. Warning center personnel supplement the WCM's efforts.

As potential tsunami generating events are infrequent, warning center personnel have time to develop new processes, software, and other items which improve the warning system. Analysts are the interface between researchers who develop new techniques and the operational system by taking these new techniques and porting them into operational practice. There are literally an endless number of projects which analysts pursue to improve the system. These include implementing new seismological and sea level processes and techniques, tsunami forecasting improvements, new product generation, historical data base studies, and developing improved outreach materials.

Another critical aspect of tsunami warning center operation is testing processes and communication routes. Multiple times daily, analysts review system functionality. In addition to these internal checks, monthly communication tests must be conducted to verify communication routes between the warning centers and their primary recipients.

Each warning center must have a backup plan. Outages could be due to fire, communications/electrical outages, or one of many other causes. In the United States, the PTWC and WCATWC provide mutual backup capabilities. If either center is out of commission, the other center can issue the appropriate products.

4.   Warning Center Messages

Once an analyst has decided on the proper course of action, the appropriate message is composed and transmitted. Centers must work with their primary recipients to ensure that statements are understandable. U. S. tsunami warning centers accomplish this coordination through state tsunami committees, the U. S. National Tsunami Hazard Mitigation Program, and IOC working groups for international coordination.

U.S. tsunami warning center messages are progressing from effectively a three-level suite to a four-level suite. The products issued by the centers are warning, watch, advisory, and information statement. Each has a distinct meaning relating to local emergency response. In summary:

Product		Threat		Action
Warning	→	Inundating tsunami possible	→	Full evacuation suggested
Watch	→	Danger level not yet known	→	Stay alert for more info
Advisory	→	Strong currents likely	→	Stay away from the shore
Information	→	Minor tsunamis at most	→	No action suggested

Based on seismic data analysis or forecasted amplitude (dependent on whether the center has obtained sea level data), the appropriate product is issued. Warnings and Advisories suggest that action be taken. Watches are issued to provide an early alert for areas that are distant from the wave front, but may have danger. Once the danger level is determined, the watch is upgraded to a warning or advisory, or canceled. The full definition of each message is given below:

Tsunami Warning: A tsunami warning is issued when a potential tsunami with significant widespread inundation is imminent or expected. Warnings alert the public that widespread, dangerous coastal flooding accompanied by powerful currents is possible and may continue for several hours after arrival of the initial wave. Warnings also alert emergency management officials to take action for the entire tsunami hazard zone. Appropriate actions to be taken by local officials may include the evacuation of low-lying coastal areas and repositioning ships to deep waters when there is time to safely do so. Warnings may be updated, adjusted geographically, downgraded, or canceled. To provide the earliest possible alert, initial warnings are normally based only on seismic information.

Tsunami Watch: A tsunami watch is issued to alert emergency management officials and the public of an event that may later impact the watch area. The watch area may be upgraded to a warning or advisory—or canceled—based on updated information and analysis. Therefore emergency management officials and the public should prepare to take action. Watches are normally issued based on seismic information without confirmation that a destructive tsunami is underway.

Tsunami Advisory: A tsunami advisory is issued for the threat of a potential tsunami that may produce strong currents or waves dangerous to those in or near the water. Coastal regions historically prone to damage due to strong currents induced by tsunamis are at the greatest risk. The threat may continue for several hours after the arrival of the initial wave, but significant widespread inundation is not expected for areas under an advisory. Appropriate actions to be taken by local officials may include closing beaches, evacuating harbors and marinas, and repositioning ships to deep waters when there is time to safely do so. Advisories are normally updated to continue the advisory, expand/contract affected areas, upgrade to a warning, or cancel the advisory.

Information Statement: A tsunami information statement is issued to inform emergency management officials and the public that an earthquake has occurred, or that a tsunami warning, watch or advisory has been issued for another section of the ocean. In most cases information statements are issued to indicate there is no threat of a destructive tsunami and to prevent unnecessary evacuations because the earthquake may have been felt in coastal areas. An information statement may, in appropriate situations, caution about the possibility of destructive local tsunamis. Information statements may be reissued with additional information, though normally these messages are not updated. However, a watch, advisory or warning may be issued for the area, if necessary, after analysis and/or updated information becomes available.

U.S. tsunami warning center messages are official National Weather Service (NWS) products and as such must be formatted according to NWS rules. In many respects these rules are based on World Meteorological Organization (WMO) standards. Products issued at the WCATWC and PTWC are listed in Table 4. Public format messages were recently introduced at the centers to supplement the standard products. Standard products include NWS Universal Generic Codes (UGC) and segmented format necessary for automated broadcasting systems. These components can render the standard products difficult to read. The newer public products remove segmentation, tables, and UGCs which results in an easier-to-read product. Examples of public and standard format warnings are given in Table 5. Warning centers also issue messages using internet-based technology which allow the use of better graphics and easier-to-read messages.

TABLE 4.
Product types issued by U.S. Tsunami Warning Centers. NWS AWIPS IDs are National Weather Service product Identifiers. WMO Headers are World Meteorological Organization codes.

WMO Header	NWS AWIPS ID	Explanation
WEPA41 PAAQ	TSUWCA	Tsunami Warnings, Watches, and Advisories *AK, BC, and US West Coast*
WEPA43 PAAQ	TIBWCA	Tsunami Information Statements *AK, BC, and US West Coast*
WEAK51 PAAQ	TSUAK1	"Public" Tsunami Warnings, Watches, and Advisories *AK, BC, and US West Coast*
WEAK53 PAAQ	TIBAK1	"Public" Tsunami Information Statements *AK, BC, and US West Coast*
SEAK71 PAAQ	EQIAKX	Tsunami Information Statements *Alaska*
SEUS71 PAAQ	EQIWOC	Tsunami Information Statements *BC and US West Coast*
WEXX20 PAAQ	TSUAT1	Tsunami Warnings, Watches, and Advisories *US East coast, Gulf, Puerto Rico/Virgin Islands, and Canadian Maritime Provinces*
WEXX22 PAAQ	TIBAT1	Tsunami Information Statements *US East coast, Gulf, Puerto Rico/Virgin Islands, and Canadian Maritime Provinces*
WEXX30 PAAQ	TSUATE	"Public" Tsunami Warnings, Watches, and Advisories *US East coast, Gulf, Puerto Rico/Virgin Islands, and Canadian Maritime Provinces*
WEXX32 PAAQ	TIBATE	"Public" Tsunami Information Statements *US East coast, Gulf, Puerto Rico/Virgin Islands, and Canadian Maritime Provinces*

WMO Header	NWS AWIPS ID	Explanation
SEXX60 PAAQ	EQIAT1	Tsunami Information Statements *US East coast, Gulf, Puerto Rico/Virgin Islands,* *and Canadian Maritime Provinces*
WEPA40 PHEB	TSUPAC	Tsunami Warnings, Watches, and Advisories *PTWC Pacific Area-of-Responsibility*
WEPA42 PHEB	TIBPAC	Tsunami Information Statements *PTWC Pacific Area-of-Responsibility*
WEHW40 PHEB	TSUHWX	Tsunami Warnings, Watches, and Advisories *Hawaii*
WEHW42 PHEB	TIBHWX	Tsunami Information Statements *Hawaii*
SEHW70 PHEB	EQIHWX	Tsunami Information Statements *PTWC Pacific Area-of-Responsibility*
WECA41 PHEB	TSUCAX	Tsunami Watches *Non-U.S. Caribbean Region*
WECA43 PHEB	TIBCAX	Tsunami Information Statements *Non-U.S. Caribbean Region*
WEIO21 PHEB	TSUIOX	Tsunami Watches *Indian Ocean Region*
WEIO23 PHEB	TIBIOX	Tsunami Information Statements *Indian Ocean Region*

TABLE 5.
Example public and standard tsunami warning/watch messages.

Public Warning/Watch Example

```
WEAK51 PAAQ 151210
TSUAK1

BULLETIN
PUBLIC TSUNAMI MESSAGE NUMBER 2
NWS WEST COAST/ALASKA TSUNAMI WARNING CENTER PALMER AK
410 AM PST WED NOV 15 2006

. . . A TSUNAMI WARNING IS IN EFFECT WHICH INCLUDES THE ALASKA
      COASTAL AREAS FROM DUTCH HARBOR ALASKA TO ATTU ALASKA . . .

. . . A TSUNAMI WATCH IS IN EFFECT FOR THE BRITISH COLUMBIA
      AND ALASKA COASTAL AREAS FROM THE NORTH TIP OF VANCOUVER I.
      BRITISH COLUMBIA TO DUTCH HARBOR ALASKA . . .

A TSUNAMI WARNING MEANS . . . ALL COASTAL RESIDENTS IN THE WARNING
AREA WHO ARE NEAR THE BEACH OR IN LOW-LYING REGIONS SHOULD MOVE
IMMEDIATELY INLAND TO HIGHER GROUND AND AWAY FROM ALL HARBORS AND
INLETS INCLUDING THOSE SHELTERED DIRECTLY FROM THE SEA. THOSE
FEELING THE EARTH SHAKE . . . SEEING UNUSUAL WAVE ACTION . . . OR THE
WATER LEVEL RISING OR RECEDING MAY HAVE ONLY A FEW MINUTES BEFORE
THE TSUNAMI ARRIVAL AND SHOULD EVACUATE IMMEDIATELY. HOMES AND
SMALL BUILDINGS ARE NOT DESIGNED TO WITHSTAND TSUNAMI IMPACTS.
DO NOT STAY IN THESE STRUCTURES.

ALL RESIDENTS WITHIN THE WARNED AREA SHOULD BE ALERT FOR
INSTRUCTIONS BROADCAST FROM THEIR LOCAL CIVIL AUTHORITIES. THIS
TSUNAMI WARNING IS BASED SOLELY ON EARTHQUAKE INFORMATION - THE
TSUNAMI HAS NOT YET BEEN CONFIRMED.
```

A TSUNAMI WATCH MEANS . . . ALL COASTAL RESIDENTS IN THE WATCH AREA
SHOULD PREPARE FOR POSSIBLE EVACUATION. A TSUNAMI WATCH IS ISSUED
TO AREAS WHICH WILL NOT BE IMPACTED BY THE TSUNAMI FOR AT LEAST
THREE HOURS. WATCH AREAS WILL EITHER BE UPGRADED TO WARNING STATUS
OR CANCELED.

AT 314 AM PACIFIC STANDARD TIME ON NOVEMBER 15 AN EARTHQUAKE WITH
PRELIMINARY MAGNITUDE 8.1 OCCURRED
NEAR THE KURIL ISLANDS RUSSIA.
THIS EARTHQUAKE MAY HAVE GENERATED A TSUNAMI. IF A TSUNAMI
HAS BEEN GENERATED THE WAVES WILL FIRST REACH
SHEMYA ALASKA AT 424 AM AKST ON NOVEMBER 15.
ESTIMATED TSUNAMI ARRIVAL TIMES AND MAPS ALONG WITH SAFETY RULES
AND OTHER INFORMATION CAN BE FOUND ON THE WEB SITE
WCATWC.ARH.NOAA.GOV.

TSUNAMIS CAN BE DANGEROUS WAVES THAT ARE NOT SURVIVABLE. WAVE
HEIGHTS ARE AMPLIFIED BY IRREGULAR SHORELINE AND ARE DIFFICULT TO
PREDICT. TSUNAMIS OFTEN APPEAR AS A STRONG SURGE AND MAY BE
PRECEDED BY A RECEDING WATER LEVEL. MARINERS IN WATER DEEPER
THAN 600 FEET SHOULD NOT BE AFFECTED BY A TSUNAMI. WAVE HEIGHTS
WILL INCREASE RAPIDLY AS WATER SHALLOWS. TSUNAMIS ARE A SERIES OF
OCEAN WAVES WHICH CAN BE DANGEROUS FOR SEVERAL HOURS AFTER THE
INITIAL WAVE ARRIVAL. DO NOT RETURN TO EVACUATED AREAS UNTIL AN
ALL CLEAR IS GIVEN BY LOCAL CIVIL AUTHORITIES.

THE PACIFIC TSUNAMI WARNING CENTER IN EWA BEACH HAWAII WILL ISSUE
MESSAGES FOR HAWAII AND OTHER AREAS OF THE PACIFIC OUTSIDE
CALIFORNIA/ OREGON/ WASHINGTON/ BRITISH COLUMBIA AND ALASKA.

ADDITIONAL MESSAGES WILL BE ISSUED EVERY HALF HOUR OR SOONER IF
CONDITIONS WARRANT. THIS TSUNAMI WARNING AND WATCH WILL REMAIN
IN EFFECT UNTIL FURTHER NOTICE. FOR FURTHER INFORMATION STAY TUNED
TO NOAA WEATHER RADIO . . . YOUR LOCAL TV OR RADIO STATIONS . . . OR SEE
THE WEB SITE WCATWC.ARH.NOAA.GOV.
$$

Standard Warning/Watch Example
WEPA41 PAAQ 151210
TSUWCA

BULLETIN
TSUNAMI MESSAGE NUMBER 2
NWS WEST COAST/ALASKA TSUNAMI WARNING CENTER PALMER AK
410 AM PST WED NOV 15 2006

. . .A TSUNAMI WARNING IS IN EFFECT WHICH INCLUDES THE ALASKA
     COASTAL AREAS FROM DUTCH HARBOR ALASKA TO ATTU ALASKA . . .

. . .A TSUNAMI WATCH IS IN EFFECT FOR THE BRITISH COLUMBIA
     AND ALASKA COASTAL AREAS FROM THE NORTH TIP OF VANCOUVER I.
     BRITISH COLUMBIA TO DUTCH HARBOR ALASKA . . .

. . .AT THIS TIME THIS MESSAGE IS INFORMATION ONLY FOR OTHER
     AREAS OF CALIFORNIA - OREGON - WASHINGTON - AND BRITISH
     COLUMBIA . . .
EVALUATION
  IT IS NOT KNOWN - REPEAT NOT KNOWN - IF A TSUNAMI EXISTS BUT A
  TSUNAMI MAY HAVE BEEN GENERATED. THEREFORE PERSONS IN LOW-
  LYING COASTAL AREAS SHOULD BE ALERT TO INSTRUCTIONS FROM THEIR
  LOCAL EMERGENCY OFFICIALS. PERSONS ON THE BEACH SHOULD MOVE TO
  HIGHER GROUND IF IN A WARNED AREA. TSUNAMIS ARE A SERIES OF
  WAVES WHICH COULD BE DANGEROUS FOR SEVERAL HOURS AFTER THE
  INITIAL WAVE ARRIVAL.

```
PRELIMINARY EARTHQUAKE PARAMETERS
 MAGNITUDE - 8.1
 TIME      - 0214 AKST NOV 15 2006
             0314 PST NOV 15 2006
             1114 UTC NOV 15 2006
 LOCATION  - 46.7 NORTH 153.5 EAST
           - KURIL ISLANDS
 DEPTH     - 21 MILES
```

PKZ176-175-172-170-171-AKZ191-185-151310-
COASTAL AREAS BETWEEN AND INCLUDING DUTCH HARBOR ALASKA TO
ATTU ALASKA

. . . A TSUNAMI WARNING IS IN EFFECT WHICH INCLUDES THE ALASKA
        COASTAL AREAS FROM DUTCH HARBOR ALASKA TO ATTU ALASKA . . .

ESTIMATED TIMES OF INITIAL WAVE ARRIVAL
SHEMYA-AK   0424 AKST NOV 15   DUTCH HARBOR-AK   0633 AKST NOV 15
ADAK-AK   0523 AKST NOV 15
$$
PKZ170-171-155-150-132-136-138-137-130-141-140-120-121-129-
127-125-126-128-052-051-053-022-012-043-013-011-021-032-
031-042-034-033-035-041-036-AKZ185-181-171-145-111-101-121-
125-131-135-017-020-018-019-021-022-023-024-025-026-028-
029-027-151310-
COASTAL AREAS FROM THE NORTH TIP OF VANCOUVER I. BRITISH
COLUMBIA TO DUTCH HARBOR ALASKA

. . . A TSUNAMI WATCH IS IN EFFECT FOR THE BRITISH COLUMBIA
        AND ALASKA COASTAL AREAS FROM THE NORTH TIP OF VANCOUVER I.
        BRITISH COLUMBIA TO DUTCH HARBOR ALASKA . . .

ESTIMATED TIMES OF INITIAL WAVE ARRIVAL
SAND PT.-AK   0715 AKST NOV 15   LANGARA-BC   0950 PST NOV 15
COLD BAY-AK   0741 AKST NOV 15   VALDEZ-AK   0852 AKST NOV 15
KODIAK-AK   0802 AKST NOV 15   CORDOVA-AK   0901 AKST NOV 15
SEWARD-AK   0829 AKST NOV 15   HOMER-AK   0925 AKST NOV 15
YAKUTAT-AK   0842 AKST NOV 15   KETCHIKAN AK   1002 AKST NOV 15
SITKA-AK   0843 AKST NOV 15   JUNEAU-AK   1004 AKST NOV 15
$$
PZZ130-131-133-134-132-135-150-153-156-110-250-210-255-350-
353-356-450-455-550-530-535-555-670-673-650-655-750-WAZ001-
503-506-507-007-508-509-510-511-514-515-516-517-021-ORZ001-
002-021-022-CAZ001-002-505-508-006-509-514-515-034-035-039-
044-040-045-046-041-087-042-043-151310-
COASTAL AREAS FROM THE CALIFORNIA-MEXICO BORDER TO THE
NORTH TIP OF VANCOUVER I. BRITISH COLUMBIA

. . . TSUNAMI INFORMATION STATEMENT . . .
NO - REPEAT NO - WATCH OR WARNING IS IN EFFECT FOR THE
    COASTAL AREAS FROM THE CALIFORNIA-MEXICO BORDER TO THE
    NORTH TIP OF VANCOUVER I. BRITISH COLUMBIA

FOR INFORMATION ONLY - ESTIMATED TIMES OF INITIAL WAVE ARRIVAL
TOFINO-BC   1117 PST NOV 15   ASTORIA-OR   1203 PST NOV 15
NEAH BAY-WA   1127 PST NOV 15   SAN FRANCISCO-CA 1223 PST NOV 15
CHARLESTON-OR   1131 PST NOV 15   SAN PEDRO-CA   1245 PST NOV 15
CRESCENT CITY-CA   1138 PST NOV 15   LA JOLLA-CA   1254 PST NOV 15
SEASIDE-OR   1139 PST NOV 15
$$
TSUNAMI WARNINGS ARE ISSUED DUE TO THE IMMINENT THREAT OF A TSUNAMI.
WARNINGS CAN BE BASED SOLELY ON SEISMIC INFORMATION . . . OR BASED
ON CONFIRMATION THAT A POTENTIALLY DESTRUCTIVE WAVE HAS OCCURRED.
COASTAL RESIDENTS IN THE WARNING AREA WHO ARE NEAR THE BEACH OR
IN LOW-LYING REGIONS SHOULD MOVE IMMEDIATELY INLAND TO HIGHER
GROUND.
```
```

```
TSUNAMI WATCHES ARE ISSUED AS AN ADVANCE ALERT TO AREAS THAT
COULD BE IMPACTED BY A TSUNAMI. THE WATCH WILL BE EITHER
CANCELLED OR UPGRADED TO A WARNING. POPULATIONS IN A WATCH AREA
SHOULD CLOSELY FOLLOW THE PROGRESS OF THIS EVENT.

THE PACIFIC TSUNAMI WARNING CENTER IN EWA BEACH HAWAII WILL
ISSUE MESSAGES FOR HAWAII AND OTHER AREAS OF THE PACIFIC
OUTSIDE THE STATES AND PROVINCES LISTED ABOVE.
MESSAGES WILL BE ISSUED EVERY HALF HOUR OR SOONER IF CONDITIONS
WARRANT. THIS TSUNAMI WARNING AND WATCH WILL REMAIN IN EFFECT
UNTIL FURTHER NOTICE. REFER TO THE INTERNET SITE
WCATWC.ARH.NOAA.GOV FOR MORE INFORMATION AND EXPECTED ARRIVAL
TIMES.
$$
```

Tsunami messages are broken down into several basic sections:

- Header
- Headline (who is in warning/watch/advisory)
- Evaluation including expected impact level
- Observed tsunami amplitudes
- Earthquake information
- Expected tsunami arrival times
- Background information (e.g., watch/warning definitions)
- Follow-up message information

Warning/watch/advisory cancellation bulletins are created and disseminated using the same product types as the initial warnings, watches and advisories. Cancellations can take the form of an all clear if no tsunami was generated. Or, warnings can be cancelled with an all clear left up to local emergency management officials if a minor to moderate tsunami was generated. In a major tsunami, ongoing danger can vary significantly from site to site along a coast and the warning center may not have observations from each location. This danger can last for over 24 hours in some harbors. Since the danger can be localized, emergency management may have better information about the ongoing danger level for their region than a tsunami warning center; especially if no coastal sea level gages exist in the area or the gages were destroyed by the tsunami.

Message composition must be automated to facilitate warning centers' response requirements. An analyst needs the ability to modify messages, but the majority of content must be pre-planned and programmed for quick composition. The message content is based on several inputs: type of message to disseminate, bulletin number, source location, earthquake magnitude, estimated tsunami arrival times, pre-defined procedures, observed tsunami amplitudes, and evaluation options selected by the analyst.

5.   Warning Dissemination

Tsunami warning dissemination is accomplished by transmitting messages described in Section 4 to coastal populations and emergency management through telecommunication networks. The communication networks provide redundant pathways to ensure connectivity during an event. Fig. 13.20 is a schematic diagram showing communication paths used to disseminate messages from WCATWC and PTWC.

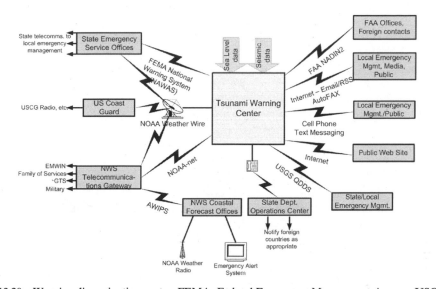

Fig. 13.20 – Warning dissemination routes. FEMA=Federal Emergency Management Agency; USCG = United States Coast Guard; EMWIN = Emergency Managers Weather Information Network; GTS = Global Telecommunications System; AWIPS = Advanced Weather Interactive Processing System; FAA = Federal Aviation Administration; NADIN2 = National Airspace Data Interchange Network; QDDS = Quake Data Dissemination System; RSS = Really Simple Syndication.

Primary communication routes at the TWCs include the FEMA National Warning System, the NOAA Weather Wire satellite system, the FAA NADIN2 system, and NOAA-net connections to the NWS Telecommunications Gateway. Secondary paths include email, FAX, text messaging, web sites, RSS feeds, and the USGS Quake Data Dissemination System.

Warning centers must define those agencies considered primary contacts. These agencies include:

- Designated national contact points
- State/Province Warning Points (WP)
- Weather Forecast Offices
- Coast Guard
- Military Commands

Primary contacts forward the message to the local emergency management agencies, the media, and the public. A typical forwarding sequence is:

TWC → State WP → County Emergency Mgmt. → Communities → Public

Several steps are necessary to propagate the message from a warning center to the public. Other pathways must also be activated in case one of the above steps fails.

TWC → Weather Forecast Offices → Public via Weather Radio/Emergency Alert System

TWC → Coast Guard → Mariners via radio

TWC → NWS Telecommunications Gateway → Media → Public

These dissemination paths provide emergency management offices with several methods to receive warning messages. Some of the more common receipt methods are:

- NOAA Weather Radio
- Emergency Alert System
- State-wide telecommunication systems
- National Warning System
- Email/FAX from warning centers
- Emergency Managers Weather Information Network
- Cell phone text messaging

With advances in communication technology, it is becoming more feasible for the warning centers to contact local emergency management and the public directly. Methods such as the Emergency Manager's Weather Information Network (EMWIN) and internet based methods such as email, RSS feeds, cell phone text messaging, auto-FAX, and the USGS's Quake Data Dissemination System (QDDS) are providing this functionality.

Testing the communication routes is critical to proper functioning during an event. Several times each day, tsunami warning centers must verify that dissemination circuits are functional. Monthly, the centers must check connectivity and response times through these links to primary contacts. To fully test the communications network, at least annually an end-to-end test must be originated at the tsunami warning centers and forwarded through the primary contacts to their recipients. Without these end-to-end tests, there can be no confidence that the communications system will function when absolutely necessary.

6. Emergency Management Response

Emergency management response organizations are responsible for: instituting response plans and procedures; preparing the public and itself to respond properly during an event; receiving warning messages from the warning center; transmitting the warning to local populations; and conducting evacuation and recovery as necessary.

A response plan is critical for all types of potential emergency situations. Tsunami response plans overlap with others in many regards. For example, warning point procedures, emergency operations center setup criteria and procedures, warning reception and local dissemination protocol, and recovery plans will be similar to weather-related hazards. Two aspects of response plans singular to tsunami response are hazard zone definition and evacuation plans. The hazard zone is the region which could be inundated during a tsunami. Hazard zones can be defined by several methods. The simplest is to set an elevation/inundation below which is considered threatened (inundation is the horizontal distance a wave travels over land). Early hazard maps in Alaska were based on the minimum of the 100 foot elevation contour or one mile inundation. A more advanced technique is to set hazard zones based on observed inundations or geologic studies which define pre-historic inundation. The method most commonly used today is to define hazard zones by mathematical inundation models (Chapter 10 this volume). Gonzalez *et al.* (2005b) provides an overview of recent inundation modeling efforts. Inundation models, especially when coupled with historic information or geologic studies providing constraints, can provide good estimates of maximum likely inundation. These studies are particularly useful when the maximum expected events are well known (one good example is when a subduction zone is directly offshore as in the U.S. Pacific Northwest and Alaska). When maximum events are not well understood, the accuracy of the computed hazard zone decreases. However the hazard zone is defined, once it is determined evacuation routes and shelter locations can be planned.

Once response plans are established, local populations must be educated regarding proper actions during a tsunami. Many coastal regions host a large number of tourists. Steps must be taken to alert these visitors to tsunami danger, nature's warnings, and proper response. Many methods are used to provide response plan information to the public. Beach signs, brochures, telephone directory postings, public meetings and door-to-door campaigns have been used to spread the word (e.g., Connor, 2005).

Emergency response organizations are essentially middle-men in the warning dissemination process. They must be able to reliably receive information from tsunami warning centers (using methods described in Section 5) and forward it to residents. Some examples of methods used to transmit the warnings locally are:

- Sirens
- Loudspeakers
- Patrol car bullhorns
- Reverse 911 phone systems
- Local CB radio networks

Local notification systems are described further by Darienzo *et al.* (2005).

An automated notification system implemented in the state of Washington has shown great promise. This system, known as All Hazard Alert Broadcasting (AHAB), broadcasts verbal response information over loudspeakers strategically located along the coast (Crawford, 2005). The speakers are activated either by: 1) NOAA Weather Radio, 2) satellite feed from the state emergency operations center, or 3) manually. One advantage to this system is that loudspeakers which broadcast verbal information are not subject to the ambiguity of siren tones (Fig. 13.21).

Fig. 13.21 – AHAB Loudspeaker located in Ocean Shores, Washington (photo provided by Washington Emergency Management Division).

To assist coastal counties and communities prepare for tsunamis, the NWS in partnership with the NTHMP initiated the TsunamiReady program in 2000. This program sets guidelines for communities to properly prepare for a tsunami. If all guidelines are followed, the NWS publicly recognizes the community as being TsunamiReady. Program guidelines are shown in Table 6. While the program can not guarantee that all citizens and visitors will respond properly during a warning, it does provide local response organizations a protocol to follow to ensure that the community government has taken appropriate actions to prepare for a tsunami.

TABLE 6.
TsunamReady guidelines.

Guideline 1 – **Communications and Coordination**	Establish 24-hour warning point (WP)
	Establish Emergency Operations Center (EOC)
Guideline 2 – **Tsunami Warning Reception**	WP/EOC must have X number of ways to receive tsunami messages where X is based on population (X = 3 or 4).
Guideline 3 – **Local Warning Dissemination**	WP/EOC must have X number of ways to disseminate tsunami messages to local residents where X is based on population (X = 1 to 4).
	NOAA Weather Radio receivers in public facilities
	County warning points must have county-wide communications network that ensures information flow.
Guideline 4 – **Community Preparedness**	Annual tsunami awareness programs (number dependent upon population).
	Designated tsunami shelter or area in a non-hazard zone.
	Provide written, locally-specific, tsunami hazard response material to the public.
	Encourage tsunami hazard curriculum in schools, and provide safety material to staff and students.
Guideline 5 – **Administrative**	Formal tsunami hazard operations plan.
	Biennial meeting/discussion between emergency manager and NWS representative.
	Visit to community by NWS official at least every other year.

In addition to warnings provided by tsunami warning centers, nature provides indicators about potential tsunami impact. Especially for those caught in the immediate generation vicinity, nature's warnings should trigger tsunami response action. A good general rule of thumb is that if the ground shakes hard in a coastal region for 20 seconds or more, a tsunami may be imminent. Upon feeling this sustained strong shaking, emergency management in tsunami-threatened regions should begin tsunami warning response procedures. Warning centers will quickly send out a notification if there is no tsunami threat. Tsunamis can strike the nearest coasts within minutes, so waiting for an official warning is not prudent. Another natural warning of an impending tsunami is a strong withdrawal of water leaving the sea floor exposed. There may be only minutes to impact following a withdrawal.

7. Summary

A tsunami warning system is an integrated network consisting of observational networks, warning centers, communication links, and emergency response organizations. Only with all of these components fully functional will the tsunami warning system be able to fulfill its mission in the event of a destructive tsunami. Tsunami warning dissemination can be thought of as a continually bifurcating series of falling dominoes with the warning centers plucking the first domino. If any domino is missing, dissemination stops on that path.

All tsunami warning communication and emergency management systems must be regularly tested and exercised. This is especially true for tsunami warning systems due to the infrequency of major tsunamis. Regular in-house training for tsunami warning center analysts is critical to both hone analyst's skills and check in-house systems by rehearsing various scenarios. Tsunami warning center processes must be checked multiple times per day, and primary contact message receipt must be verified monthly. An end-to-end test of the entire communication pathway, from the warning centers through the primary contacts to their customers, must also be tested annually. Only with vigorous and routine testing will the tsunami warning system perform properly in an event.

Acknowledgments

Thanks to Slava Gusiakov and Eddie Bernard for their thoughtful reviews which improved the content of this chapter, and to Cindi Preller of the WCATWC for assistance with graphics. Thanks to Laura Furgione and Jeff LaDouce, NWS Alaska and Pacific Region Directors, for their continued support of the tsunami program. Thanks to Thomas Sokolowski, former PTWC Scientist and WCATWC Director, for laying the initial groundwork necessary to modernize the U.S. tsunami warning centers. Finally, thanks to all the past and present tsunami warning center staff that have made the advancements cited in this chapter possible.

References

Acharya, H. K., 1979. Regional variations in the rupture-length magnitude relationships and their dynamical significance, *Bull. Seism. Soc. Am.*, **69**, 2063–2084.

Bernard, E. N., 2005. The U.S. National Tsunami Hazard Mitigation Program: A successful state-federal partnership, *Natural Hazards,* **35,** 5–24.

Blewitt, G., C. Kreemer, W. C. Hammond, H. P. Plag, S. Stein, and E. Okal, 2006. Rapid determination of earthquake magnitude using GPS for tsunami warning systems, *Geophys. Res. Lett.,* **33,** L11309, doi:10.1029/2006GL026145.

Blackford, M. E., 1984. Use of the Abe magnitude scale by the tsunami warning system, *Sci. Tsu. Hazards,* **2,** 27–30.

Burgy, M. C. and D. K. Bolton, 2006. New Coastal Tsunami Gauges: Application at Augustine Volcano, Cook Inlet, Alaska, *EOS Trans. AGU,* 87(52), Fall Meet. Suppl., Abstract V51C-1693.

Connor, D., 2005. The city of Seaside's tsunami awareness program – Outreach assessment: How to implement an effective tsunami preparedness outreach program, Oregon Department of Geology and Mineral Industries Open File Report O – 05–10, pp. 78.

Crawford, G. L., 2005. NOAA Weather Radio (NWR) – A coastal solution to tsunami alert and notification, *Natural Hazards,* **35,** 163–171.

Crowley, H. A., W. Knight, and P. Whitmore, 2007. Computation and application of tsunami travel times, GSA *Abstracts with Programs,* **39,** 156.

Darienzo, M. A. Aya, G. L. Crawford, D. Gibbs, P. M. Whitmore, T. Wilde, and B. S. Yanagi, 2005. Local tsunami warning in the Pacific coastal United States, *Natural Hazards,* **35,** 111–119.

Furumoto, A. S., H. Tatehata, and C. Morioka, 1999. Japanese tsunami warning system, *Sci. Tsu. Hazards,* **17,** 85–106.

Gerard, F., 2007. Framework document of the ad hoc working group for global tsunami and other ocean-related hazards early warning system (GOHWMS), IOC, 24th Session of the Assembly, IOC-XXIV/2 Annex 10, 14 pp.

Gonzalez, F. I., E. N. Bernard, C. Meinig, M. C. Eble, H. O. Mofjeld, and S. Stalin, 2005a. The NTHMP tsunameter network, *Natural Hazards,* **35,** 25–39.

Gonzalez, F. I., V. V. Titov, H. O. Mofjeld, A. J. Venturato, R. S. Simmons, R. Hanson, R. Combellick, R. K. Eisner, D. F. Hoirup, B. S. Yanagi, S. Yong, M. Darienzo, G. R. Priest, G. L. Crawford, and T. J. Walsh, 2005b. Progress in NTHMP Hazard Assessment, *Natural Hazards,* **35,** 89–110.

Gusiakov, V. K., A. G. Marchuk, and A. V. Osipova, 1997. Expert tsunami database for the Pacific: motivation, design and proof-of-concept demonstration, in *Perspectives on Tsunami Hazard Reduction: Observations, Theory and Planning,* G. Hebenstreit (ed.), Kluwer Academic Publisher, Dordrecht-Boston-London, 21–24.

Iida, K., D. C. Cox and G. Pararas-Carayannis, 1967. Preliminary catalog of tsunamis occurring in the Pacific Ocean, Data Report Mo. 5, HIG-67-10, U. of Hawaii, Honolulu, 274 pp.

Johnson, C. E., A. Bittenbinder, B. Bogaert, L. Dietz, and W. Kohler, 1995. Earthworm: a flexible approach to seismic network processing, *IRIS Newsletter,* **14,** 1–4.

Kanamori, H., 1977. The energy release in great earthquakes, *J. Geoph Res.,* **82,** 2981–2987.

Johnson, J. M., Y. Tanioka, L. J. Ruff, K. Satake, H. Kanamori, and L. R. Sykes, 1994. The 1957 great Aleutian earthquake, *Pure and Applied Geophysics,* **142,** 3–38.

Knight, W., 2006. Model predictions of Gulf and southern Atlantic coast tsunami impacts from a distribution of sources, *Sci. Tsu. Hazards,* **24,** 304–312.

Kowalik, Z. and P. M. Whitmore, 1991. An investigation of two tsunamis recorded at Adak, Alaska, *Sci. Tsu. Hazards,* **11,** 216–226.

Kowalik, Z., W. Knight, T. Logan, and P. Whitmore, 2005. The tsunami of 26 December 2004: Numerical modeling and energy considerations, in *Proceedings of the International Tsunami Symposium,* Eds.: G. A. Papadopoulos and K. Satake, Chania, Greece, 27–29 June, 2005, 140–150.

Lander, J. and P. A. Lockridge, 1989. United States tsunamis (including United States possessions) 1690–1988, Publication 41-2, National Geophysical Data Center, Boulder, CO, USA, 265 pp.

Lander, J., P. A. Lockridge and J. Kozuch, 1993. Tsunamis affecting the west coast of the United States 1806–1992, NGDC Key to Geophysical Research Documentation No. 29, USDOC/NOAA/NESDIS/NGDC, Boulder, CO, USA, 242 pp.

Lander, J., 1996. Tsunamis affecting Alaska 1737–1996, NGDC Key to Geophysical Research Documentation No. 31, USDOC/NOAA/NESDIS/NGDC, Boulder, CO, USA, 195 pp.

McCreery, C. S., 2005. Impact of the National Tsunami Hazards Mitigation Program on operations of the Richard H. Hagemeyer Pacific Tsunami Warning Center, *Natural Hazards,* **35,** 73–88.

McMillan, J. R., 2002. Methods of installing United States National Seismographic Network (USNSN) stations – a construction manual, United States Geological Survey Open-File Report 02-144, 23 pp.

National Oceanographic and Atmospheric Administration National Geophysical Data Center Tsunami Data Base (2006). Retrieved October, 2006 from http://www.ngdc.noaa.gov/seg/hazard/tsu.shtml.

Okada, Y., 1985. Surface deformation due to shear and tensile faults in a half-space, *Bull.Seism. Soc. Am.,* **75,** 1135–1154.

Okal, E. A. and J. Talandier, 1987. M_m: Theory of a variable-period mantle magnitude, *Geoph. Res. Lett.,* **14,** 836–839.

Oppenheimer, D. H., A. N. Bittenbinder, B. M. Bogaert, R. P. Buland, L. D. Dietz, R. A. Hansen, S. D. Malone, C. S. McCreery, T. J. Sokolowski, P. M. Whitmore, and C. S. Weaver, 2005. The seismic project of the National Tsunami Hazards Mitigation Program, *Natural Hazards,* **35,** 59–72.

Pelayo, A. M. and D. A. Wiens, 1991. The April 1, 1946 Aleutian tsunami earthquake: The largest known slow seismic event, Abstract, *Eos Trans AGU,* **72,** 292–293.

Reymond, D., O. Hyvernaud, and J. Talandier, 1993. An integrated system for real time estimation of seismic source parameters and its application to tsunami warning, in *Tsunamis in the World,* Ed.: S. Tinti, Kluwer Academic Publishers, 177–195.

Sokolowski, T. J., G. W. Fuller, M. E. Blackford, and W. J. Jorgensen, 1983. The Alaska Tsunami Warning Center's automatic earthquake processing system, in *Proceedings, 1983 Tsunami Symposium, Hamburg, FRG, August, 1983,* 131–147.

Sokolowski, T. J., P. M. Whitmore, and W. J. Jorgensen, 1990. Alaska Tsunami Warning Center's automatic and interactive computer processing system, *Pure and Applied Geophysics,* **134,** 163–174.

Sokolowski, T. J., 1999. The U.S. west coast and Alaska tsunami warning center, *Sci. Tsu. Hazards,* **17,** 49–56.

Soloviev, S. L. and M. D. Ferchev, 1961. Summary of data on tsunamis in the USSR, *Bulletin of the Council for Seismology,* Academy of Sciences of the USSR, No. 9, 1961, translation by W.G. van Campen, Hawaii Inst. of Geophysics, Translations Series 9, 37 pp.

Spaeth, M. G. and S. C. Berkman, 1967. The tsunami of March 28, 1964, as recorded at tide stations, ESSA Technical Report Coast and Geodetic Survey Technical Bulletin No. 33, 86 pp.

Spence, W., S. A. Sipkin, and G. L. Choy, 1989. Measuring the size of an earthquake, *Earthquakes and Volcanoes,* **21,** 58–63.

Stein, S. and E. A. Okal, 2005. Size and speed of the Sumatra earthquake, *Nature,* **434,** 581–582.

Talandier, J., D. Reymond, and E. A. Okal, 1987. M_m: Use of a variable-period mantle magnitude for the rapid one-station estimation of teleseismic moments, *Geoph. Res. Lett.,* **14,** 840–843.

Tappin, D. R., P. Watts, G. M. McMurty, Y. Lafoy, and T. Matsumoto, 2001. The Sissano, Papua New Guinea tsunami of July 1998 – offshore evidence on the source mechanism, *Marine Geology,* **175,** 1–23.

Tatehata, H., 1998. The new tsunami warning system of the Japan Meteorological Agency, *Sci. Tsu. Hazards,* **16,** 39–50.

Titov, V.V. and F.I. Gonzalez, 1997. Implementation and testing of the Method of Splitting Tsunami (MOST) model, NOAA Technical Memorandum ERL PMEL-112, 11 pp.

Titov V. V., F. I. Gonzalez, E. N. Bernard, M. C. Eble, H. O. Mofjeld, J. C. Newman, and A. J. Venturato, 2005. Real-time tsunami forecasting: Challenges and Solutions, *Natural Hazards,* **35,** 41–58.

Trnkoczy, A., P. Bormann, W. Hanka, L. G. Holcomb, and R. L. Nigbor, 2002. IASPEI New manual of seismological observatory practice, Volume 1, Ch. 7, Ed.: P. Bormann, Publisher GeoForschungsZentrum, Potsdam, Germany, 108 pp.

Tsuboi, S. K., K. Abe, K. Takano, and Y. Yamanaka, 1995. Rapid determination of M_w from broadband P waveforms, *Bull.Seism. Soc. Am.,* **83,** 606–613.

Tsuboi, S. K., P. M. Whitmore, and T. J. Sokolowski, 1999. Application of M_{wp} to deep and teleseismic earthquakes, *Bull.Seism. Soc. Am.,* **89,** 1345–1351.

UNESCO Intergovernmental Oceanographic Commission, 1999. ITSU Master Plan – Second Edition April 1999, Retrieved October, 2006 from ioc3.unesco.org/ptws/documents/itsu_master_plan _eng.pdf, 40 pp.

United States Geological Survey Earthquake Hazards Program Global Earthquake Data Base (2006). http://wwwneic.cr.usgs.gov/neis/epic/epic.html.

Urban, G. W., A. H. Medbery, and M. C. Burgy, 2003. New approach to sea level monitoring tested at Shemya, Alaska, *Eos Trans AGU,* **84,** Fall Meeting Suppl., Abstract OS21D-04.

Walker, D. A. and R. K. Cessaro, 2002. Locally generated tsunamis in Hawaii: a low cost, real time warning system with world wide applications, *Sci. Tsu. Hazards,* **20,** 177–182.

Whitmore, P. M. and T. J. Sokolowski, 1996. Predicting tsunami amplitudes along the North American coast from tsunamis generated in the northwest Pacific Ocean during tsunami warnings, *Sci. Tsu. Hazards,* **14,** 147–166.

Whitmore, P.M. and T.J. Sokolowski, 2002. Automatic earthquake processing developments at the U.S. West Coast/Alaska Tsunami Warning Center, in *Recent Research Developments in Seismology,* Transworld Research Network, Kervala, India, 1–13.

Whitmore, P. M., S. K. Tsuboi, B. Hirshorn, and T. J. Sokolowski, 2002. Magnitude-dependent correction for M_{wp}, *Sci. Tsu. Hazards,* **20,** 187–192.

Whitmore, P. M., 2003. Tsunami amplitude prediction during events: a test based on previous tsunamis, *Sci. Tsu. Hazards,* **21,** 135–143.

Whitmore, P. M., J. C. Ferris, and S. A. Weinstein, 2007. Seismic data monitoring at U.S. Tsunami Warning Centers, *EOS Trans. AGU,* 88 (23), Jt. Assem. Suppl., Abstract S33C-01.

Wielandt, E., 2002. IASPEI New manual of seismological observatory practice, Volume 1, Ch. 5, Ed.: P. Bormann, Publisher GeoForschungsZentrum, Potsdam, Germany, 46 pp.

Zitek, W. O., A. H. Medbery, and T. J. Sokolowski, 1990. Concurrent seismic data acquisition and processing using a single IBM PS/2 computer, NOAA Technical Memorandum NWS AR-41, 20 pp.

INDEX